Characterization and Control of Interfaces for High Quality Advanced Materials II

Characterization and Control of Interfaces for High Quality Advanced Materials II

Ceramic Transactions, Volume 198

Proceedings of the Second International Conference on Characterization and Control of Interfaces for High Quality Advanced Materials, and Joining Technology for New Metallic Glasses and Inorganic Materials, Kurashiki, Japan (2006)

Edited by
Kevin Ewsuk, Kiyoshi Nogi, Rolf Waesche, Yukichi Umakoshi, Tom Hinklin, Keizo Uematsu, Tony Tomsia, Hiroya Abe, Hidehiro Kamiya, Makio Naito

The American Ceramic Society

BICENTENNIAL 1807 WILEY 2007 BICENTENNIAL

A JOHN WILEY & SONS, INC., PUBLICATION

Published by John Wiley & Sons, Inc., Hoboken, New Jersey
Published simultaneously in Canada.

Library of Congress Cataloging-in-Publication Data is available.

ISBN 978-0-470-18414-1

Printed in the United States of America.

10 9 8 7 6 5 4 3 2 1

Contents

Preface

Interfaces are critically important to a broad spectrum of materials and technologies. In 2003, the first International Conference on Characterization and Control of Interfaces for High Quality Advanced Materials (ICCCI2003) established an international forum for materials interface science and technology. Joining technology for new metallic glasses and inorganic materials was added to the dialog at the second International Conference on Characterization and Control of Interfaces for High Quality Advanced Materials, and Joining Technology for New Metallic Glasses and Inorganic Materials (ICCCI2006) in Kurashiki in Japan in 2006, which also featured sessions on nanoparticles and powders, energy and environment, smart processing technology, and materials design.

ICCCI2006 was supported by the Institute of Materials Research (IMR) of Tohoku University, the Materials and Structures Laboratory (MSL) of the Tokyo Institute of Technology, the Japan JSPS 124th Committee, the Hosokawa Powder Technology Foundation, the Project on Joining Technology for New Metallic Glasses and Inorganic Materials, the 21st Century COE Program "Center of Excellence for Advanced Structural and Functional Materials Design" lead by Professor Yukichi Umakoshi of Osaka University, and the Joining and Welding Research Institute (JWRI) of Osaka University.

Over 200 scientists and engineers from academia and industry from 16 different countries attended ICCCI2006 to see and discuss 165 invited and contributed presentations and posters on the state-of-the art of interface characterization and control for particulate materials, joining, and nanotechnology. The Proceedings of ICCCI 2006 features 67 peer–reviewed papers on interface characterization and control technology for materials synthesis, powder processing, composite processing, joining, and to control airborne particulates.

Kevin Ewsuk
Kiyoshi Nogi
Rolf Waesche
Yukichi Umakoshi
Tom Hinklin
Keizo Uematsu
Tony Tomsia
Hiroya Abe
Hidehiro Kamiya
Makio Naito

JOINING TECHNOLOGY FOR NEW METALLIC GLASSES AND INORGANIC MATERIALS

THE STRUCTURE OF THE WELDED ZONE AND PHASE TRANSFORMATION BEHAVIOR OF Ni-BASED BULK-GLASS FORMING ALLOY

D. V. Louzguine-Luzgin[1,*], G. Xie[1], T. Tsumura[2], K. Nakata[2], Y. Murakami[1], H. M. Kimura[1] and A. Inoue[1]

[1]*Institute for Materials Research, Tohoku University, Katahira 2-1-1, Aoba-Ku, Sendai 980-8577, Japan*
[2]*Joining & Welding Research Institute, Osaka University, Mihogaoka, Ibaraki, Osaka 567-0047, Japan*

ABSTRACT

The crystallization behavior and the structure of the welded samples of the $Ni_{53}Nb_{20}Ti_{10}Zr_8Co_6Cu_3$ glassy alloy are studied. A large supercooled liquid region of about 50 K observed enables its successful welding. On heating the supercooled liquid crystallizes by polymorphous mechanism forming a single metastable hR14 Ni_4Ti_3 phase which further transforms at higher temperature forming several equilibrium phases. Its composition was found to be very close (equal within the confidential interval) to that of the glassy phase. The kinetic analysis confirmed the interface-controlled polymorphous transformation mechanism observed.

Good bonding between two glassy samples was obtained after welding. The results indicate that with an increase in the welding speed from 2 to 3 m/min the structure of the welded zone changes from the amorphous + hR14 Ni_4Ti_3 crystalline phase to an almost single amorphous one. By adjusting the welding speed one can obtain the desired microstructure. The samples electron-beam welded at the beam current of 4 mA and voltage of 60 kV at a speed of 4 m/min were found to be almost amorphous. The results indicate successful welding of the Ni-based glassy samples.

INTRODUCTION

Bulk glassy alloys with a size exceeding 1 mm in three dimensional space were obtained by stabilization of the supercooled liquid against crystallization at the relatively low cooling rate of less than 100 K/s from liquid state in the various, mostly multi-component, metallic alloys.[1-3] These alloys are promising materials for structural applications as they exhibit high mechanical strength, high hardness, good fracture toughness, good corrosion resistance[2], etc.

In particular, bulk glass formation was achieved in various Ni-based alloy systems: Ni–Zr–Ti–(Si,Sn)[4],[5], Ni-Nb-Ti-Zr[6], Ni-Nb-Ti[7], Ni-Nb-Ti-Hf[8,9], Ni-Nb-Sn[10] and so on, which are actually a group of LTM-ETM or LTM-ETM-M alloys (where LTM is(are) late transition metal(s), ETM is(are) early transition metal(s) and M is metalloid or Sn). These alloys are promising candidates for structural applications owing to their high strength attaining 3 GPa. $Ni_{53}Nb_{20}Ti_{10}Zr_8Co_6Cu_3$ bulk glassy alloy exhibits a high tensile fracture strength of 2700 MPa and a critical diameter of 3 mm[11]. $Ni_{59}Zr_{16}Ti_{13}Si_3Sn_2Nb_7$ bulk glassy alloy which showed a significant plastic deformation of 6.5 % to failure[12] may contain nanoparticles which are not detectable by an X-ray diffraction (XRD) technique. Devitrification behavior of Ni-Nb-Ti[13], Ni-Nb-Ti-Zr, Ni-Nb-Ti-Zr and Ni-Nb-Ti-Zr-Pt[14] glassy alloys has been studied recently.

* Corresponding author. Tel: +081(22)215-2592, Fax: +081(22)215-2381, e-mail: dml@imr.tohoku.ac.jp

At the same time a limited size of bulk metallic glassy samples requires bonding technology to obtain industrial-size samples. In the present work we study the crystallization and welding behavior of the $Ni_{53}Nb_{20}Ti_{10}Zr_8Co_6Cu_3$ glassy alloy. This alloy is a good candidate for study as it has one of the highest GFAs among Ni-based alloys and a large supercooled liquid region.

EXPERIMENTAL PROCEDURE

An ingot of the $Ni_{53}Nb_{20}Ti_{10}Zr_8Co_6Cu_3$ alloy was prepared by arc-melting using pure elements of 99.9 mass% purities in an argon atmosphere. Ni-based metallic glassy ribbon with a width of 40 mm and thickness of 25 μm used in this study was produced by rapid solidification of the melt on a single copper roller at a roller tangential velocity of 42 m/s. The structure of the samples was examined by X-ray diffractometry (XRD) with monochromatic CuK_α radiation. The structure of the welded bead was studied using a micro-area XRD (CuK_α radiation) with a collimator size of 30 μm. The phase transformations were studied by differential scanning calorimetry (DSC) at a heating rate of 0.67 K/s and differential isothermal calorimetry (DIC).

As the laser welding heat source, diode laser welding apparatus KATAOKA LD-YAG HP300a was used. The welding was carried out at the diameter of the laser beam of 0.3 mm, the laser beam power of 26 W, and the welding speed range of 2 to 3 m/min. MITSUBISHI EM-9HB election beam welding apparatus was used as the election welding heat source. The welding was carried out at the accelerating voltage of 60 kV, beam current range of 3 to 5 mA, welding speed of 4 m/min under an evacuated atmosphere of about 0.1 Pa. In order to prevent crystallization at the heat affected zone copper plates were used as the cover and backing plates. The cover plate has had a slit of 0.6 mm in width which is similar in size to the laser/electron spot diameter.

Transmission electron microscopy (TEM) investigation was carried out using a JEM 2010 (JEOL) microscope operating at 200 kV equipped with an energy dispersive X-ray (EDX) spectrometer of 0.1 keV resolution. The samples for TEM were prepared by the ion-polishing technique.

RESULTS

Thermal stability and crystallization behavior of the glassy alloy.

First of all the crystallization behaviour of the $Ni_{53}Nb_{20}Ti_{10}Zr_8Co_6Cu_3$ alloy on heating was studied in order to determine the temperature interval of welding. A DSC trace taken within the present work is shown in Fig. 1 (a). The sample undergoes glass transition, followed by a large supercooled liquid region and then two exothermic reactions due to crystallization. The glass transition temperature (T_g), temperature of crystallization (T_x), and melting temperature (T_m) are 845 K, 897 K, and 1265 K, respectively. This alloy has a very high T_{rg} of 0.67.

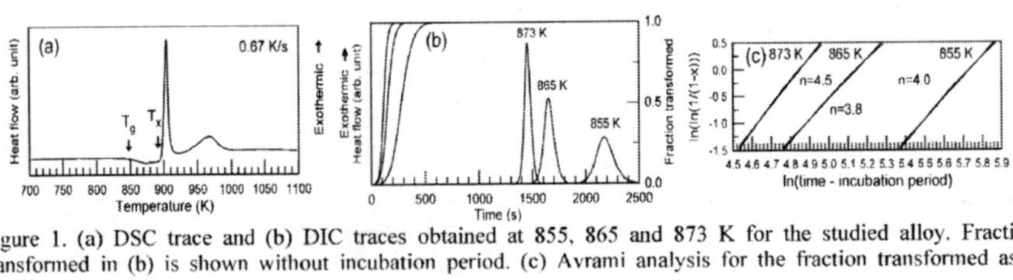

Figure 1. (a) DSC trace and (b) DIC traces obtained at 855, 865 and 873 K for the studied alloy. Fraction transformed in (b) is shown without incubation period. (c) Avrami analysis for the fraction transformed as a function of time.

Upon heating the studied glassy alloy initially devitrifies through a supercooled liquid producing a multicomponent hR14 Ni_4Ti_3 phase (the lattice parameters a=1.12nm, c=0.51 nm) starting at about 900 K and hR14 $Ni_{6.5}Nb_{6.5}$ phase at about 950 K (Fig. 2).

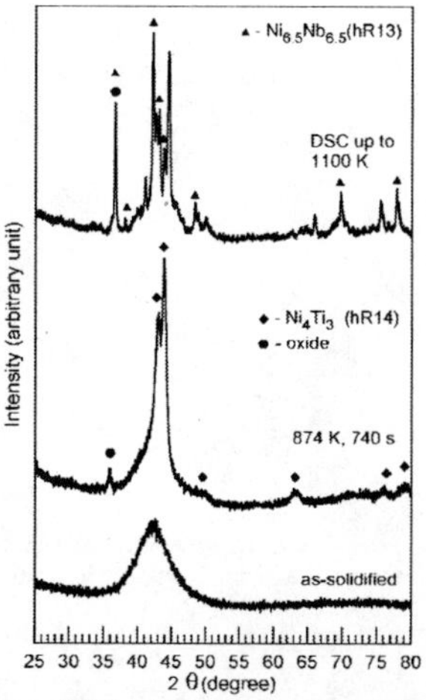

Figure 2. XRD patterns of the studied alloy in as-solidified and annealed states as indicated.

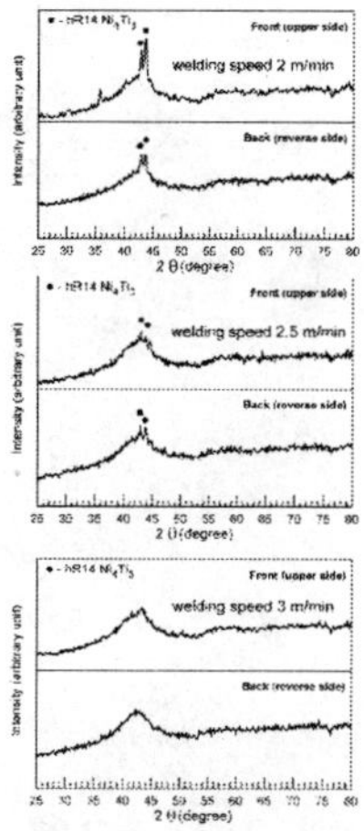

Figure 3. Micro-area XRD patterns of the laser-beam welded samples. Welded bead, upper and bottom side.

The isothermal calorimetry traces obtained at different temperatures shown in Fig. 1 (b) were analyzed using an Avrami method (including the least squares fitting of the Avrami plot) for the fraction transformed as a function of time (Fig. 1 (c)) according to the following kinetic law[15] for the volume fraction (x) transformed as a function of time (t):

$$x(t) = 1 - \exp[-Kt^n] \qquad (1)$$

The existence of the incubation period and the Avrami exponent values of about 4 indicate that the studied alloy undergoes nucleation and growth-type transformation by polymorphous reaction on heating above the heating-rate dependent crystallization temperature.

Laser beam welding of glassy samples.

XRD patterns of the samples laser-beam welded at different speed are shown in Fig. 3. As well as in the heat treated sample the hR14 Ni_4Ti_3-type phase was found to precipitate within the central area of the sample laser beam welded at 2 and 2.5 m/min (see Figs. 2 and 3). One can see that the volume fraction of the crystalline phase decreases with an increase in the welding speed. The appearance of the welded sample is shown in Fig. 4.

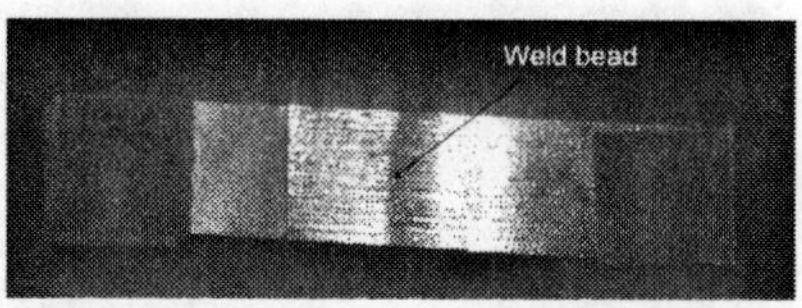

Figure 4. Optical micrograph of the sample subjected for laser beam welding.

The TEM images of the welded zone in the sample welded at 2.5 m/min are shown in Fig. 5. The existence of the submicron hR14 Ni_4Ti_3 solid solution type particles is observed in some parts of the sample (Fig. 5 (a)). Fig. 5 (b,c) shows the selected-area electron diffraction (SAED) patterns indexed according to hR14 lattice. The chemical composition of these multicomponent crystalline particles and a residual amorphous phase after welding are shown in Table 1.

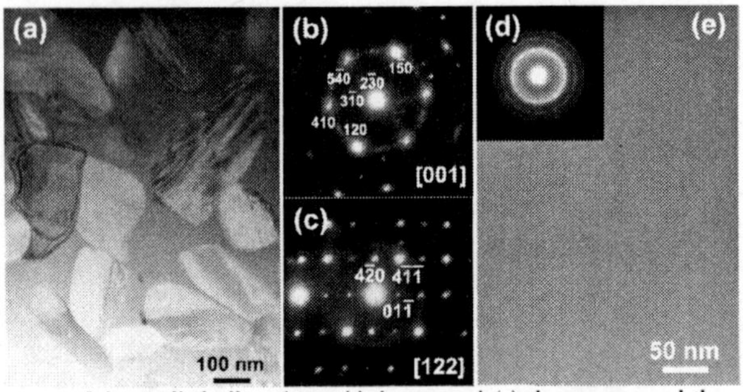

Figure 5. TEM images of the studied alloy. (a) welded zone and (e) the area around the welded zone (heat affected zone). (b and c) SAED patterns taken from the crystalline particles in (a) and indexed according to hR14 Ni_4Ti_3 phase lattice. (d) SAED from the area in (e).

Table 1. Chemical composition of the hR14 crystalline particles and a residual amorphous phase after welding (at%).

	Ni	Nb	Ti	Zr	Co	Cu
Crystal 1	47.6	29.3	10.2	5.2	4.8	2.9
Crystal 2	46.2	30.1	6.0	10.3	4.7	2.8
Amorph.	45.5	30.9	6.9	9.0	4.8	3.0

The areas around the welded zone with a large volume fraction of the glassy phase (Fig. 5 (e)) contain nanocrystals of Ni_4Ti_3. The laser beam welded sample showed a tensile mechanical strength of 410 MPa.

Electron-beam welding of glassy samples.

The electron-beam welded samples shown in Fig. 6 were found to be almost amorphous (Fig. 7). In other words the volume fraction of crystalline particles is lower compared to those in the laser-beam welded samples. A low fraction of the crystalline particles were found in the welded zone of the sample electron-beam welded at the welding current of 4 mA. However, in some places the surface of the welded sample was oxidized. According to XRD data no crystals were found in the thermally affected area around the welded zone.

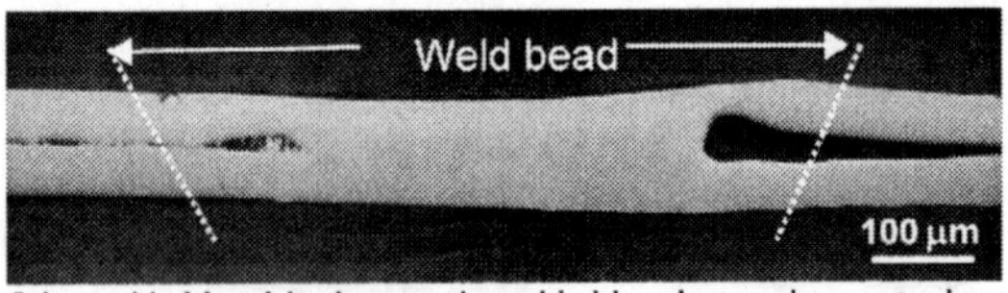

Figure 6. The edge of the welded bead in the sample welded by electron-beam at a beam current of 3.9 mA.

SEM.

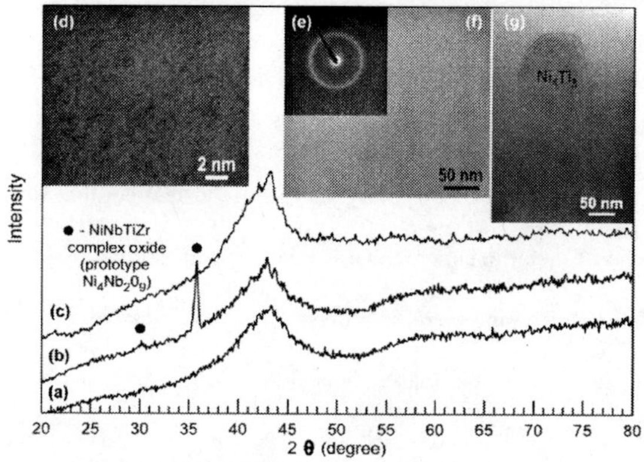

Figure 7. (a-c) XRD patterns and (d-g) TEM images of the electron-welded sample of the studied alloy. XRD of the welded zone: (a) clear area and (b) oxidized area and (c) thermally affected area. (d) HRTEM image inside the welded zone, (e) SAED pattern, (f) bright-field image of the amorphous area. (g) a single particle of the hR14 Ni_4Ti_3 phase.

The electron beam welded samples showed much higher tensile strength (960 MPa) compared to the laser beam welded samples. No oxide particles were found in the body of the sample, which allows us to conclude that the oxidation took place on the surface only.

DISCUSSION

The studied glassy alloy exhibits a large supercooled liquid region on heating and initially crystallizes by polymorphous reaction using nucleation and growth mechanism forming uniaxial grains of a metastable hR14 Ni_4Ti_3 solid solution phase (LTM Co and Cu may substitute Ni while ETM Zr and Nb are presumed to substitute Ti). The kinetic analysis confirmed the interface-controlled polymorphous transformation mechanism observed. An Avrami exponent value of about 4 suggests the interface-controlled growth. The lattice parameters of the multicomponent hR14 Ni_4Ti_3 phase are slightly different from those[16] of a binary phase. hR14 $Ni_{6.5}Nb_{6.5}$ phase and an unidentified phase form at higher temperature while the hR14 Ni_4Ti_3 phase disappears.

Successful bonding is achieved after both laser-beam and electron beam welding. Electron-beam welding produces somewhat better results compared to laser beam one. Almost fully amorphous samples were obtained with a small volume fraction of the hR14 Ni_4Ti_3 solid solution crystalline phase upon electron-beam welding, though some surface oxidation was observed due to insufficient vacuum. A complex oxide formed in some parts of the surface has a structure of $Ni_4Nb_2O_9$ one. The obtained results indicate high stability of the sample upon heating. It is also found that with an increase in the laser-beam welding speed from 2 to 3 m/min the structure of the welded zone changes from the amorphous + hR14 Ni_4Ti_3 crystalline phase to an almost single amorphous one. By adjusting the welding speed one can obtain the desired microstructure. Within the sample welded at 2.5 m/min the metastable Ni_4Ti_3 particle size varied from a few nanometres in the outer area of the welded zone to a few hundreds of nanometres in the center. The electron-beam samples welded at the beam current of 4 mA, voltage of 60 kV and a speed of 4 m/min were found to be almost amorphous.

The crystallization behavior and the structure of the welded samples of the $Ni_{53}Nb_{20}Ti_{10}Zr_8Co_6Cu_3$ alloy were studied. The results indicate successful welding of the glassy alloy by the electron-beam welding. The obtained data are promising for future applications of the laser and electron-beam welding techniques to glassy alloys.

REFERENCES:

[1] A. Inoue, "High-strength bulk amorphous alloys with low critical cooling rates" Mater. Trans. JIM 36, 866-875 (1995).
[2] A. Inoue, "Stabilization of metallic supercooled liquid and bulk amorphous alloys" Acta Mater. 48, 279-306 (2000).
[3] W. L. Johnson, "Bulk glass forming metallic alloys: science and technology" MRS Bull 24, 42-56 (1999).
[4] S. Yi, T.G. Park and D.H. Kim, "Ni-based bulk amorphous alloys in the Ni-Ti-Zr-(Si, Sn) system" J. Mater. Res. 15, 2425-2429 (2000).
[5] J.K. Lee, S.H. Kim, S. Yi, W.T. Kim and D.H. Kim, „A Study on the Development of Ni-Based Alloys with Wide Supercooled Liquid Region" Mater. Trans. JIM 42, 592-596 (2001).
[6] A. Inoue, W. Zhang and T. Zhang, "Thermal Stability and Mechanical Strength of Bulk Glassy Ni-Nb-Ti-Zr Alloys" Mater. Trans. 43, 1952-1956 (2002).
[7] W. Zhang and A. Inoue, "Effects of Ti on the Thermal Stability and Glass-Forming Ability of Ni-Nb Glassy Alloy" Mater. Trans. 43, 2342-2345 (2002).
[8] T. C. Hufnagel, C. Fan, R. T. Ott, J. Li, S. Brennan, "Controlling shear band behavior in metallic glasses through microstructural design" Intermetallics 10, 1163-1168 (2002).
[9] W. Zhang, A. Inoue, "Formation and Mechanical Properties of Ni-based Ni-Nb-Ti-Hf Bulk Glassy Alloys" Scripta Mater. 48, 641-645 (2003).
[10] H. Choi-Yim, D. Xu, and W. L. Johnson, "Ni-based Bulk Metallic Glass Formation in the Ni-Nb-Sn and Ni-Nb-Sn-X (X=B,Fe,Cu) Alloy System" Appl. Phys. Lett. 82, 1030-1032 (2003).
[11] T. Zhang and A. Inoue, "New Bulk Glassy Ni-Based Alloys with High Strength of 3000 MPa" Mater. Trans. JIM 43, 708-711 (2002).
[12] M.H. Lee, J.Y. Lee, D.H. Bae, W.T. Kim, D.J. Sordelet, D.H. Kim, "A development of Ni-based alloys with enhanced plasticity" Intermetallics 12, 1133-1137 (2004).
[13] T. Shimada, D. V. Louzguine, J. Saida and A. Inoue "Thermal Stability and Devitrification Behavior of Ternary Ni-Nb-Ti and Quaternary Glassy Alloys Containing Noble Metals" Materials Transactions, 46, 675-680 (2005).
[14] D. V. Louzguine-Luzgin, T. Shimada and A. Inoue "Ni-based bulk glassy alloys with large supercooled liquid region exceeding 90 K" Intermetallics, 13, 1166-1171 (2005)
[15] J.W. Christian "The Theory of Transformation in Metals and Alloys" Oxford: Pergamon Press Ltd; 1975. p. 542.
[16] T. Saburi, T. Tatsumi and S. Nenno "Effects of heat treatment on mechanical behavior of Ti-Ni alloys" Journal de Physique C4, 261-266 (1982).

BRAZING OF ADVANCED CERAMIC COMPOSITES: ISSUES AND CHALLENGES

Mrityunjay Singh[1] and Rajiv Asthana[2]

[1]Ohio Aerospace Institute
NASA Glenn Research Center, Cleveland, OH 44135

[2]Department of Engineering & Technology,
University of Wisconsin-Stout, Menomonie, WI 54751

ABSTRACT

The relative effectiveness of several Ag-, Cu-, and Ni-base brazes for joining of C-C, C-SiC and SiC-SiC composites to Ti, Inconel 625, Hastealloy and Cu-clad Mo, is evaluated. Titanium-containing active metal brazes led to sound metallurgical microstructure and good joints. The surface preparation and fiber architecture at the mating surface influence the shear strength of the composite/metal joints. The need to characterize molten braze contact angles and thermoelastic compatibility issues between the composite and the base metal to fabricate robust composite-to-metal joints is highlighted.

INTRODUCTION

Recently, there has been considerable interest in development and testing of advanced ceramic composites for a wide variety of aerospace and ground-based applications. Among the composites currently being developed and evaluated are C-C, C-SiC and SiC-SiC composites. These composites must be joined to themselves and to other materials such as metals and alloys for a number of applications. Robust assembly and integration technologies, therefore, play an essential role in fabricating composite components [1].

Active metal brazing is a simple and cost-effective method to join ceramics and ceramic composites to metals using braze powders, pastes, ribbons, or wires that are melted, allowed to wet and spread in the joint region, and solidified to form joint interfaces. Ceramic brazing utilizes either premetallized ceramic surfaces or braze foils, pastes or wires that usually contain an reactive filler metal such as Ti that promotes braze wettability and flow characteristics by inducing chemical reactions with the ceramic. Generally, joining becomes feasible and joint strength improves when either a wettability-enhancing chemical reaction (e.g., oxide dissolution, oxide reduction, compound formation) is induced in the joint region, as in active metal brazing, metal substrate is oxidized (oxide/ceramic bond), or ceramic surface is metallized (metal/metal bond). Braze alloys containing Ti or another reactive metal yield very high bond strengths although pre-coating of the ceramic with a Ti-bearing compound (e.g., TiH_2) that forms a Ti layer on the ceramic may also be used.

The braze materials for ceramic composite-to-metal joints must meet a large number of criteria. The design of the braze composition must simultaneously meet several of these criteria, which essentially represents a compromise. For example, braze must show good wetting and adherence to the substrates, be resistant to grain growth, creep and oxidation, be ductile (or have closely-matched coefficient of thermal expansion (CTE) with the joined materials), and have liquidus temperature greater than the operating temperature of the joint, but lower than the metal's melting temperature. For ceramic composites, as a first step, we empirically evaluated the joining response of several commercial brazes.

In this paper, observations on the joining of C-C, C-SiC and SiC-SiC composites to metals and alloys (Ti, Inconel 625, Hastealloy and Cu-clad Mo) using various Cu-, Ag-, and Ni-base braze alloys with and without active metal additions are presented. Initial evaluation of braze effectiveness to join the composites to metals, and the microstructure and strength of the joint are presented. Issues and challenges in designing and fabricating robust composite-to-metal joints are highlighted.

BRAZE EVALUATION

Table 1 summarizes initial results on the effectiveness of commercial brazes for joining C-C, C-SiC, and SiC-SiC composites to metals and alloys. The filler metals included Ag and Cu-based

Table 1. Effectiveness of Ag, Cu and Ni-Base Braze Alloys in Joining C-C and Ceramic Composites

Composite	Metallic Substrate	Braze	Bonding
C-C[1,6]	Ti	Silcoro-75[8], Palcusil-15[8]	Weak
C-C and SiC-SiC	Ti	Ticuni, Cu-ABA, Ticusil	Good
C-SiC[9]	Ti and Hastealloy	MBF-20[8]	Good (Ti), Fair (Hastealloy)
C-C[1,6]	Ti and Hastealloy	MBF-20[8], MBF-30[8]	Good (Ti), Fair (Hastealloy)
SiC-SiC	Hastealloy	MBF-20[8], MBF-30[8]	Fair (MBF20), Good (MBF30)
C-SiC[9]	Hastealloy	MBF-30[8]	Good
C-SiC[1,9]	Ti, Inconel 625,	MBF-20[8], MBF-30[8]	Good
SiC-SiC	Ti	MBF-20[8]	Good
SiC-SiC	Ti	MBF-30[8]	Good
C-SiC[1,9]	Inconel 625	Incusil-ABA[8], Ticusil[8]	Good
C-SiC[1,9]	Inconel 625	Cu-ABA[8], Cusil[8], Cusil-ABA[8]	Good
C-SiC[1,2,9]	Ti	Cusil-ABA[7,8]	Good
SiC-SiC[1,2,10]	Ti	Cusil-ABA[7,8]	Good[7], Fair[8]
C-SiC[2,9]	Ti, Inconel 625, Cu-clad Mo[11]	Ticusil[7]	Fair (Ti), Good
SiC-SiC[2,10], C-SiC[1,9], C-C[6] (T-300)	Ti, Inconel 625, Cu-clad Mo[11]	Ticusil[7]	Good
C-C[3,4,5]	Ti, Inconel 625, Cu-clad Mo[11]	Ticusil[7]	Good[4], Fair[5]
C-C[3,4,5], C-SiC[1,2,9], SiC-SiC[1,2,10]	Cu-clad Mo[11]	Cusil-ABA[7]	Good
C-C[3,4,5]	Ti and Inconel 625	Cusil-ABA[7]	Good
SiC-SiC[1,2,10]	Ti	Cusil-ABA[7]	Good
C-SiC[1,2,9]	Ti	Cusil-ABA[7]	Good[1], Weak[2]

[1]*Polished;* [2]*Not-polished;* [3]*3-D composite (P120 C fiber, CVI C matrix), Goodrich Corp., CA;* [4]*Oriented fiber at the joint (3-D composite);* [5]*Non-oriented side at the joint;* [6]*T-300 C fibers in resin-derived C matrix, C-CAT, TX;* [7]*Braze paste;* [8]*Braze foil;* [9]*T300 C fiber in CVI SiC matrix, GE Power System Composites, DE;* [10]*Sylramic SiC fiber in MI-SiC matrix, GE Power System Composites, DE;* [11]*H.C. Starck, Inc., MA.*

Table II. Selected Physical and Mechanical Properties of Braze Alloys used in the Study

Braze	Comp., %	T_L, K	T_S, K	E, GPa	YS, MPa	CTE, x10^{-6} K^{-1}	K, W/m.K	%El
Cu-ABA[a]	92.8Cu-3Si-2Al-2.25Ti	1297	1231	96	279	19.5	38	42
MBF-30[b]	Ni-4.61Si-2.8B-0.02Fe-0.02Co	1257	1327	--	--	--	--	--
MBF-20[b]	Ni-6.48Cr-3.13Fe-4.38Si-3.13B	1242	1297	--	--	--	--	--
TiCuNi[a]	15Cu-15Ni-70Ti	1233	1183	144	--	20.3	--	--
TiCuSil[a]	68.8Ag-26.7Cu-4.5Ti	1173	1053	85	292	18.5	219	28
Palcusil-15[a]	65Ag-20Cu-15Pd	1173	1123		379	--	98	23
Silcoro 75[a]	75Au-20Cu-5Ag	1168	1158	--	--	--	--	--
Cusil-ABA[a]	63Ag-35.3Cu-1.75Ti	1088	1053	83	271	18.5	180	42
Cusil[a]	72Ag-28Cu	1053	1053	83	272	19.6	371	19
Incusil-ABA[a]	59Ag-27.3Cu-12.5In-1.25Ti	988	878	76	338	18.2	70	21

[a]Morgan Advanced Ceramics; [b]Metglass, Inc., T_L: liquidus temperature, T_S: solidus temperature, E: Young's modulus, YS: yield strength, CTE: coefficient of thermal expansion, K: thermal conductivity, %El: percent elongation.

alloys with or without Ti, and Ni-base metallic glasses, MBF-20 and MBF-30, containing boron and/or Cr. The brazing process consisted of placing the braze foil or paste as an interlayer between the metal and the composite substrates, heating the assembly to 15-20° above the braze liquidus in vacuum (10^{-6} torr) under a normal load of 0.30-0.4 N, isothermally holding for 5 min. at the brazing temperature, and slowly cooling the brazed joint to room temperature. Table I lists a qualitative comparison of joint quality based primarily on metallurgical soundness of the joint microstructures produced; further details on joint microstructures and strength are presented in ref. [2-7]. Table I also shows that Ag and Cu filler metals containing Ti produced sound joints whereas those without Ti yielded weak joints. The MBF brazes also generally produced metallurgically sound joints although cracking was noted in some joints. The compositions and selected properties of the braze filler metals used are given in Table II.

JOINT MICROSTRUCTURE

Figure 1 shows examples of the interfaces produced between ceramic composites and metals via brazing. Detailed SEM and EDS analyses of the joint region were done to examine the distribution of elements across the joint [2-7]. The results confirmed that Ti in braze preferentially segregated at the composite/braze interface which promoted braze spreading in the contact region and improved the joint strength. Chemical interactions between composites and molten braze led to dissolution and diffusion, which caused compositional changes in the joint.

JOINT STRENGTH

The shear strength of the Ti/composite joints extracted from Butt-Strap Tensile (BST) test is displayed in Fig. 2(a) for C-C, CVI C-SiC and melt-infiltrated (MI) SiC-SiC composites in as-

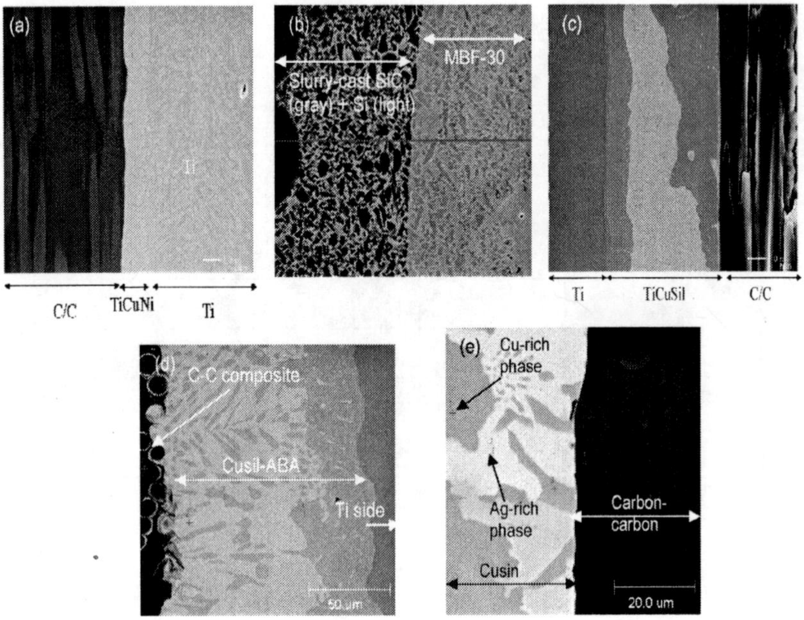

Fig. 1. Brazed composite/metal joints: (a) C-C/Ticuni/Ti, (b) SiC-SiC/MBF-30/Hastealloy, (c) C-C/Ticusil/Ti, (d) C-C/Cusil-ABA/Ti, and (e) C-C/Cusin/Cu-clad Mo.

received and polished conditions. The C-C composite is made from P120 C fiber in a resin-derived carbon plus CVI C matrix, the CVI C-SiC composite (GE Power Systems) is made from T300 C fiber in CVI SiC matrix, and SiC-SiC composite (GE Power Systems) is made from Sylramic SiC fiber in MI SiC matrix. The average joint strengths are in the range of 1.5-9.0 MPa (standard deviation: ±0.09 to 1.62 MPa). Generally, polished composite samples yielded better joint strength than unpolished (as-fabricated) composites (the results for CVI C-SiC are somewhat inconclusive because ofdelamination failure within the composite) [5]. Polishing of the composite prior to brazing led to slightly stronger joints. Although polishing will reduce the surface defect population, it may also introduce new subsurface damage (grain pull-out, porosity, cracking); as a result, the shear strength increments due to polishing were moderate.

Figure 2(b) shows the average shear strength of Ti tubes bonded to C-C plates of T-300 C fibers in resin-derived C matrix (C-CAT, TX). This figure shows the effect of braze type and orientation of the exterior fiber tow on the joint strength of C-C to Ti. The standard deviation around the strength is in the range ±0.04-0.13 MPa. Figure 2(b) shows that Ticuni yielded the highest joint strength because of its good wetting and spreading. The fiber tow orientation at the mating surface significantly influenced the joint strength; joints, with fiber tows perpendicular to the Ti tube axis exhibited higher failure loads than joints with parallel fibers. Details of the tube tensile testing procedure and the resulting strength data are presented in recent publications [2,3].

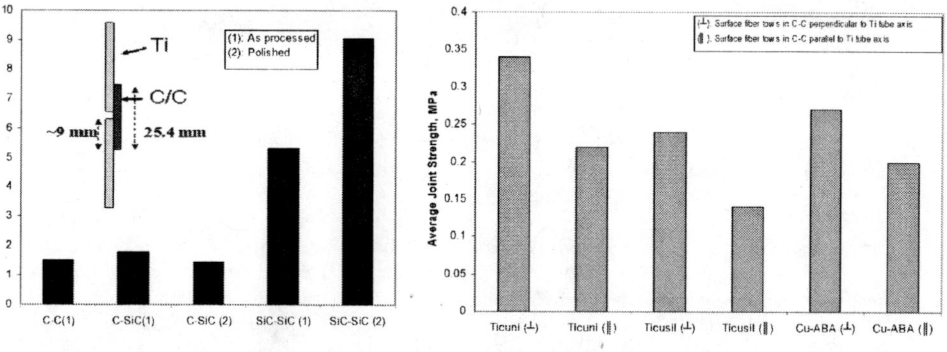

Fig. 2(a) Butt-strap lap tensile test (inset) and test outcomes for polished and as-fabricated (un-polished) composites jointed to Ti using Cusil-ABA, and (b) Average joint strength for Ti tubes bonded to C-C plates showing the effect of braze type and exterior fiber tow orientation.

WETTABILITY

The wetting of Ag and Cu braze fillers containing Ti on ceramics is excellent and the equilibrium contact angles approach near-zero values from the initially large obtuse values. The spreading rate varies depending upon the system and test conditions. The radius, R(t), of the contact region between the ceramic and the molten braze is empirically described by $R(t) = R_0 \exp(t/3\tau)$, where R_0 is the initial contact radius and τ is a characteristic time (on the order of a few minutes in most cases). Thus, to achieve a final radius R(t) just 10% larger than R_0 and with τ~10 min., a contact time of ~3 min. is estimated which is close to the brazing times of 4-5 minutes used in this work.

Ceramic composites are multi-phase materials and their surface consists of regions of both chemical and structural inhomogeneity. Therefore, different fiber architecture and regions of different chemical phases at the mating interface at a composite/metal joint will influence the contact angle and braze spreading behavior [9]. Scant information, however, is available on measurement of contact angle of brazes on ceramic-ceramic composites, and this represents an important area for future investigation.

THERMAL EXPANSION MISMATCH

The strength of a ceramic/metal joint essentially represents a convolution of several factors that include wetting, residual stress, phase transformation, and segregation. An important source of residual stresses in joints between dissimilar materials is the mismatch of the coefficient of thermal expansion (CTE) [8]. In brazed ceramic composite/metal joints, metal will be under tension, and the composite under compression during cooling from the brazing temperature. Depending upon the system under consideration, the elastic thermal strain, $\Delta\alpha\Delta T$ ($\Delta\alpha$: CTE mismatch, ΔT: temperature interval) during brazing may exceed the yield strain of the braze interlayer, causing the latter to plastically deform to accommodate the stress. Most Ag and Cu base filler metals are ductile (Table II), and they may plastically deform to reduce the stress and cracking tendency as was actually observed. It is well-known that metal interlayers reduce the stresses at the bonded region, and joint strength generally increases with increasing interlayer thickness. In this context, brazed joints of C-C composites to Cu-clad Mo (Fig. 1(e)) for thermal management applications can be considered as an example [7]. The CTE of Cu-clad Mo is ~5.7×10^{-6}/K and the CTE of C-C composites is ~$0\text{-}1.0\times10^{-6}$/K over 20-250 °C, and ~$2.0\text{-}4.0\times10^{-6}$/K over 20-2500°C [10]. Thus, CTE mismatch between C-C and Cu-clad Mo is moderate rather than excessive, and may lead to moderate thermal stresses during temperature excursions, which may be accommodated by the ductile Cu interlayer. More complete characterization of the joining behavior of ceramic composites to metals is currently underway.

CONCLUSIONS

The relative effectiveness of several Ag, Cu- and Ni-base brazes for joining C-C, C-SiC, and SiC-SiC composites to pure Ti, Inconel 625, Hastealloy and Cu-clad Mo was evaluated. Ti-containing active metal brazes generally led to sound joint microstructure and strong joints. The surface preparation and fiber architecture at the mating surface appear to influence the shear strength. Contact angles and thermoelastic compatibility in joints made from ceramic composites/high-temperature metals and alloys need to be characterized in detail.

ACKNOWLEDGEMENT

The authors would like to thank T.P. Shpargel and G.N. Morscher for their help. R. Asthana acknowledges research support from NASA Glenn Research Center, Cleveland, OH.

REFERENCES

1. M. Singh and E. Lara-Curzio, 'Design, fabrication and testing of ceramic joints for high-temperature SiC/SiC composites', *Trans ASME*, 288-292, 123 (2001).
2. M. Singh, T. Shpargel, G.N. Morscher, R. Asthana, 'Active metal brazing and characterization of brazed joints in Ti to carbon-carbon composites', *Mater. Sci. Eng.*, 412-413, 123-28 (2005).
3. G.N. Morscher, M. Singh, T. Shpargel, R. Asthana, 'A test to determine the effectiveness of braze composition for joining Ti tubes to C/C composite plates', *Mater. Sci. Eng.*, 19-24, A418 (2006).
4. M. Singh, T. Shpargel, G.N. Morscher and R. Asthana, 'Active metal brazing of carbon-carbon composites to titanium', *Proc. 5th Int. Conf. on High-Temperature Ceramic-Matrix Composites (HTCMC-5),* M. Singh et al (Eds.), 2005, pp. 457-462, Amer. Ceramic Society.
5. M. Singh, T. Shpargel, R. Asthana, G.N. Morscher, 'Effect of composite substrate properties on the mechanical behavior of brazed joints in metal-composite system', in *Brazing and Soldering*, J.J. Stephens and K. Scott Weil (eds.), ASM International and AWS, 246-51 (2006).
6. R. Asthana, M. Singh T. Shpargel, 'Brazing of ceramic-matrix composites to Ti using metallic glass interlayers', in *Proc. 30th Int. Conf. on Adv. Ceramics and Compos.*, Cocoa Beach (2006).
7. M. Singh, R. Asthana, T. Shpargel, 'Brazing of C-C composites to Cu-clad Mo for thermal management applications', submitted to *Mater. Sci. Eng. A (2006).*
8. G.N. Morscher, M. Singh, T. Shpargel, Comparison of different braze and solder materials for joining Ti to high-conductivity C/C composites', in *Brazing and Soldering*, J.J. Stephens and K. Scott Weil (eds.), ASM International and AWS, 257-261 (2006).
9. N. Sobczak, M. Singh, R. Asthana, 'High-temperature wettability measurements in ceramic-metal systems: Some methodological issues', *Current Opinion in Solid State & Mater. Sci. (2006).*
10. R. Taylor, 'Carbon-matrix composites,' in *Comprehensive Composite Materials*, 387-426, 4 (2000) Elsevier Science Ltd.

DEVELOPMENT OF A VARIABLE TEMPERATURE UHV-LASER MICROSCOPE COMBINED WITH A PULSED LASER DEPOSITION FOR IN SITU STUDIES OF CERAMICS/METALLIC GLASS INTERFACES

Yuji Matsumoto, Takuma Obata, Miki Hiraoka and Masao Katayama

Materials and Structures Laboratory, Tokyo Institute of Technology
4259 Nagatsuta Midori-ku Yokohama 226-8503 Japan

ABSTRACT

A new laser microscope (LM) has been designed and developed for in situ observation of metallic glass thin films and their interfaces with oxides. It can operate at various gas pressure and temperature conditions, from UHV(1×10^{-8}Torr) to atmosphere and from RT to 1000°C, respectively. The LM system can be connected with a pulsed laser deposition (PLD) system that allows us to directly transfer the sample prepared in the PLD to the LM system without exposing it to air. In order to verify the capability of this LM system, the surface morphology of metallic glass thin films was characterized at various temperature and atmospheric conditions. As a consequence, we succeeded in making the first observation of the dynamic behavior of a metallic glass in liquid droplet form, and also found different behavior of transitions in vacuum.

INTRODUCTION

Laser microscope (LM) has been often used to directly observe the behavior of *bulk* metallic glasses at high temperature because the LM technique can overcome the effect of thermal radiation from the melts on microscope viewing. Actually, in situ observation of the crystal growth of Fe-C alloys and $Pd_{40}Cu_{30}Ni_{10}P_{20}$ alloy in a B_2O_3 have previously reported so far [1, 2].

For wider applications of metallic glasses involving some hetero-interfaces with different materials such as oxide ceramics, studies on metallic glasses in *thin film* form have also become important today. As thin film properties of metallic glasses, however, are likely to subject to surface effects such as oxidation, the conventional laser microscope technique is not available for the studies on metallic glass thin films and their interfaces with oxides, owing to the observation in air or insufficient low vacuum condition.

Here, we report on a new LM system we developed with an aid of ultra high vacuum (UHV) technology: it can operate at various gas pressure and temperature conditions, from UHV to atmosphere and from RT to 1000°C, respectively. In addition, this LM system can be connected with other vacuum system through the gate valve. We describe the detailed implementation of the variable temperature UHV-LM together with some of the observation results obtained for the metallic glass thin films.

EXPERIMENTAL

Fig. 1 shows a schematic drawing of the developed variable temperature UHV-LM system, together with photographs. We used a confocal scanning laser microscope system (1LM21P from lasertec Co.; with 0.61 μm resolution at 875X magnification) with a 1.5 mW

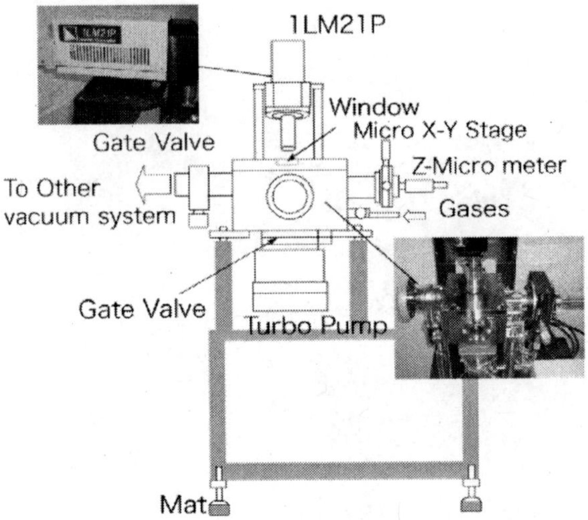

Figure 1. Schematic drawing of the developed variable temperature UHV-LM system, together with photographs.

laser diode. The acquisition rate of LM images is 30 images/sec at maximum and the digital data of LM images are directly recorded in the DVD. A UHV furnace is about 230x160x140mm^3 in size (from Kitano Co.) and evacuated by a turbo molecular pump system (V70 from Varian Co.) the base pressure of which is less than $3x10^{-8}$ Torr even before baking the system. In this pressure range, it is true that almost metal should be thermodynamically oxidized, but the bulk oxide formation would be slow enough for our purpose to avoid serious oxidizing effect on the high-temperature behavior of metallic thin films, in spite of a possible formation of the ultra-thin surface oxide. The LM is fixed on the UHV furnace and the sample in the furnace can be observed through the quartz window. The sample is mounted on the xyz stage and one can control the position of the sample on a micro-scale with respect to the objective of the LM and observe a wide sample area of 10x20mm^2. The SiC block heater is used for heating the sample at

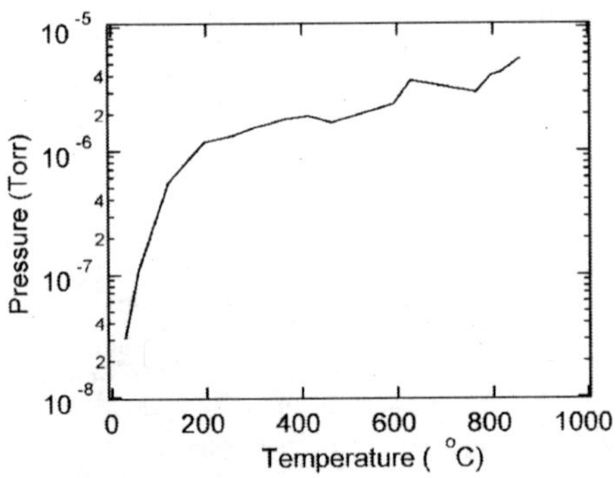

Figure 2. Behavior of the vacuum condition vs. the sample

as high as 1000°C.

Fig. 2 shows a behavior of the vacuum condition vs. the sample temperature. The pressure increases with the sample temperature due to the degas process. Even at a sample temperature of 800°C, the pressure is kept as low as $4x10^{-6}$ Torr enough to *kinetically* avoid undesirable complete oxidation of the metallic glass thin film.

Another advantage of this UHV furnace system is that the UHV furnace also has an ICF70 port and it can be connected through the gate valve to other vacuum system, *e.g.* a pulsed laser deposition (PLD) system [3] that enables us to deposit thin films of ceramics materials as well as metals in the present study. As shown in Fig. 3, the gas pressure of the LM system is thus controlled independently of that of the PLD system: one can operate the LM system in a high pressure condition up to atmosphere (760Torr) keeping an UHV condition of the PLD system. In the PLD system, a focused pulsed laser beam of KrF excimer laser is impinged with an energy density of ~$1J/cm^2$ through a quartz window onto a sintered ceramics target to ablate atomic spices of the ceramics material. Films of the target material are deposited under UHV or oxidizing atmosphere of O_2 gas. The sample prepared in the PLD can be transferred into the LM system without taking it in air. To reduce vibrations in LM measurement, whole the LM system is set on special rubber sheets to cut the vibrations from the pump system through the floor.

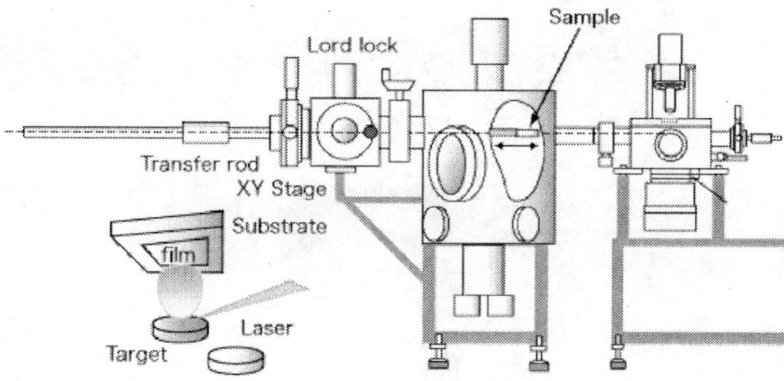

Figure 3. Schematic illustration of *in situ* LM-PLD system and the PLD growth.

RESULTS AND DISCUSSION

Observation of metallic glass thin films in atmospheric N_2 condition

We first observed metallic glass thin films of $Pd_{76}Cu_7Si_{17}$ grown on SiO_2 substrates with ~3μm thickness by arc plasma deposition [4]. The Tg, Tx and Tm reported in the previous literature [5] are 364, 396 and 746°C, respectively. Once the LM chamber was evacuated to reach a background pressure of $3x10^{-8}$ Torr, an atmospheric N_2 gas (~615 Torr) was introduced into the chamber at room temperature. The sample temperature was increased up to 1000 °C at a rate of 0.4°C/s. Fig. 4 shows a series of the LM images of $Pd_{76}Cu_7Si_{17}$ film surface at temperature of

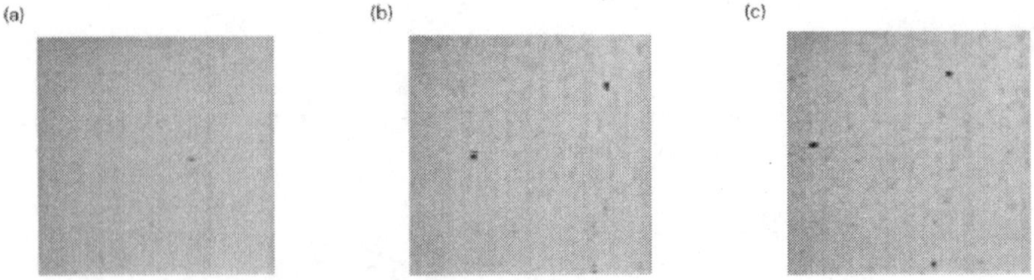

Figure 4. A series of the LM images (65μmx65μm) of $Pd_{76}Cu_7Si_{17}$ film surface at temperature of 401 to 447 °C in N_2 atmospheric condition. (a) 401 °C, (b) 435 °C, and (c) 447 °C

401 to 447 °C. There appeared many voids and creases on the film surface above 400 °C, which is closer to the Tx, but not Tg; the change in the surface morphology may thus correspond to the crystallization of the metallic glass film. In general, the molar volume of metallic glass decreases when the crystallization occurs from the undercooled liquid and the voids and creases found on the surface are therefore considered to compensate the reduction in the molar volume. At temperature around the melting point, we could also observe a dynamic behavior of the melt to be liquid droplets as shown in Fig. 5. From (c) to (d), there are found some droplets move and grow by the coalescences and they are stable at least up to 985°C. Here, we would like to emphasize that an oxygen free condition by using our new LM system allows us to observe the liquid droplet of metallic glass for the first time.

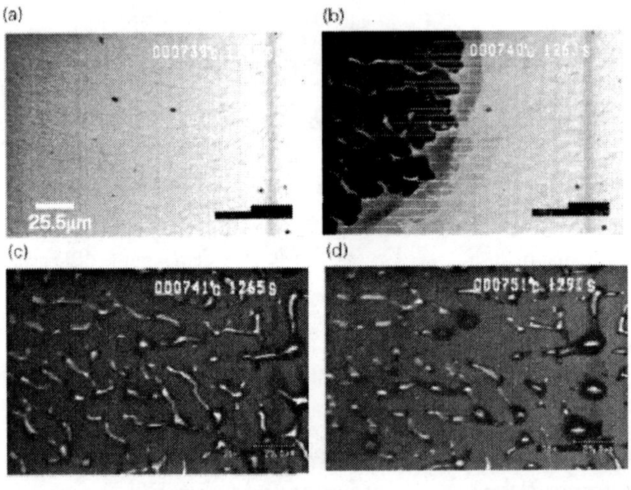

Figure 5. A series of the LM images of $Pd_{76}Cu_7Si_{17}$ film surface around at melting point of 746 °C in N_2 atmospheric condition. (a) 739 °C, (b) 740 °C, (c) 741 °C and (d) 751 °C

Observation of metallic glass thin films in vacuum condition

Next, we tried to observe the behavior of the metallic glass thin film of $Pd_{76}Cu_7Si_{17}$ in vacuum condition (~10^{-6} Torr). Being different from the case observed in N_2 atmospheric condition, we did not find a clear transition at around Tx. Instead, the drastic change in the LM image was observed at temperature of 648 to 728 °C as shown in Fig. 6. The LM image suddenly became darker and darker, and then slowly recovered its brightness. In general, the brightness and contrast of LM image are very sensitive to the surface conditions of being flat or rough in addition to the reflectance property of materials itself, owing to the different scattering behavior of the probing laser on the sample surface. The observed change of the LM image in brightness thus might reflect on the surface roughness of the metallic glass film; when the surface became rough, the LM image was dark and it turned to be bright when the surface recovered its flatness. It should be pointed out that this behavior of the metallic glass thin film is found to greatly depend on the local composition and/or defect structures. For example, the local areas indicated by the white arrows in Fig. 6(a), where something impurities or defects might exist showed less change in their brightness and contrast, *i.e.* surface roughness as compared to that around them. The origin of the different behavior of the metallic glass thin film observed in vacuum will be studied further in our later work.

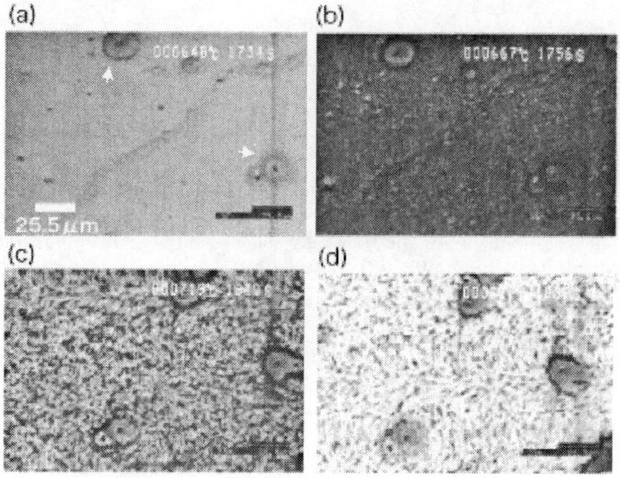

Figure 6. A series of the LM images of $Pd_{76}Cu_7Si_{17}$ film surface at temperature of 648 to 728 °C in vacuum condition. (a) 648 °C, (b) 667 °C, (c) 719 °C and (d) 728 °C

CONCLUSION

We successfully developed a new LM designed for in situ observation of metallic glass thin films under various atmospheric conditions. The transitions at Tx and Tm in the $Pd_{76}Cu_7Si_{17}$ metallic glass films could be detected by the LM, but not at Tg in atmospheric N_2 gas. In contrast, the metallic glass film exhibited a different behavior in vacuum, which is characterized by the drastic change of the image contrast and brightness at around 700°C, while maintaining a stabile the thin film form at temperature as high as 920°C.

ACKNOWLEDGEMENTS

The authors would like to thank Prof. Seiichi Hata (Tokyo Inst. of Tech.) for providing metallic glass thin film samples. This work was supported by Ministry of Education, Culture, Sports, Science and Technology Grant Program of Metallic Glass Inorganic Materials Joining Technology Development.

REFERENCES

[1]H. Chikama, H. Shibata, T. Emi and M. Suzuki, "'In-situ' real time observation of planar to cellular and cellular to dendritic transition of crystals growing in Fe-C alloy metals", *Mater. Trans. JIM*, **37** 620-626 (1996)

[2] J.-H. Kim, S. –G. Kim and A. Inoue, "In situ observation of solidification behavior in undercooled Pd-Cu-Ni-P alloy by using a confocal scanning laser microscope", *Acta Mater*. **49** (2001) 615-622

[3] R. Takahashi, Y. Yonezawa, M. Ohtani, M. Kawasaki, K. Nakajima, T. Chikyow, H. Koinuma and Y. Matsumoto, "Perfect $Bi_4Ti_3O_{12}$ Single-Crystal Films via Flux-Mediated Epitaxy", *Adv. Funct. Mater.* **16** 485-491 (2006)

[4] S. Hata, R. Yamaguchi, J. Sakurai and A. Shimokobe, "Combinatorial Arc Plasma Deposition of Thin Films", *Jpn. J. Appl. Phys*. **45** (2006) 2708-2713

[5] H. S. Chen and D. Turnbull, "Formation, stability and structure of palladium-silicon based alloy glasses", Acta Metallurgica 17 (1969) 1021-31

STRESS ANALYSIS OF GEOMETRICALLY COMPLEX AND ULTRA LARGE SCALE MODEL BY FRACTAL MULTI-GRID METHOD

H. Murakawa and H. Serizawa
Joining and Welding Research Institute, Osaka University
11-1 Mihogaoka, Ibaraki, Osaka 567-0047, Japan

M. Tejima, K. Taguchi and S. Itoh
Graduate School of Engineering, Osaka University
2-1 Yamadaoka, Suita, Osaka 565-0871, Japan

ABSTRACT

Finite element method is a powerful tool to predict not only the mechanical behavior of structures but also the residual stresses and distortions caused in the fabrication processes. In most problems of engineering interest, the phenomena are of three-dimensional nature. Generally, the three-dimensional elastic-plastic FE analysis requires very long computational time and large memory size. To overcome this problem, a Fractal Multi-Grid method is proposed. Its potential capability is demonstrated through application to very simple mechanical problems.

INTRODUCTION

The technique of computer simulation with finite element method (FEM) has been widely used to evaluate stress, strain and deformation of complex structures. R & D efforts in both computer hardware and software make it possible to solve large scale FE models with more than 10^8 degree of freedom. As for the software improvement, various methods for solving large classes of matrix equations have been developed to decrease the computational time. These are, for example, ICCG (Incomplete Cholesky Conjugate Gradient) method [1], multifrontal method [2], multi-grid method [3] and AMG (algebraic multi-grid) method [4]. Also, DDM (Domain Decomposition Method) [5] and BDD (Balancing Domain Decomposition) method [6] have been developed for parallel and grid computing. The size of FE model becomes much larger, when fabrication processes and fracture behavior of complex and large real products are to be simulated with reasonable accuracy. Simulations of welding distortions and residual stresses in dissimilar material joints or crash simulation of automobiles can be mentioned as examples. In such cases, a great amount of computational time and memory are necessary although the capacity of computers has been improved drastically. However, it is still necessary to speed up the computation and increase its efficiency greatly. In this research, as one of new methods for solving large problems, the authors developed a Fractal Multi-Grid method, which may be categorized as a geometrical multi-grid method and tested its potential capabilities for two dimensional elastic problem and linear elastic plate bending problem.

FRACTAL MULTI-GRID METHOD

Solution Procedure

The idea of the Fractal Multi-Grid method is illustrated using a two dimensional simple elastic problem as shown in Fig. 1. A square sheet is stretched at its four vertexes. When the model is subdivided into 8×8 elements, deformation and stress are computed by solving basic cells consisting of 2×2 elements under the prescribed displacement at four vertexes as shown in Fig. 2. Such basic operation is repeated hierarchically from the top level to the lowest level. In

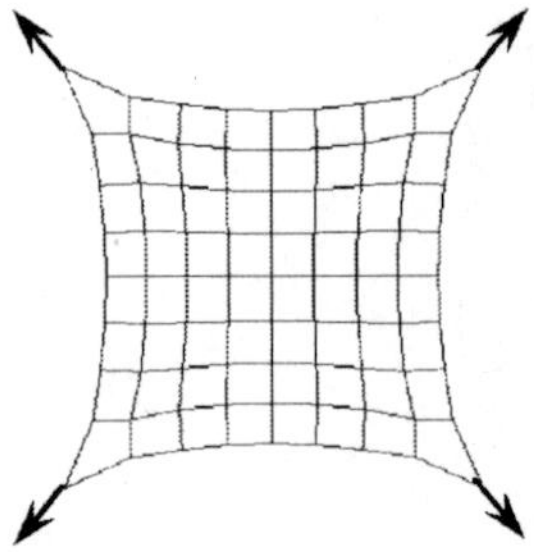

Fig. 1 Elastic sheet stretched at four corners.

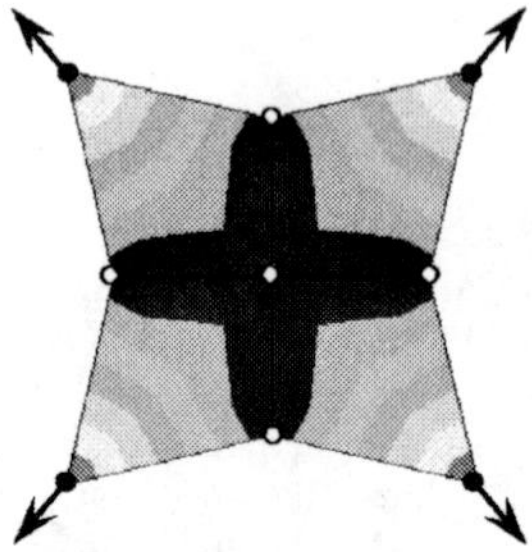

Fig. 2 Basic cell consist of four elements.

this process, the continuity between the neighboring cells is ignored as illustrated in Fig. 3. The continuity of the displacement can be recovered at the lowest level by interpolating the displacements at two nodes sharing the same cell boundary. In this way, the continuity of the traction is not guaranteed. It is retained through the iteration. Since all the governing equations and boundary conditions must be satisfied at the lowest or the finest level, the error in stress field is evaluated at the lowest level and it is transferred to the upper level.

The detail of the scheme is as follows,

Step-(1) Solve the cell at the 1st-level under the given boundary condition, not necessarily prescribed displacements at four vertexes.

Step-(2) Use the displacements at 9 nodes obtained for the 1st-level as the prescribed displacements at four vertexes, compute displacements of four cells belonging to the 2nd-level.

Step-(3) Repeat the same computation as in Step-(2) until the lowest level.

Step-(4) Force the continuity of the displacements at four nodes on the edge of the cell in the lowest level by interpolating the displacements at the two nodes belonging to the two cells sharing the same edge.

Step-(5) Compute the unbalanced force at the lowest level.

Step-(6) Redistribute the unbalanced force at the four middle nodes and one center node of the cell to the four nodes at vertexes according to the following rule:

Middle node: Redistribute one half of the unbalanced force to the two corner nodes on the same edge.

Center node: Redistribute one fourth of the unbalanced force to the four corner nodes of the cell.

Step-(7) Sprit the above unbalanced nodal forces to the two nodes belonging to the cells sharing the same edge according to the ratio of the stiffness.

Step-(8) Compute the correction of the nodal displacement of the cell using the redistributed unbalanced force with the prescribed correction displacements at the four corner nodes.

Step-(9) Repeat the Steps-(2) through (8) until the norm of the unbalanced force at the lowest level becomes small enough.

The details of the computational scheme may be different when the problem to be solved is different, such as heterogeneous or nonlinear problems. If the problem is a linear isotropic two or three-dimensional problem and a regular uniform mesh is used, the stiffness matrix of all cells becomes the same. Therefore, it is not necessary to save the stiffness matrixes of all cells. Once

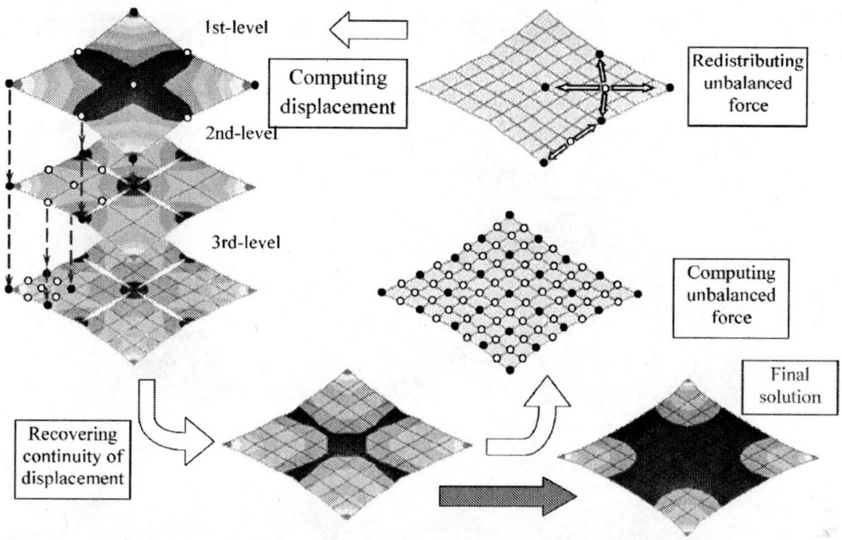

Fig. 3 Procedure of computation in FMG method.

the inverse of the stiffness matrix of a cell is computed, it can be used repeatedly. From the aspect of memory space, the required memory size is only three or four times of the degree of freedom.

The advantage of the proposed FMG method is that the stiffness matrices are exactly identical throughout the hierarchical levels as long as the domain is a rectangle or a rectangular parallelepiped and the problem is homogeneous. However, this becomes disadvantage when the problem is inhomogeneous such as an elastic-plastic problem. In such cases, the stiffness matrix of the element can be derived only for the element on the lowest level. The stiffness of the elements on the higher level can not be defined.

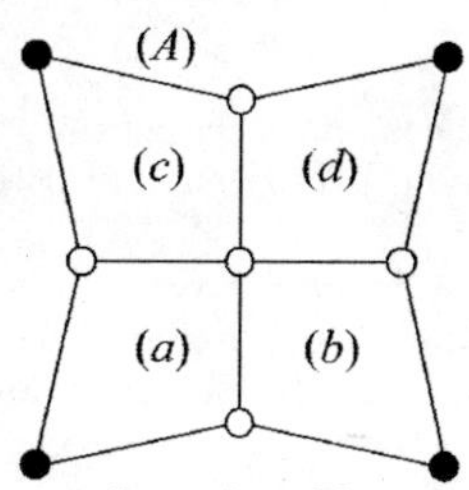

Fig. 4 A cell consist of four elements.

This problem can be solved by introducing the idea similar to the substructure. To explain the idea, a cell (*A*) shown in Fig. 4 is taken as an example. The cell (*A*) consists of four elements, namely (*a*), (*b*), (*c*) and (*d*). Denoting,

$[k_i]$, $\{f_i\}$: stiffness matrix and load vector of element (i) (i = *a*, *b*, *c*, *d*) on lower level

$[k_{abcd}]$: stiffness matrix of cell composed of 4 elements

$\{f_{abcd}\}$: load vector matrix of cell composed of 4 elements

the stiffness matrix $[k_A]$ and the load vector $\{f_A\}$ for the higher level can be given by the following equation:

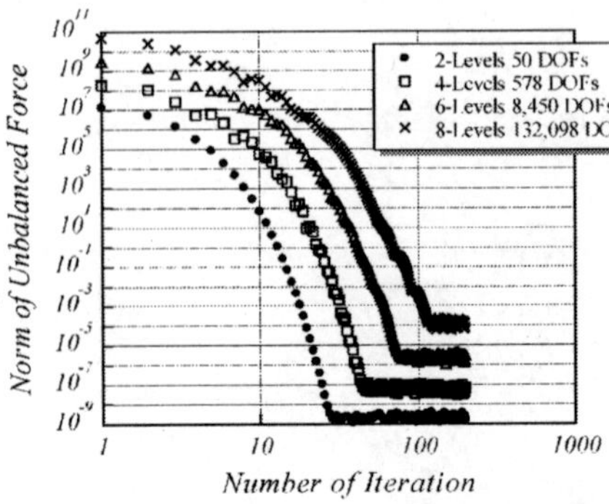

Fig. 5 Convergence of norm of unbalanced force

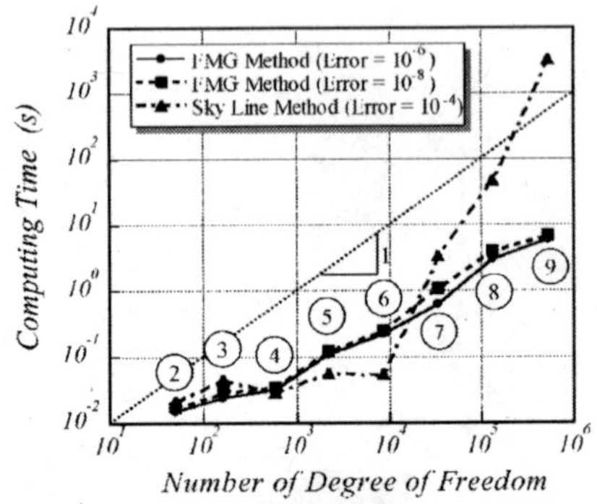

Fig. 6 Relation between computing time and degree of freedom.

$$[k_A]=[k_{abcd}^{22}]-[k_{abcd}^{21}][k_{abcd}^{11}]^{-1}[k_{abcd}^{21}]^{T},\ \{f_A\}=\{f_{abcd}^{2}\}-[k_{abcd}^{21}][k_{abcd}^{11}]^{-1}\{f_{abcd}^{1}\} \tag{1}$$

Where,

$$[k_{abcd}]=\begin{bmatrix} k_{abcd}^{11} & k_{abcd}^{12} \\ k_{abcd}^{21} & k_{abcd}^{22} \end{bmatrix},\ \{f_{abcd}\}=\begin{Bmatrix} f_{abcd}^{1} \\ f_{abcd}^{2} \end{Bmatrix} \tag{2}$$

and the superscript 2 corresponds to the displacements at corner nodes where the displacements are given and the superscript 1 corresponds to those at other nodes without restraint.

Modified Solution Procedure Using Coarse Global Mesh

As it is noticed, the solution procedure in the basic form can deal with only a cubic or a square domain. This limitation can be overcome by introducing course global mesh. The finer grid with fractal structure is built in each global element.

Example Problems

Elastic Square Sheet Stretched at Four Vertexes

One of the example problems is the elastic square sheet stretched at four vertexes as shown in Fig. 1. This problem is solved by FMG in which the stiffness of the element on higher hierarchy is defined by the substructure method. Figure 5 shows the convergence of the norm of an unbalanced force with iterations for 4 cases, in which the number of hierarchy changes from 2 to 8. The number of degrees of freedom (DOFs) for 8-level is 132,098. The rate of convergence becomes slightly smaller as the number of hierarchy increases but good convergence is observed generally. The relation between the computing time and the degree of freedom is summarized in Fig. 6. The computing time to achieve the relative error of 10^{-6} or 10^{-8} is plotted. The error in this figure is the relative value which is normalized by that of the first iteration. As seen from the slope of the curve, the computational time increases almost linearly with the degree of freedom up to 9-level (526,338 DOFs). In case of conventional direct solution method (Sky Line method), the computational time is proportional to $n^{2.0}$ (for two dimensional problem) or $n^{2.3}$ (for three

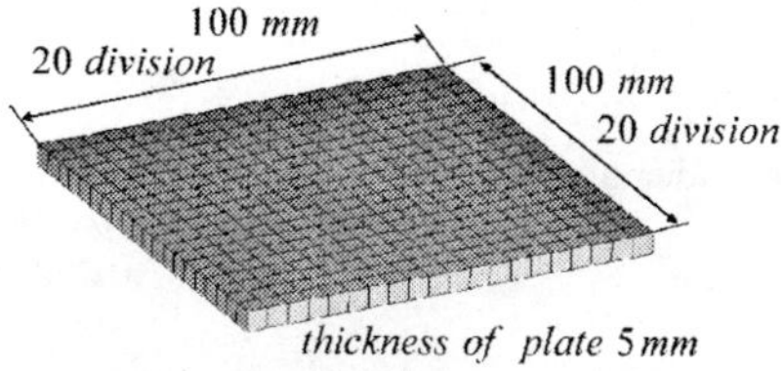

Fig. 7 Coarse global mesh of a square plate.

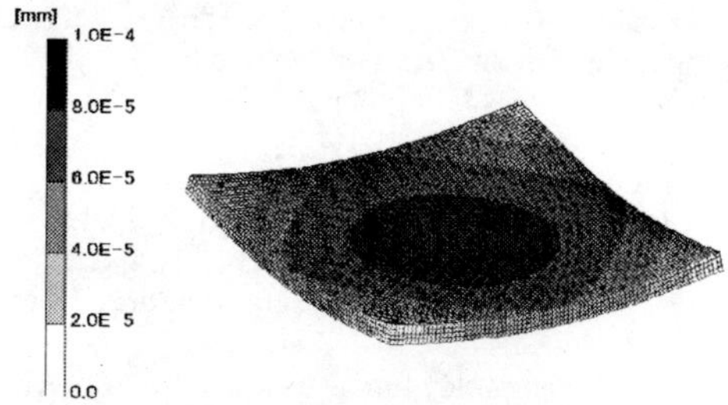

Fig. 8 Deformation of plate.

dimensional problem). Thus, the computational speed in FMG becomes about 500 faster than that in the Sky Line method.

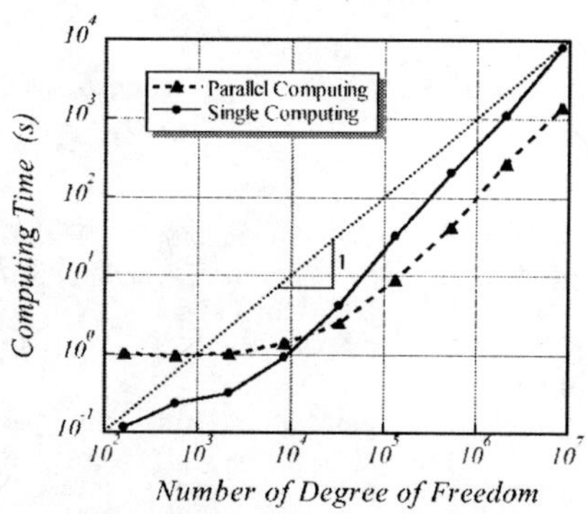

Fig. 9 Effect of parallel computing on computing time.

Plate Modeled Using Solid Elements

When the coarse global mesh is employed, the deformation of a plate can be analyzed. Figure 7 shows the coarse global mesh of a square plate. The size of the plate is 100 mm in width and length and 5 mm in thickness. The plate is divided into 20 x 20 elements. Each element is further divided into FMG of two-level or subdivided into 64 elements. Thus, the total number of elements becomes 25,600 and the degree of freedom is 98,415. The plate is assumed to be simply supported at four corners and the center of the plate is pulled down 0.0001 mm. The computed deflection is shown in Fig. 8 using contour lines. Although this example can be also solved by the Sky Line method, this is the largest which can be solved using single PC (Pentium 4 3.8GHz, 2GB RAM, MS-Windows XP) while much larger model can be solved by FMG.

Parallel Computing

Since the proposed FMG method is a kind of multi-grid method, it can be applied to parallel computing. One PC is used as mother computer for analyzing the global mesh and controlling the whole computing, and eight PCs compute the internal region of global element. Then, by exchanging only the information on the boundary of the each global mesh between the mother PC and the eight PCs, the problem can be solved in the parallel computing using the FMG. As one example for the parallel computing, the elastic square plate stretched at four vertexes is analyzed. The plate is divided into 4 x 4 global meshes and each element is divided into the fractal mesh with different hierarchical level from 1 to 9. This corresponds to 162 to 8,396,802 DOFs. Figure 9 shows the relation between the computational time and the number of degree of freedom. When the degree of freedom is smaller, the parallel computational time is larger than the time computed by one PC. This comes from the time for transferring of information. However, when the degree of freedom is larger than 20,000, the parallel computation becomes faster than that by single PC. When the degree of freedom is 8,396,802, it becomes 5.6 times faster than one PC.

CONCLUSIONS

Problems of practical interest in industry are generally large in size and complex in geometry. To solve such large problems, the authors proposed a Fractal Multi-Grid method. Its effectiveness and potential usefulness are demonstrated through several examples and the following conclusions are drawn.

(1) The computing time is roughly proportional to the degree of freedom.

(2) Ultra large problem such as a problem with 8,396,802 degree of freedom can be solved by FMG method on the parallel computers in 2,000 seconds.

(3) Using coarse global mesh, models with arbitrary geometry such as a plate can be analyzed by FMG method.

(4) The computational speed becomes faster by applying FMG method to the parallel computing when the degree of freedom is larger than 20 thousands.

REFERENCES

[1]D. Kershaw, "The incomplete Cholesky conjugate gradient method for iterative solution of systems of linear equations," *Journal of Computatinal Physics*, **26**, 43-65 (1978).

[2]P. Geng, J.T. Oden and R.A. van de Geijn, "A Parallel Multifrontal Algorithm and its Imprementation," *Computer Methods in Applied Mechanics and Engineering*, **149**, 289-301 (1997).

[3]W.L. Briggs, "Multi-Grid Methods and Applications," *Springer-Verlag*, Berlin (1985).

[4]K. Stuben, "A Review of Algebraic Multigrid," *Journal of Computational and Applied Mechanics*, **128**, 281-309 (2001).

[5]Q.V. Dinh Glowinski and J. Periaux, "Domain Decomposition Methods for Nonlinear Problems in Fluid Dynamics," *Computer Methods in Applied Mechanics and Engineering*, **40**, 27-109 (1983).

[6]J. Mandel, "Balancing Domain Decomposition," *Communications on Numerical Methods in Engineering*, **9**, 233-341 (1993).

INFLUENCE OF GRAIN-GRAIN INTERFACES ON HEAT TRANSFER IN DENSE AND POROUS OXIDE CERAMICS

D.S. Smith, C. Poulier, B. Naït-Ali, A. Michot, J. Absi
Groupe d'Etude des Matériaux Hétérogènes, ENSCI
47 - 73 avenue Albert Thomas, 87065 Limoges cedex, France

ABSTRACT

The influence of grain-grain interfaces on the effective thermal conductivity is examined in three different cases: (i) dense oxide ceramics, (ii) porous oxide ceramics, (iii) green bodies. Laser flash measurements, on dense alumina and dense tin oxide with significant grain size variation, indicate that the thermal resistivity of these materials increases with the number of grain boundaries per unit length of heat path. Values of the average grain boundary thermal resistance were calculated: 1.0×10^{-8} $m^2.K.W^{-1}$ for alumina and 1.2×10^{-8} $m^2.K.W^{-1}$ for tin oxide. Numerical simulations at the local scale were then used to show there is a strong temperature drop at a grain boundary in a porous polycrystalline solid. At the macroscopic scale, experimental thermal conductivity values for porous zirconia and porous alumina agree closely with predictions by Landauer's effective medium expression providing the grain boundary resistance is taken into account. Finally, the high thermal resistivity of a green body of pure tin oxide is explained by the small grain-grain contact area.

INTRODUCTION

Ceramic materials are generally made with a high temperature process and are often used in high temperature situations. Heat transfer is consequently a critical mechanism controlled principally by the thermal conductivity. In transient situations the heat capacity must also be taken into account. Heat conduction through a polycrystalline ceramic involves the crossing of a large number of interfaces. This paper focuses on the influence of grain-grain interfaces on the effective thermal conductivity in three different cases: (i) dense oxide ceramics, (ii) porous oxide ceramics, (iii) non consolidated ceramic materials such as a green body.

In a non-metallic crystalline solid, heat is carried essentially by lattice vibrations, termed phonons. The phonon – phonon interaction controls the thermal conductivity through a wide range of temperature and this can be sufficient understanding for many practical purposes. However in polycrystalline material grain boundaries act as additional scattering sites and hence decrease the thermal conductivity compared to the equivalent single crystal. When the intrinsic mean free path for phonon scattering is significantly less than the grain size, that is at room temperature and above, the effect of the grain boundary can be assimilated to an additional thermal resistance in the heat path. In the first part of the paper, we present thermal conductivity measurements for dense ceramic oxides. Modification to the firing cycle has been used to produce strong variations in grain size and hence in the number of grain boundaries per unit length of heat path. The data is analysed to deduce the average grain boundary thermal resistance for alumina and tin oxide.

The presence of porosity in a solid reduces the effective heat carrying cross section. Using a simplified geometry of 2 hemispherical grains in contact and an associated neighbouring pore volume, numerical simulations are used to study the effect of grain boundary thermal resistance on the thermal response at the local scale. At the macroscopic scale several analytical relations can be used to predict the thermal conductivity of the material as a function of pore volume fraction. In particular Landauer's effective medium expression is able to describe the transition of behaviour from a ceramic containing a small amount of closed porosity to a ceramic containing a larger pore volume fraction which is essentially open. Experimental data for zirconia ceramics containing upto 70% of pore volume fraction show good agreement to this approach. For more thermally conducting oxides, like alumina and tin oxide, it is necessary to account for grain boundary thermal resistance as well. This is confirmed with data for porous alumina ceramics.

In a final situation, we examine the thermal conductivity of non consolidated or only slightly consolidated porous ceramic materials such as a pressed powder compact (green body). Because of the small contact area, the role of the interface between grains should be even more extreme than for the case of consolidated porous materials. During firing the essential transformations of the microstructure involve the nature and area of the grain – grain contacts (cohesion of powder compact), elimination of porosity, and grain growth. The influence of these factors on the equivalent thermal conductivity has been studied for tin oxide. Experimental measurements for green bodies are compared to measurements for samples subjected to thermal treatments which progressively go to higher temperatures.

EXPERIMENTAL

Sample preparation

Dense samples of alumina and tin oxide were made in the form of disks. Uniaxial pressing of a dry powder was followed by sintering of the green pellet in an electrical furnace. Variation of the firing duration at 1600°C yielded alumina ceramics with 95% relative density and average grain sizes from 2 to 30 µm. For tin oxide, MnO_2 was added as a sintering aid to a 99.9% pure powder in an amount of 0.5 wt%. Variation of the firing duration at 1150°C (0.2 to 24h) yielded 95 - 96% dense ceramics with average grain sizes from 1.3 to 6.4 µm.

Porosity was generated in the ceramics by a variety of techniques. The simplest consists of reducing the firing temperature and duration; the case for a batch of alumina samples. Another method involves the use of a pore former such as potato starch which is mixed with the powder and eliminates itself during thermal treatment. Batches of porous alumina and porous zirconia supplied by the ICT Prague were made in this way. Alternatively a latex former can be used and this was done for zirconia powder supplied by AGH Crackovie to obtain upto 70% porosity. The approach also has the advantage of maintaining constant grain size despite changes in pore volume fraction. Finally if pure tin oxide (no MnO_2 additive) is used, virtually no densification occurs during firing and the initial pore volume fraction of the green body is conserved.

Microstructure characterization and thermal conductivity measurements

Sample densities were evaluated using the method based on Archimedes' principle. Results were checked with the dry weight of the disc sample divided by an estimate of its volume. This gives a slightly lower value to the first method but typically within 2%. On micrographs the linear intercept method was used to estimate n the average number of grain boundaries per unit length crossed by a heat flow path, given by the expression

$$n = \frac{N}{(L-P)} \tag{1}$$

where N is the number of grains, L is the line length and P the pore length crossed by the line. The thermal conductivity (λ) was evaluated with the laser-flash technique.

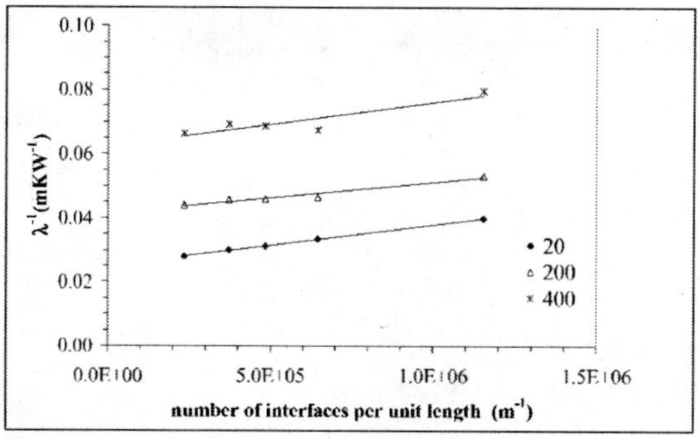

Figure 1. Inverse thermal conductivity as a function of the number of interfaces per unit length for dense tin oxide (0.5 wt.% MnO_2) at 20, 200 and 400°C.

RESULTS AND DISCUSSION

Dense ceramics

The microstructure of a ceramic material can be simplified to a 3-dimensional lattice of cubic cells. For convenience all the grain boundaries can be grouped together as one resistance and all the crystallite cells can be grouped as another resistance. Such an approach (the brick layer model) is well known in a.c. impedance studies of polycrystalline ceramics [1]. The thermal conductivity can then be written

$$\lambda^{-1}_{poly} = \frac{1}{\lambda_{crystal}} + n\,R^{*} \qquad (2)$$

where $\lambda_{crystal}$ is the single crystal thermal conductivity, n is as defined in equation (1), and R^* is the thermal resistance for a grain boundary of unit area. Figure 1 shows a plot of the inverse thermal conductivity values versus n for dense samples of tin oxide at 3 different temperatures. These data sets confirm the validity of Eq. (2). The average grain boundary thermal resistance for a polycrystalline material can be calculated directly from the slope and in this case yields 1.3×10^{-8} $m^2.K.W^{-1}$ at 20°C and after a small correction for the porosity, 1.2×10^{-8} $m^2.K.W^{-1}$. Furthermore the grain boundary thermal resistance is essentially constant as a function of temperature, consistent with a theoretical study [2]. Similar behavior is observed for alumina and a value of 1.0×10^{-8} $m^2.K.W^{-1}$ for 100% dense material at 20°C is obtained. Other workers have reported a value of 0.5×10^{-8} $m^2.K.W^{-1}$ for yttria stabilized zirconia [3].

Porous ceramics

The effect of an interface on the heat flow through a porous solid was first studied at the local scale using a numerical model. Numerical simulations were performed with a finite element analysis package, Abaqus. A difference of temperature is applied across the model system until thermal equilibrium is achieved. Then the effective conductivity is obtained from the calculated average heat flow density using Fourier's equation. The chosen geometry for the model consists of two hemispheres in contact (Coble geometry) and the resulting temperature distribution along the central axis was calculated with and without a grain boundary resistance, Figure 2. On the curve "No resistance", the effect of the constriction on the heat flow is revealed by the increase in the thermal gradient in the vicinity of the neck. With a grain boundary thermal resistance of $1x10^{-8}$ $m^2.K.W^{-1}$, corresponding to literature values for dense polycrystalline ceramics, this becomes a sharp temperature drop at the interface. The influence of grain size variation was studied by changing the length of the axis of revolution with the values: 2 μm, 20 μm, 2 mm and 20 mm. It can be noted that for a perfect contact, despite the constriction, there is no effect on the value of effective thermal conductivity. In contrast the imperfect contact ($1x10^{-8}$ $m^2.K.W^{-1}$) yields a decrease in effective thermal conductivity for smaller grain sizes.

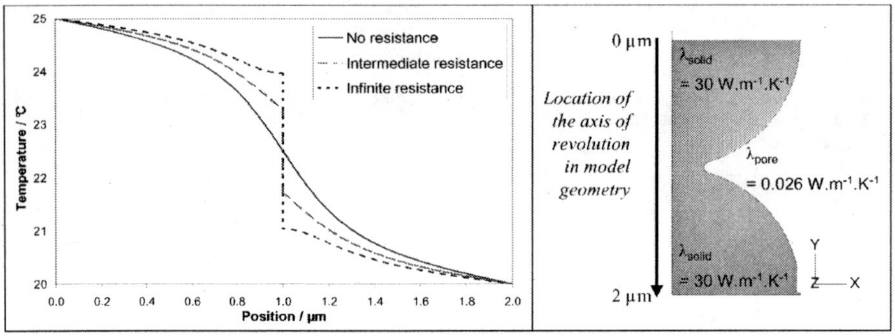

Figure 2. Two hemispheres in contact with a grain boundary. Temperature distributions along the axis of revolution for different values of the interface resistance; the intermediate one is $1x10^{-8}$ $m^2.K.W^{-1}$

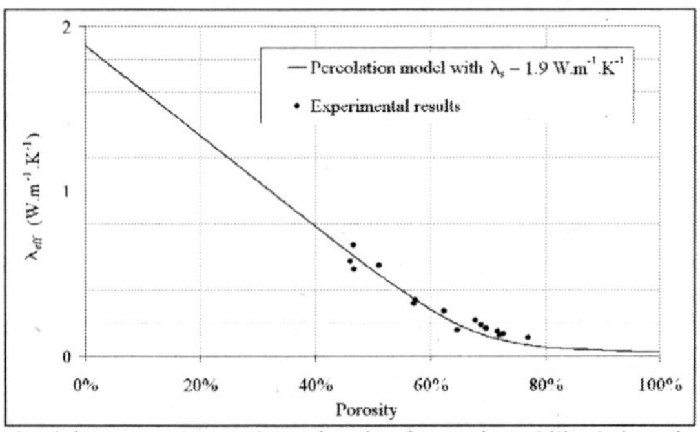

Figure 3. Thermal conductivity versus pore volume fraction for yttria stabilized zirconia at 20°C. Calculation uses Eq. 3, λ_s = 1.9 $W.m^{-1}.K^{-1}$ and λ_p = 0.026 $W.m^{-1}.K^{-1}$.

We now examine the thermal conductivity of a porous ceramic that would be measured at a macroscopic scale much larger than the pore size. For materials containing randomly dispersed pores, corresponding to the case for ceramics obtained by pressing of a powder compact in a die, Landauer's effective medium expression (percolation model) can be used to describe the effective thermal conductivity of such a two-phase system [4]. This is given by:

$$\lambda = \frac{1}{4}\left[\lambda_p(3v_p - 1) + \lambda_s(2 - 3v_p) + \left\{\left[\lambda_p(3v_p - 1) + \lambda_s(2 - 3v_p)\right]^2 + 8\lambda_s\lambda_p\right\}^{1/2}\right] \quad (3)$$

where λ_s is the solid phase thermal conductivity, λ_p is the thermal conductivity of the pore phase, and v_p is the pore volume fraction. The predictions agree closely to experimental data for 8 mol.% yttria stabilized zirconia ceramics containing up to 70% porosity, Figure 3. One reason is that the low thermal conductivity of the zirconia grains masks the effect of the grain boundaries on λ_s.

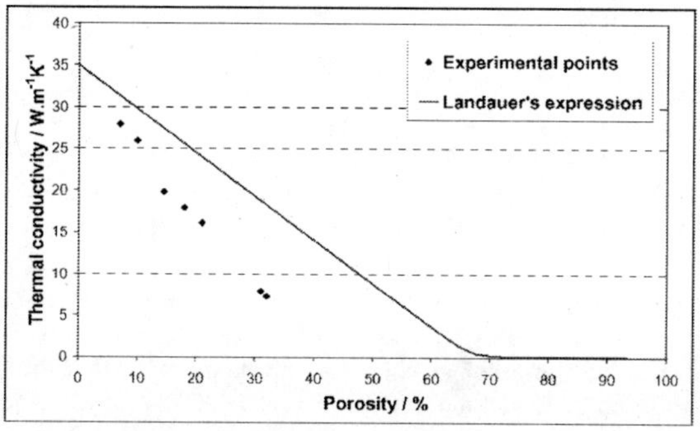

Figure 4. Thermal conductivity versus pore volume fraction for alumina ceramics at 20°C. Average grain size decreases with porosity from 2 μm (7%) to 0.6 μm (30%). Calculation uses Eq. 3, see text for values.

Alumina has a significantly higher single crystal conductivity compared to zirconia. Divergence, from predictions made with Landauer's expression with $\lambda_s = 35\ W.m^{-1}.K^{-1}$ and $\lambda_p = 0.026\ W.m^{-1}.K^{-1}$, can be seen for alumina ceramics with porosities upto 30% but also decreasing grain size, Figure 4. This can be explained by the greater relative effect of the grain boundary resistance. However if the grain size is fixed at a constant value by the preparation method, good agreement between Landauer's expression and experimental results can be achieved when λ_s is calculated with Eq. (2).

Green bodies

The discussion, up to this point, implies that the thermal resistance of grain boundaries can affect the overall conductivity of a porous ceramic. We now consider the more extreme case of a green body where the grain boundaries have not yet been formed through thermal treatment and for which also the contact area of grain – grain interfaces will be small. Pure tin oxide without a sintering additive has the particularity of maintaining a virtually constant pore volume fraction ($v_p = 0.47$ for samples in Figure 5) under different firing cycles. In contrast the grain size varies strongly for thermal treatments above 1000°C. It can be seen that the thermal resistivity decreases with the increase of grain size, Figure 5. This behaviour is well described by Eq. (2) where the contribution of the grain boundary term becomes progressively smaller as n decreases. The predominance of the grain boundary resistance in the small grain samples (0.2 μm) is also supported by the observation of an almost constant value of λ^{-1} with temperature. Theoretical studies predict that the thermal resistance of a grain boundary approaches a constant value for $T > \theta_D/2$ where θ_D is the Debye temperature [2]. In contrast the large grain sample (4.9 μm) yields a steady linear increase with temperature corresponding to phonon-phonon scattering and therefore predominance of the intrinsic grain thermal conductivity.

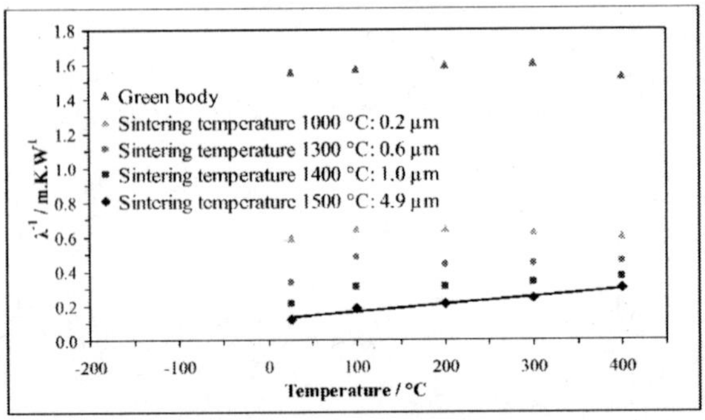

Figure 5. Thermal resistivity of a green body of pure tin oxide treated at 500°C compared with the samples sintered above 1000°C - Grain sizes are also indicated.

In fact there are complications in the initial behaviour of a green body under thermal treatment and so the measurements were made after a heating cycle to 500°C to obtain reproducible values. It could be assumed as a first approximation that the grain size variation between the tin oxide green body and the sample fired at 1000°C is small. The much higher value of thermal resistivity for the green body (factor of 2.5) compared to the sintered sample can then be explained by a much smaller contact area between the grains in the pressed powder compact. This is consistent with the microscopic numerical models. We note also that λ^{-1} does not vary significantly with temperature. Any contribution to heat transfer due to radiation is assumed to be attenuated by the grain boundaries and pores and hence negligible.

CONCLUSIONS

The average grain boundary thermal resistance is of the order of 10^{-8} $m^2.K.W^{-1}$ in a dense oxide ceramic. This contribution to the overall thermal resistance of a porous polycrystalline solid is amplified by the reduction in heat carrying cross section. The thermal resistivity becomes particularly high in the case of a green body where the grain-grain interfaces have a very small contact area.

REFERENCES

[1]J. Fleig and J. Maier, "A finite-element study on the grain boundary impedance of different microstructures", J. Electrochem. Soc. **145**, 2081-89 (1998).
[2]D.A. Young and H.J. Maris, "Lattice dynamical calculation of the Kapitza resistance between fcc lattices", Phys. Rev. B: Condens Matter, **40**, 3685-93 (1989).
[3]H.S. Yang, G.R. Bai, L.J. Thompson, and J.A. Eastman, "Interfacial thermal resistance in nanocrystalline yttria stabilized zirconia", Acta Mater., **50**, 2309-17 (2002).
[4]R. Landauer, "The electrical resistance of binary metallic mixtures", J. Appl. Phys., **23**, 779-84 (1952).

FABRICATION OF COLLOIDAL PHOTONIC CRYSTALS FROM SUBMICRON-SIZED SPHERES

Daisuke Nagao, Mitsuaki Hirose, Ryoji Kameyama, Hideki Matsumoto, Yoshio Kobayashi, Mikio Konno*
Department of Chemical Engineering, Graduate School of Engineering, Tohoku University
6-6-07, Aoba, Aramaki-aza, Aoba-ku
Sendai, 980-8579, Japan

ABSTRACT

A number of colloidal methods were examined to rapidly fabricate photonic crystals from well-ordered polystyrene (PSt) or silica particles with controlled layers. Using evaporation methods, colloidal solvent was evaporated under a condition to control the deposition of the well-ordered particles on a substrate. However, the time required to produce well-ordered particle arrays was long, with the deposition rates for PSt particles less than 0.02 μm/sec. To increase the deposition rate, the substrates were lifted up from the suspensions at a constant speed. Although slow lift up of the substrates resulted in the fabrication of a close-packed structure on the substrates, the number of layers on the substrates was relatively high. A simple dip-coating method proposed by the authors can fabricate close-packed particles in a single layer. Repetition of the dip-coating can fabricate a double-layer of PSt particles on a substrate and a PSt particle mono-layer on a silica particle mono-layer. The fabrication of well-ordered structures shows that the dip-coating is a possible way to rapidly fabricate well-ordered particle arrays with a small number of layers.

INTRODUCTION

Colloidal photonic crystals that have photonic band gaps located in the visible and near-infrared region have attracted increasing attention. Silica or polystyrene (PSt) particles have frequently been used for fabrication of the colloidal crystals, because monodisperse and spherical particles are easily prepared with a conventional synthetic method[1-3]. On the other hand, colloidal crystals have been fabricated with various methods. The sedimentation method can simply fabricate well-ordered structure of spherical particles in solutions. However, the time needed for fabricating the colloidal crystals is long: more than a week for silica particles and more than a month for PSt particles. Another problem in the method is the difficulty in controlling of the number of layers in the colloidal crystals.

Fabrication methods enabling control of the number of layers on a substrate can be classified into two groups. One is an evaporation method that controls evaporation rate of colloidal solvent to deposit the well-ordered particles on a substrate. Another is a lift up method that lifts the substrate from the colloidal solution via control of the evaporation rate and the lift up speed. Particle concentration, evaporation rate, lift up speed and particle diameter have been regarded as parameters to quantify the number of layers of colloidal crystals[4].

To obtain the required property for a photonic crystal, both the regularity of colloidal crystals and the number of layers have to be simultaneously controlled. In the present work, various types of fabrication methods were studied to control regularity and number of layers. Methods that can rapidly fabricate well-ordered colloidal crystals with controllable layers were examined.

EXPERIMENTAL

All chemicals used were special grades of Wako Pure Chemical Industries, Ltd. (Osaka, Japan). Ethanol (99.5%), ammonia (25% aqueous solution), methylamine (40% aqueous solution) and tetraethyl orthosilicate (95%) were used for the synthesis of silica particles. Styrene (99%) and potassium persulfate (95%) were used for the synthesis of polystyrene particles. Styrene was purified by inhibiter removal and distillation under a nitrogen atmosphere. The other chemicals were used as received. Water was deionized and distilled to have an electric resistance higher than 18 MΩ cm.

The colloidal suspension of PSt particles was prepared according to our previous work[8]. Soap-free emulsion polymerization was performed in a batch reactor (11 cm in inside diameter, 12 cm in height) equipped with four baffles and a six-blade pitched paddle impeller with a 6 cm diameter, which was located one-third of liquid height from the bottom. The polymerization was conducted in water with potassium persulfate (KPS) initiator at 70 °C. PSt particles with two different number-averaged sizes (d_n) of 520 and 280 nm were prepared at styrene concentrations of 1.1 and 0.022 kmol/m^3 H_2O, respectively, for a KPS concentration of 4 mol/m^3 H_2O. Coefficient of variation of particle size distributions (C_V) was 2.0 % for the 520-nm particles and 2.3 % for the 280-nm particles.

The colloidal suspension of silica particles was prepared with a sol-gel method according to our previous papers[9, 10]. Silica particles with d_n = 268 and 233 nm were synthesized by the use of methylamine catalyst at 0.2 and 0.1 kmol/m^3, respectively, for 0.2 kmol/m^3 TEOS. Values of C_V were 3.8 % for 268-nm particles and 3.5 % for 233-nm particles.

The colloidal suspensions were washed by repeating centrifugation (Tomy SRX-201 high-speed refrigerated centrifuge) of the suspensions, removal of supernatant and addition of pure water. The content of the particles in the suspension was varied in a concentration range of 0.1 - 2 vol. %.

Glass substrates were immersed in the piranha solution, rinsed with water and dried by nitrogen blowing. In the evaporation method, the substrate was vertically inserted into a suspension, and the position of the substrate was kept under conditions of 20 – 23 °C and 20 – 40 relative humidity (RH) %. In the lift up method, the substrate immersed in a suspension, was lifted up at constant speeds of 0.38 – 0.89 μm/sec. In a dip-coating method proposed by the authors[12], the substrate was wetted by dropping the suspension on the surface. Then, the substrate was inclined vertically until the suspension was dried.

RESULTS AND DISCUSSION

Fabrication of PSt colloidal crystals

Colloidal photonic crystals of polystyrene (PSt) particles were fabricated by evaporation of 0.2 vol% aqueous PSt suspension according to the previous report[5]. Temperature and relative humidity were adjusted to 60 °C and 90 RH%, respectively. The condition enabled the fabrication of well-ordered PSt colloidal crystal with layers more than 40. However, it took more than a day to form a 1 mm-length PSt colloidal crystal, corresponding to a lift up speed lower than 0.02 μm/sec.

To increase the formation rate of colloidal crystal, the substrate was lifted up at different speeds. **Figure 1** shows SEM images of PSt colloidal crystals fabricated with the lift up method. The number of particle layers in the images of Fig.1 (A) – (C) are 10, 5 and 1, respectively. The decrease in the number of layers with lift-up speed was also reported by previous researchers[4].

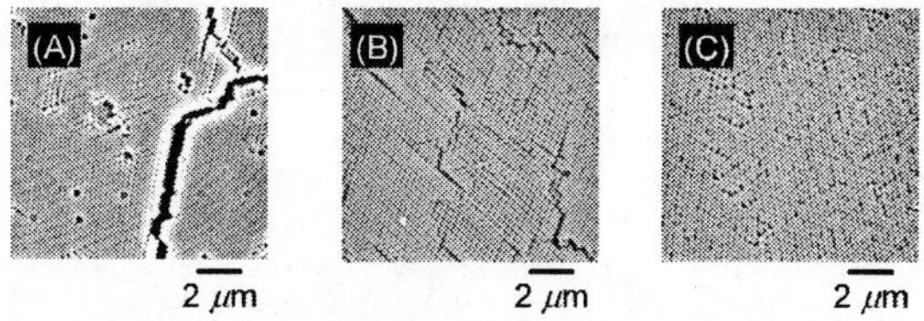

Figure 1 SEM images of PSt colloidal crystals fabricated with a lift up method at different speeds of (A) 0.38, (B) 0.65 and (C) 0.89 μm/s. The concentration of PSt particles was 1.0 %.

Most particles in the colloidal crystal fabricated at a low speed of 0.38 μm/s (Fig.1 (A)) were hexagonally close-packed, although a small number of cracks or point defects, which were reported in other lift-up methods[7, 11], were observed. In contrast, in the images of (B) and (C), a large number of lines and dots comprising voids among particles were observed, and the number of close-packed particles were much less than that in the image of (A).

The concentration of particles in solution was varied to control the number of particle layers in the lift up method[4]. **Figure 2** shows SEM images of PSt colloidal crystals fabricated in a particle concentration range of 0.3 – 2.0 vol% at 0.38 μm/s. The SEM images indicate that more than 90 % of particles on the substrates lifted up at concentrations of 1.0 and 2.0 vol% were hexagonally close-packed.

The number of particle layers for the colloidal crystals in Fig. 2 are shown in **Figure 3**, where the number of layers for the other lift up speeds are also shown. Open symbols in Fig. 3 indicate that close-packed structures like Figs. 2 (C) and (D) were obtained at the high concentration of PSt particles . High concentrations of PSt particles can produce hexagonally close-packed structures with more than 10 layers, while the lower concentration of particles leads to the formation of a large number of voids in the particle array. It is concluded that the lift up method is effective for fabrication of close-packed particle arrays with a large number of layers.

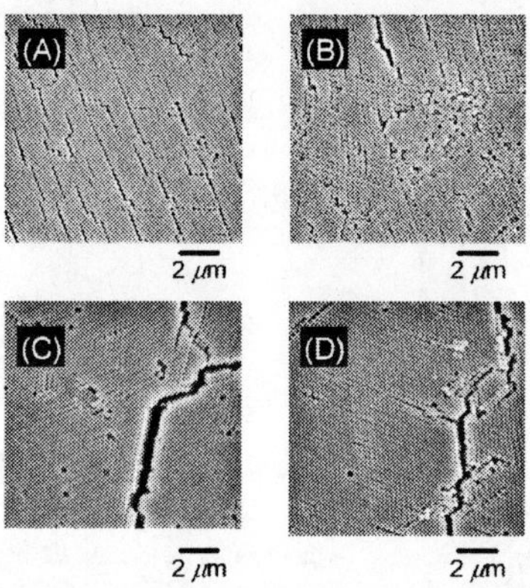

Figure 2 SEM images of PSt colloidal crystals fabricated with a lift up method at 0.38 μm/s for the particle concentrations of (A) 0.3, (B) 0.5, (C) 1.0 and (D) 2.0 vol%.

Fabrication of Silica colloidal crystals

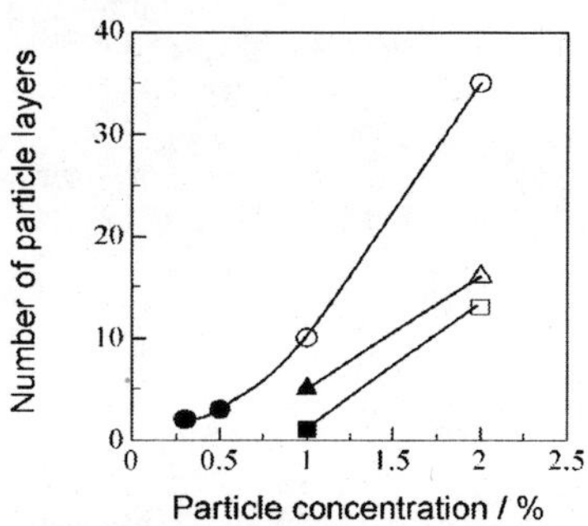

Figure 3 Number of particle layers of colloidal crystals fabricated at different concentrations of PSt particles. The lift up speeds of 0.38, 0.65 and 0.89 μm/s were shown with symbols of circles, triangles and squares. Open symbols indicate close-packed structure of colloidal crystal.

To fabricate the colloidal crystal of silica particles with an evaporation method, silica particles were suspended in water at 20 °C and 40 %RH. However, the low evaporation rate of water caused sedimentation of particles before the deposition of silica particles on a substrate.

Alternatively, ethanol was used as a solvent to promote the formation of colloidal crystals. **Figure 4** shows a SEM image of colloidal crystals fabricated with an ethanolic solution of silica particles at the concentration of 1.0 vol%. Temperature and relative humidity in the fabrication were adjusted to 20 °C and 40 RH%, respectively. A closed-packed structure of silica particles was obtained in a wide area of the substrate. However, the formation rate of colloidal was <0.02 μm/sec that was calculated from the evaporation rate of ethanol[7].

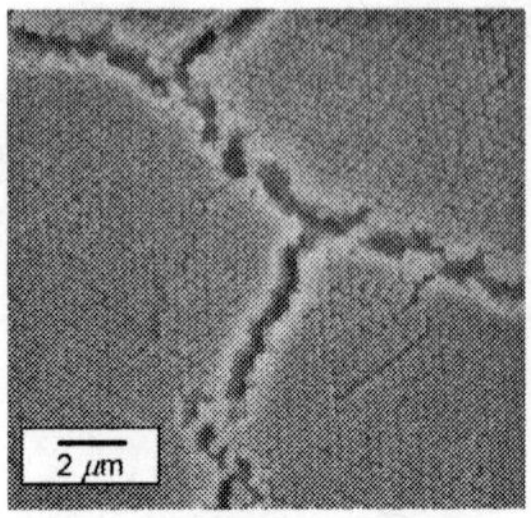

Figure 4 SEM images of Silica colloidal crystals fabricated with an evaporation method at the particle concentration of 0.5 vol%. Temperature and relative humidity were adjusted to 20 ℃ and 40 %

To further promote the formation rate of colloidal crystals, the substrate was lifted up at a slow speed of 0.38 μm/sec. **Figure 5** shows SEM images of silica colloidal crystals with the lift up method at particle concentrations of 0.5, 1.0 and 2.0 vol%. A large number of voids between silica particles is observed in Fig. 5 (A). Although an increase in the particle concentration seems to decrease the number of voids, complete close-packed structures are scarcely observed in Fig.5 (B). On the other hand, a highly close-packed structure was obtained for the concentration of 2.0 vol %.

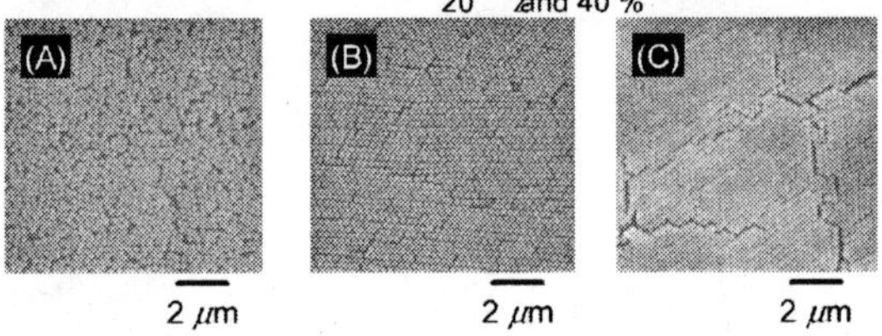

Figure 5 SEM images of Silica colloidal crystals fabricated with a lift up method at 0.38 μm/s for the particle concentrations of (A) 0.5, (B) 1.0 and (C) 2.0 vol%.

Figure 6 shows the number of layers of colloidal crystals at the different concentrations of silica particles. For comparison, the numbers of layers formed with the evaporation method were also shown in the same figure. Close-packed structure of silica particles was illustrated with open symbols in Fig. 6. The open symbols show that close-packed structures were obtained only for layers more than 8. Close-packed structure of silica particles with a small number of layers could not be obtained with the lift up method.

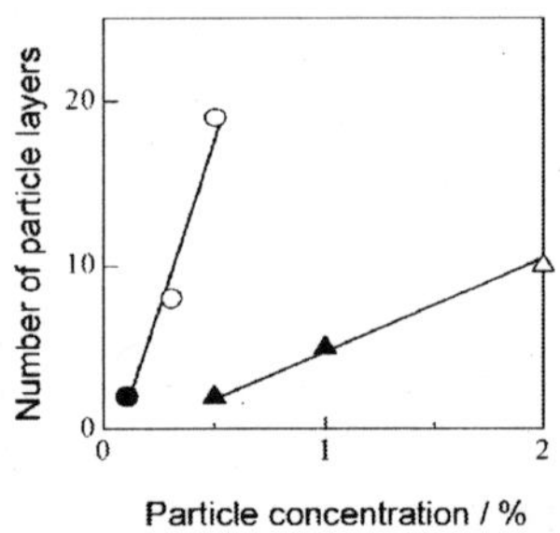

Figure 6 The number of layers in silica colloidal crystals fabricated with and without lifting up of substrate. The substrate was lifted up at 0.38 μm/s. Open symbols indicate close-packed structure for a colloidal crystal

Fabrication of well-ordered particles with a small number of layers

The authors recently proposed an alternative method that was a simple dip-coating for monolayer of particles with close-packed structure[12]. The dip-coating can fabricate the structure on a single side of the substrate within an hour, and achieve mulilayers of close-packed particles by repeated iteration of dip-coating and drying. **Figure 7** shows an example of an image of double-layered PSt particles (520 nm) by the double dip-coatings. In the center of SEM image, the layer of well-ordered PSt particles was observed under the surface layer of PSt particles, and the PSt particles in both layers were generally close-packed.

Figure 7 SEM image of PSt particle array prepared by double dip-coatings (d_n = 520 nm)

Figure 8 Top view of the particle array prepared by the double coatings with PSt particles (d_n = 520 nm) and silica particles (d_n = 233 nm)

Figure 8 shows another example image of close-packed particles obtained with the dip-coating method. A colloidal solution of 233-nm silica particles were dropped onto a monolayer of close-packed 520-nm PSt particles. PSt particle concentration in the first dip-coating was 10 vol% and silica particle concentration in the second coating was 2 vol%. Interestingly, the second layer of silica particle is located between the substrate and the monolayer of PSt particles. Thus, the simple dip-coating could be applied to the fabrication of close-packed structure consisting of different particles with different diameters. PSt is hydrophobic with low affinity to water while silica is hydrophilic with a high affinity to water. This difference in the affinity might create the interesting structure.

CONCLUSION

The lift up method is effective for fabricating close-packed PSt and silica particles with a large number of layers. Dip-coating can fabricate a close-packed structure on a single side of a substrate in a relatively short time, and is an effective way to fabricate colloidal crystals with a small number of layers.

REFERENCES

[1]C. Lopéz, A. Blanco, H. Míguez, and F. Meseguer, "Photonic Crystals for Laser Action," *Opt. Mater.*, **13**, 187-192 (1999).

[2]M. Trau, D. A. Saville, and I. A. Aksay, "Field-Induced Layering of Colloidal Crystals," *Science*, **272**, 706-709 (1998).
[3]L.M. Goldenberg, J. Wagner, J. Stumpe, B.-R. Paulke, and E. Görnitz, "Ordered Arrays of Large Latex Particles Organised by Vertical Deposition," *Mater. Sci. Eng. C*, **22**, 405-408 (2002).
[4]A. S. Dimitrov, and K. Nagayama, "Continuous Convective Assembling of Fine Particles into Two-Dimension Arrays on Solid Surfaces," *Langmuir*, **12**, 1303-1311 (1996).
[5]S.-L. Kuai, X.-F. Hu, A. Haché, and V.-V. Truong, "High-Quality Colloidal Photonic Crystals Obtained by Optimizing Growth Parametes in a Vertical Deposition Technique," *J. Cryst. Growth*, **267**, 317-324 (2004).
[6]K. P. Velikov, C. G. Christva, R. P. A. Dullens, and A. v. Blaaderen, "Layer-by-Layer Growth of Binary Colloidal Crystals," *Science*, **296**, 106-109 (2002).
[7]P. Jiang, J. F. Bertone, K. S. Hwang, and V. L. Colvin, "Single-Crystal Colloidal Multilayers of Controlled Thickness," *Chem. Mater.*, **11**, 2132-2140 (1999).
[8]S. Orihara, and M. Konno, "A Simple Method for Producing Micrometer-Sized Monodisperse Polystyrene Particles in Aqueous Media," *J. Colloid Interface Sci.*, **230**, 210-212 (2000).
[9]D. Nagao, E. Mine, Y. Kobayashi, and M. Konno, "Synthesis of Silica Particles in the Hydrolysis of Tetraethyl Orthosilicate with Amine Catalysts," *J. Chem. Eng. Japan*, **37** 905-907 (2004).
[10]D. Nagao, H. Osuzu, A. Yamada, E. Mine, Y. Kobayashi, and M. Konno, "Particle Formation in the Hydrolysis of Tetraethyl Orthosilicate in pH Buffer Solution," *J. Colloid Interface Sci.*, **279** 143-149 (2004).
[11]Z.-Z. Gu, A. Fujishima, and O. Sato, "Fabrication of High-Quality Opal Films with Controllable Thickness," *Chem. Mater.*, **14** 760-765 (2002).
[12]D. Nagao, R. Kameyama, Y. Kobayashi, and M. Konno, "Multiformity of Particles Arrays Assembled with a Simple Dip-Coating," *Colloid Surf. A*, accepted.

SPARK PLASMA SINTERING OF Al_2O_3 PARTICULATE DISPERSED $Zr_{55}Cu_{30}Al_{10}Ni_5$ METALLIC GLASSY MATRIX COMPOSITE

Guoqiang Xie, Dmitri V. Louzguine-Luzgin, Akira Okubo, Hisamichi Kimura, and Akihisa Inoue
Institute for Materials Research, Tohoku University
Katahira 2-1-1, Aoba-Ku, Sendai 980-8577, Japan

ABSTRACT

A spark plasma sintering technique was successfully used to consolidate $Zr_{55}Cu_{30}Al_{10}Ni_5$ bulk metallic glassy matrix composite containing Al_2O_3 particulates. No crystallization of the glassy matrix occurred during the spark plasma sintering process. The Al_2O_3 particles were homogeneously dispersed in the $Zr_{55}Cu_{30}Al_{10}Ni_5$ glassy matrix. The crystallization behavior of the sintered composite was similar to that of the metallic glassy powder. A large compressive strength of the sintered composite specimens was exhibited.

INTRODUCTION

Bulk metallic glasses (BMGs) exhibit many interesting physical and chemical properties such as high strength, high corrosion resistance and good soft magnetic properties. Such materials have attracted increasing attention in recent years due to their scientific and engineering significance, and are considered as promising candidates for the next generation of structural and functional materials.[1] However, monolithic BMGs usually exhibit poor ductility (inability to deform plastically) due to the formation of highly localized shear bands under loading at room temperature, which leads to brittle fracture features and catastrophic failure of the materials. Such absence of the plasticity limits the extensive practical applications of the BMGs.

In order to improve ductility of metallic glasses, a number of studies have been carried out. It has been demonstrated that the BMGs containing hard particles or fibres, such as ZrC, SiC, WC, Ta and W exhibit either higher strength or larger plastic strain as compared to monolithic glassy phase.[2-5] Recently the glassy matrix plus crystalline ceramic particles mixed structure alloys have been prepared by a copper mould cast technique. For example, $Cu_{60}Zr_{30}Ti_{10}$ metallic glassy matrix composite material containing 10 vol.% of ZrC dispersoid exhibited 3 % plastic deformation in the as cast state.[6] The composite structure was prepared by in situ reaction between the melt and graphite micro-powder. However, by this method it is difficult to obtain a homogeneous distribution of the ceramics particles. Furthermore, in order to prevent the formation of the more thermodynamically stable crystalline phases, a sufficiently high cooling rate is always required during casting. Therefore, the dimensions of the BMGs are limited, and the shape is also rather simple.

On the other hand, the spark plasma sintering (SPS) process, as a novel technique developed for rapid sintering metals, ceramics and composite materials, has a great potential for producing specimens with nearly no crystallization for amorphous materials or minimal grain growth for crystalline materials in a short sintering time.[7] The SPS process is a type of solid compression sintering technique which is similar to hot-pressing (HP) sintering process, so that the sintered specimens with the large-size and complicated shape can be easily achieved.[8] In the SPS process, a pulse electrical current flows directly in the sintered materials. A very high heating efficiency is achieved. Furthermore, it is also postulated that the high electric-pulse power might generate plasma in the gaps between powder particles,[8] although the plasma generation during the SPS process has still not been experimentally verified. The plasma can partially eliminate adsorbed gases and oxide films on nonoxide particles, and activate the surface of the powders. The sintering can be carried out at a lower temperature with a shorter time than that of conventional sintering processes. Therefore, crystallization of the metallic glassy alloy and coarsening of the dispersed particles can be avoided. However, up to now, the investigations about metallic glass powders sintered by the SPS process are relatively limited. In particular, no report has been presented about the SPS consolidation of metallic glassy matrix composite.

In the present study, we will focus on sinterability of metallic glassy matrix composite by the SPS process. The Zr-Cu-Al-Ni alloy is known to be one of the best glass formers. The alloy is very robust against heterogeneous nucleation at surfaces or interfaces, making it desirable for use as the matrix in the development of novel bulk metallic glassy matrix composites. In traditional metal matrix composites, the most widely used reinforcements are ceramics, such as Al_2O_3. Another useful feature is that ceramic particles have a high strength and low density, and thus, the materials with high specific strength (strength to density ratio) can be expected. Therefore, the $Zr_{55}Cu_{30}Al_{10}Ni_5$ alloy powder and the Al_2O_3 powder were used in this study.

EXPERIMENTAL PROCEDURES

Master ingots of the $Zr_{55}Cu_{30}Al_{10}Ni_5$ alloy (composition is given in nominal atomic percentages) were prepared by arc melting a mixture of high purity Zr, Cu, Al and Ni in an argon atmosphere purified using Ti getter. The $Zr_{55}Cu_{30}Al_{10}Ni_5$ metallic glassy alloy powders were produced by a high pressure argon gas atomization method. The ingots were melted again under an evacuated state in a quartz tube using an induction heating coil, followed by injection through a nozzle with a diameter of 0.8 mm and then atomization by high pressure argon gas with a dynamic pressure of about 8.0 MPa.

Following atomization, size distribution of the gas-atomized powders was measured by a conventional sieving method, and the powders were classified. The glassy alloy powder with the size in the range of 25 μm < d < 37 μm (d is the diameter of the powder particles) was used in the present study. Then the $Zr_{55}Cu_{30}Al_{10}Ni_5$ glassy alloy powder and 10 vol.% Al_2O_3 powder having a purity of 99.99% were uniformly blended. The mixed powders were pre-compacted, and then sintered. The sintering of the mixed powders was carried out in a vacuum using a SPS system (Model SPS-1050, Sumitomo Coal Mining Co. Ltd., Japan). The schematic illustration of the SPS system is shown in Fig. 1. This system is composed of control unit, mechanical driving unit, vacuum system and the chamber where the sample and the dies are inside. The sintering temperature used was 623 K, which is lower than the glass transition temperature (T_g = 685 K) of the $Zr_{55}Cu_{30}Al_{10}Ni_5$ glassy alloy powders. The heating rate was 50 K/min (from room temperature to T_S-30 K) and 5 K/min (T_S-30 K to T_S) (T_S: sintering temperature). The holding time at sintering temperature was 10 min. A

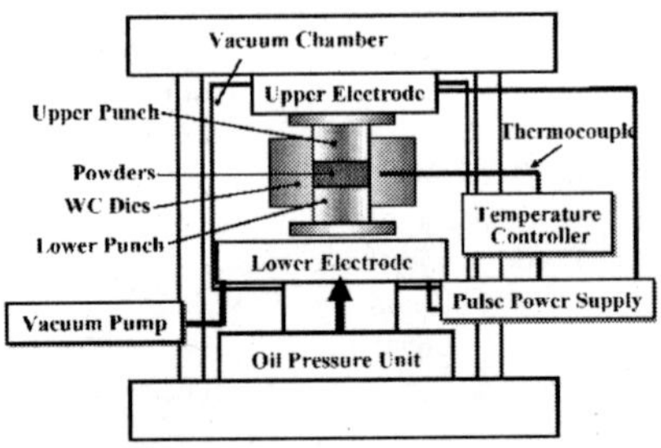

Fig. 1 Schematic illustration of a spark plasma sintering equipment used in the present study.

uniaxial pressuring method was conducted using top and bottom WC hard metal punches. The loading pressure was 600 MPa in the present study. The details of temperature and loading pressure control model of the SPS process are shown in Fig. 2. The sintered specimens obtained were a cylindrical shape with a diameter of 10 mm and a height of about 5 mm.

The structures of the powders and sintered BMG composite specimens were examined by X-ray diffractometry (XRD) in reflection with a monochromatic Cu Kα radiation, and the microstructure was characterized by scanning electron microscopy (SEM). Thermal stability associated with the glass transition temperature (T_g), supercooled liquid region (ΔT_X), and crystallization temperature (T_X) was examined by differential scanning calorimetry (DSC) at a heating rate of 0.67 K s^{-1}. The density of the sintered BMG composite specimens was determined by measuring their mass and dimension of the samples (mass per volume). Mechanical properties under a compressive load were measured for three or four samples of each condition at room temperature using a conventional mechanical testing system (Instron 4201). The test specimens with a rectangular shape of 2.5 mm in width, 2.5 mm in thickness and 5.0 mm in height were used in the present study. The initial strain rate used was 5×10^{-4} s^{-1}.

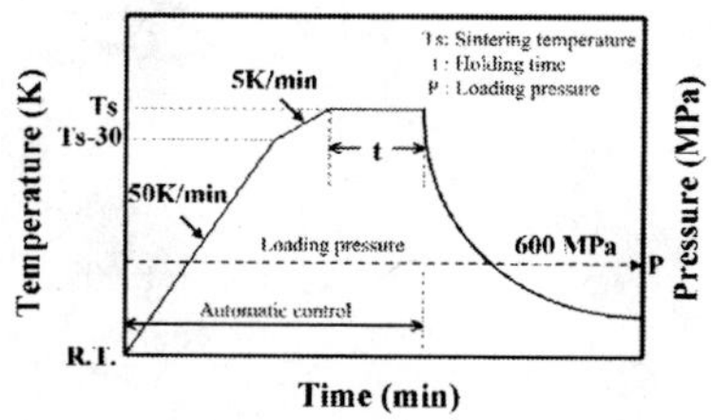

Fig. 2 Temperature and loading pressure control model of the SPS process.

RESULTS

Using the mixed powder of the gas-atomized $Zr_{55}Cu_{30}Al_{10}Ni_5$ glassy alloy powder and 10 vol.% Al_2O_3 powder, the sintering was performed by the SPS process at a sintering temperature of 623 K with a loading pressure of 600 MPa. The density of the sintered composite was measured and was found to be 6.17 Mg/m^3, which was about 94.2 pct of the theoretical density, 6.547 Mg/m^3. Microstructure of the sintered composite specimen was observed with a SEM. Figure 3 shows a typical SEM micrograph of the cross section. The white parts in Fig. 2 indicate the Al_2O_3 particulates, the dark parts are pores, and the gray parts are $Zr_{55}Cu_{30}Al_{10}Ni_5$ glassy matrix. It is seen that the Al_2O_3 powders are relatively homogeneously dispersed in the $Zr_{55}Cu_{30}Al_{10}Ni_5$ glassy matrix. Except for the Al_2O_3 crystalline phase, no appreciable contrast revealing the formation of other crystalline phase in the $Zr_{55}Cu_{30}Al_{10}Ni_5$ glassy matrix is observed.

Figure 4 shows the XRD patterns of the sintered composite specimen and the monolithic gas-atomized $Zr_{55}Cu_{30}Al_{10}Ni_5$ glassy powder, as well as the monolithic Al_2O_3 powder. The monolithic $Zr_{55}Cu_{30}Al_{10}Ni_5$ glassy powder shows only a diffuse diffraction pattern typical for a glassy phase, and no diffraction peak corresponding to crystalline phases is observed, while the XRD pattern of the sintered composite specimen shows sharp peaks diffracted from the Al_2O_3 superimposed on a broad halo peak, indicating that the matrix of the sintered composite consists of a fully glassy phase.

Fig. 3 SEM image of the sintered metallic glassy matrix composite with 10 vol.% Al_2O_3 powder.

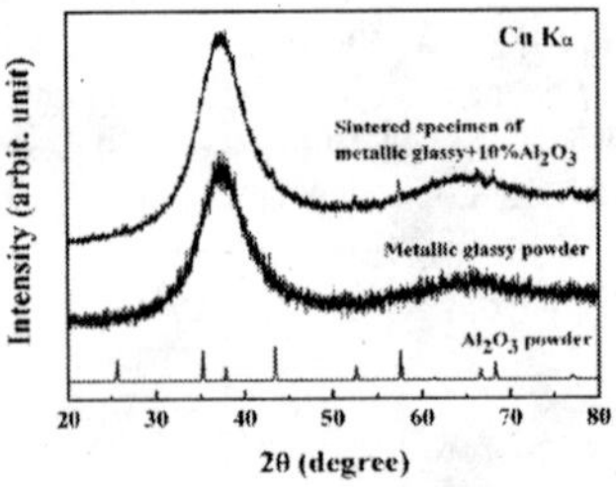

Fig. 4 XRD patterns of the sintered composite, the monolithic glassy alloy powder and Al_2O_3 powder.

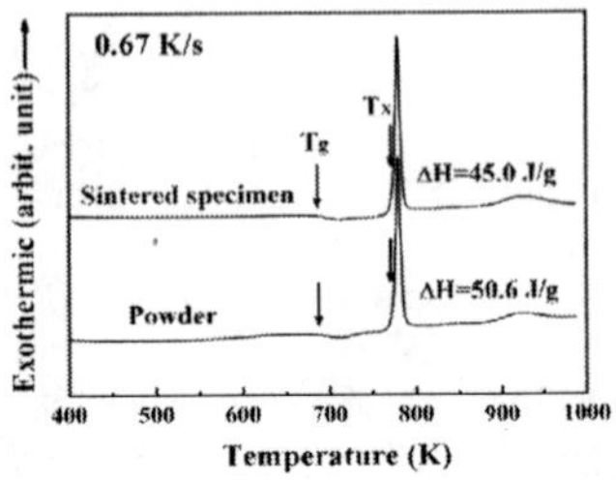

Fig. 5 DSC curves of the sintered composite and the monolithic glassy alloy powder.

Figure 5 shows the DSC curves of the sintered composite specimen and the monolithic gas-atomized $Zr_{55}Cu_{30}Al_{10}Ni_5$ glassy powder taken at a heating rate of 0.67 K s^{-1}. The features of the DSC curve of the sintered composite specimen are similar to that of the original powder, namely, an endothermic reaction due to the glass transition, followed by a large supercooled liquid region and two exothermic reactions due to crystallization events. The sintered composite specimen shows glass transition temperature (T_g) of 685 K, the onset temperature of the first-stage crystallization (T_X) of 774 K, and the extent of the supercooled liquid region ($\Delta T = T_X - T_g$) of 89 K are same for those from the original powder. Furthermore, the crystallization enthalpy (ΔH) of the sintered composite specimen is 45.0 J/g. It is around 90% of that of the monolithic gas-atomized $Zr_{55}Cu_{30}Al_{10}Ni_5$ glassy powder, which is 50.6 J/g, as indicated in Fig. 5.

Figure 6 shows nominal compressive stress–strain curves of the sintered composite specimen, together with that of the as-cast $Zr_{55}Cu_{30}Al_{10}Ni_5$ bulk glassy alloy rod sample. The mechanical properties were improved after the introduction of the Al_2O_3 particles. For the monolithic $Zr_{55}Cu_{30}Al_{10}Ni_5$ as-cast rod sample, the compressive fracture strength is 1640 MPa, but with no plastic deformation. In contrast, for the sintered composite specimen, the fracture strength reaches 1470 MPa and the sample exhibits macroscopic compressive plastic deformation after the elastic strain.

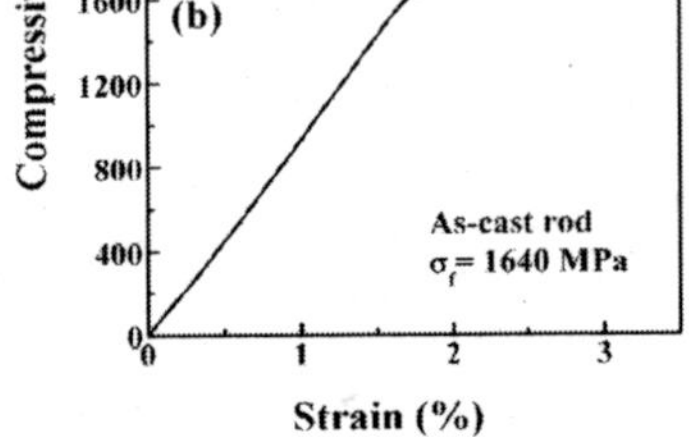

Fig. 6 Compressive stress-strain curves of the sintered composite (a) and the as-cast rod (b).

DISCUSSION

It is known that the SPS process is an electrical sintering technique which applies an ON-OFF DC pulse voltage. The pulse electric current directly flows through the sintered powder materials in the SPS process, and a very high heating efficiency is offered. The application of pulsed DC voltage induces various phenomena caused by electrical and thermal effects, providing advantages which could not be obtained using conventional sintering processes.[8] These phenomena can be summarized as follows:[8-11] (a) electrical breakdown of surface oxide film and removal of contaminated layer on particle surface by spark generation and sputtering effect; (b) destruction of surface oxide film and neck formation between powder particles; (c) focused current and Joule heat at the neck, and (d) enhanced migration of atom or ion by temperature difference between neck and particle core as well as by electrical field, and enhanced the neck growth. Therefore, the sintering can be carried out at a lower temperature and in a shorter time than that with conventional processes. In the present study, the $Zr_{55}Cu_{30}Al_{10}Ni_5$ bulk glassy matrix composite specimen has been successfully fabricated in a total period only 22 min including the SPS heating-up and holding time and at a sintering temperature of 623 K, which is far lower than the crystallization temperature (T_x) as well as the glass transition temperature (T_g) of the $Zr_{55}Cu_{30}Al_{10}Ni_5$ glassy alloy. They were 774 K and 685 K, respectively, at a heating rate of 0.67 K s^{-1}.

During the SPS process, the temperature at the contact interfaces between powder particles should be higher than the average temperature for the sintered specimens due to the focused current and Joule heat at the bonded interface between powder particles, which has been demonstrated in the investigation of Al-Mg alloy powders sintered by the SPS process.[12] However, in the present study, no crystallization of the metallic glassy matrix was observed by SEM (Fig. 3) and XRD pattern analyses (Fig. 4). DSC trace analyses (Fig. 5) also demonstrated that the sintered composite specimen shows the same curve features as the original glassy alloy powder. The glass transition temperature (T_g), the onset temperature of the first-stage crystallization (T_X), and the extent of the supercooled liquid region ($\Delta T = T_X - T_g$) are also same as those from the original glassy alloy powder. It indicates that no crystallization of the metallic glassy matrix occurred during the SPS process. Furthermore, the crystallization enthalpy (ΔH) of the sintered composite specimen is around 90% of that of the monolithic gas-atomized $Zr_{55}Cu_{30}Al_{10}Ni_5$ glassy powder. This further demonstrated that the crystallization of the $Zr_{55}Cu_{30}Al_{10}Ni_5$ glassy matrix in the sintered composite specimen did not take place during the SPS process. The reason of no crystallization should be originated from the low sintering temperature and rapid cooling in the SPS process. Tokita suggested that the application of the pulse voltage in the SPS process enhances thermal diffusion.[8] The heat is transferred immediately and diffused to whole powder particle so that intergranular bonding portion is quickly cooled. This is in favor of the sintering of glassy materials or metastable phase materials. Kawamura et al.[13] demonstrated that the crystallization was prevented because the cooling rate was higher than the critical cooling rate of the $Zr_{55}Cu_{30}Al_{10}Ni_5$ glassy alloy prepared by spark welding.

Based on the results and the aforementioned discussion, it is clear that the metallic glassy matrix composites are sinterable by the SPS process, and the crystallization of the metallic glassy matrix can be prevented during the SPS process. Furthermore, redistribution of the glassy matrix and the dispersed particles does not exist during solid state sintering, which results in more homogeneous products than those prepared by the casting method. Hence, this is a promising technique for the fabrication of metallic glassy matrix composite without size and shape limitations.

CONCLUSIONS

Using a mixed powder of the argon gas-atomized $Zr_{55}Cu_{30}Al_{10}Ni_5$ glassy alloy powder blended with 10 vol.% Al_2O_3 powder, $Zr_{55}Cu_{30}Al_{10}Ni_5$ bulk metallic glass matrix composite was successfully fabricated by the SPS process. The results indicated that the Al_2O_3 particles were homogeneously dispersed in the $Zr_{55}Cu_{30}Al_{10}Ni_5$ metallic glassy matrix. The crystallization of the metallic glassy matrix was prevented during the SPS process. The crystallization behavior of the sintered composite was similar to that of the metallic glassy powder. The sintered composite specimen exhibited a large compressive strength.

ACKNOWLEDGEMENTS

This work was supported by a Grant-In-Aid for Science Research in a priority Area "Research and Development Project on Advanced Metallic Glasses, Inorganic Materials, and Joining Technology" from the Ministry of Education, Science, Sports and Culture, Japan. A part of this work was carried out at Advanced Research Center of Metallic Glasses, Institute for Materials Research, Tohoku University, Japan.

REFERENCES

[1] A. Inoue, "Stabilization of metallic supercooled liquid and bulk amorphous alloys," *Acta Mater.*, **48,** 279-306 (2000).

[2] D.H. Bae, M.H. Lee, D.H. Kim, and D.J. Sordelet, "Plasticity in $Ni_{59}Zr_{20}Ti_{16}Si_2Sn_3$ metallic glass matrix composites containing brass fibers synthesized by warm extrusion of powders," *Appl. Phys. Lett.*, **83,** 2312-2314 (2003).

[3] H. Choi-Yim, R.D. Conner, F. Szuecs, and W.L. Johnson, "Processing, microstructure and properties of ductile metal particulate reinforced $Zr_{57}Nb_5Al_{10}Cu_{15.4}Ni_{12.6}$ bulk metallic glass composites," *Acta Mater.*, **50,** 2737-2745 (2002).

[4] H. Kato, T. Hirano, A. Matsuo, Y. Kawamura, and A. Inoue, "High strength and good ductility of $Zr_{55}Al_{10}Ni_5Cu_{30}$ bulk glass containing ZrC particles," *Scripta Mater.*, **43,** 503-507 (2000).

[5] H. Choi-Yim, R. Busch, U. Köster, and W.L. Johnson, "Synthesis and characterization of particulate reinforced $Zr_{57}Nb_5Al_{10}Cu_{15.4}Ni_{12.6}$ bulk metallic glass composites." *Acta Mater.*, **47,** 2455-2462 (1999).

[6] H. Kato, K. Yubuta, D.V. Louzguine, A. Inoue, and H.S. Kim, "Influence of nanoprecipitation on strength of $Cu_{60}Zr_{30}Ti_{10}$ glass containing μm-ZrC particle reinforcements," *Scripta Mater.*, **51,** 577-581 (2004).

[7] V. Mamedov, "Spark plasma sintering as advanced PM sintering method," *Powder Metall.*, **45,** 322-328 (2002).

[8] M. Tokita, "Trends in advanced SPS systems and FGM technology," *Proceedings of NEDO International Symposium on Functionally Graded Materials*, New Energy and Industrial Technology Development Organization, Tokyo, Japan, 1999, pp.23-33.

[9] M. Omori, "SPS for the preparation of advanced materials and functionally graded material," *Proceedings of NEDO International Symposium on Functionally Graded Materials*, New Energy and Industrial Technology Development Organization, Tokyo, Japan, 1999, pp.75-84.

[10] O. Yanagisawa, T. Hatayama, and K. Matsugi, "Recent research on spark plasma sintering," *Materia Japan*, **33,** 1489-1496 (1994).

[11] D.K. Kim, H.R. Park, and K. Okazaki, "Electrodischarge compaction of nickel powders," *Mater. Sci. Eng. A*, **104,** 191-200 (1988).

[12] G.Q. Xie, O. Ohashi, N. Yamaguchi, M.Song, K. Mitsuishi, K. Furuya and T. Noda, "Reduction of surface oxide films in Al-Mg alloy powders by pulse electric current sintering," *J. Mater. Res.*, **19,** 815-819 (2004).

[13] Y. Kawamura and Y. Ohno, "Spark welding of $Zr_{55}Al_{10}Ni_5Cu_{30}$ bulk metallic glasses," *Scripta Mater.*, **45,** 127-132 (2001).

POROUS BULK METALLIC GLASS PRODUCED BY SPARK PLASMA SINTERING OF GAS ATOMIZED $Zr_{55}Cu_{30}Al_{10}Ni_5$ GLASSY POWDERS

Guoqiang Xie, Wei Zhang, Qingsheng Zhang, Dmitri V. Louzguine-Luzgin, Akira Okubo, Hisamichi Kimura, and Akihisa Inoue
Institute for Materials Research, Tohoku University
Katahira 2-1-1, Aoba-Ku, Sendai 980-8577, Japan

ABSTRACT

Porous bulk metallic glasses recently attract increasing scientific and industrial attention due to their unique mechanical, physical, chemical, thermal, electrical and acoustic properties. In the present study, porous $Zr_{55}Cu_{30}Al_{10}Ni_5$ bulk metallic glass specimens were fabricated by a spark plasma sintering process using the $Zr_{55}Cu_{30}Al_{10}Ni_5$ glassy alloy powders made by argon gas atomization method. No crystallization was observed in the sintered porous bulk glassy alloy. The thermal stability of the sintered porous glassy alloy specimen was similar to that of the original powder. The sintered porous glassy alloy specimen exhibited larger plastic ductility and lower Young's modulus than those of the as-cast alloy specimen.

INTRODUCTION

Bulk metallic glasses (BMGs) have been produced in a large number of alloy systems and many studies are focused on the development of new alloy compositions due to their attractive technological promises and excellent properties such as high strength, high corrosion resistance and good soft magnetic properties.[1] They are considered as promising candidates for the next generation of structural and functional materials. However, there are still some disadvantages of the BMGs, which limit their large-scale industrial applications. A sufficiently high cooling rate is always required during casting in order to prevent the formation of the more thermodynamically stable crystalline phases, and the dimensions of the metallic glasses are limited. Therefore, appropriate casting techniques are necessary to acquire the high cooling rate required for glass formation. Although it is now possible to fabricate the metallic glasses with diameters of several centimeters at certain chemical compositions, such as $Pd_{40}Cu_{30}Ni_{10}P_{20}$ and $Zr_{55}Al_{10}Ni_5Cu_{30}$,[2,3] for most of the other metallic glasses, their dimensions are still limited to millimeter scale and their shapes are also rather simple. In order to overcome the limitations of the dimension and the shape, the development of appropriate techniques is an important research subject.

On the other hand, porous materials are of significant interest due to their wide applications in filters, catalysis, separation, as well as lightweight structural components and thermal structural materials. A number of investigations have been carried out in the fabrication of the porous metals and alloys.[4-7] However, only limited researches dealing with the porous BMGs have been reported so far.

Spark plasma sintering (SPS), as a newly developed rapid sintering technique, has a great potential for producing specimens with nearly no crystallization for amorphous materials or minimal grain growth for crystalline materials in a short sintering time.[8] The SPS process is a type of solid compression sintering technique which is similar to hot-pressing (HP) sintering process, so that the sintered specimens with the large-size and complicated shape can be easily achieved.[9] In the SPS process, a pulse electrical current flows directly in the sintered materials. A very high heating efficiency is offered. The sintering can be carried out at a lower temperature with a much shorter time than that of conventional sintering processes. Hence the SPS process can be applied to materials that are requested to suppress crystallization and grain growth.

In the present study, $Zr_{55}Cu_{30}Al_{10}Ni_5$ glassy alloy powders were prepared by the argon gas atomization method. Using the obtained glassy alloy powders, porous bulk metallic glassy specimens were produced by the SPS process. The structure, thermal stability and mechanical properties of the sintered porous glassy alloy specimens were investigated.

EXPERIMENTAL PROCEDURES

Master ingots of the $Zr_{55}Cu_{30}Al_{10}Ni_5$ alloy (composition is given in nominal atomic percentages) were prepared by arc melting a mixture of high purity Zr, Cu, Al and Ni in an argon atmosphere purified using Ti getter. The $Zr_{55}Cu_{30}Al_{10}Ni_5$ metallic glassy alloy powders were produced by a high pressure argon gas atomization method. The ingots were melted again under an evacuated state in a quartz tube using an induction heating coil, followed by injection through a nozzle with a diameter of 0.8 mm and then atomization by high pressure argon gas with a dynamic pressure of about 8.0 MPa.

Following atomization, size distribution of the atomized powders was measured by a conventional sieving method, and the powders were classified. The glassy alloy powders with the size in the range of 37 μm < d < 53 μm (d is the diameter of the powder particles) were used for the subsequent sintering experiment. The sintering of the powders was carried out in a vacuum using a SPS system (Model SPS-1050, Sumitomo Coal Mining Co. Ltd., Japan). The sintering temperature used was 613 K. The heating rate was 50 K/min (from room temperature to T_S-30 K) and 5 K/min (T_S-30 K to T_S) (T_S: sintering temperature). The holding time at sintering temperature was 10 min. A uniaxial pressuring method was conducted using top and bottom graphite punches. The loading pressure was 40 MPa in the present study. The details of heating, temperature measurement and control, and loading pressure control in the SPS process have been described in a previous paper.[10] The sintered specimens obtained were a cylindrical shape with a diameter of 10 mm and a height of about 5 mm.

The structures of the powders and the sintered porous bulk specimens were examined by X-ray diffractometry (XRD) in reflection with a monochromatic Cu Kα radiation, and the microstructure was observed with a scanning electron microscope (SEM). Thermal stability associated with the glass transition temperature (T_g), crystallization temperature (T_X), and supercooled liquid region ($\Delta T = T_X - T_g$) was examined by differential scanning calorimetry (DSC) at a heating rate of 0.67 K s^{-1}. The density of the sintered porous bulk specimens was determined by measuring their mass and dimension of the samples (mass per volume). The porosity was evaluated from the obtained density. Mechanical properties under a compressive load were measured using a conventional mechanical testing system (Instron 4201). The test specimens with a rectangular shape of 2.5 mm in width, 2.5 mm in thickness and 5.0 mm in height were used in the present study. The initial strain rate used was 5×10^{-4} s^{-1}.

RESULTS AND DISCUSSION

Preparation of glassy alloy powder

The as-prepared powders were classified. The particle sizes of the gas-atomized powders are below 125 μm. The volume fraction of the particles with a size in the range of 37□53 μm is about 50%. The observation with the SEM indicated that, irrespective of size, spherical morphology as well as clean surfaces was shown in all sizes of as-prepared gas-atomized alloy powders. Figure 1 presents a SEM micrograph of the as-prepared $Zr_{55}Cu_{30}Al_{10}Ni_5$ powder with a particle size in the range of 37□53 μm. No appreciable contrast revealing the formation of a crystalline phase is observed on the outer surface of the particles.

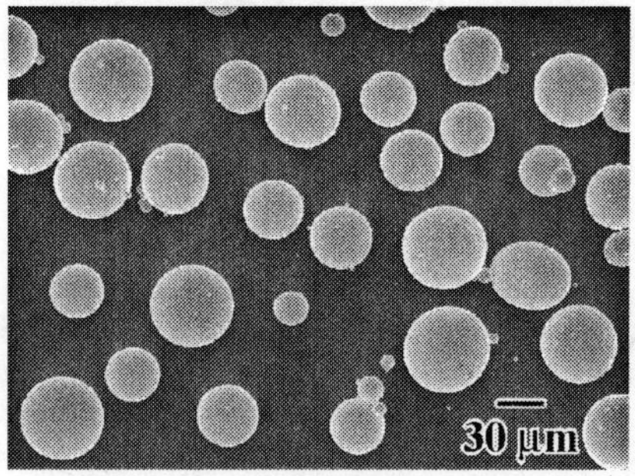

Fig. 1 SEM image of the argon gas-atomized $Zr_{55}Cu_{30}Al_{10}Ni_5$ powder with a particle size fraction of 37□53 μm.

Figure 2 shows the XRD patterns of the as-atomized powders as a function of particle sizes. Only a diffuse diffraction pattern typical for a glassy phase is seen. Neither diffraction peaks corresponding to crystalline phases are observed, nor any apparent effect of powder size on the glassy phase stability is observed (Fig. 3). This indicates that a single glassy phase without crystallinity is formed in the particle size range below 125 μm. Figure 3 shows the DSC curves of the gas-atomized powders taken at a heating rate of 0.67 K s^{-1} for various powder sizes. The glass transition temperature (T_g), the onset temperature of the first-stage crystallization (T_X) corresponding to the primary exothermic peak, the crystallization enthalpy (ΔH) and the extent of the supercooled liquid region ($\Delta T = T_X - T_g$) are almost same for all powder sizes. The average values are 685 K, 774 K, 50.3 J/g and 89 K, respectively. These values are close to the previously reported ones for the $Zr_{55}Cu_{30}Al_{10}Ni_5$ metallic glassy alloy.[3,11] Therefore, it is demonstrated that the gas atomization process is a useful method for preparation of metallic glassy alloy powders.

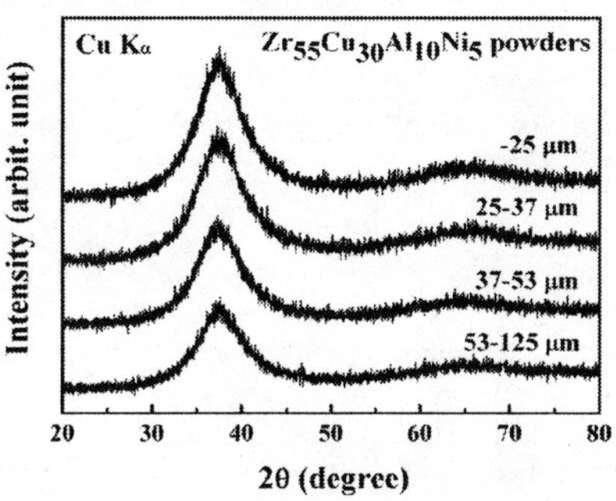

Fig. 2 XRD patterns of the gas-atomized powders with different particle sizes.

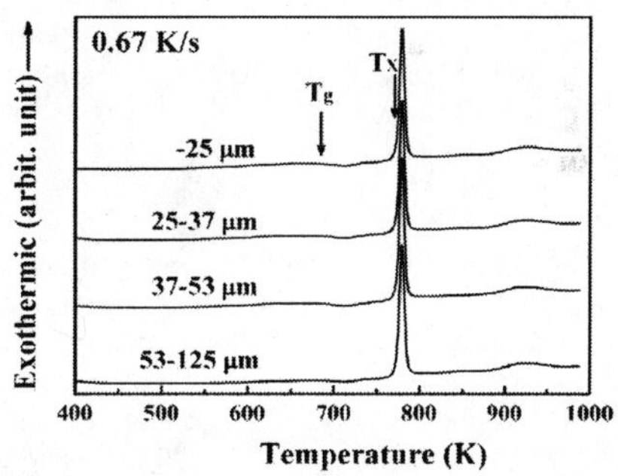

Fig. 3 DSC curves of the gas-atomized powders with different particle sizes.

Fabrication and properties of porous bulk metallic glass

Using the as-prepared $Zr_{55}Cu_{30}Al_{10}Ni_5$ powder with a particle size in the range of 37□53 μm, the sintering was performed by the SPS process at a sintering temperature of 613 K with a loading pressure of 40 MPa. The density of the sintered specimen was measured. The porosity, which was evaluated from the obtained density, was 25.2%. The structure and thermal stability of the sintered porous specimen were examined by XRD and DSC. Figure 4 shows the XRD patterns of the sintered porous bulk glassy alloy specimen as well as that of the original $Zr_{55}Cu_{30}Al_{10}Ni_5$ powder for comparison. The XRD patterns of the as-sintered porous specimens consist only of a broad diffraction peak and no sharp peak corresponding to crystalline phase is observed, in agreement with the original glassy alloy powder. Figure 5 shows the DSC curve of the sintered porous bulk specimen, together with the curve of the original powder. The feature of the DSC curve of the sintered porous specimen is similar to that of the original powder. No obvious difference in T_g, T_X, and the heat of exothermic reaction is observed between the sintered porous specimen and the original powder. These results indicate that the sintered porous $Zr_{55}Al_{10}Ni_5Cu_{30}$ bulk glassy alloy is composed of a glassy phase.

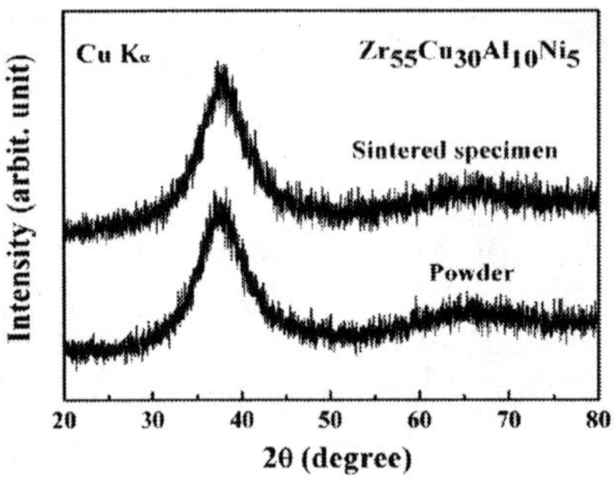

Fig. 4 XRD pattern of the sintered porous bulk glassy alloy. The pattern of the original powder is also shown for comparison.

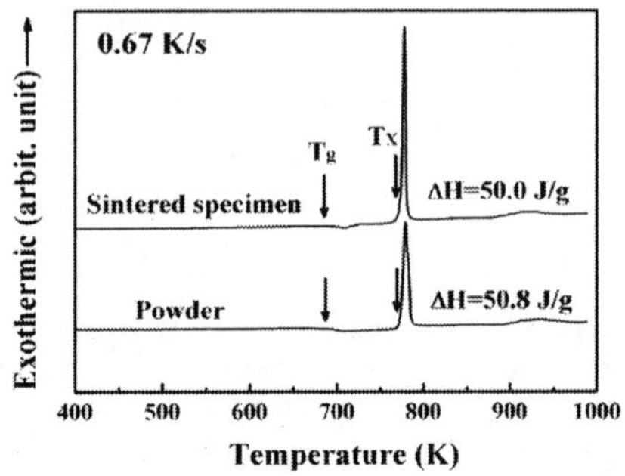

Fig. 5 DSC curve of the sintered porous bulk glassy alloy. The trace of the original powder is also shown for comparison.

Figure 6 shows nominal compressive stress–strain curves of the sintered porous $Zr_{55}Cu_{30}Al_{10}Ni_5$ bulk glassy alloy specimen, together with that of the as-cast rod sample. A distinct difference between the porous sample and the as-cast rod sample can be seen. That is, the as-cast rod sample ruptures instantly after the elastic strain limit, while the porous bulk glassy alloy specimen shows a large plastic ductility after the elastic strain. The 0.2% proof stress ($\sigma_{0.2}$) and Young's modulus (E) are 480 MPa and 23 GPa, respectively, for the sintered porous glassy specimen, while those of the as-cast rod specimen are 1640 MPa and 94 GPa, respectively. Although the compressive strength of the sintered porous bulk glassy specimen was lower than that of the as-cast specimen, it is noted that plastic ductility increased for the porous sample. The increase in the plastic ductility is proposed to originate from the polyhedral shape of the pores in the porous glassy alloys. The polyhedral cell walls with different cross sectional areas and various directions to the applied load axis are suffered to multiple stress conditions even under a uniaxial compressive applied load.[12] The enhancement of the plastic ductility results from the multiple generation of shear bands under the multi-axis stress condition in the cold-rolled bulk glassy alloy, as has been demonstrated.[13] It is also shown that the Young's modulus of the sintered porous $Zr_{45}Cu_{30}Al_{10}Ni_5$ bulk glassy alloy specimen is 23 GPa, which is consistent with the Young's modulus of human bone of 10-40 GPa.[14] Such mechanical properties of the porous bulk glassy alloy specimens which are significantly different from those of the as-cast bulk glassy alloys indicate the possibility of future applications as new structural and functional materials. Furthermore, for the sintering of the powders, the SPS process is similar to the conventional sintering process such as hot pressing, and hence it is easy to achieve the sintered specimens with the large-size and complicated shape.

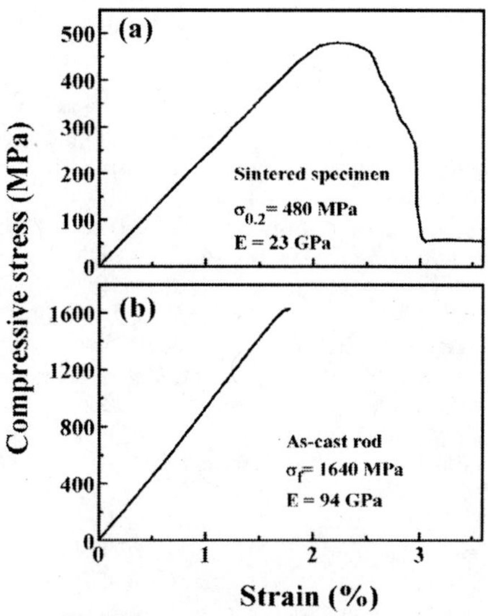

Fig. 6 Compressive stress-strain curves of the sintered porous bulk glassy alloy (a) and the as-cast specimen (b).

CONCLUSIONS

We have prepared the $Zr_{55}Cu_{30}Al_{10}Ni_5$ glassy alloy powders by the argon gas atomization method. The thermal properties of the powders are independent of the particles size. Using the obtained glassy alloy powders, the porous bulk glassy specimen has been successfully fabricated by the SPS process. No crystallization is observed in the sintered porous bulk glassy alloy. The thermal stability of the sintered porous glassy alloy specimen is similar to that of the original powder. The sintered porous glassy alloy specimen exhibits larger plastic ductility, lower Young's modulus and lower fracture strength than those of the as-cast alloy specimen.

ACKNOWLEDGEMENTS

This work was supported by a Grant-In-Aid for Science Research in a priority Area "Research and Development Project on Advanced Metallic Glasses, Inorganic Materials, and Joining Technology" from the Ministry of Education, Science, Sports and Culture, Japan. A part of this work was carried out at Advanced Research Center of Metallic Glasses, Institute for Materials Research, Tohoku University, Japan.

REFERENCES

[1] A. Inoue, "Stabilization of metallic supercooled liquid and bulk amorphous alloys," *Acta Mater.*, **48,** 279-306 (2000).
[2] A. Inoue, N. Nishiyama, and H. Kimura, "Preparation and thermal stability of bulk amorphous $Pd_{40}Cu_{30}Ni_{10}P_{20}$ alloy cylinder of 72 mm in diameter," *Mater. Trans.*, **38,** 179-183 (1997).
[3] A. Inoue and T. Zhang, "Fabrication of bulk glassy $Zr_{55}Al_{10}Ni_5Cu_{30}$ alloy of 30 mm in diameter by a suction casting method", *Mater. Trans.*, **37,** 185-187 (1996).
[4] M. Hakamada, Y. Yamada, T. Nomura, Y. Chen, H. Kusuda, and M. Mabuchi, "Fabrication of porous aluminum by spacer method consisting of spark plasma sintering and sodium chloride dissolution," *Mater. Trans.*, **46,** 2624-2628 (2005).
[5] C.E. Wen, M. Mabuchi, Y. Yamada, K. Shimojima, Y. Chino, and T. Asahina, "Processing of biocompatible porous Ti and Mg," *Scripta Mater.*, **45,** 1147-1153 (2001).
[6] C. Hong, X. Zhang, J. Han, and B. Wang, "Fabrication and mechanical properties of porous TiB_2 ceramic," *J. Mater. Sci.*, **41,** 4790-4794 (2006).
[7] Y. Zhao, M. Taya, Y. Kang, and A. Kawasaki, "Compression behavior of porous NiTi shape memory alloy," *Acta Mater.*, **53,** 337-343 (2005).
[8] V. Mamedov, "Spark plasma sintering as advanced PM sintering method," *Powder Metall.*, **45,** 322-328 (2002).
[9] M. Tokita, "Trends in advanced SPS systems and FGM technology," *Proceedings of NEDO International Symposium on Functionally Graded Materials*, New Energy and Industrial Technology Development Organization, Tokyo, Japan, 1999, pp.23-33.
[10] G.Q. Xie, O. Ohashi, and N. Yamaguchi, "Sintering behavior of aluminum powder by spark plasma sintering," *Trans. Mater. Res. Soc. Jpn.*, **27,** 743-746 (2002).
[11] A. Castellero, S. Bossuyt, M. Stoica, S. Deledda, J. Eckert, G.Z. Chen, D.J. Fray, and A.L. Greer, "Improvement of the glass-forming ability of $Zr_{55}Cu_{30}Al_{10}Ni_5$ and $Cu_{47}Ti_{34}NZr_{11}Ni_8$ alloys by electro-deoxidation of the melts," *Scripta Mater.*, **55,** 87-90 (2006).
[12] T. Wada, K. Takenaka, N. Nishiyama, and A. Inoue, "Formation and mechanical properties of porous Pd-Pt-Cu-P bulk glassy alloys," *Mater. Trans.*, **46,** 2777-2780 (2005).
[13] Y. Yokoyama, K. Yamano, K. Fukaura, H. Sunada, and A. Inoue, "Enhancement of ductility and plasticity of $Zr_{55}Cu_{30}Al_{10}Ni_5$ bulk glassy alloy by cold rolling," *Mater. Trans.*, **42,** 623-632 (2001).
[14] M. Long and H.J. Rack, "Titanium alloys in total joint replacement: a materials science perspective," Biomaterials 19, 1621-1639 (1998).

Mechanical Properties of Friction-Stir Welded Titanium Joint

Hidetoshi Fujii, Hideaki Kato, Kazuhiro Nakata and Kiyoshi Nogi
Joining and Welding Research Institute, Osaka University
11-1 Mihogaoka, Ibaraki, Osaka, 567-0047, Japan

ABSTRACT

The Commercial purity titanium (cp-Ti) plates with a 2 mm thickness, were butt-welded by friction-stir welding (FSW). Using a W alloy tool, FSW was performed at 200rpm and 50-300 mm/min. The mechanical properties of the joints were evaluated using tensile tests and Vickers micro-hardness tests. The microstructure was observed by optical microscopy (OM), field emission scanning electron microscopy (FE-SEM) and transmission electron microscopy (TEM). An electron back-scattering pattern (EBSP) technique was also used in the FE-SEM to clarify the crystallographic features of the microstructure. The tensile strength of joints significantly increases with the increasing welding speed because the grain size decrease and the dislocation density increases. The tensile strength of the specimen welded at 200 mm/min exceeds that of the base metal, and therefore the specimen fractured in the base metal during the tensile tests. However, the defects were formed in the specimen at 300 mm/min leading to decrease in tensile strength. The friction-stir welded cp-Ti microstructure consists of recrystallized grains with many dislocations, and the dislocation density increased with the increasing welding speed (decreasing heat input). For pure Al or IF steel, few dislocations are observed because the recovery occurs during cooling. Accordingly, the change in strength of the stir zone by varying heat inputs is greater for cp-Ti than for pure Al or IF steel.

INTRODUCTION

The friction-stir welds of low melting point materials such as Al[1-4], Mg[5-6], Cu[7] have been widely investigated since its development by The Welding Institute (TWI) in 1991. Recently, welding research activities on high melting point materials such as IF steel[8], and carbon steel[9-12], and stainless steel[13-16] have also started. However, there are few reports on welding of Ti[17-20] whose melting point is higher than that of steel. Lee[17] and Ramirez[18] succeeded in welding pure Ti plates (α-Ti) and jointed Ti-6Al-4V (α-β Ti), respectively. John[19] investigated the residual stress of FSWed Ti-6Al-4V and Reynolds[20] reported the texture of the FSWed Ti metal 21S (β Ti).

Ti is widely used in aerospace industry because it is not only a light weight and high strength material but also has superior heat resistance. Moreover, Ti is used in the sea water environment and for the petrochemistry due to its high corrosion resistance. In addition, Ti is also used for biomedical implants materials due to its good compatibility to the human body, and more applications are expected in the future. When FSW is performed for Ti alloys, a tool with a high temperature strength is necessary because of its excellent mechanical properties as in the previous study[17-18, 20]. An atmosphere control is also needed because Ti is very easily oxidized. The crystal structure of Ti is a hexagonal closed packed (hcp) structure under 880°C and a body centered cubic (bcc) structure over 880°C. A smaller number of slip systems in the hcp structure than in the bcc or fcc[20] structures may affect the deformation behavior, although friction-stir welding has been already used for Mg[5-6], a hcp metal.

In this study, the cp-Ti plates were successfully friction-stir welded. The effect of the welding conditions on the mechanical properties and microstructure of the joints is also investigates so that the best welding conditions could be determined.

EXPERIMENTAL PROCEDURE

Commercial purity Ti plates with 300 mm length, 50 mm width, and 2 mm thickness, were butt-welded by friction-stir welding. The chemical composition (wt %) of the base material was 0.007% C, 0.0013% B, 0.08% O, 0.004% N, 0.05% Fe and balance Ti. The ultimate tensile strength, 0.2% proof strength, elongation and hardness of base material were 418 MPa, 319 MPa, 39%, 146 Hv respectively. A W alloy tool with shoulder diameter of 15 mm, probe diameter of 6 mm, and probe length of 1.8 mm, was used at a rotation rate of 200 rpm. The welding speed was varied at 50-300 mm/min to control the heat input. During welding, an Ar gas was thrown around the tool for the atmosphere to prevent Ti from being oxidized.

After welding, an electrical discharge machine was used to form the tensile test specimens and cross-sections perpendicular to the welding direction for the hardness test. The configuration and size of the transverse tensile specimens were prepared according to Figure 1, where RS and AS denote the retreating side and advancing side, respectively. The type (a) in Fig.1 was used for all specimens, and type (b) was used for specimens 200 mm/min because the type (A) specimen was broken at the base metal during the tensile tests. The crosshead speed was 1 mm/min during the tensile tests. The Vickers micro hardnesses were measured under a load of 0.98 N for 15 seconds along the centerlines of the cross-section with a spacing of 0.5 mm.

The microstructure was observed by field emission scanning electron microscopy (FE-SEM), transmission electron microscopy (TEM). An electron back scattering pattern (EBSP) technipue was also used in the FE-SEM to clarify the crystallographic features of the microstructure. After cross-sections of the stir zone perpendicular to the welding direction were cut off for the EBSP, they were polished by emery papers and electrochemical polished using a twin jet electro polishing machine.

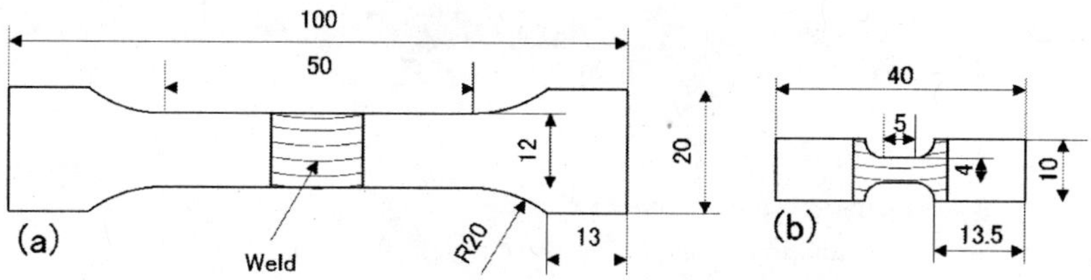

Figure 1. Configuration and size of the tensile tests used for (a) all specimens (b) only specimens welded at 200 mm/min.

RESULTS AND DISCUSSION

Tensile Strength of Joints

Figure 2 shows the tensile strength of FSWed cp-Ti at different welding speeds. Two specimens were tested under each condition. The broken line indicates the tensile strength of the base metal. The tensile strength increases with the increasing welding speed (decreasing heat input). Especially, the tensile strength of the joint exceeded that of the base metal at 200 mm/min, and therefore, the specimen was fractured in the base metal during the tensile tests. The other specimens were fractured in the stir zone during the tensile tests. However, the tensile strength of the specimens welded at 300 mm/min were decreased due to the defect formed at butt interface.

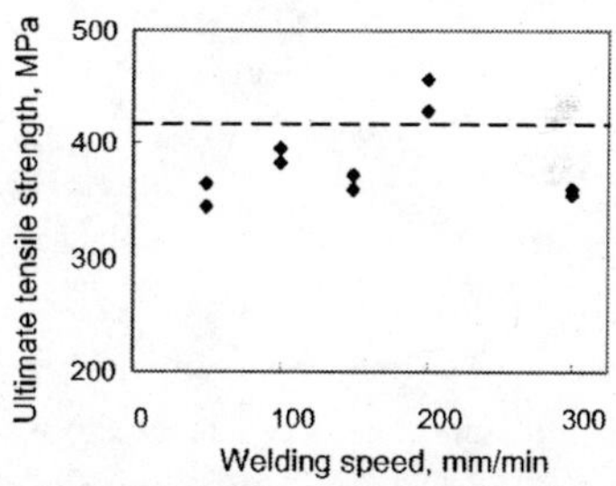

Fig. 2 Tensile strength of FSWed cp-Ti joints.

Hardness Distributions of Joints

Figure 3 shows the Vickers micro hardnesses along the centerlines on the cross-sections. The broken line indicates hardness of the base metal. In common with the tensile strength, the hardness increases with the increasing welding speed (decreasing heat input). The hardest value of about 200 Hv was obtained at 300 mm/min.

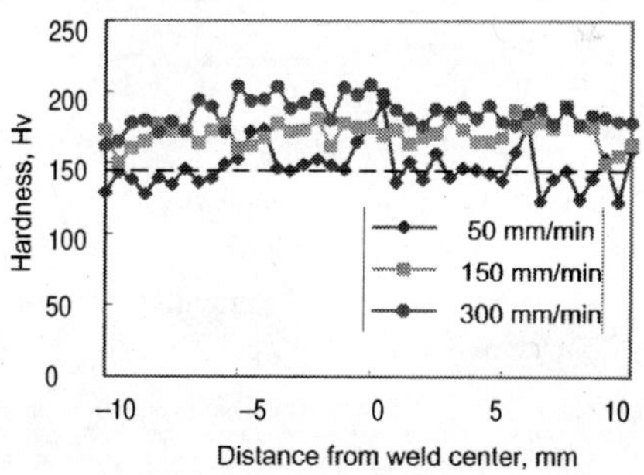

Figure 3. Hardness distribution of cp-Ti.

TEM observation

Figure 4 shows TEM bright field images at the center of stir zone. In common with the EBSP analysis, the crystal grains become finer with the increasing welding speed. Furthermore, the dislocation density increased with the increasing welding speed. It is considered that the dislocations at a high welding speed were not recognized as the low angle boundary due to the small difference in the orientation. This result clearly indicates that cp-Ti was dynamic recrestallized during welding by getting great deal of strain.

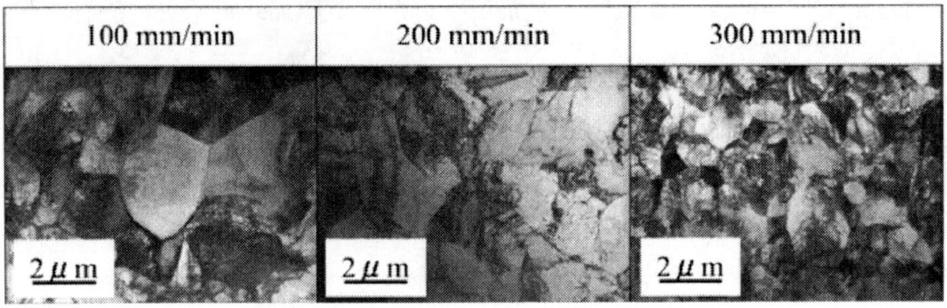

Figure 4. TEM photographs of FSWed cp-Ti

CONCLUSIONS

(1) Commercial purity Ti (cp-Ti) was successfully friction-stir welded at 50 mm/min to 300 mm/min. The mechanical properties of the FSWed cp-Ti increases with the increasing welding speed because the grain size decreases and the dislocation density increases due to the plastic deformation under the welding conditions. In this study, the highest tensile strength was obtained at 200 mm/min, because a small defect was formed at 300 mm/min.

(2) In the stir zone of FSWed, cp-Ti equiaxial grains with lots of dislocations were observed, indicating that dynamic recrestallization occurred.

ACKNOWLEDGEMENTS

The authors wish to acknowledge the financial support of the Toray Science Foundation, a Grant-in-Aid for the Cooperative Research Project of Nationwide Joint-Use Research Institutes on Development Base of Joining Technology for New Metallic Glasses and Inorganic Materials, "Priority Assistance of the Formation of Worldwide Renowned Centers of Research - The 21st Century COE Program (Project: Center of Excellence for Advanced Structural and Functional Materials Design) "from the Ministry of Education, Sports, Culture, Science and Technology of Japan and a Grant-in-Aid for Science Research from Scientific Research from the Japan Society for Promotion of Science.

REFERENCES

[1] C. G. Rhodes, M. W. ahoney, W. H. Bingel, R. A. Spurling, and C. C. Bampton, EFFECTS OF FRICTION STIR WELDING ON MICROSTRUCTURE OF 7075 ALUMINUM, *Scripta Mater.*, **36**, 69-75 (1997).

[2] S. Benavides, Y. Li, L. E. Murr, D. Brown, and J. C. McClure, LOW-TEMPERATURE FRICTION-STIR WELDING OF 2024 ALUMINUM, *Scripta Mater.*, **41**, 809-15 (1999).

[3] R. S. Mishra, M.W. Mahoney, S. X. McFadden, N. A. Mara, and A. K. Mukherjee, HIGH STRAIN RATE SUPERPLASTICITY IN A FRICTION STAR PROCESSED 7075 Al ALLOY, *Scripta Mater.*, **42**, 163-68 (2000).

[4] H. J. Liu, H. Fujii, M. Maeda, and K. Nogi, Tensile properties and fracture locations of friction-stir-welded joints of 2017-T351 aluminum alloy, *Mater. Proce. Techno.*, **142**, 692-69 (2003).

[5] S. H. C. Park, Y. S. Sato, and H. Kokawa, Effect of micro-texture on fracture location in friction stir weld of Mg alloy AZ61 during tensile test, *Scripta Mater.*, **49**, 161-66 (2003).

[6] D. Z. Zhang, M. Suzuki, and K. Maruyama, Microstructural evolution of a heat-resistant magnesium alloy due to friction stir welding, *Scripta Mater.*, **52**, 899-03 (2005).

[7] H. S. Park , T. Kimura , T. Murakami , Y. Nagano , K. Nakata, and M. Ushio, Microstructures and mechanical properties of friction stir welds of 60% Cu–40% Zn copper alloy, *Mater. Sci. and Eng. A*, **371,** 160-69 (2004).

[8] H. Fujii, R. Ueji, Y. Takada, H. Kitahara, N. Tsuji, K. Nakata, and N. Nogi, Friction Stir Welding of Ultrafine Grained Interstitial Free Steels, *Mater. Trans.*, **47**, 239-42 (2006).

[9] W. M. Thomas, P.L. Threadgill, and E. D. Nicholas, Feasibility of friction stir welding steel, *Sci. Technol. Weld. Join.*, **4**, 365-72 (1999).

[10] A. P. Reynolds, W. Tang, M. Posada, and J. DeLoach, Friction stir welding of DH36 steel, *Sci. Technol. Weld. Join.*, **8**, 455-60 (2003)

[11] H. Fujii, L. Cui, N. Tsuji, M. Maeda, K. Nakata, and K. Nogi, Friction stir welding of carbon steels, *Mater. Sci. and Eng. A*, **429,** 50-57 (2006).

[12] R. Ueji, H. Fujii, L. Cui, A. Nishioka, K. Kunishige, and K. Nogi, Friction stir welding of ultrafine grained plain low-carbon steel formed by the martensite process, *Mater. Sci. and Eng. A*, **423,** 324-30 (2006).

[13] A. P. Reynolds, W. Tang, T. Gnaupel-Herold, and H. Park, Structure, properties, and residual stress of 304L stainless steel friction stir welds, *Scripta Metar.*, **48**, 1289-94 (2003).

[14] Y. S. Sato, T. W. Nelson, and C. Sterling, Recrystallization in type 304L stainless steel during friction stirring, *Acta Mater.*, **53,** 637-45 (2005).

[15] J. H. Cho, D. E. Boyee, and P. R. Dawson, Modeling strain hardening and texture evolution in friction stir welding of stainless steel, *Mater. Sci. and Eng. A*, **398,** 146-63 (2005).

[16] T. Ishikawa, H. Fujii, K. Genchi, L. Cui, S. Matsuoka, and K. Nogi, Joint Properties of Friction Stir Welded Austenitic Stainless Steels, Q. J. Jpn. Weld. Soc., **24** 174-80 (2006). (in Japanese)

[17] W. B. Lee, C. Y. Lee, W. S. Chang, Y. M. Yeon, and S. B. Jung, Microstructural investigation of friction stir welded pure titanium, *Mater. Letters,* **59,** 3315-18 (2005).

[18] A. J. Ramirez, and M. C. Juhas, MICROSTRUCTUAL EVOLUTION IN Ti-6Al-4V FRICTION STIR WELDS, *Mater. Sci. Forum,* **426-432,** 2999-04 (2003).

[19] R. John, K. V. Jata, and K. Sadananda, Residual stress effects on near-threshold fatigue crack growth in friction stir welds in aerospace alloys, *Inter. Jour. of Fatigue,* **25,** 939-48 (2003).

[20] A. P. Reynolds, E. Hood, and W. Tang, Texture in friction stir welds of Timetal 21S, *Scripta Mater.*, **52,** 491-94 (2005).

[21] H. R. Wenk, and P. V. Houtte, Texture and anisotropy, *Rep. Prog. Phys.*, **67,** 1367-28 (2004).

SILICA-COATING OF BARIUM TITANATE PARTICLES

Hideki Matsumoto, Daisuke Nagao, Yoshio Kobayashi, Mikio Konno*
Department of Chemical Engineering, Graduate School of Engineering, Tohoku University
6-6-07, Aoba, Aramaki-aza, Aoba-ku, Sendai, Japan, 980-8579

ABSTRACT

Barium titanate (BT) particles were coated with silica to improve their stability in water. The BT particles were prepared using a sol-gel method in a mixed solvent of ethanol and acetonitrile (weight ratio = 25/75) with additives of methylamine (MA) and acetylacetone. The silica coating was produced with the hydrolysis of tetraethylorthosilicate (TEOS) in the mixed solvent using a MA catalyst in the presence of the BT particles. TEM observation showed that the BT particles are spherical and have silica shells. The Zeta-potential measured for the coated BT particles with thick silica shells revealed that isoelectric point of the coated particles was almost consistent with that of silica particles. The size distribution of the coated particles was measured with dynamic light scattering (DLS) to examine the colloidal stability of the coated particles in water. The high stability of the coated particles was confirmed in the DLS measurement, and the averaged diameter of the coated particles obtained in this measurement was close to that obtained by TEM. Annealing of the coated particles produced fresnoite ($Ba_2TiSi_2O_8$) crystallite particles.

INTRODUCTION

Barium titanate (BT) particles have been used in electric devices such as multilayer capacitors (MLCCs), because of high dielectric constants.[1-4] The capacity of MLCC strongly depends on the thickness and roughness of the dielectric thin films. Organic solvents are often used as dispersive media in the fabrication of MLCC, since BT particles are unstable in water due to the dissolution of barium ions from the BT surfaces. Therefore, stability of BT particles in suspension is required for increasing the MLCC capacity.

Some researchers have studied on stabilization of BT particles in aqueous solution to reduce environmental loading.[2-5] Silica coating of BT particles is a candidate to stabilize the BT particles in water. Chen et al.[6] coated BT particles with 5 nm-silica thin layers deposited from a sodium silicate solution. Their work did not examine stability of coated particles in water. Park et al.[1] applied hydrolysis of tetraethylorthosilicate (TEOS) to silica-coating of BT particles. However, low stability of BT particles in the hydrolysis caused aggregation of BT particles, and coated BT particles with core-shell structure were scarcely formed.

In a present work, we proposed an alternative method for preparing silica-coated BT particles with core-shell structure, and examined stability of silica-coated particles with dynamic light scattering and zeta-potential measurement.

EXPERIMENTAL

Synthesis of BT core particles

A complex alkoxide solution of barium-titanium was prepared as a precursor of BT particles. At first, titanium tetraisopropoxide (TTIP, 9mmol) was added to an ethanol (73cm^3)

that dissolved metallic barium (9mmol). The ethanolic solution was refluxed at 70°C for 2h to form the complex alkoxide. Those procedures were conducted under a nitrogen atmosphere. Then, an acetonitrile (221ml) containing methylamine (18mmol) and acetylacetone (9mmol) was mixed with the alkoxide solution in a hermetically sealed flask reactor at 0°C. The reaction was performed under stirring. The initial concentrations of barium, TTIP, methylamine, acetylacetone and water were 30, 30, 60, 30 and 200mM, respectively. The mixed solvent weight ratio of ethanol/acetonitrile was 25/75. After a reaction time of 6h, paste of BT particles was obtained by centrifugation at 20000rpm for 10min and removal of supernatant.

Silica-coating of BT core particles

TEOS was hydrolyzed in the presence of the BT core particles for silica-coating. Solution of 25/75 (wt/wt) ethanol/acetonitrile containing water and methylamine was added to the paste of BT particles. Then, TEOS was added to the solution. The concentrations of methylamine, water, and BT particles in the mixture were 0.1, 1M and 0.23vol%, respectively. To control thickness of silica shells, TEOS was added in the range of 10 – 200mM. The hydrolysis was conducted in a hermetically sealed flask reactor at 35°C for 12h. The silica-coated BT particles were annealed in air at 800°C for 4h for crystallization.

Characterization

Particles formed were characterized with transmission electron microscopy (Zeiss, LEO 912 OMEGA), X-ray diffraction (Rigaku RU-200A), dynamic light scattering (Otsuka Electronics, ELS8000) and zeta potential (ELS-8000).

RESULT AND DISCUSSION

Figure 1 shows a TEM image of BT particles used as cores. The BT particles formed were spherical, and had a volume-averaged diameter of 234nm. According to XRD measurement, the BT particles were amorphous.

Figure 2 shows photographs of the BT particles in water and the ethanol/acetonitrile solvent. The BT particles were dispersively stable in the ethanol/acetonitrile, while the particles in water were rapidly aggregated to form precipitates as shown in the bottom of container.

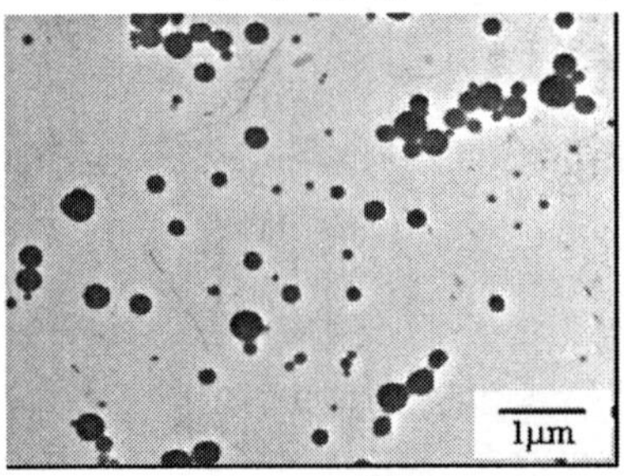

Figure 1 TEM image of the BT core particles. The core particles were prepared at concentrations of 30mM barium, 30mM TTIP, and 200mM H_2O.

Figure 2 Photographs of the BT core particles in ethanol/acetonitrile (A) and in water (B).

Figure 3 shows TEM images of silica-coated BT particles formed in a TEOS concentration range of 10 - 200mM. Aggregation of particles formed in the TEOS concentration range of 10 - 100mM was scarcely observed, while a small amount of precipitates was observed on the reactor wall at the TEOS concentration of 200mM. High magnification images of (C) and (D) demonstrated that the particles produced had core-shell structure. The average diameters of the coated particles in Figs (A) - (C) were close to those calculated from SiO_2/BT volume ratios. The average diameter in Fig. (D) was slightly higher than that calculated from the ratio. Possibly the difference between the measured and calculated diameters occurred because of aggregation of BT core particles.

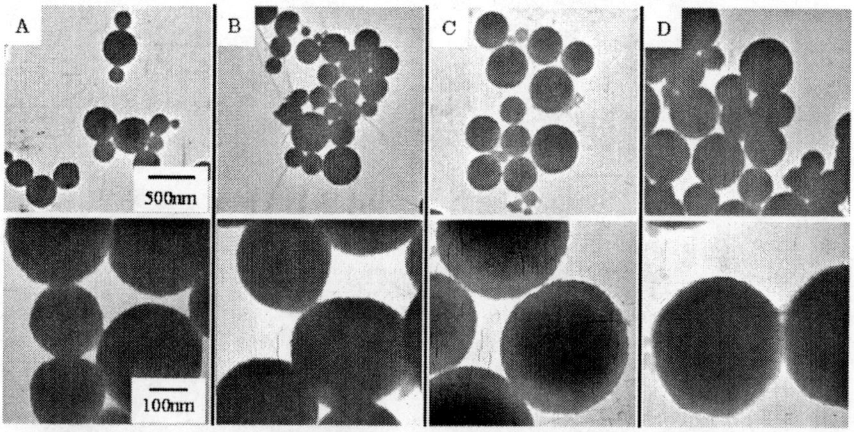

Figure3 TEM images of the silica-coated BT particles. Their high magnification images are shown in lower column. The silica-coating was performed at TEOS concentrations of 10mM (A), 20mM (B), 100mM (C) and 200mM (D). These TEOS concentrations corresponded to volume ratios of SiO_2/BT of 0.14 (A), 0.28 (B), 1.42 (C) and 2.83 (D), respectively.

Figure 4 shows photographs of the silica-coated BT particles in water. Both of pH in the suspensions were approximately 10. No precipitates were observed even after more than a month, indicating that the coated particles had high colloidal stabilities in water.

Figure 4 Photographs of the silica coated BT particles in water. BT particles were coated at TEOS concentrations of 10mM (A) and 100mM (B).

Figure 5 shows the size distributions of the coated particles obtained with dynamic light scattering. Distribution of non-coated BT particles could not be measured because of extremely low stability. More than 95% of the particles in Figs (A) and (B) had sizes less than 1μm, indicating that the coated particles were scarcely aggregated in water. Volume-averaged diameters for the main peaks in Fig. 5 (A) and (B) were close to those obtained from the TEM images in Fig. 3 (A) and (C).

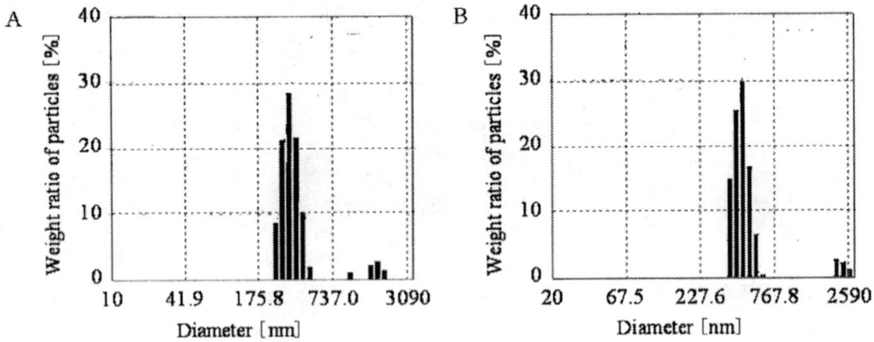

Figure 5 Diameter distributions of the silica-coated BT particles in water. Volume ratios of SiO_2/BT were 0.14 (A) and 1.42 (B).

Figure 6 shows zeta potentials of silica and silica-coated BT particles as a function of pH. Isoelectric point (IEP) for the volume ratio 0.14(A) was much higher than that of silica particles (C), and was in the range of IEP of BT (PH 6-9)[3, 4]. The difference in IEP might be caused by incomplete coverage of BT core particles. In contrast, IEP of the silica-coated particles for the volume ratio 1.42(B) was almost consistent with that of silica particles.

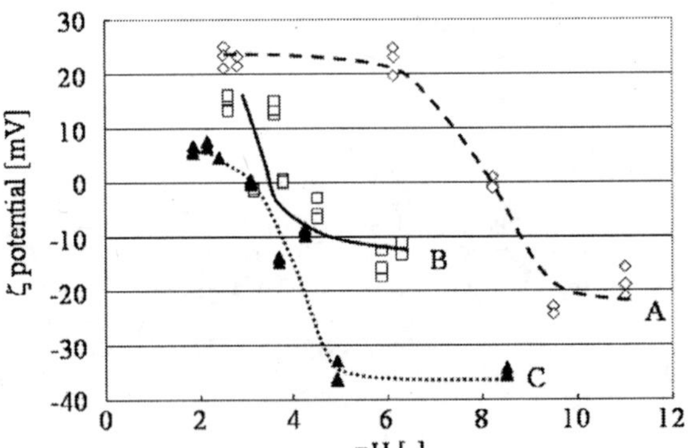

Figure 6 ζ-potential curves of the silica coated BT particles prepared at SiO_2 /BT volume ratios of 0.14 (A) and 1.42 (B). The curve of (C) shows the ζ-potential of silica particles.

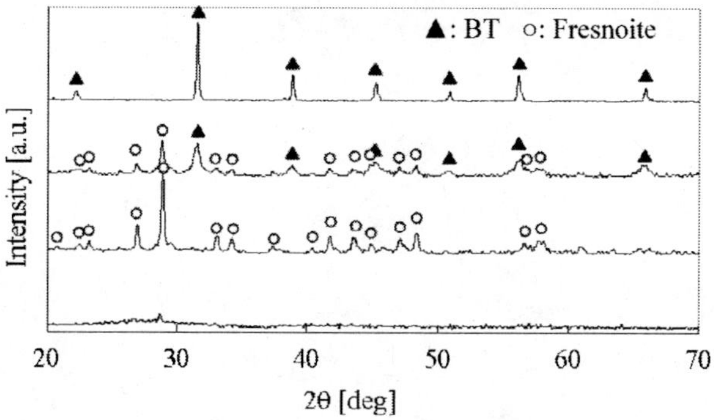

Figure 7 XRD patterns of the BT core (A) and silica-coated particles (B) – (D) after the heat treatment (800°C, 4h). The volume ratios of SiO_2/BT were 0.14 (B), 0.28 (C) and 1.42(D).

Figure 7 shows XRD patterns of the silica-coated and non-coated BT particles annealed at 800°C for 4h. The coated particles formed at the SiO_2/BT volume ratio of 0.14 (Fig. 7 (B)) revealed a pattern that included peaks of BT (cubic). For the ratio of 0.14, peaks attributed to fresnoite ($Ba_2TiSi_2O_8$) appeared. An increase in SiO_2/BT volume ratio to 0.28 (Fig. 7(C)) provided no formation of BT but only fresnoite. Further increase in the SiO_2/BT ratio to 1.42 suppressed the crystallization of the coated BT particles.

CONCLUSION

Hydrolysis of TEOS in a solvent of ethanol and acetonitrile produces silica-coated BT particles with core-shell structure. The coated particles prepared in TEOS in the concentration range of 10 - 100mM (SiO_2/BT volume ratio 0.14-1.42) are colloidally dispersed in water. Annealing the coated BT particles produces crystallized particles of fresnoite.

REFERENCES

[1]J. S. Park and Y. H. Han, "Nano Size $BaTiO_3$ Powder Coated with Silica," *Ceramics International*, **31**, 777-82 (2005)

[2]X. Y. Wang, S. W. Lu, B. I. Lee, and L. A. Mann, "Dispersion and Aging Behavior of $BaTiO_3$ and PZT in Water," *Mater. Res. Bull.*, **35**, 2555-63 (2000)

[3]M. C. Blanco López, B. Rand, and F. L. Riley, "The Isoelectric Point of $BaTiO_3$," *J. the Eur. Ceram. Soc.*, **20**, 107-18 (2000)

[4]C.-W. Chiang and J.-H. Jean, "Effects of Barium Dissolution on Dispersing Aqueous Barium Titanate Suspensions," *Mater. Chem. Phys.*, 80, 647-55 (2003)

[5]M. C. Blanco López, G. Fourlaris, and F. L. Riley, "Interaction of Barium Titanate Powders with an Aqueous Suspending Medium," *J. Eur. Ceram. Soc.*, **18**, 2183-92 (1998)

[6]R.Z. Chen, A.L. Cui, X.H. Wang, Z.L. Gui, and L.T. Li, "Structure, Sintering Behavior and Dielectric Properties of Silica-Coated $BaTiO_3$." *Mater. Lett.*, **54**, 314-17 (2002)

DISSOLUTION OF HYDROGEN INTO TUNGSTEN PHOSPHATE GLASSES THROUGH PALLADIUM COATING

Hiromasa Tawarayama* and Hiroshi Kawazoe
Kawazoe Frontier Technologies Corporation
Kuden-cho 931-113, Sakae-ku, Yokohama 247-0014, Japan

Shouichi Sugata, Futoshi Utsuno** and Hiroyuki Inoue
Institute of Industrial Science, The University of Tokyo
Komaba 4-6-1, Meguro-ku, Tokyo 153-8505, Japan

Hideo Hosono
Materials and Structures Laboratory & Frontier Collaborative Research Center, Tokyo Institute of Technology, Nagatsuta 4259, Midori-ku, Yokohama 226-8503, Japan

ABSTRACT

Dissolution of hydrogen into tungsten phosphate glasses through coated Pd layers was observed upon heating below the glass transition temperatures. The color of the glasses was varied from pale yellow to dark blue during the heat-treatments under hydrogen atmosphere, and the center responsible for the coloration was found to be W^{5+} ions formed by electron trapping from the optical and ESR absorption measurements. Here, a proton works as the charge compensator of W^{5+}. Hydrogen was reversibly released from the glasses through coated Pd under low partial pressures of hydrogen such as N_2 atmosphere. Proposed was mechanisms of hydrogen dissolution into the glasses; 1) decomposition of H_2 to 2H by coated Pd, 2) dissolution of the resulting hydrogen atom through the reaction with W^{6+} ions to form W^{5+}/H^+ redox pair in the glasses, and 3) diffusion of hydrogen in the form of proton and electron trapped on W^{6+} via hopping between W^{6+} and W^{5+} into the glass. The diffusion coefficient of hydrogen was enhanced with decreasing ion radius of modifiers such as alkali and alkali-earth ions. The electrical conductivity increased with increasing the concentration of hydrogen. The W-containing glass dissolving hydrogen was found to be a mixed conductor of proton and electron.

INTRODUCTION

Hydrogen is expected to be an energy carrier in hydrogen energy technology. Large scale production of hydrogen, at present, is depending on steam reforming of hydrocarbons because of its cost competitiveness. Therefore, process of hydrogen separation from byproducts gasses, especially carbon monoxide which is poisonous for polymer electrolyte fuel cells, is indispensable.

Pd-Ag alloy is an excellent filter material for production of high purity hydrogen because of its high thermal and chemical stability and hydrogen selectivity. However, noble metal Pd is expensive for widespread use. Consequently, hydrogen filters composed of thin film of Pd or Pd-Ag alloy on porous materials or non-Pd based alloys have been studied by many researchers.[1,2]

A protonic and electronic mixed conductor is also a candidate for hydrogen filter materials. It is well known that some acceptor doped perovskites such as $BaCe_{0.9}Y_{0.1}O_{3-\alpha}$ are high temperature proton conductors. Recently, Matsumoto et. al. reported that Ru doped

$BaCe_{0.9}Y_{0.1}O_{3-\alpha}$ was a mixed conductor of proton and electron in a moist atmosphere and showed good hydrogen permeability at 800 °C based on the ambipolar diffusion.[3]

Here we will report that hydrogen is dissolved into tungsten phosphate glasses by reacting with W^{6+} ions through coated Pd layers, the hydrogen diffuses into the glass and the dissolved hydrogen is recovered to the ambient with low partial pressure of hydrogen through the coated Pd. Hydrogen permeation rate through the glass is estimated. In addition, the glass is found to be a mixed conductor of proton and electron.

REACTION OF THE GLASS WITH HYDROGEN

Chemical composition of glass samples used in this experiment was $30PO_{5/2}\cdot 10WO_3\cdot 25NbO_{5/2}\cdot 10BaO\cdot 25NaO_{1/2}$ in mole. A stoichiometric mixture of raw materials was melted with a platinum crucible in air. The batch weight was 100 g. The melt was poured into a mold and was placed in a furnace pre-heated at 570 °C, which is the glass transition temperature (T_g) of the glass determined by differential thermal analysis. The glass obtained was cut and mirror-polished. Thin layer of Pd was deposited by sputtering on the opposite surfaces of the glass plates with thickness of 1 mm. The Pd coated and naked plates were heat-treated in $3.5\%H_2$-$96.5\%N_2$ or N_2. Optical and ESR absorption spectra of the samples were measured before and after the heat-treatments. Pd coating on the samples was removed when the measurements were performed. The absolute spin concentration of the color center was estimated by comparing the intensities for the glasses with that of $CuSO_4\cdot 5H_2O$ reference.

Although the as-prepared glass was pale yellow, the Pd coated glass was strongly colored dark blue even the inside by the heat-treatment in the forming gas. As shown in Fig.1 the induced optical absorption band peaking around 17000 cm^{-1} was thought to be due to two strong intra-ionic excitations and a weak intervalence excitation of W^{5+}.[4,5] The same absorption band was detectable in the naked glass heated in the forming gas, but the intensity was considerably weak than that of the Pd coated glass. In addition, the optical absorption intensity of the Pd coated glass heated in hydrogen containing gas was found to decrease by the additional heat-

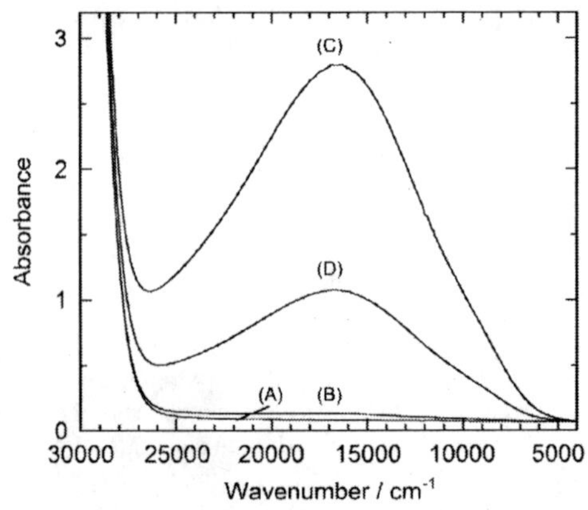

Figure 1. Optical absorption spectra of (A) the as-prepared glass and the (B) naked and (C) Pd coated glasses heated at 400 °C for 1 h in $3.5\%H_2$-$96.5\%N_2$. Glass D was obtained by heat-treating glass C at 400 °C for 1 h in N_2.

treatment under N_2 which is the ambient with lower partial pressure of molecular hydrogen (D in Fig. 1).

Simultaneously with the appearance of the optical absorption, a broad and anisotropic ESR singlet assignable to W^{5+} (g=1.7)[5] was also observed. No signal assigned as Nb^{4+} was detected. Several glasses with the different concentrations of W^{5+} were prepared similarly, and the optical absorption intensities were found to be proportional to the absolute spin concentrations of ESR absorption with g=1.7. In addition the linear plot passes through the origin. Therefore, the center responsible for the coloration was concluded to be W^{5+}.

The reversible reaction of the glass with hydrogen was thought to be expressed by the following equations,

$$H_2 \leftrightarrow 2H_S \quad (1)$$
$$W^{6+}_{G} + H_S \bullet W^{5+}_{G} + H^{+}_{G}, \quad (2)$$

where H_S is hydrogen atom formed at the surface of the glass and W^{6+}_{G}, W^{5+}_{G} and H^{+}_{G} denote the species dissolved in the glass. Hydrogen atom on the glass surface dissolved in the glass by reducing W^{6+} to W^{5+}. A proton is thought to be located around a nearby W^{5+} for charge compensation like tungsten hydrogen bronze.[6] A pair of the charge compensating proton and electron trapped on W^{6+} diffuses into inside of the glass via hopping between W^{6+} and W^{5+}. Chemical reaction by Eq. 1 is extremely promoted by coated Pd, which works as a catalyst for dissociation and recombination of hydrogen, on the surface of the glass.

COMPOSITION DEPENDENCE OF DIFFUSION COEFFICIENT

It is well known that transport characteristics of alkali ion in glasses are affected by introducing second alkali and alkaline-earth ions. Therefore, we examined effects of types of modifier ions in the tungsten phosphate glasses on the diffusion coefficient of hydrogen. The $30PO_{5/2}\cdot25NbO_{5/2}\cdot10WO_3\cdot35xRO_y$ (xRO_y=$LiO_{1/2}$, $NaO_{1/2}$, $KO_{1/2}$, 0.5BaO and 0.5ZnO) glasses were prepared with the same manner. Details of measuring the diffusion coefficient of hydrogen in the glass was reported in previous paper.[7] The mirror-polished glass plates with 20 mm×20 mm×10 mm in size heated in air or dry air below the T_g until those were completely decolored. Thin layer of Pd was deposited by sputtering on a single surface of the bleached glass plates, and the glass plates were heat-treated at 500 °C in 3.5%H_2-96.5%N_2. The heat-treated samples were cut and mirror-polished, and the peak intensities of the optical absorption at different positions were continuously measured by moving the glass sample. The diffusion coefficient was calculated from the obtained depth profile.

Figure 2 shows the relationship between the diffusion coefficient of hydrogen in the glasses and the radius of the coexisting modifier ions. In case of alkali ions the diffusion coefficient exponentially increased from 8.6×10^{-7} cm^2 s^{-1} to 2.0×10^{-5} cm^2 s^{-1} with decreasing the radius of the alkali ion. The surface concentration of the color center was also increased in the order of Li > Na > K, and that of the Li containing glass was approximately 10^{19} cm^{-3}. On the other hand, when R was alkaline-earth ions, the diffusion coefficients were drastically reduced and were a few orders of magnitude lower than those of the alkali ions having almost the same ion radii. It was found that the coexisting modifier such as alkali and alkaline-earth ions suppressed movement of hydrogen in the glass.

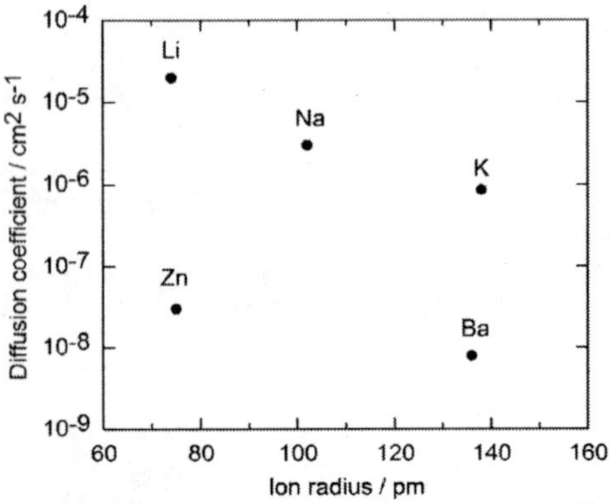

Figure 2. Relationship between the diffusion coefficient of hydrogen at 500 °C and the radius of the coexisting modifier ions (R) in the $30PO_{5/2}\cdot 25NbO_{5/2}\cdot 10WO_3\cdot 35xRO_y$ (xRO_y=$LiO_{1/2}$, $NaO_{1/2}$, $KO_{1/2}$, 0.5BaO and 0.5ZnO) glasses.

The flux of hydrogen F passed through the glass with thickness x is given by Fick's first low,

$$F = -D\frac{\partial C}{\partial x}, \tag{3}$$

where D and $\partial C/\partial x$ denote the diffusion coefficient of hydrogen and the concentration gradient of hydrogen, respectively. If D and $\partial C/\partial x$ are respectively 2×10^{-5} cm^2 s^{-1} and 5×10^{19} H atom cm^{-3} / 0.01 cm, the flux of hydrogen is calculated to be 0.08 μmol s^{-1} cm^{-2}, which is comparable with that of $BaCe_{0.9-x}Y_{0.1}Ru_xO_{3-\alpha}$.[3]

ELECTRICAL PROPERTIES

The glasses dissolving hydrogen in the form of proton and electron are expected to be a protonic and electronic mixed conductor. We measured AC and DC electrical conductivities of the $37PO_{5/2}\cdot 10WO_3\cdot 30NbO_{5/2}\cdot 23ZnO_{1/2}$ glass which shows no conduction due to alkali ions. Thin layer of Pd was deposited by sputtering on the opposite surfaces of the bleached glass plates with 10 mm ×10 mm × 1 mm in size, and the glass plates were heat-treated at 500 °C in 3.5%H_2-96.5%N_2. Impedance measurement of the samples before and after the heat-treatments was carried out at different temperatures under N_2 by using an impedance analyzer where sputtered Au electrodes were used. The applied bias was 1 V. The conductivity was determined from the Cole-Cole plot.

From the optical absorption measurements, the concentration of hydrogen in the samples was increased rapidly up to heat-treatment for 6 h and slowly after that. Figure 3 shows the electrical conductivity of the glass samples before and after the heat-treatments as a function of reciprocal absolute temperature (1/T). The symbol (●) is the experimentally observed electrical conductivity and the solid line is the calculated by the Arrhenius equation. With increasing duration of the heat-treatment, the isothermal electrical conductivity of the glass samples

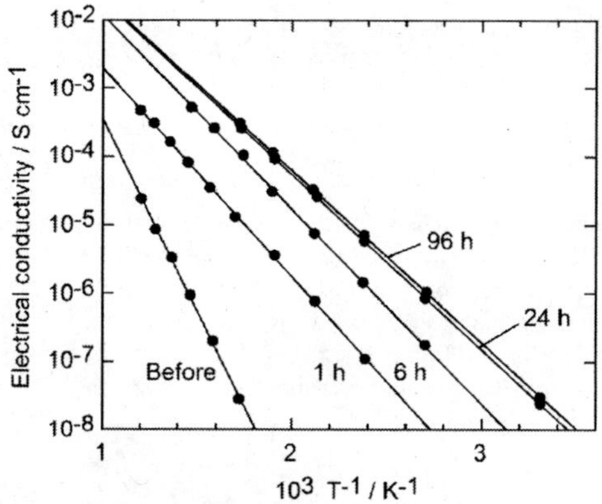

Figure 3. Temperature dependence of electrical conductivity for the glasses before and after heat-treatment in 3.5%H_2-96.5%N_2 at 500 °C. Duration of heat-treatment was described in the figure.

increased and the activation energies for the conduction decreased from 107 kJ mol^{-1} to 48 kJ mol^{-1}. The result was thought to be due to dissolution of hydrogen.

The DC electrical conductivity at 300 °C was measured in a two-probe configuration using Pd electrodes in 3.5%H_2-96.5%N_2. The glass heat-treated for 96 h was used for the measurement. Similarly, DC polarization measurement of the same glass was also performed in N_2 atmosphere by using Au electrodes to evaluate the electronic conductivity. Applied bias in this measurement was set to 0.1 V and 3 V, respectively. Optical absorption of the glasses was measured before and after the measurement.

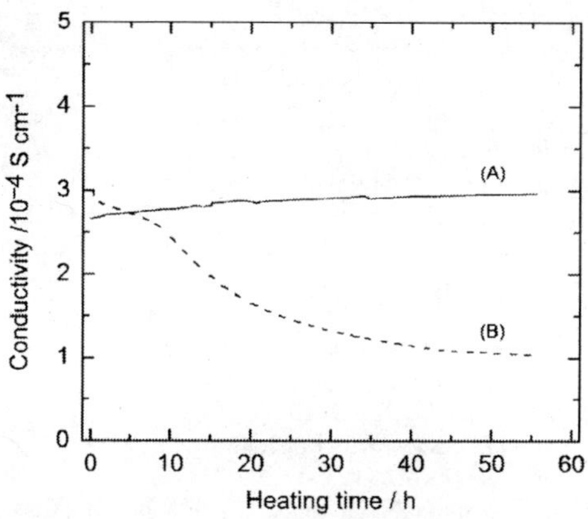

Figure 4. The change of electrical conductivity measured with increasing time. Measurement conditions are: (A) Pd electrodes in the forming gas and (B) Au electrodes in N_2.

The change of electrical conductivity measured with increasing time is shown in Figure 4. In case of using Pd electrodes under the forming gas, the conductivity slightly increased with increasing the measuring time (curve A), which is thought to be due to small concentration change of the charge compensating pair of proton and electron from the optical absorption measurement. Here the obtained DC electrical conductivity at 55 h after starting of the measurement was 3.0×10^{-4} S cm^{-1}, which is the summation of protonic and electronic conductivities. In DC polarization measurement using Au electrodes in N_2, the conductivity decreased after the electric field was applied and became 1.0×10^{-4} S cm^{-1} after 55 h (curve B). The value is thought to be the electronic conductivity because the optical absorption intensities due to W^{5+} in the glass before and after the measurement were almost the same. From the obtained conductivities the transport number of proton at 300 °C in the heat-treated glass is calculated as 0.66. The result strongly suggests that the glass is a protonic and electronic mixed conductor.

CONCLUSION

It was found that hydrogen was dissolved in tungsten phosphate glasses by reacting W^{6+} to form a pair of W^{5+} and H^+ through coated Pd. layers. The hydrogen in the form of proton and electron trapped on W^{6+} ion diffused into the glass. The modifier ions such as alkali and alkaline-earth ions work as obstacle for movement of hydrogen in the glass. The glass dissolving hydrogen was a protonic and electronic mixed conductor.

FOOTNOTES

*Present address: Materials and Structures Laboratory, Tokyo Institute of Technology, Nagatsuta 4259, Midori-ku, Yokohama 226-8503, Japan.
**Present address: Central Research Laboratories, Idemitsu Kosan Co.'Ltd., Kamiizumi 1280, Sodegaura, Chiba 299-0293, Japan.

REFERENCES

[1]S. Yamaura, M. Sakurai, M. Hasegawa, K. Wakoh, Y. Shimpo, M. Nishida, H. Kimura, E. Matsubara, and A. Inoue, Hydrogen permeation and structural features of melt-spun Ni-Nb-Zr amorphous alloys, *Acta Mater.*, **53**, 3703-11 (2005).
[2]L. Yang, Z. Zhang, X. Gao, Y. Guo, B. Wang, O. Sakai, and T. Takahashi, Changes in hydrogen permeability and surface state of Pd-Ag/ceramic composite membrane after thermal treatment, *J. Membr. Sci.*, **252**, 145-54 (2005).
[3]H. Matsumoto, T. Shimura, T. Higuchi, H. Tanaka, K Katahira, T. Otake, T. Kudo, K. Yashiro, A. Kaimai, T. Kawada, and J. Mizusaki, Protonic and Electronic Mixed Conduction and Hydrogen Permeation in $BaCe_{0.9-x}Y_{0.1}RuxO_{3-\alpha}$, *J. Electrochem. Soc.*, **152**, A488-92 (2005).
[4]G. H. Hollinger, T. M. Duc, and A. Deneuville, Charge Transfer in Amorphous Colored WO_3 Films Observed by X-Ray Photoelectron Spectroscopy, *Phys. Rev. Lett.*, **37**, 1564-7 (1976).
[5]P. Gerard, A. Deneuville, and R. Courths, Characterization of a-"WO_3" thin films before and after colouration, *Thin Solid Films*, **71**, 221-36 (1980).
[6]M. Miyakawa, H. Hosono, and H. Kawazoe, Formation of hydrogen tungsten bronze by proton implantation, *Mat. Res. Bull.*, **34**, 115-21 (1999).
[7]H. Tawarayama, F. Utsuno, H. Inoue, H. Hosono, and H. Kawazoe, Coloration and Decoloration of Tungsten Phosphate Glasses by Heat Treatments at the Temperature Far below Tg under a Controlled Ambient, *Chem. Mater.*, **18**, 2810-6 (2006).

PROPERTIES OF METALLIC GLASS COATINGS ON AN ALUMINUM ALLOY SUBSTRATE PRODUCED USING A HVOF SPRAYING PROCESS

Hyun-Guen Kim
Graduate School of Engineering, Osaka University
11-1, Mihogaoka, Ibaraki, Osaka, Japan, 567-0047

Kazuhiro Nakata, Takuya Tsumura
Joining and Welding Research Institute, Osaka University
11-1, Mihogaoka, Ibaraki, Osaka, Japan, 567-0047

Masaharu Sugiyama, Takanori Igarashi
TOPY Industries, Limited
5-9, Yonban-cho, Chiyoda-ku, Tokyo, Japan, 102-8448

Masahiro Fukumoto
Toyohashi University of Technology
1-1, Tempaku-cho, Toyohashi, Aichi, Japan, 441-8580

Hisamichi Kimura, Akihisa Inoue
Institute for Materials Research, Tohoku University
2-1-1, Katahiro, Aoba-ku, Sendai, Japan, 980-8577

ABSTRACT

An 170μm thick, Fe-based bulk metallic glass ($Fe_{43}Cr_{16}Mo_{16}C_{15}B_{10}$) coating was successfully formed on an aluminum alloy (A5052) metal substrate using an HVOF (High Velocity Oxygen Fuel) spraying process. X-ray diffraction analysis determined that the sprayed remains amorphous after spraying. The amorphous coating shows good adhesion to the metal substrate, and has a high hardness of HMV 900-1000, very low wear loss, and a comparably high friction coefficient of 0.5 to 0.6 determined using a ball-on-disc wear tester.

INTRODUCTION

Aluminum alloys are widely used in industry for their lighter weight, energy efficient and recyclable goods, the need to replace many materials, both with technical and economical considerations, and so on. However, aluminum materials in general exhibit poor wear resistance because of their softness. To avoid this disadvantage, the implementation of wear resistance coatings deposited by spraying process has been proposed [1-5].

Significant interest has been generated in amorphous alloys due to their unique physical, mechanical, and chemical properties [6]. Consequently, a variety of technologies have been used to produce such coatings. Recently, it has been discovered that special multi-component alloys can form amorphous structure with much slower cooling rate (around 100K/sec) during solidification than conventional alloys [6-7]. It is known that the cooling rate required for the formation of amorphous structure of conventional alloys is usually higher than 10^5K/sec. These alloys with high glass forming ability are called BMG (Bulk Metallic Glass) alloys.

It is known that the Fe-based metallic glass is a promising candidate due to high hardness and corrosion resistance [7-10]. Because of the absence of crystalline anisotropy the Fe-based buck metallic glass exhibits higher strength and hardness as compared with the corresponding crystalline phase.

Thermal spraying is probably the most economical method for the production of thick amorphous surface layers. However, few successful thermal sprayed coatings have been developed and commercialized since it is difficult to produce fully amorphous structure in air. Furthermore, thermal sprayed coatings with partially amorphous structure do not give anticipated excellent protection on wear and corrosion since defects such as lamellar structure and pores which are unique in thermal sprayed coatings overwhelm the beneficial effects of amorphous structure.

Most of previous investigations have worked with Fe- and Ni-based alloys with eutectic composition containing metalloid such as B, Si, and C [7-8, 11-12]. Although it was reported that the initial powder state has little or no influence on the amorphous content of the coatings, utilizing the fully amorphous content of the coatings since the HVOF spraying frequently contains unmelted particles. It is also expected that utilizing the HVOF spraying may prevent the formation of the oxide phase in the coating since the atmosphere of the HVOF spraying is reducing rather than oxidizing.

This article reports on the amorphous phase formation of Fe-based BMG forming alloy by HVOF spraying. This is a new approach that the BMG forming alloys with high hardness and good wear resistance praying process.

EXPERIMENTAL DETAILS

Preparation of the powders for spraying

One of the well-known BMG alloys, $Fe_{43}Cr_{16}Mo_{16}C_{15}B_{10}$ (at. %) was used for this investigation [7-10]. The preparation of the powders for spraying was manufactured by water atomization. The shape of the powders is spherical as shown in Fig. 1. Table 1 shows the compositions of the powder prepared for spraying material.

Table 1 Chemical composition (at. %) of the powder.

Composition (at. %)					Size (μm)
Fe	Cr	Mo	C	B	
43	16	16	15	10	25□45

Spraying conditions

HVOF spraying experiments were conducted using a TAFA JP-5000 system. Fuel was Kerosene and the substrate was A5052 with a thickness 5mm. The substrate was blasted with Al_2O_3 grit prior to spraying in order to enhance the adhesive strength between coatings and substrate. A typical set of spray parameters used is presented in Table 2. Deposition thicknesses of 150-200μm were achieved from 12 vertical passes of the gun.

Characterization of the coatings

The microstructure of the coatings was examined by optical microscopy and SEM (Scanning Electron Microscopy) equipped with EDS (Energy Dispersive Spectroscopy). The porosity (% area) of the coatings was measured by image analyzer at a magnification of 500 (KEYENCE Digital microscope VHX-200). About 10 measurements of each sample were conducted to have the average value of the coatings. Phase identification was conducted by XRD on the as-received powder and as-sprayed coatings using Cu K_{α} radiation at 40kV, 40mA. The scanning rate was 0.02°/sec. for 2θ range of 20-80°. The hardness of the sprayed coatings was measured using a microvickers hardness tester (AKASHI AAV-500 series automatics hardness testing machine) at the load of 0.98N.

Tribological behavior of $Fe_{43}Cr_{16}Mo_{16}C_{15}B_{10}$ bulk metallic glass HVOF sprayed coating on the aluminum alloy substrate was evaluated using a ball-on-disk tribometer under room temperature. Ball of alumina with a diameter of 4.8 mm was used as the counterface materials. The normal load was 10N. The sliding distance was 47.48m and the sliding speed was controlled at 0.21m/s. Coefficient of friction was recorded during each test.

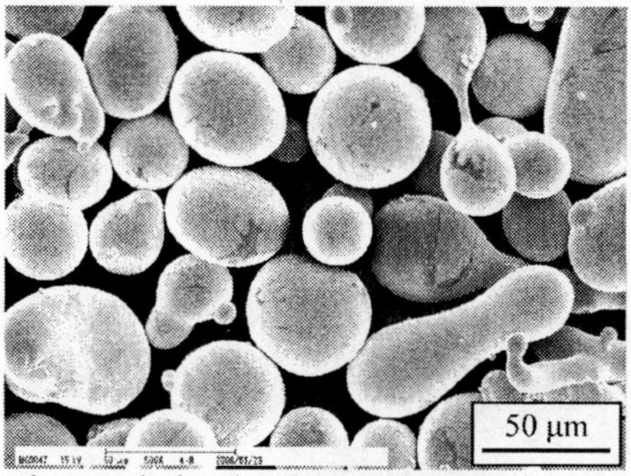

Fig. 1 Scanning electron micrograph showing the morphology of the powder.

Table 2 Experimental conditions of the HVOF spraying using a TAFA JP-5000 spraying gun.

Spray gun	TAFA JP-5000HP
Fuel flow rate	5.1, 6.1, 7.1 GPH*
Oxygen gas flow rate	1800 SCFH**
Pass	12 Pass
Step	5 mm
Gun traverse speed	1000 mm/sec.
Spray distance	380 mm

* GPH: gallon per hour, **SCFH: standard cubic feet per hour

RESULTS AND DISCUSSION

Microstructure of the coatings

Fig. 2 shows the cross sectional microstructures of the sprayed coatings with a variation of the fuel flow rates. As the fuel flow rate is increased, the microstructure of the coatings tends to be denser, that is, less pores and thinner lamellae thickness. Fig.3 shows the SEM morphology of the surface of the coating sprayed with different fuel flow rate. According to the fuel flew rates, cross sectional microstructure changes of the as-sprayed Fe-based metallic glass coating can be seen Fig. 2 and Fig. 3. Similar to the other thermal spraying process, HVOF sprayed coating microstructure is largely dependent on the impacting particle energy and resulting deposition phenomena. Energy state of particle at the moment of impact could be estimated from the microstructural features according to the fuel flow rates. As the fuel flow rates are increased, unmelted particle size and number density are decreased and then it disappears. The flattening ratio is considered to be increased with the increased of the fuel flow rates from the viewpoint of splat thickness.

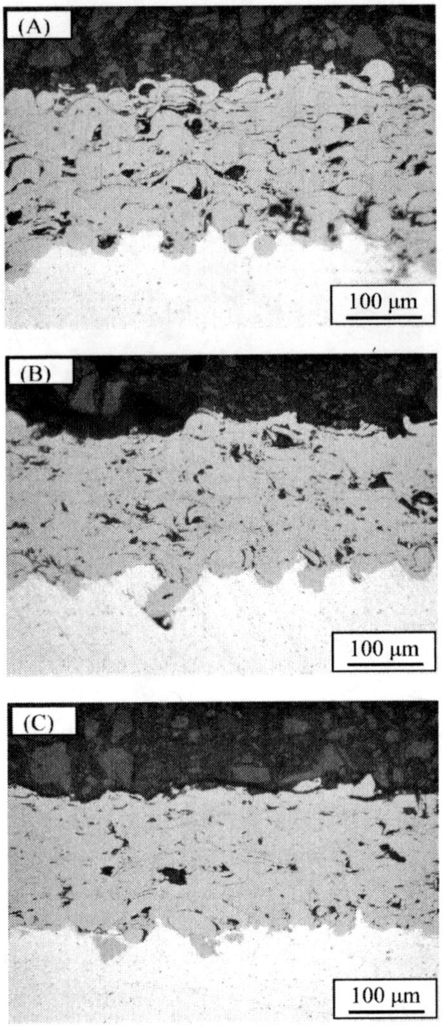

Fig. 2 Optical micrographs of the cross section of the coatings sprayed with different fuel flow rates; (A) 5.1GPH, (B) 6.1 GPH, (C) 7.1GPH.

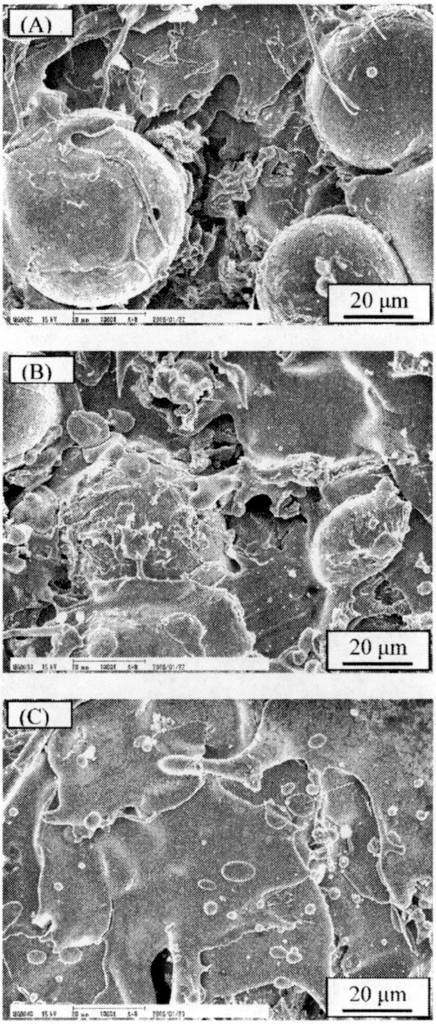

Fig. 3 SEM morphology of the surface of the coatings sprayed with different fuel flow rates; (A) 5.1GPH, (B) 6.1 GPH, (C) 7.1GPH.

The results of the quantitative measurements of porosity by image analysis are shown in Fig. 4. It is shown that as the fuel flow rate is increased; the porosity of the coatings is decreased. Fig. 5 shows the XRD patterns of the as-sprayed coating surface in compare with used spraying powder. All of the diffraction patterns show hallow patterns that are typical of amorphous structure.

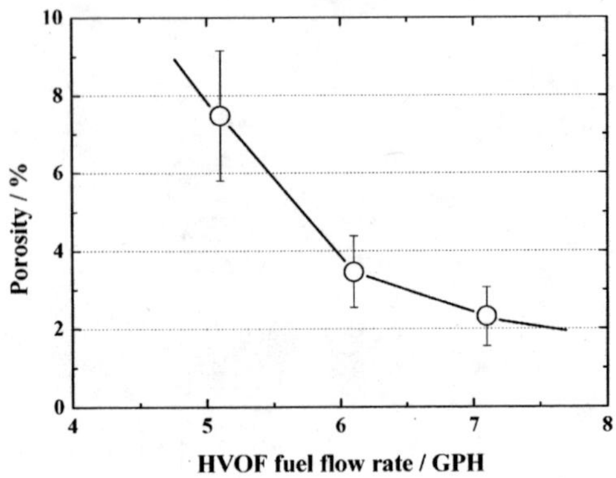

Fig. 4 Porosity variations of the sprayed coatings with different fuel flow rates.

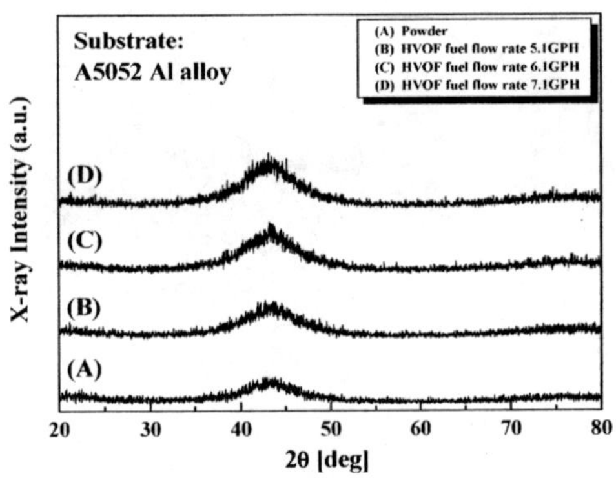

Fig. 5 XRD spectra of the sprayed coatings with different fuel flow rates.

Mechanical properties of sprayed coating

The microvickers hardness profiles of the cross section of the sprayed coatings with the fuel flow rates are shown in Fig. 6. The average hardness of the sprayed coating is slightly increased with the increased fuel flow rates. Fig. 7 shows the weight loss of the sprayed coating after ball-on-disc wear test. Fig.8 indicates the relationship between microhardness and wear loss of the coatings.

Fig.9 shows the evolution of the friction coefficient as a function of the friction time when sliding against the Al_2O_3 ball. It is shown that as the fuel flow rate is increase, the friction coefficient of the coatings is decreased. Friction coefficient of a certain material system can be affected by a great number of factors such as phase composition, porosity, microhardness, and so on. Friction coefficient was decreased with the increased of porosity in the as-sprayed coating.

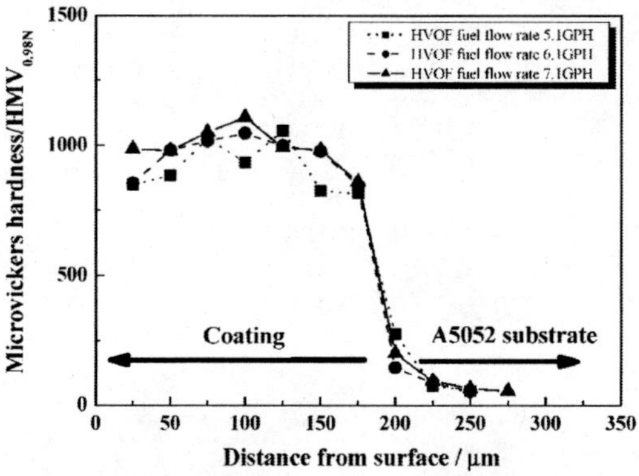

Fig. 6 Microvickers hardness profiles of the cross section of the sprayed coatings with different fuel flow rates.

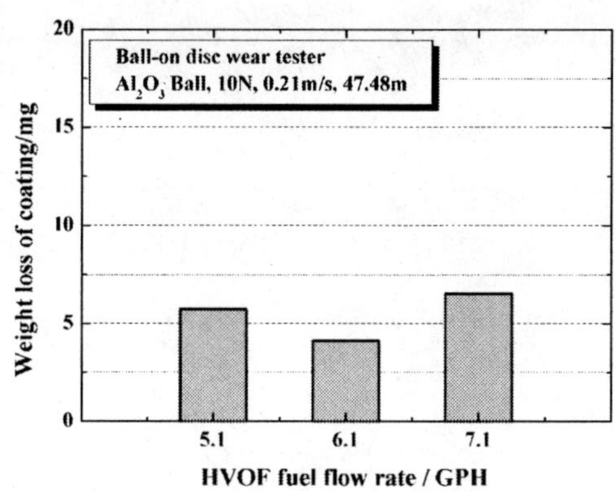

Fig. 7 Weight loss of the sprayed coatings with different fuel flow rates.

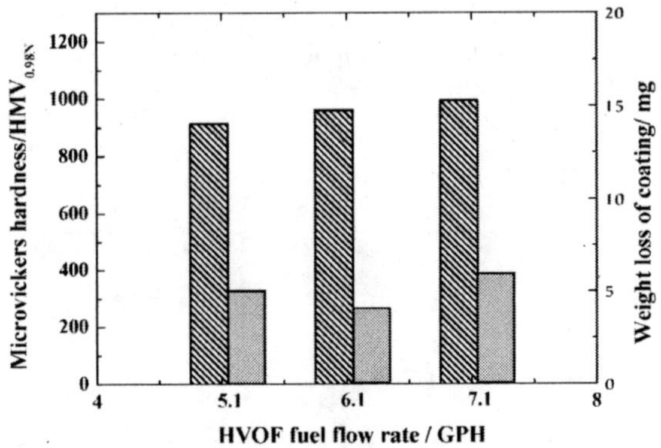

Fig.8 Relationship between microvickers hardness and weight loss of coatings with different fuel flow rates.

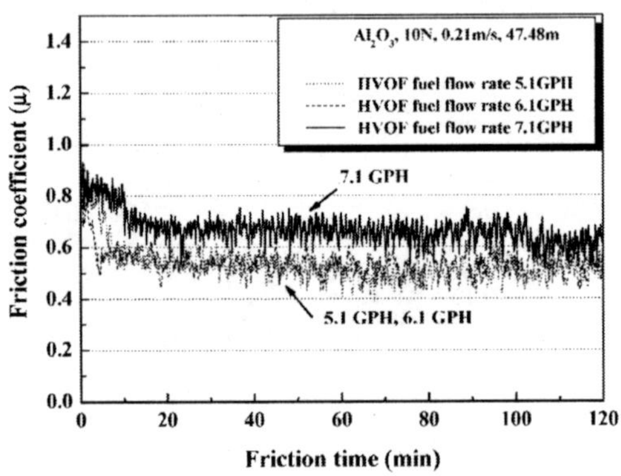

Fig. 9 Variation of friction coefficient of the sprayed coatings during ball-on-disc wear test.

CONCLUSION

A Fe-based metallic glass powder was sprayed using an HVOF process, and the effects of the fuel flow rates on the microstructure of the sprayed coating were investigated. Additionally, microhardness and porosity effects on the friction and wear loss of the metallic glass coating were investigated. The results of the study can be summarized as follow:

1. An amorphous coating was produced using the HVOF process, and, as the fuel flow rates are increased, the microstructure of the coatings tends to be denser.

2. The sprayed coating has a high hardness of HMV 900-1000, very low wear loss, and a comparably high friction coefficient of 0.5 to 0.6 as determined using a ball-on-disc wear tester.
3. The friction coefficient decreases with the increasing porosity of the as-sprayed coating.

REFERENCES

[1]K. Nakata and M. Ushio, "Al rich hard coatings on Al alloy by low pressure plasma spraying," *Surf. Eng.*, **13**, 45-49 (1997).
[2]K. Nakata and M. Ushio, "Wear resistance of plasma sprayed Al-Si binary alloy coatings on A6063 Al alloy substrate," *Surf. Coat. Technol.*, **142-144**, 277-82 (2001).
[3]K. Nakata and M. Ushio, "Effect of Fe content on wear resistance of thermal-sprayed Al-17Si-XFe alloy coating on A6063 Al alloy substrate," *Surf. Coat. Technol.*, **169-170**, 443-46 (2003).
[4]S. Uozato, K. Nakata and M. Ushio, "Corrosion and wear behaviors of ferrous powder thermal spray coatings on aluminum alloy," *Surf. Coat. Technol.*, **169-170**, 691-94 (2003).
[5]S. Uozato, K. Nakata and M. Ushio, "Evaluation of ferrous powder thermal spray coatings on diesel engine cylinder bores," *Surf. Coat. Technol.*, **200**, 2580-86 (2005).
[6]A. Inoue, "Stabilization of metallic supercooled liquid and bulk amorphous alloys," *Acta mater.*, **48**, 279-306 (2000).
[7]K. Hildal, N. Sekido, J.H. Perepezko, "Critical cooling rate for $Fe_{48}Cr_{15}Mo_{14}Y_2C_{15}B_6$ bulk metallic glass formation," *Intermetalicsl.*, **14**, 898-902 (2006).
[8]S.J. Pang, T. Zhang, K. Asami, A. Inoue, "Synthesis of Fe-Cr-Mo-C-B-P bulk metallic glasses with high corrosion resistance," *Acta matar.*, **50**, 489-97 (1999).
[9]K. Kishitake, H. Era, and F. Otsubo, "Thermal-Sprayed Fe-10Cr-13P-7C Amorphous Coatings Possessing Excellent Corrosion Resistance," *J. Therm. Spray Technol.*, **5**, 476-82 (1996).
[10]M. Stoica, J. Eckert, S. Roth, Z.F. Zhang, L. Schultz, W.H. Wang, "Mechanical behavior of $Fe_{65.5}Cr_4Mo_4Ga_4P_{12}C_5B_{5.5}$ bulk metallic glass," *Intermatallics.*, **13**, 764-69 (2005).
[11]H.J. Kim, K.M. Lim, B.G. Seong, C.G. Park, "Amorphous phase formation of Zr-based alloy by HVOF spraying process," *J. Mater. Sci.*, **36**, 49-54 (2001).
[12]S. Yoon, C. Lee, H. Choi H. Jo, "Kinetic spraying deposition behavior of bulk amorphous NiTiZrSiSn feedstock," *Mater. Sci. Eng A.*, **415**, 45-52 (2006).

PHOTOELECTRON SPECTROSCOPIC STUDY OF ENERGY LEVEL ALIGNMENT AT C12A7:e^- / Alq_3 INTERFACES

Ki-Beom Kim[1], Maiko Kikuchi[1], Masashi Miyakawa[3], Hiroshi Yanagi[1], Toshio Kamiya[1,2], Masahiro Hirano[2], Hideo Hosono[1,2,3]

Materials and Structures Laboratory, Tokyo Institute of Technology, Mail-box R3-1, 4259 Nagatsuta-cho, Midori-ku, Yokohama 226-8503, Japan[1]
ERATO-SORST, Japan Science and Technology Agency (JST), in the Frontier Collaborative Research Center (FCRC), Tokyo Institute of Technology, Mail-box S2-13, 4259 Nagatsuta-cho, Midori-ku, Yokohama 226-8503, Japan[2]
Frontier Collaborative Research Center (FCRC), Tokyo Institute of Technology, Mail-box S2-13, 4259 Nagatsuta-cho, Midori-ku, Yokohama 226-8503, Japan[3]

Phone: +81-45-924-5628, E-mail: kevin@lucid.msl.titech.ac.jp

ABSTRACT

Alq_3 (tris 8-hydroxyquinoline aluminum) is the most widely used electron transporting material for organic light-emitting diodes (OLEDs). However, because of small electron affinity of Alq_3, a low work function cathode is required to reduce the electron injection barrier (E_{IB}). From this point of view, the electride of $12CaO{\cdot}7Al_2O_3$ (C12A7:e^-) is a good candidate as a cathode material for OLED because of its small work function (~2.4eV) and chemical stability in an ambient atmosphere. In this study, the energy level alignment at the interfaces between polycrystalline C12A7:e^- thin films and Alq_3 layers was studied by ultraviolet photoelectron spectroscopy (UPS, He I) and X-ray photoelectron spectroscopy (XPS, Mg Kα). As a result of this study, it was found that the E_{IB}, i.e., the distance from E_F of cathode to $LUMO^*$ (LUMO level after occupying an electron) of Alq_3, does not correspond to the value expected from the work functions of C12A7:e^-. This result indicates that only adopting a low work function cathode material is not sufficient to form a low E_{IB} interface: the interface energy shift should also be counted as well.

INTRODUCTION

The electride of 12CaO·7Al2O3 (C12A7:e^-) is an interesting material because of its low work function (Φ~2.4eV) and chemical stability in air [1]. Due to its unique properties, C12A7:e^- has possibility of various applications such as a tip for an electron gun and an electron emitter of field emission display (FED)[2]. On the other hand, Alq_3 (tris 8-hydroxyquinoline aluminum) is the most widely used electron transporting and emitting material for organic light-emitting diodes (OLEDs). [3, 4] Because of the small electron affinity of Alq_3, a low work function cathode is required to reduce the E_{IB}. Most of these studies are might be classified into the two categories. One is to insert an ultra thin (below 0.5 nm) insulating layer between a cathode and an organic layer. [5-7] Although this technique is presently used as a standard process of OLED manufacturing, uniform deposition at the sub-nanometer scale is very difficult. Therefore, it has potential possibility to be a bottle neck of large area fabrication process in the near future. The other is to use low work function metals such as Ca, Mg, and Yb for the cathode [8, 9] or to dope alkali metals such as Li, Sr and Sm into the organic layer by co-depositing with organic material. [10] In the case of using low work function metals, these chemically active elements are tending to react with organic material and form gap state. [11] This should result in a significant change in the electronic structure of the organic molecule. Considering these consequences, low work function materials that possess chemical stability may be an ideal solution for the cathode of organic electronic devices. From this point of view, C12A7:e^- has advantages of low work function and chemical stability at the same time. In this work, we examined the possibility of C12A7:e^- as a cathode for OLED by investigating the energy level alignment at the interface with Alq_3.

EXPERIMENT

Polycrystalline C12A7:e^- thin films were fabricated on MgO(100) substrates by the PLD process. Detailed fabrication condition is described in ref. [12]. After deposition of thin films, several kinds of surface treatment processes were examined to find out appropriate surface condition. The conditions of the processes are shown in Table 1.

Table 1. Process conditions of surface treatments

Surface treatment	Process condition
Vacuum annealing	600°C, 5×10^{-5}Pa, 10min
UV-ozone cleaning	100°C, O_2 10lpm, 5min
Ar plasma	50W, 6×10^{-1}Pa, 1min

In this experiment, samples were prepared by three kinds of process schemes (1) vacuum annealing, (2) vacuum annealing and UV-ozone treatment and (3) vacuum annealing and Ar plasma treatment. After surface treatment processes, samples were analyzed by X-ray photoelectron spectroscopy (XPS) and ultraviolet photoelectron spectroscopy (UPS) to measure the chemical composition and work function. Alq_3 over-layers were deposited by thermal evaporation in the deposition chamber (2×10^{-6}Pa) and the deposition rate (0.02 - 0.05 nm/s) was monitored with a pre-calibrated quartz micro balance. After finishing each deposition step, samples were transferred in the ultra-high vacuum to analysis chamber to be measured with UPS (He I, 21.2eV). The highest occupied molecular orbital (HOMO) level and work function of C12A7:e^-/Alq_3 interface were measured until they were saturated as a function of the Alq_3 thickness (0 - 5nm).

RESULTS AND DISCUSSION

Fig. 1 shows the chemical composition and work function of the samples after surface treatment process. These are measured by XPS and UPS, respectively. In Fig. 1 (a), the chemical compositions were compared as the peak area ratios of Al 2p, Ca 2p, C 1s and O 1s peaks. For the simplicity of comparison, data were normalized to the values of the vacuum annealed sample. When compared with the VA sample, VA & UV-ozone and VA & Ar plasma treatment drastically decreased C 1s/O 1s ratio. This suggests UV-ozone and Ar plasma treatment effectively remove residual carbon contaminant on the surface. UV-ozone treatment decreased the Ca 2p/O 1s ratio and Al 2p/O 1s ratio together. It is considered to be related with the formation of an oxygen rich surface by the ozone during the process. On the other hand, Ar plasma treatment decreased the Ca 2p/O 1s ratio more than the Al 2p/O 1s ratio. It can be deduced that Ca deficient surface was formed by the selective sputtering during the Ar plasma treatment. The work functions of the samples are shown in the Fig. 1 (b). Work functions of the samples were determined as the difference between the excitation energy (21.2 eV) and the secondary electron emission cut-off energy in binding energy scale. Work functions of the VA, VA & UV-ozone treated and VA & Ar plasma treated samples were measured to be 2.6eV, 3.9eV and 4.2eV, respectively. This result shows that the vacuum-annealed C12A7:e^- had the lowest work function while the work function was significantly increased by the other treatments followed by vacuum annealing. The work function increase of the UV-ozone treatment and Ar plasma treatment can be explained by the formation of oxygen-rich layer and the non-stoichiometry of the Ca/Al ratio, respectively. To solve this side effect of the surface treatments, it is required to develop a "damage-free" surface treatment which can make the surface of C12A7:e^- clean with maintaining low work function.

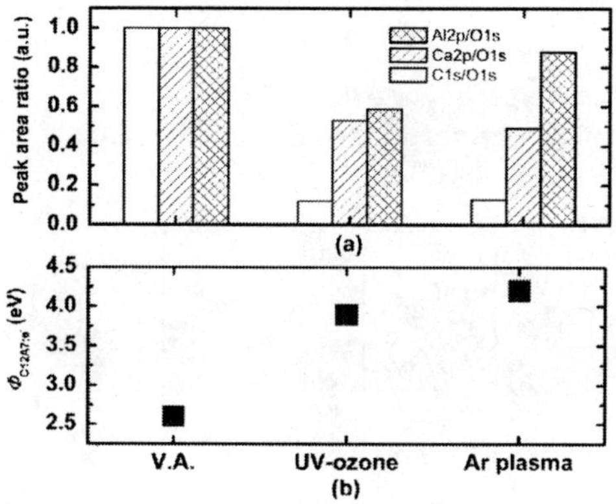

Figure 1. Effects of various surface treatments on (a) chemical composition and (b) work function of C12A7:e^- surfaces.

Fig. 2 is the UPS spectra of the VA-C12A7:e^- / Alq_3 interface measured with the function of the Alq_3 over-layer thickness (d). When d = 0 nm, the spectrum shows that of the C12A7:e^- bare surface. Fig. 2 (a) shows that vacuum level shifted to a lower binding energy by 0.3 eV when the Alq_3 thickness was 0.3 nm. There was no further change at $d > 0.3$nm. Fig. 2 (b) shows the valance band spectra of the VA- C12A7:e^- / Alq_3 interface. The onset of HOMO ascribed peak is located at 1.6eV below the Fermi level (E_F) of C12A7:e^-.

Fig. 3 is the UPS spectra at the interface of VA & UV-ozone treated C12A7:e^-/Alq_3. In the Fig. 3 (a), unlike to the result of the former, vacuum level was shifted to a higher binding energy by 0.2 eV when the Alq_3 thickness was 0.3 nm. In Fig. 3(b), when the thickness of Alq_3 is 0nm, oxygen-originated structure is observed around 7eV. It is regarded that this structure is due to UV-ozone treatment after vacuum annealing and that is the reason of difference between spectra of $d_{Alq3}=0$ in Fig. 2 (b). The HOMO level of the Alq_3 of was observed at 1.9 eV. And as can be seen in the valance band spectra of Fig. 2(b) and Fig. 3(b), there were no sign of an interfacial gap state at the C12A7:e$^-$/Alq_3 interface, which are reported by Yokoyama et al. in the case of Al/Alq_3 interface. [13] This may suggest that there is no chemical reaction between C12A7:e$^-$ and Alq_3.

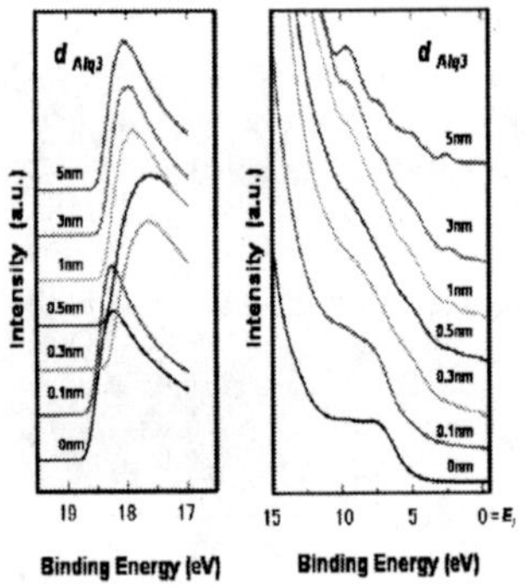

(a) (b)

Figure 2. UPS spectra of the interface and near-interface region of the vacuum annealed C12A7:e$^-$ / Alq_3. (a) cut-off region and (b) valance band region.

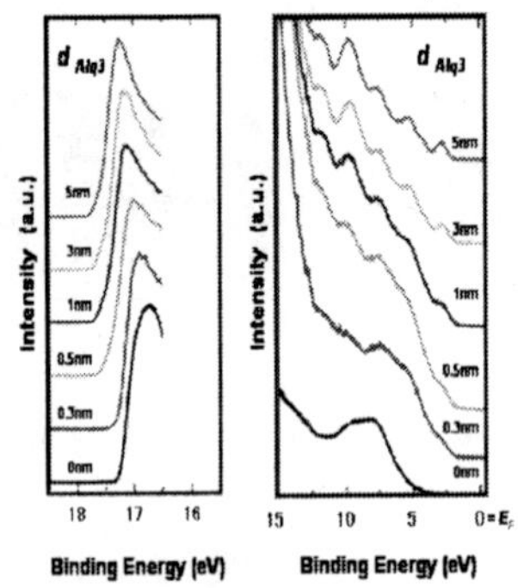

(a) (b)

Figure 3. UPS spectra of the interface and near interface region of the VA & UV ozone treated C12A7:e$^-$/Alq_3. (a) cut-off region and (b) valance band region.

In Fig. 4, the energy level alignments of the VA C12A7:e$^-$/Alq_3 and the VA & UV-ozone treated C12A7:e$^-$/Alq_3 interfaces were compared. These energy diagrams are based on the UPS spectra of Fig. 2 and Fig. 3. Because of severe change of surface chemical elements, the VA & Ar treated sample was excluded from this consideration. The values of the lowest unoccupied molecular orbital (LUMO) and LUMO* levels were deduced by using the reported optical band gap (E_{opt} = 2.9 eV) and transport gap (E_t = 4.6 eV) of Alq_3. [14]. From these diagrams, the actual

E_{IB} can be determined as the distance from E_F of C12A7:e$^-$ to LUMO* of Alq$_3$. The E_{IB} of the VA sample and of the VA & UV-ozone treated sample correspond to 3.0eV and 2.6eV, respectively. In this calculation, the height of the E_{IB} was determined by the position of onset HOMO peak in the UPS spectra. Consequently, the shift of HOMO level toward high binding energy is desirable to lower the E_{IB}. The comparison between these two interfaces makes us know that the small work function of an electrode material does not mean a small E_{IB}. Ishii et al. [15] have reported dependence of vacuum level shift (Δ) on the work function of electrode (Φ_m). According to their study, in most cases Δ values are negative as observed in the UV-ozone treated C12A7:e$^-$/Alq$_3$ interface, and it brings lowering of vacuum level. However, the results of this study suggest that Δ is not the only factor to determine the height of E_{IB} at the C12A7:e$^-$/Alq$_3$ interface.

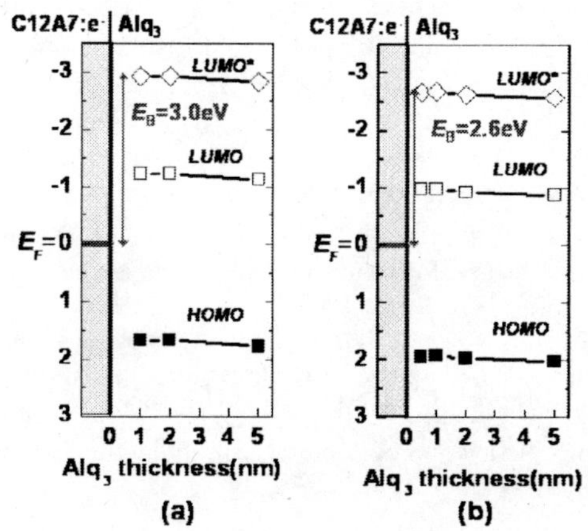

Figure 4. Band diagram of C12A7:e$^-$/Alq$_3$ interfaces. (a) Vacuum annealed (b) VA & UV-ozone treated.

SUMMARY

Until now, the low work function of a cathode material was believed to be a key to efficient electron injection in OLED.[16] However, the result of this study shows us that only adopting a low work function cathode material is not sufficient to form a interface with low E_{IB}. Since the actual E_{IB} is determined by the energy difference between E_F of cathode and LUMO* of organic material, the position of HOMO level is directly responsible for deciding injection barrier. Thus, just adopting low work function cathode is not enough to form a low E_{IB}. The energy level alignment at the interface should be counted as well.

REFERENCES

[1]S. Matsuishi, Y. Toda, M. Miyakawa, K. Hayashi, T. Kamiya, M. Hirano, I. Tanaka, H. Hosono, High-Density Electron Anions in a Nanoporous Single Crystal: $[Ca_{24}Al_{28}O_{64}]^{4+}(4e^-)$. Science 301, 626, (2003).

[2]Y. Toda, S. Matsuishi, K. Hayashi, K. Ueda, T. Kamiya, M. Hirano, and H. Hosono, Field Emission of Electron Anions Clathrated in Subnanometer-Sized Cages in $[Ca_{24}Al_{28}O_{64}]^{4+}(4e^-)$, Adv. Mater. **16**, 685 (2004).

[3]C.W. Tang, S.A. Van Slyke, Organic electroluminescent diodes, Appl. Phys. Lett. **51**, 913 (1987).

[4]C.W. Tang, S.A. Van Slyke, Electroluminescence of doped organic thin films, C.H. Chen., J. Appl. Phys. **65**, 3610 (1989).

[5]F. Li, H. Tang, J. Anderegg, J. Shinar, Fabrication and electroluminescence of double-layered organic light-emitting diodes with the Al_2O_3/Al cathode, Appl. Phys. Lett., **70**, 1233, (1997).

[6]G. E. Jabbour, Y. Kawabe, S. E. Shaheen, J. F. Wang, M. M. Morrell, B. Kippelen, N. Peyghambarian, Highly Efficient and Bright Organic Electroluminescent Devices with an Aluminum Cathode, Appl. Phys. Lett., 71, 1762, (1997).

[7]Y.-E. Kim, H. Park, and J.-J. Kim, Enhanced quantum efficiency in polymer electroluminescence devices by inserting a tunneling barrier formed by Langmuir–Blodgett films, Appl. Phys. Lett. **69**, 599, (1996).

[8]M. Stossel, J. Staudigel, F. Steuber, J. Simmerer, A. Winnacker, Impact of the cathode metal work function on the performance of vacuum-deposited organic light emitting-devices, Appl. Phys. A, 68, 387, (1999).

[9]M. Stoßel, J. Staudigel, F. Steuber, J. Blassing, J. Simmerer, A. Winnacker, H. Neuner, D. Metzdorf, H.-H. Johannes, W. Kowalsky, Electron injection and transport in 8-hydroxyquinoline aluminum, Synth. Met. 111–112, 19, (1999).

[10]J. Kido, T. Matsumoto, Efficient and stable organic light-emitting diodes with a sputter-deposited cathode, Appl. Phys. Lett., **73**, 20, (1998).

[11]G. Parthasarathy, C. Shen, A. Kahn, S. R. Forrest, Lithium doping of semiconducting organic charge transport materials, J. Appl. Phys. **89**, 4986, (2001).

[12]M. Miyakawa, T. Kamiya, M. Hirano, H. Hosono, To be submitted.

[13]T. Yokoyama, D. Yoshimura, E. Ito, H. Ishiii, Y. Ouchi, K. Seki, Energy Level Alignment at Alq3/LiF/Al Interfaces Studied by Electron Spectroscopies: Island Growth of LiF and Size-Dependence of the Electronic Structures, Jpn. J. Appl. Phys. **42**, 3666, (2003).

[14]I. G. Hill, A. Kahn, Z.G. Soos, R.A. Pascal. Jr., Charge-separation energy in films of π-conjugated organic molecules, Chemical Physics Letters **327**, 181 (2000).

[15]H. Ishii, K. Sugiyama, E. Ito, K. Seki, Energy Level Alignment and Interfacial Electronic Structures at Organic/Metal and Organic/Organic Interfaces, Adv. Mater. **11**, 8, (1999).

[16]I. D. Parker, Carrier tunneling and device characteristics in polymer light-emitting diodes, J. Appl. Phys. **75**, 3, (1994).

Development of Nanoceramics; Application to Diffusion Bonding

Michiyuki Yoshida, Yutaka Shinoda, Takashi Akatsu and Fumihiro Wakai

Cecure Materials Center, Materials and Structures Laboratory, Tokyo Institute of Technology, 4259 Nagatsuta, Midori, Yokohama, 226-8503, Japan

ABSTRACT

Nanoceramics are experiencing a rapid development in recent years. Nanoceramics exhibit superplasticity at homologous temperatures lower than the deformation temperature of their coarser-grained counterparts, because of the enhanced diffusion. The phenomenon of low-temperature-superplasticity is of technological importance owing to its cost effectiveness of near-net–shape forming preparation. The enhanced diffusion in nanoceramics should lend itself nicely to the process of diffusion bonding, then, the superplastic nanoceramics can be used as interlayers in diffusion bonding. As the first step, we performed diffusion bonding of conventional Y_2O_3 stabilized tetragonal ZrO_2 (Y-TZP, d = 300 nm) by using an interlayer of nano Y-TZP (d = 60 nm), which was fabricated by hot isostatic pressing (HIP). The joining was accomplished by uniaxially applying a compressive load (50 MPa) at temperature of 1423 K, which was 300 K lower than the sintering temperature of the conventional Y-TZP.

INTRODUCTION

As the market for ceramic materials grows, methods of fabricating larger and more complexly shaped products must be developed. At present, slip casting and injection molding are the main techniques for such shaping. However, these techniques are limited in the size and shape of the component that can be formed. The obvious alternative to these fabrication techniques is to use method of joining to build up complex or large shapes from a series of smaller and simply shaped components. Bonding techniques applied to ceramics include (1) brazing, in which oxide solder, metal solder, or metal leaf is inserted at an interface, and (2) diffusion bonding. Diffusion bonding has an advantage in the mechanical performance, especially at elevated temperatures at which a metallic interlayer will exhibit diminishment of its mechanical properties. Therefore, diffusion bonding has a potential to produce strong joints in complex-shaped components for high-temperature applications.

Diffusion bonding of the superplastic ceramics has been studied, since the superplasticity of Y_2O_3 stabilized tetragonal ZrO_2 (Y-TZP) was discovered.[1-3] Superplastic joining required minimal surface preparation and generally occurs at lower temperatures than does conventional diffusion bonding. Recently, we developed dense nanocrystalline Y-TZP (nano Y-TZP) by hot isostatically pressing.[4] Nano Y-TZP exhibited superplasticity at homologous temperatures lower than the deformation temperature of their coarser-grained counterparts, because of the enhanced diffusion. The enhanced diffusion in nano Y-TZP should lend itself nicely to the process of diffusion bonding, then, the nano Y-TZP can be used as interlayers in diffusion bonding.

In this study, we performed diffusion bonding of conventional Y_2O_3 stabilized tetragonal ZrO_2 (Y-TZP, d = 300 nm) by using an interlayer of nano Y-TZP (d = 60 nm). The pore-free joining was accomplished by uniaxially applying a compressive load (50 MPa) at temperature of 1423 K, which was 300 K lower than the sintering temperature of the conventional Y-TZP.

EXPERIMENTAL PROCEDURE

The materials used in the bonding tests were conventional Y-TZP and nanocrystalline Y-TZP. The Y-TZP sample was prepared from a tetragonal ZrO_2 powder containing 3 mol% Y_2O_3 (3Y-TZP, Tosoh Co., Japan: <0.005 Al_2O_3, <0.007 SiO_2, <0.003 Fe_2O_3, <0.017 Na_2O in wt%). The powder compacts were formed by die-pressing at 20 MPa, followed by cold-isostatically pressing (CIP) at 200 MPa. The conventional Y-TZP specimens were prepared by pressureless sintering at 1773 K for 2 h in Air. The nano Y-TZP interlayer was prepared by hot isostatic pressing (HIP). The green compacts (13 mm × 13 mm × 48 mm) were coated with BN powder (GP, Denka Co., Japan), and then encapsulated by borosilicate glass. The glass-encapsulated specimens were hot isostatically pressed (HIP) under Ar gas. They were heated up to 1073 K, then the pressure was applied to 200 MPa. The temperature for HIP was increased up to 1373 K. They were kept at the sintering temperature for 30 min. The bulk density of both conventional Y-TZP and nano Y-TZP were 6.05 g/cm^3. Both materials had 99% of the theoretical density. The sintered bodies were polished and thermally etched for microstructural observation by scanning electron microscope (Model S4500, Hitachi Co., Japan). The average grain size d was defined as the linear intercept length. The average grain sizes of conventional Y-TZP and nano Y-TZP were 300 and 60 nm, respectively.

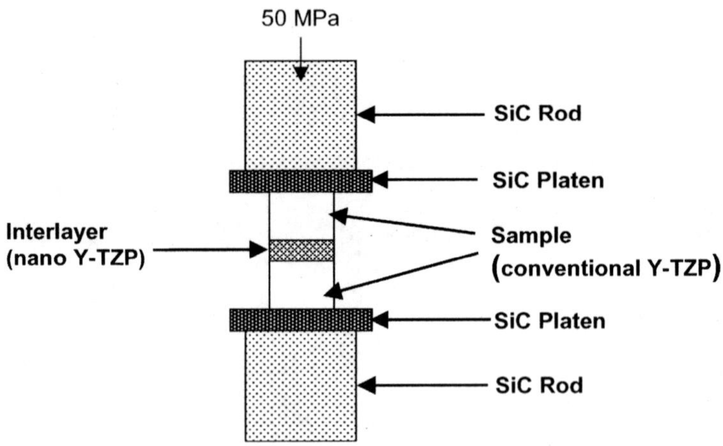

Fig. 1. Schematic of the joining set.

The bulk conventional Y-TZP and nano Y-TZP were cut to the dimensions of 3 mm × 3 mm × 4 mm and 3 mm × 3 mm × 0.6 mm, respectively. The specimens were ground with a 200-mesh diamond wheel. The joining method is schematically illustrated in Fig. 1. Two conventional Y-TZP specimens were joined either directory or with an interlayer. The sandwiched sample was produced by inserting the interlayer of nano Y-TZP between the two conventional Y-TZP specimens. Compressive deformation and joining experiments were conducted by using the universal testing machine (AG-I, Shimadzu Co., Japan) in air at 1323 K, 1373 K and 1423 K. When the temperature of the furnace reached a joining temperature, the specimens were pressed in the SiC jig at a speed of 1.1 MPa/sec up to a stress of 50 MPa. They were kept at the joining stress for 30 min. The compression tests were conducted at constant crosshead speed.

RESULTS AND DISCUSSION

Deformation behavior of nano Y-TZP

The strain rate $\dot{\varepsilon}$ is conveniently expressed as a function of the applied stress σ, the absolute temperature T, and the grain size d as follows,

$$\dot{\varepsilon} = A\frac{Gb}{RT}\left(\frac{\sigma}{G}\right)^n\left(\frac{b}{d}\right)^p D_0 \exp(-Q/RT) \qquad (1)$$

where D_0 is the frequency factor, Q the activation energy, R the gas constant, G the shear modulus, b the Burger's vector, p the exponent of the inverse grain size, n the stress exponent and A is a dimensionless constant.

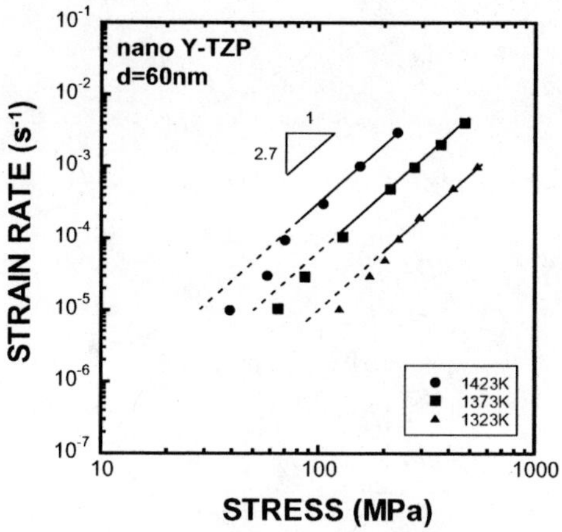

Fig. 2. Variation in strain rate with stress for the nano Y-TZP deformed at 1323-1423 K.

Table 1. Bonding results*

Temperature (K)	1323	1373	1423
With interlayer	B	B	B
No interlayer	U	U	U

*B: bonded. U: unbonded.

Figure 2 illustrates the variation in strain rate with stress for nano Y-TZP at 1323-1423 K. The slope of the curve is the stress exponent. In the deformation of nano Y-TZP, the strain rate decreased with decreasing deformation temperature. The stress exponent of nano Y-TZP was 2.7 at high stress region and increased with decreasing flow stress. Therefore, the stress exponent of nano Y-TZP is expressed as a function of the applied stress. This transition of stress exponent is peculiar to the superplasticity of high-purity Y-TZP.[5]

Bonding

The results of the bonding tests are summarized in Table 1. Under the present conditions, all sandwich specimens (with interlayer) could be bonded to each other, while no successful joints could be made without the interlayer of nano Y-TZP.

Fig. 3 shows the thermally etched interface between the conventional Y-TZP and the interlayer of nano Y-TZP bonded at 1423 K. In this figure, it was difficult to identify the interface of bonding. No interfacial residual pore was found. After bonding, no grain growth was observed in conventional Y-TZP (upper part), while the grain growth of nano Y-TZP (lower part) was remarkable. The grain size of nano Y-TZP after bonding at 1423 K was 110 nm, which was two times larger than that before the bonding (d = 60 nm). The grains at the interface retained equi-axed shape. These observations indicate complete bonding and suggest that grain boundary sliding and grain boundary migration are activated in the interlayer of nano Y-TZP during the bonding.

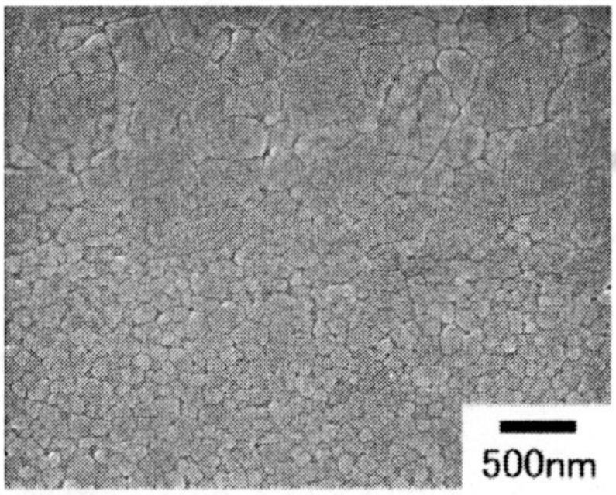

Fig. 3. SEM photograph of thermal etched interface between conventional Y-TZP (upper part) and nano Y-TZP (lower part).

SEM photographs of the bonding interfaces of the conventional Y-TZP and the interlayer of nano Y-TZP at 1323-1423 K are shown in Fig. 4. The seamless bonding was obtained at 1423 K, while interfacial residual pores were observed at 1373 and 1323 K. The size of interfacial residual pore increased with decreasing bonding temperature. The externally applied stress acting on the sandwiched sample forced the interlayer of nano Y-TZP to deform and accommodate to the surface of bulk conventional Y-TZP. The elimination of interfacial voids could therefore occur by the superplastic deformation and diffusion at the interface. The higher temperature greatly promoted the deformation of the interlayer of nano Y-TZP, and improved the mass transport (diffusion) at the interface, leading to good adhesion and pore-free joints.

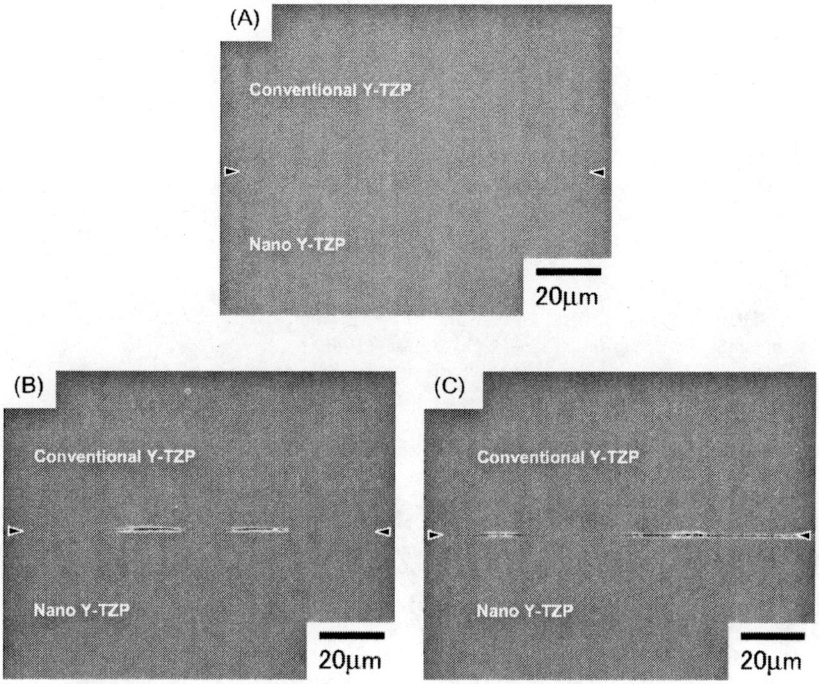

Fig. 4. SEM photographs at bonding interface of (A) 1423 K, (B) 1373 K and (C) 1323 K. The arrows indicate the bonding interface.

In this study, the seamless bonding was accomplished by using the interlayer of dense nanocrystalline Y-TZP at the temperature of 1423 K. Cross et al.[6] bonded conventional Y-TZP using the interlayer of pre-sintered Y-TZP at 1363 K. However residual pores were still remained in the interlayer of pre-sintered Y-TZP after bonding. Gutierrez-Mora et al.[7] also performed the bonding of conventional Y-TZP with interlayer of nanocrystalline Y-TZP ($d = 20$ nm) at 1423 K. The grain size of their interlayer ($d = 20$ nm) was smaller than that used in this study ($d = 60$ nm). But the stress at which the seamless bonding was accomplished (240 MPa) was larger than that of our study (50 MPa). Unfortunately, they did not measure the impurity level of their nanocrystalline Y-TZP. The deformation of nanocrystalline ZrO_2 is significantly affected by the addition of small amount of impurity.[8] It is suggested that the difference in the observed results in bonding between Gutierrez-Mora and this work is attributed to the difference in impurity concentration.

CONCLUSION

Diffusion bonding was performed on conventional Y-TZP ($d = 300$ nm) by using an interlayer of nano Y-TZP ($d = 60$ nm). The joining was accomplished at temperatures between 1323 and 1423 K, while no successful joints could be made without the interlayer of nano Y-TZP. Pore-free joint was achieved by applying a compressive stress (50MPa) at temperature of 1423 K, which was 300 K lower than the sintering temperature of the conventional Y-TZP.

REFERENCES

[1]F. Wakai, S. Sakaguchi and Y. Matsuno, "Superplasticity of Yttria-stabilized tetragonal ZrO_2 polycrystals," *Adv. Ceram. Mater.*, **1** [3], 259-63 (1986).

[2]T. Nagano, H. Kato and F. Wakai, "Diffusion Bonding of Zirconia / Alumina Composites," *J. Am. Ceram. Soc.*, **73** [11], 3476-3480 (1990).

[3]Y. Motohashi, K. Jyumonji and T. Sakuma," Mechanical Property of Joined Alumina Polycrystals Using Superplastic Ceramics as an Insert," *Mater. Sci. Forum*, **170-172**, 427-432 (1994).

[4]M. Yoshida, Y. Shinoda, T. Akatsu, and F. Wakai, "Deformation of Monoclinic ZrO_2 Polycrystals and Y_2O_3-Stabilized Tetragonal ZrO_2 Polycrystals below the Monoclinic-Tetragonal Transition Temperature," *J. Am. Ceram. Soc.*, **85** [11] 2834-2836 (2002).

[5]M. Jiménez-Melendo, A. Domínguez-Rodríguez and A. Bravo-León," Superplastic Flow of Fine-Grained Yttria-Stabilized Zirconia Polycrystals: Constitutive Equation and Deformation Mechanisms," *J. Am. Ceram. Soc.*, **81**, 2761-76 (1998).

[6]T. H. Cross and M. J. Mayo, "Ceramic-Ceramic Diffusion Bonding Using Nanocrystalline Interlayers," *Nanostruct. Mater.*, **3**, 163-168 (1993).

[7]F. Gutierrez-Mora, A. Dominguez-Rodriguez, J. L. Routbort, R. Chaim and F. Guiberteau," Joining of Yttria- Tetragonal Stabilized Zirconia Polycrystals Using Nanocrystals," *Scripta Matel.* **41** [5] 455-460 (1999).

[8]M. Yoshida, Y. Shinoda, T. Akatsu, and F. Wakai," Deformation behavior of SiO_2 doped nanocrystalline monoclinic zirconia at low temperatures, " *Key Eng. Mat.* **317-318**, 433-436 (2006).

INFLUENCE OF FRICTION STIR WELDING PARAMETERS ON MICROSTRUCTURE AND MECHANICAL PROPERTIES OF AA2024-T3 ALUMINUM ALLOY

Saad Ahmed Khodir, Toshiya Shibayanagi and Masaaki Naka
Joining and Welding Research Institute, Osaka University
11-1,Mihogaoka, Ibaraki, Osaka, Japan, 567-0047

ABSTRACT

In this study, 3mm thick AA 2024-T3 Aluminum alloy plates were friction stir butt welded. Effects of welding parameters such as rotation speeds (6.67-25 sec^{-1}), welding speeds (0.83, 1.67 mm/sec) and backing plates on microstructures, hardness distributions, and tensile properties of the joints were investigated. Three types of backing materials were used, including stainless steel, copper block, and a combination of the two. Orientation distributions in the weld zone were investigated using electron backscatter diffraction patterns (EBSP). The grain size in the stir zone (SZ) increased with increasing rotation speed and decreased from 7.4μm with stainless steel to 4.4 and 1.7 μm with a combination backing type and copper-block, respectively. The orientation distribution changes gradually in the SZ, suggesting a complicated metal flow during FSW. Hardness increased both in the SZ and the thermo- mechanically affected zone (TMAZ) as the rotation speed increased. A hardness maximum appeared in heat affected zone (HAZ) for a higher welding speed and when using a combination of stainless steel and copper block plate (backing type 3). The tensile strength of the joints increased with increasing rotation speed and a maximum value of 427MPa was achieved with type 3 backing plate.

INTRODUCTION

Over the last decade, friction stir welding (FSW) has offered excellent welding quality to the joining of Al alloys specially for high strength Al alloys (2xxx, 7xxx) which are difficult to be welded using conventional fusion welding[1]. FSW joint is known to generally possess four zones such as (i) intensively deformed zone called stir zone (SZ), (ii) thermomechanically affected zone (TMAZ), (iii) heat affected zone (HAZ), and (iv) base metal (BM) [2]. According to previous studies[3-8], dynamic recrystallization takes place in SZ to produce equiaxed fine grain structure. No recrystallization occurs in TMAZ. HAZ has experienced only thermal cycle which affects mechanical properties of welded joint. The present study aims to investigate the effect of welding parameters such as rotation speed, welding speed, backing materials on microstructure and mechanical properties of AA 2024-T3 aluminum alloy joints friction stir welded.

EXPERIMENTAL PROCEDURES

Friction stir welds was performed on 3.0mm thick 2024-T3AA (Al- 4.1Cu- 1.4Mg- 0.5Mn- 0.6Si- 0.08Fe wt. Pct.). The yield, tensile strengths and elongation of BM are 322, 455MPa, and 21.5% respectively. Welding speed was kept constant at 0.83mm/sec and rotation speeds were set at 6.67, 10, 13.3, 16.67, 20.8, and 25sec^{-1} using backing material SUS304 of 10mm thick (type 1). Then rotation speed was kept constant at 20.8sec^{-1} and two welding speeds as 0.83 and 1.67 mm/sec were adopted at three types of backing materials such as SUS304 (type 1), pure copper-block of 12mm thick (type 2), and combination of 0.5mm thick SUS304 plate with 12mm thick copper block (type 3). Temperatures were measured during FSW by utilizing K-type thermocouples of 0.5mm diameter at welding centerline and HAZ on the advancing side. The observation of microstructure was performed on a cross-section in the weld region after mechanical grinding and polishing followed by colloidal silica suspension. Crystallographic data collection by Electron Back-Scattered Pattern (EBSP) for BM and SZ were performed in scanning electron microscope operating at 20 kV. Microhardness measurements were carried out along the center line of cross section with the internal spacing of 0.5mm under the load of 0.98N for 15s loading time. The tensile test was carried out at room temperature at a constant crosshead speed of 1mm/min.

RESULTS AND DISCUSSION

Microstructures of Joints

Figure 1 shows microstructures of BM and recrystallized grains in the center region of SZ for joints welded at 0.83mm/sec of welding speed and different rotation speeds. Microstructure of BM as shown in Fig. 1 (a) represents elongated grains containing randomly distributed second phase particles and inclusions. On the other hand, microstructure of SZs as shown in Fig. 1 (b, c, d) appear to consists of equiaxed grains with much smaller size compared to the large elongated grains of BM. Grain size increases with increasing rotation speed as shown clearly in (b), (c). Table I represents the average grain size in SZ for all welding parameters. A maximum grain size of 7.4μm was obtained at the rotation speed of 25sec^{-1}. Grain size in SZ decreases from 7.4 μm (Fig. 1 c) for backing type 1 to 1.7μm for type 2 (Fig. 1 d) at the same welding and rotation speeds. Table I also shows the temperature measurements in SZ and HAZ for most welded joints. Temperature increases both in SZ and HAZ as rotation speed increases. On the other hand, the joint produced using backing type 2 has the lowest peak temperature in SZ (683K) and HAZ (671K). For the case of backing type 3 the temperatures were between these two cases.

In general, average grain size in SZ depends mainly on the peak temperature[3]. Peak temperature in SZ depends on heat input which increases with increasing rotation speed and decreasing welding speed. Thus lower grain size in SZ was brought about by decreasing rotation speed, increasing welding speed or using suitable backing plates. In case of backing type 1 the largest grain size is obtained because of the highest peak temperature measured in SZ as shown in Table I. In case of backing type 2, reduction of grain size in SZ is explained by an increase in the amount of frictional heat extracted by pure copper block due to its higher thermal conductivity than SUS304. For the backing type 3, the 0.5mm-thick SUS304 plate added onto the pure copper bock slightly decreases the amount of heat extraction through the copper block which in turn increases the peak temperature resulting in larger grain size than that pure copper only. Therefore, grain size in SZ is expected to be controlled easily by controlling welding parameters.

Fig.2 shows orientation maps and pole figures of BM and central region in SZ of joint welded at 20.8sec^{-1} and 1.67mm/sec using backing type 3. We select this welding condition for OIM analysis because maximum mechanical properties were achieved as shown later when talking about the hardness and tensile properties of joint. These images in Fig.2 were constructed such that the misorientation angle was greater than 15^0. The orientations are colored according to the color key of the inverse pole figure (IPF) // WD as shown in the middle of upper side of Fig. 2. BM in Fig. 2(a) shows large grains with high angles grain boundaries. On the other hand, Fig. 2 (b) shows that SZ is composed of equiaxed very fine grain structure with high fraction of high angle grain boundaries as a result of dynamic recrystallization. A (100) pole figure of BM in Fig. 2(c)shows that the most dominant component is close to strong rotated cube orientation

{001}<110>. The cube texture is well known as a typical recrystallization texture in aluminum alloys. On the other hand the (111) pole figure of SZ in Fig.2 (d) shows that the texture in this region roughly takes the orientations of {114} <221> and {110} <001>. This (111) pole figure was produced by rotation of -16, 10, and 17 degrees about ND, WD and TD, respectively. These two texture components in the central region of SZ were also observed in 6063-T5 aluminum alloy by Sato et al[9]. The change of rotated cube texture of the BM to the{114} <221> and {110} <001> textures in SZ suggesting a complicated metal flow during FSW.

FSW is considered as a shearing deformation process. Shearing and compressive stresses generated at the surface between rotating shoulder and metal surface during rotation and forward motion of tool shoulder. Also shear stresses arise along the probe surface and during its forward motion. The effect of shoulder is limited to the surface region while the effect of probe extends from the area under the shoulder to the all center area of SZ[9]. The (111) plane located almost parallel to the axial direction of the rotating probe. Therefore the week shearing texture <111>// ND in middle region of SZ resulted from the shearing stresses arise from the probe surface and its forward motion.

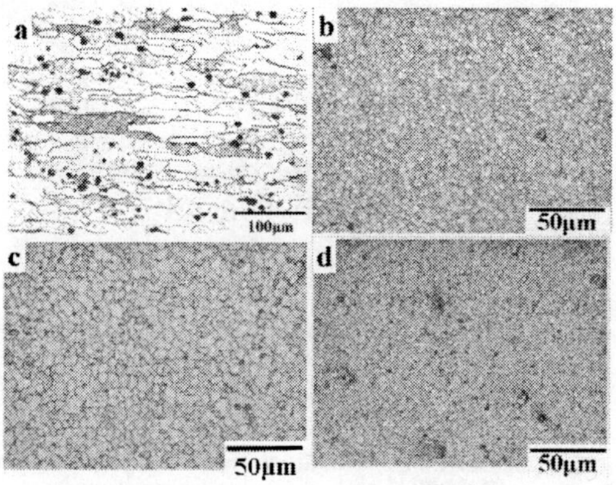

Fig.1 Microstructures of BM and SZ of joints welded at 0.83mm/sec: (a) BM, (b) SZ at 10sec^{-1}, (c) SZ at 20.8sec^{-1} using backing type1, and (d) SZ at 20.8sec^{-1} using backing type 2.

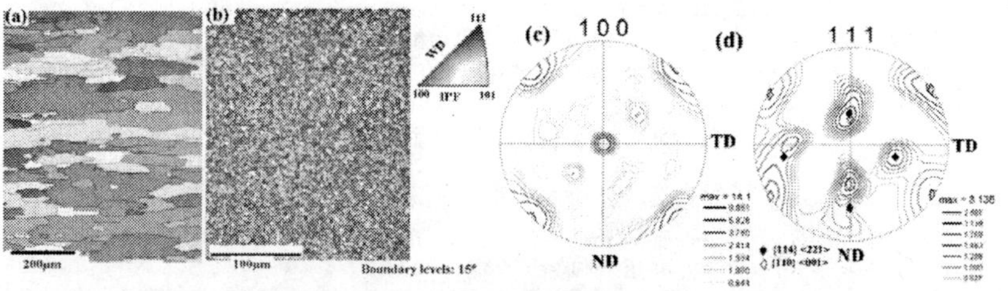

Fig.2 Orientation images and pole figures of BM and SZ of joint welded at 20.8sec^{-1} and 1.67mm/sec using backing type 3: (a) Orientation image of BM, (b) Orientation images of SZ, (c) 100pole figure of BM, (c) 111 pole figure in central region of SZ

Hardness of Joints

Figure 3 shows hardness profiles after 4 hours for the cases 6.67sec^{-1} (a), 25sec^{-1} (b) and 4 months of natural aging for the cases of 6.67, 13.3, 20.8, and 25sec^{-1} (c), (d), (e), and (f) respectively welded at 0.83mm/sec using baking type 1. In case of 6.67sec^{-1} as shown in (a), prolonged aging time of 4 months as shown in (c) brought about further increase of hardness in the length of shoulder diameter and higher value of increment was observed in the region comparable to the length of probe diameter. On the other hand, Fig. 3 (b, f) for the case of 25sec^{-1} represents a significant increase of hardness in SZ and TMAZ after natural aging of both 4 hours and 4 months. Particularly 4-months aging made SZ equal to or slightly harder than BM. HAZ showed less increment of hardness and no change occurred in BM. Also, as shown in Fig. 3 (c, d, e, f) microhardness in SZ, TMAZ, and slightly in HAZ increases with increasing rotation speed and almost the values in SZ and TMAZ reached to that of BM for a rotation speed of 20.8 sec^{-1}. Although increasing rotation speed resulted in larger grain size in SZ, hardness increased as rotation speed increased regardless of the coarsening of grains in SZ. This tendency suggests that significant increase of hardness with increasing rotation speed is not a function of grain size but a function of hardening precipitates.

Figure 4 shows three hardness profiles at 20.8sec^{-1} and 1.67mm/sec for backing type 1 (a), type 2 (b) and type 3 (c). In case of backing type 1 as shown in (a) comparing with Fig.3 (e), no significant hardness change appears in SZ both at lower and higher welding speeds. On the other hand, HAZ becomes slightly harder at higher welding speed. For backing types 2 and 3, as shown in (b) and (c), respectively, HAZ shows further increase of hardness and the maximum value increases by 10-12Hv than that welded using backing type one. On the contrary, SZ for backing types 2, 3 becomes less harder than that of type 1.

In 2024-T3 base metal, the initial hardening precipitates is Guinier Preston Bagarystskii (GPB) zones[8]. These zones dissolve during heating resulting in very fine S' or S precipitates (Al_2CuMg) without or slightly changes in hardness. Upon continuous heating to higher temperature S' or S precipitates coarse and lost their coherency. The peak temperatures measured in HAZ as shown in Table I are within coarsening and/or dissolution temperature ranges. So, softening behavior observed in HAZ could be caused by a coarsening and/or a dissolution of S' or S precipitates during welding. Hardening in SZ could be attributed to the formation of new GPB zones at room temperature due to higher peak temperature than coarsening temperature range as shown in Table I. At higher welding speed of 1.67mm/sec, hardness increased in HAZ especially for the backing types 2 and 3. Therefore, the results shown in Fig.4 suggest that improvement of hardness distribution should be realized not only by higher welding speed but also by selecting type of backing materials. Optimizing welding, rotation speeds and designing backing materials composed of suitable materials having appropriate thermal conductivity and heat capacity would be promising for obtaining little fluctuation of hardness throughout the joints friction stir welded.

Tensile Properties of Joints

Table I also shows mechanical properties of the welded joints. Tensile strength, yield strength, and elongation increased with increasing rotation speeds but still lower than those of BM. The maximum value of tensile strength and elongation to fracture were 427 MPa and 14.3% respectively, which were achieved by using backing type 3 at 1.67mm/sec of welding speed, where higher hardness is obtained. The lowest tensile properties were obtained for the joint welded at 6.67and 20.8sec^{1} using backing types 1and 2 respectively. This could be attributed to

Table I Grain size, peak temperatures, and tensile properties at all welding parameters

Welding speed (mm/sec)	Rotation speed (sec^{-1})	Backing materials	Grain size (μm)	Temperature (K)		Mechanical properties		
				SZ	HAZ	Yield strength (MPa)	Tensile strength (MPa)	Elongation (%)
0.83	6.67	Type 1	2.6	718	679	253	335	5.5
	10		4.3	735	685	285	368	8.5
	13.3		6.0	758	-	283	384	9.5
	16.67		7.1	774	701	296	386	11.3
	20.8		7.4	786	707	301	394	12.4
	25		7.8	796	711	291	384	11.7
1.67	20.8		5.5	-	-	247	355	5.6
0.83		Type 2	1.7	683	691	272	340	8.0
1.67			1.3	-	-	276	370	6.0
0.83		Type 3)	4.4	744	691	275	385	11.0
1.67			3.6	708	648	284	427	14.3

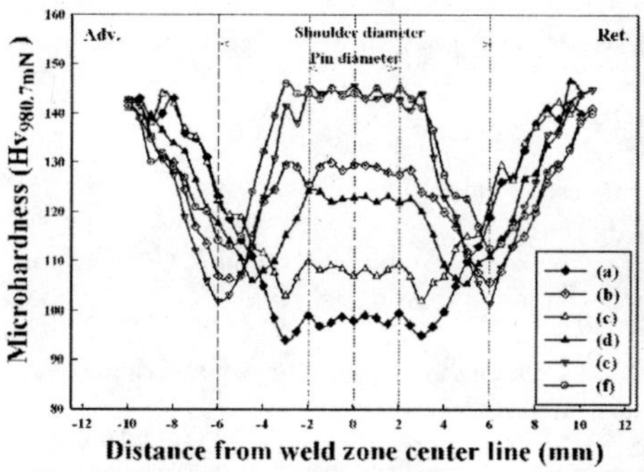

Fig. 3 hardness profiles at 0.83mm/sec using backing type1: (a) 6.67sec^{-1} and (b) 25 sec^{-1}after 4h, (c) 6.67sec^{-1}, (d) 13.3sec^{-1}, (e) 20.8sec^{-1}, and (f) 25sec^{-1} after 4months.

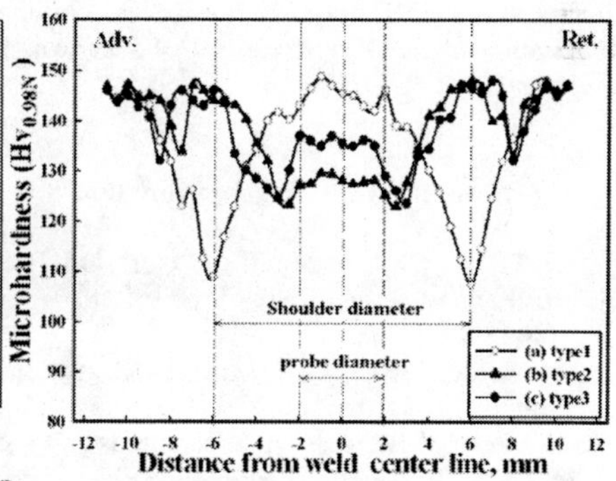

Fig.4 Hardness profiles at 1.67mm/sec and 20.8sec^{-1}: (a) backing type 1 (b) type 2, and (c) type 3.

lower hardness and kissing bond at the root of welds. Figure 5 shows the changes in fracture location of tensile tested specimens welded at three kinds of rotation speeds. Fracture occurs in SZ for the specimen welded at 6.67sec^{-1} as shown in (a). In this case, crack propagated almost at center position of SZ. As rotation speed raised to25sec^{-1}, the crack changed to propagate along the border between SZ and TMAZ on the retreating side as shown in (b). On the other hand, when the joints are fabricated at 20.8sec^{-1} where no kissing bond defect exist, crack propagated in HAZ on the retreating side as seen in (c). Tensile properties of FSW joints mainly depend on the welding defects and hardness distributions of joints, of which are functions of the welding parameters. When joints are free from these defects, their tensile properties are affected dominantly by hardness distributions.

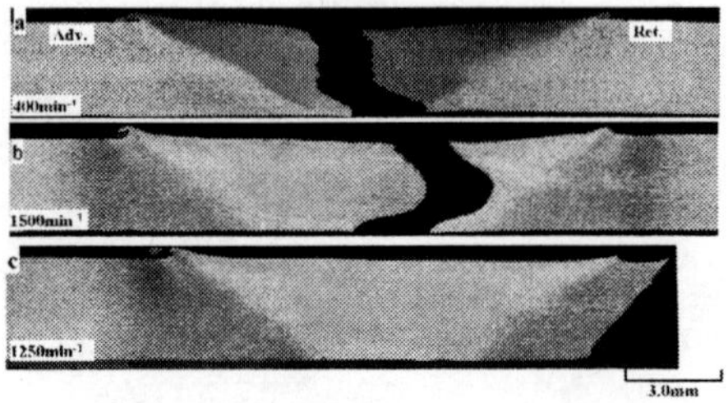

Fig. 5 Optical micrograph of ruptured tensile test specimens.

SUMMARY

The equiaxed grain size and peak temperature in the SZ increased with increasing rotation speed. A significant reduction in grain size in the SZ was realized with the type 2 backing. Hardness increased in the SZ and the TMAZ as the rotation speed increased, and it changed greatly in the SZ and the HAZ, depending on the backing material. The hardness value increased and reached the maximum with the type 3 backing type at a welding speed of 1.67mm/sec.The kissing bond-free joints fractured at the HAZ on the retreating side, while joints containing defects fractured at the SZ or on the border between the SZ and the TMAZ on the retreating side. The maximum tensile strength was 427.0 MPa, which was achieved by using the type 3 backing and a welding speed of 1.67mm/sec.

REFERENCES

[1]C.J. Dawes and W.M. Thomas, Friction Stir Process Welds Aluminum Alloys, *Weld J.*, **75,** 41–45(1996).

[2]O. V. Flores, C. Kennedy, L. E. Murr, D. Brown, S. Pappu, Brook M. N. and J. C. McClure, Microstructural Issues in a Friction-Stir-Welded Aluminum Alloy, *Scripta Materialia*, **38,** 703-08(1998).

[3]S. Benavides, Y. Li, L. E. Murr, D. Brown and J. C. McClure, Low-Temperature Friction-Stir Welding of 2024 Aluminum , *Scripta Materialia*, **41,** 809-15(1999).

[4]M. A. Sutton, B. Yang , A. P. Reynolds and J. Yan, Banded Microstructure in 2024-T351 and 2524-T351 Aluminum Friction Stir Welds: Part II. Mechanical Characterization , *Mater. Sci. Eng. A*, **364,** 66-74(2004).

[5]M. A. Sutton, B. Yang , A. P. Reynolds and R. Taylor, Microstructural Studies of Friction Stir Welds in 2024-T3 Aluminum, *Mater. Sci. Eng. A*, **323,** 160-66(2002).

[6]K. V. Jata and S. L. Semiatin, Continuous Dynamic Recrystallization During Friction Stir Welding of High Strength Aluminum Alloys, *Scripta Materialia*, **43,** 743-49(2000).

[7]C. G. Rhodes, M. W. Mahoney, W. H. Bingel and M. Calabrese, Fine-Grain Evolution in Friction-Stir Processed 7050 Aluminum, *Scripta Materialia*, **48,** 1451-55(2003).

[8]M.J. Jones, P. Heurtier, C. Desrayaud, F. Montheillet, D. Allehaux and J.H. Driver, Correlation Between Microstructure and Microhardness in A Friction Stir Welded 2024 Aluminum Alloy, *Scripta Materialia*, **52,** 693-97(2005).

[9]Y. S. Sato, H. Kokawa, K. Ikeda, M. Enomoto, S.Jogan and T. Hashimoto, Microtexture in the Friction Stir Welding of an Aluminum alloys, *Metall. Mater. Trans A*, **32,** 941-48(2001).

EFFECTS OF PROCESSING TEMPERATURE ON BONDING BEHAVIOR OF HYDROXYAPATITE CERAMICS AND TITANIUM BY HYDROTHERMAL HOT-PRESSING METHOD

Takamasa Onoki, Tomoaki Watanabe, Eiichi Yasuda, Masahiro Yoshimura
Materials and Structures Laboratory, Tokyo Institute of Technology
4259 Nagatsuta, Midori-ku,
Yokohama, 226-8503, Japan.

Toshiyuki Hashida
Fracture and Reliability Research Institute, Graduate School of Engineering, Tohoku University
6-6-11-7 Aoba, Aramaki, Aoba-ku,
Sendai, 980-8579, Japan.

Yasuhiro Tanabe
Department of Chemical Engineering, Graduate School of Engineering, Nagoya University
Furo-cho, Chikusa-ku,
Nagoya, 464-8603, Japan.

ABSTRACT

Solidificatioin of hydroxyapatite (HA:$Ca_{10}(PO_4)_6(OH)_2$), and its bonding with titanium (Ti) was achieved simultaneously by using a hydrothermal hot-pressing (HHP) method with no special surface treatment of the Ti. In this study, the relationship between the HA/Ti bonding properties and the the HHP processing temperature is examined. Mixed powders of calcium hydrogen phosphate dehydrate and calcium hydroxide were used as the starting material for the HA ceramics. The powder mixture and the Ti disk were put into an autoclave for the HHP process. The applied pressure was maintained at 40 MPa. The autoclave was heated up to 373 - 573K at a heating rate of 10K/min., and then the temperature was held constant for 2 hours. The optimal operating temperature was 423K. The bonding of HA to Ti was not achieved at temperatures over 423K.

INTRODUCTION

Titanium (Ti) and its alloy are widely used as orthopedic and dental implant materials because of their high mechanical strength, low modulus and good corrosion resistance[1]. Traditionally, Ti and its alloys have been reported as being bioinert. When embedded in the human body, a fibrous tissue capsules the implant, isolating it from the surrounding bone forms.

Some bioactive ceramics such as HA, bioglass, and glass ceramics can directly bond to living bone when used as bone replacement materials [2-4]. In the traditional method for solidifying, HA powder, it is sintered at high temperatures over 1273K [5]. The mechanical properties of bulk HA only allow applications for small non-loaded structure [6]. The possibility of depositing it into films has permitted the exploitation of its bioactive properties in structural prostheses such as for teeth root, hip, knee and shoulder joint replacement. Therefore, HA is also used as a coating material for those prostheses surface in order to prepare bioactive layers on titanium and its alloys [7]. The HA surface improves the fixation of implants by the growth of bone into the coating, forming a mechanical interlock. A plasma spraying technique has been frequently employed for the coating process of HA [8, 9]. However, this high temperature method results in some of significant problems like a poor coating-substrate adherence [10], lack of uniformity of the coating in terms of morphology and crystallinity[11, 12], that affect the long-term performance and lifetime of the implants. Despite of various investigations carried out, the produced coatings can suffer from at least one of the following problems: Lack of coating adherence to the substrate, poor structure integrity and non-stoichiometric composition of the coatings [13].

The HHP method is a possible processing route for producing a ceramic body at relatively low temperatures [14]. The compression of samples under hydrothermal conditions accelerates densification of inorganic materials. It is known that the water of crystallization in calcium hydrogen phosphate di-hydrate ($CaHPO_4 \cdot 2H_2O$; DCPD) is slowly lost below 373 K [15]. In our previous studies [16, 17], it has been demonstrated that bonding HA ceramics and pure Ti can be achieved by using the HHP method without special surface modifications of Ti, in addition to the synthesis and solidification of HA through the chemical reaction as follows:

$$6CaHPO_4 \cdot 2H_2O + 4Ca(OH)_2 \rightarrow Ca_{10}(PO_4)_6(OH)_2 + 18H_2O \quad (1)$$

In general, it is expected that the higher processing temperature can provide the higher mechanical strength through the HHP method. This report describes the relationship between bonding behavior of HA/Ti and the HHP treatment temperature.

EXPERIMANTAL METHODS

A commercially available pure Ti rod (99.5%; Nilaco, Japan), 20mm in diameter, was used in this experiment. The Ti rod was cut into disks with a thickness of 10-15mm. The disks were cleaned in deionized water and acetone by using an ultrasonic cleaner. The Ti surfaces were finished using 1500# emery paper. After the surface finish with emery paper, the titanium disks were washed again by deionized water, and then dried in air.

DCPD used as a starting powder was prepared by mixing 1.0*M* calcium nitrate solution (99.0%; $Ca(NO_3)_2 \cdot 4H_2O$, KANTO CHEMICAL, Japan) and 1.0*M* diammonium hydrogen phosphate solution (98.5%; $(NH_4)_2HPO_4$; KANTO CHEMICAL, Japan). The mixing was carried out at a room temperature (approximately 20°C). In order to control the value of pH, acetic acid and ammonia solution were added. The precipitate from the mixture was filtered and washed with deionized water and acetone. The washed filter cake was oven-dried at 50°C for 24h, and then the dried cake was ground to a powder.

The synthetic DCPD and calcium hydroxide (95.0%; $Ca(OH)_2$; KANTO CHEMICAL, Japan) were mixed in a mortar for 90min with a Ca/P ratio of 1.67. The powder mixture and Ti disks were placed into the middle of the autoclave simultaneously, as shown in Figure1. The autoclave made of steel has a pistons-cylinder structure with an inside diameter of 20mm. The pistons possess escape space for hydrothermal solution squeezed from the sample, and this space regulates the appropriate hydrothermal conditions in the sample. A grand packing of polytetrafluoroethylene (PTFE) is placed between a cast rod and push rod. The PTFE was used to prevent leakage of the hydrothermal solutions.

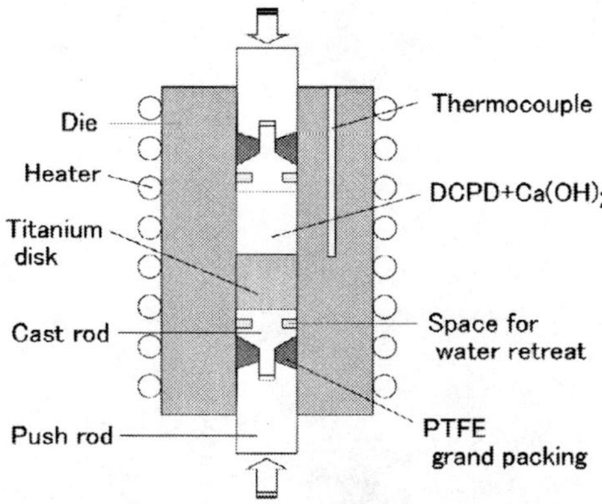

Figure1. Schematic illustration of the autoclave for Hydrothermal Hot-Pressing (HHP).

A pressure of 40MPa was initially applied to the sample through the push rods from the top and bottom at a room temperature. After the initial loading the autoclave was heated up to setting temperature (373, 423, 473 and 573K) at heating rate of 10 K/min., and then the temperature was kept constant for 2hours. The autoclave was heated with a sheath-type heater. The axial pressure was kept at 40MPa during the HHP treatment. After the treatment, the autoclave was naturally cooled to a room temperature, and the sample removed from the autoclave.

The HA ceramics were determined by X-ray diffraction analysis (XRD; MX21, Mac Science, JAPAN) with Cu$K\alpha$ radiation 40kV 40mA at a scanning speed of 3.00 degrees/min. with a scanning range (2θ) from 10 to 40 degrees. 3-point bending tests were conducted to obtain an estimate of the fracture toughness for the HA/Ti interface as well as for the HA ceramics made by the HHP method. The configuration of the core-based specimen is schematically shown in our previous literature [17, 18].

RESULTS AND DISCUSSION

Photographs of the samples treated at 423K and 573K are shown in Figure2 (a) and (b), respectively. Bonding of HA ceramics to Ti was achieved only for the specimens treated at 373K and 423K. In contrast, for the specimens treated at 473K and 573K, no bonding was observed and the almost complete separation was noticed after the removal of the specimens from the autoclave, as shown in Figure2 (b). The samples of 473K and 573K, as shown in Figure2 (b), could be found something like a thin oxide film on the surface. It is seemed that some chemical reactions occur on the Ti surface. From these observations, it is suggested that the some chemical compounds layer formed on the Ti surface inhibit the HA ceramics from bonding to Ti. It is supposed that the reaction layer, described in the literature [19], changes or disappears and that the elemental gradient around the interface is broken. Otherwise, the oxide film is formed faster than the formation of the reaction layer and the elemental gradient. Furthermore, the bonding failures at 473K and 573K are potentially due to the residual stress induced by the thermal expansion mismatch between HA and Ti.

3-point bending tests were conducted to obtain an estimate of the fracture toughness for the HA/Ti interface. The fracture toughness determined on the bonded HA/Ti specimen was close to or higher than that of the HA ceramics only (0.25-0.30MPam$^{1/2}$).

As given in Figure3, powder X-ray diffractions indicate that the HA starting powder (DCPD and $Ca(OH)_2$) materials were completely transformed into HA by the HHP treatment at 423K, 473K and 573K, respectively. On the other hand, it can be seen that starting materials are

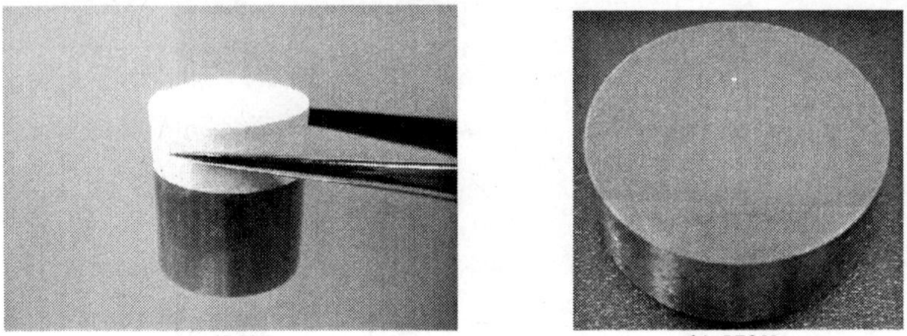

(a) 423K (b) 573K

Figure2. Photographs of the samples after the HHP treatment at 423K (a) and 573K (b).

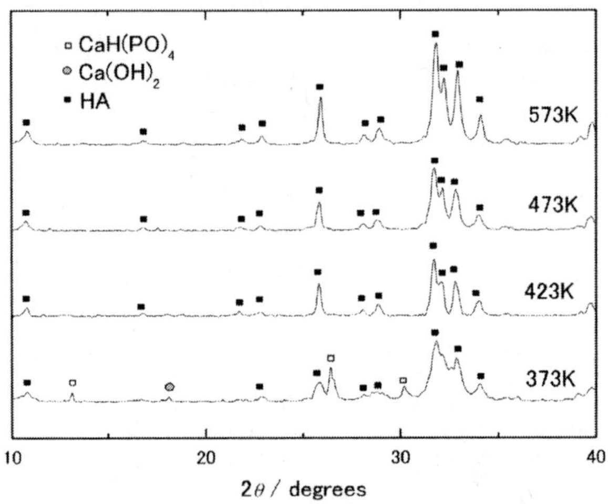

Figure3. X-ray diffraction patterns of the HA ceramics made by the HHP method at 373, 423, 473 and 573, respectively.

remained in the HA ceramics processed at 373K. If the HHP treatment time is more than 2 hours, the solidified body processed at 373K may be converted to HA entirely. When the Ti surfaces after the HHP processing at 473K and 573K respectively were examined by the XRD analysis, no significant peaks of any other calcium phosphate compounds could be observed. No bonding of HA/Ti at 473K and 573K might be affected by the crystal growth of the HA ceramics in relatively high temperature.

Based on the above results, it is judged that the appropriate temperature for bonding HA ceramics to Ti by the HHP method is 423K. Appropriate processing conditions are summarized in Table 1 in order to achieve both of HA/Ti bonding and HA yield. In order to investigate the HA/Ti interfacial structure, the interface should be observed and analyzed with transmission electron microscopy (TEM). While further development is needed to improve the fracture property of the solidified HA, the HHP treatment may have the advantage over the plasma-spraying technique in the preparation of thermodynamically stable HA without decomposition. The above results demonstrate the usefulness of the HHP method for bonding HA and Ti in order to produce a bioactive layer in bioinert materials.

Table1.HA/Ti bonding properties and HA yield conditions.

Temperature	373K	423K	473K	573K
Bonding	Good	Good	NG	NG
HA Yield	NG	Good	Good	Good

CONCLUSIONS

A mixture of calcium hydrogen phosphate dihydrate and calcium hydroxide was used as a starting powder material to solidify hydroxyapatite (HA). Solidification of HA and its bonding with titanium (Ti) was achieved simultaneously by using a hydrothermal hot-pressing (HHP) method under 40 MPa for 2 hours. It was demonstrated that the optimal operating temperature was 423K, and that the boding of HA to Ti could not be achieved at temperatures over 423K. Investigation of the bonding mechanism and *in vivo* tests of the HA/Ti bonded bodies are in progress.

ACKNOWLEDGMENT

This work was supported by Grant-in-Aid for Cooperative Research Project of Nationwide Joint-Use Research Institutes on Development Base of Joining Technology for New Metallic Glasses and Inorganic Materials and was partially supported by Grant-in-Aid for Young Scientists (B), 18760516, 2006 from The Ministry of Education, Culture, Sports, Science and Technology, Japan.

REFERENCES

[1]M. Long and H. J. Rack, "Titanium alloys in total joint replacement - a materials science perspective," Biomaterials 19, 1621- 1639 (1998).

[2]L. L. Hench, "Bioceramics," J. Am. Ceram. Soc. 81, 1705-1728 (1998).

[3]M. Jarcho, J. F. Kay, K. I. Gummaer, R. H. Doremus and H. P. Drobeck, "TISSUE, CELLULAR AND SUBCELLULAR EVENTS AT A BONE-CERAMIC HYDROXYLAPATITE INTERFACE," J. Bioengng. 1, 79-92 (1977).

[4]H.Aoki, In Medical Applications of Hydroxyapatite. Tokyo, Japan: Ishiyaku Euro-America, 1994, p. 133.

[5]J. G. J. Peelen, B. V. Rejda and K. De Groot, "Preparation and Properties of Sintered Hydroxyapatite," Ceramurgia Int. 4, 71-74 (1978).

[6]L. L. Hench, "BIOCERAMICS - FROM CONCEPT TO CLINIC," J. Am. Ceram. Soc. 74, 1487-1510 (1991).

[7]P. Ducheyne, L. L. Hench, A. Kagan, M. Martens, A. Burssens and J. C. Mulier, "EFFECT OF HYDROXYAPATITE IMPREGNATION ON SKELETAL BONDING OF POROUS COATED IMPLANTS," J. Biomed. Mater. Res. 14, 225-237 (1980).

[8]S. R. Radin and P. Ducheyne, "PLASMA SPRAYING INDUCED CHANGES OF CALCIUM-PHOSPHATE CERAMIC CHARACTERISTICS AND THE EFFECT ON INVITRO STABILITY," J. Mater. Sci. Mater. Med. 3, 33-42 (1992).

[9]K. De Groot, R. Geesink, C. P. A. T. Klein and P. Serekian, "PLASMA SPRAYED COATINGS OF HYDROXYLAPATITE," J. Biomed. Mater. Res. 21, 1375-1381 (1987).

[10]X. Zheng, M. Huang and C. Ding, "Bond strength of plasma-sprayed hydroxyapatite/Ti composite coatings," Biomaterials 21, 841-849 (2000).

[11]H. C. Gledhill, I.G. Turner and C. Doyle, "In vitro dissolution behaviour of two morphologically different thermally sprayed hydroxyapatite coatings," Biomaterials 22, 695-700 (2001).

[12]F. Fazan and P. M. Marquis, "Dissolution behavior of plasma-sprayed hydroxyapatite coatings, " J. Mater. Sci. Mater. Med. 11, 787-792 (2000).

[13]K. A. Thomas, "HYDROXYAPATITE COATINGS," Orthopaedics 17, 267-278 (1994).

[14]N. Yamasaki, K. Yanagisawa, M. Nishioka and S. Kanahara, "A HYDROTHERMAL HOT-PRESSING METHOD - APPARATUS AND APPLICATION," J. Mater. Sci. Lett. 5, 355-356 (1986).

[15]H. Monma and T. Kamiya, "PREPARATION OF HYDROXYAPATITE BY THE HYDROLYSIS OF BRUSHITE," J. Mater. Sci. 22, 4247-4250 (1987).

[16]K. Hosoi, T. Hashida, H. Takahashi, N. Yamasaki and T. Korenaga, "New processing technique for hydroxyapatite ceramics by the hydrothermal hot-pressing method," J. Am. Ceram. Soc. 79, 2771 (1996).

[17]T. Onoki, K. Hosoi and T. Hashida, "New technique for bonding hydroxyapatite ceramics and titanium by the hydrothermal hot-pressing method," Scripta Materialia, 52, 767-770 (2005).

[18]T. Hashida, "FRACTURE-TOUGHNESS TESTING OF CORE-BASED SPECIMENS BY ACOUSTIC-EMISSION," Int. J. Rock Mech. Min. Sci. Geomech. Abstr. 30, 61 (1993).

[19]T. Onoki, K. Hosoi and T. Hashida, "GD-OES Analysis of the interface between Titanium and Hydroxyapatite Ceramics Produced by the hydrothermal Hot-pressing Method," Trans. Mater. Res. Soc. Japan, 29, 2929-2932 (2004).

ORIENTATION DISTRIBUTION IN FRICTION STIR PROCESSED A6061 ALUMINUM ALLOY

Toshiya Shibayanagi
Joining and Welding Research Institute, Osaka University
11-1,Mihogaoka, Ibaraki, Osaka, 567-0047, Japan

Takamasa Matsumoto
Graduate student of Osaka University
Joining and Welding Research Institute, Osaka University
11-1,Mihogaoka, Ibaraki, Osaka, 567-0047, Japan

ABSTRACT

The present study examines the spatial crystallographic orientation distributions in the stir zone of friction stir processed 6061 Al alloy. Friction stir processing was performed with a rotation speed of 1800rpm, and a traveling speed of 100 mm/min. The microstructures of cross sections of the stir zone were characterized using scanning electron microscopy (SEM). Orientation distributions in stir zone were investigated by means of SEM together with electron back-scatter diffraction EBSD.

The orientation distribution in the stir zone fluctuated from the retreating side to the advancing side along the horizontal line at the middle thickness position. The main component in the traveling direction in the stir zone changed from (111) at the advancing side to (111) at the retreating side via (001) in the center zone of the area analyzed in the measuring direction.

INTRODUCTION

Friction stir welding (FSW) is a novel joining technology developed for welding aluminum alloys, and has been expanding its application to the joining of other metallic materials such as copper, titanium, magnesium, steel and their alloys as well as their dissimilar materials [1,2].

Many research reports on the FSW of aluminum alloys pointed out that grain refinement was successfully achieved due to both dynamic and static recrystallization during welding. Besides the uniqueness in joining technology, FSW is also utilized for fabricating peculiar microstructure in a limited area. This is called friction stir processing (FSP).

The present study aims to investigate characteristics of microstructure of 6061 aluminum alloy friction stir processed focusing on the orientation distributions.

EXPERIMENTAL PROCEDURE

The present study used an A6061-T6 aluminum alloy plate of 3mm in thickness. Chemical composition of the base metal is listed in Table 1. FSP was performed for this alloy plate by a stir-in-plate method. The processing was carried out at a rotation speed and a traveling speed of 1800rpm and 100 mm/min, respectively. The direction of tool movement was parallel to the rolling direction of base metal. The tool was inclined by 3 degrees backward from the vertical axis and was preheated by FSP of an A1050 aluminum plate before the processing. The present work utilized a backing plate made of SUS304 stainless steel of 10 mm in thickness.

Table 1 Chemical composition of A6061 Al alloy used. (mass%)

Material	Si	Fe	Cu	Mn	Mg	Cr	Zn	Ti	Al
A6061-T6	0.66	0.24	0.31	0.07	1.0	0.18	0.03	0.01	Bal.

Evaluation of microstructure was performed on a cross section of the plate processed, and this plane is usually named as "welding direction (WD)" plane. The present study mainly investigated a spatial distribution of crystallographic orientation by utilizing scanning electron microscopy (SEM) together with electron back-scatter diffraction (EBSD) method. The EBSD patterns were taken at near mid-thick positions of the plate with an accelerating voltage and a step size of 20kV and 1μm, respectively. The areas for measurement were set almost parallel to the transversal direction (TD) of the plate.

RESULTS AND DISCUSSION

Figure 1 shows a SEM image of a cross section of the friction stir processed plate. A white rectangular area highlights the areas of EBSD measurement. The abbreviated words in this figure such as "Adv." and "Ret." mean advancing side and retreating side, respectively. Slight changes of contrast in this area, which were caused by a contamination during the measurement, represent the regions of measurement; 28 measurements were taken to make one strip across the stir zone

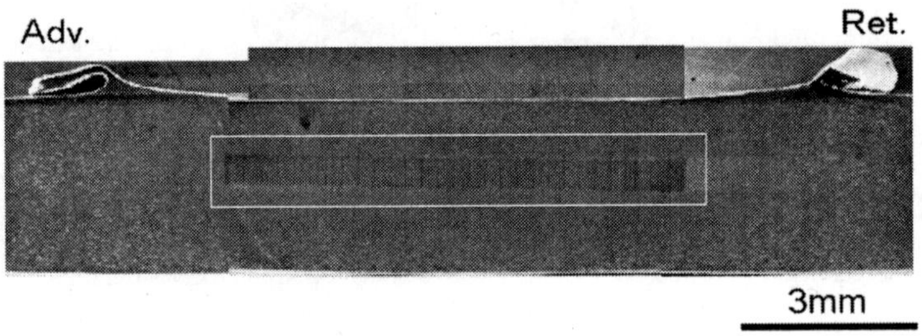

Fig.1 Cross sectional view (WD plane) of friction stir processed 6061Al alloy.

(SZ) from the advancing side to the retreating side. Each measured area was moved manually and the operation was carried out to be as parallel to the horizontal line as possible.

Figure 2 represents a result of orientation analysis based on the EBSD data taken from the area shown in Fig.1. The inverse pole figure map in Fig.2 (a) shows a fine grain structure throughout the SZ. Thermo-mechanically affected zones (TMAZ) neighboring the SZ consist of elongated grains lying along the edge of the SZ. The orientation distribution in SZ fluctuates. Some previous papers reported an onion-ring-like pattern for the orientation distribution, indicating a periodic change of orientations [4, 5]. The present study also found a heterogeneous spatial distribution of orientations, which is similar to these reports. There are orientation bands composed of grains belonging to similar orientation group in each band. There are roughly six orientation bands in this case, and each band is named as zone 1 to 6 from the advancing side to the retreating side, as indicated in the figure.

The orientation distribution using {111} pole figures corresponding to each zone are shown in Fig.2(b). Zone 1, which is located in the edge region of the SZ, represents a mixture of some orientation groups such as {111}//WD and {100}//WD. The {111} pole is close to the center of the figure, indicating that the {111} plane is nearly parallel to the WD plan; however, the intensity is low, indicating this orientation is not the main component in this region. Meanwhile, the {100}//WD-oriented grains seem dominant in this region. Zone 2 has an orientation distribution similar to that observed in zone 1, but the intensity of {111}//WD orientation becomes larger. Zone 3 clearly shows the stronger intensity of the {111} pole that is closer to the center of pole figure. These changes in orientation distribution indicate a gradual change of orientation depending on the position/area in the SZ.

Zone 4 has a larger area than in the previous three zones, and a quite different orientation distribution close to a "rotated cube" orientation. No big change in grain size was observed. Thus,

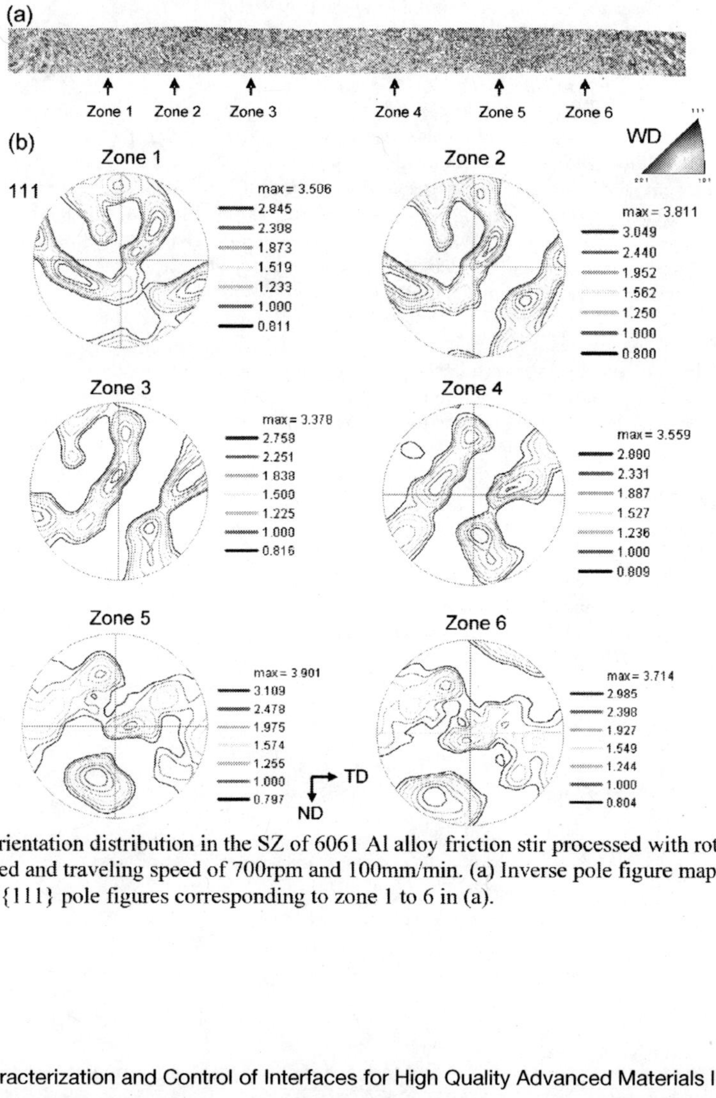

Fig.2 Orientation distribution in the SZ of 6061 Al alloy friction stir processed with rotation speed and traveling speed of 700rpm and 100mm/min. (a) Inverse pole figure map, and (b) {111} pole figures corresponding to zone 1 to 6 in (a).

a different mode of metal flow might have proceeded in this area during the processing.

Slight changes in orientation distribution become evident as the area of measurement approaches the neighboring zone 5, where the {111}//WD-oriented grains are dominant. The strongest intensity of the {111}//WD pole was obtained in zone 6, which neighbors the TMAZ on retreating side.

Throughout theses observation, the most frequently observed orientation parallel to the WD is {111}, but {100} should be recognized as another main component in the SZ. No symmetry was observed in the pole figure in the symmetric areas such as in zones 2 and 5.

The fundamental process of microstructure evolution in FSW/FSP is a high temperature deformation in a really thin region around a rotating pin. This mass flow at high strain rate yields a peculiar microstructure, mainly due to the dynamically and static recrystallization. The flowing metal is always exposed to the surrounding metal that is not rotating, and is causing a frictional stress to the flowing metal. Thus a severe shear deformation mode arises along the flow direction, and a shear texture may develop during the process. However, the flow behavior is not so simple; the probe has a thread that will cause a complicated flow, as well as a complicated flow by the traveling of the probe. Therefore, no significant development of (111) shear texture was observed in the present case.

However, it will be possible to generate strong texture in the limited region by designing suitable shape and size welding tools, and also by controlling welding parameters such as the rotation speed, welding speed, and thermal history.

The present study focuses on the spatial distribution of crystallographic orientation in the SZ to determine whether the distribution is homogeneous for this case. For this reason, one strip of measured area shown in Fig.1 is enough. But, the orientation distribution would also change along the WD and ND, thus further detailed measurements will bring about three-dimensional image of orientation distribution in friction stir processed region.

SUMMARY

Friction stir processing was carried out on 6061 Al alloy plate to investigate the crystallographic orientation distribution in the SZ. SEM-EBSP analysis revealed some "orientation bands" having weak texture, in which the main component differs in each band-like region. The (111)-orientated region, which is a feature of severely shear-deformed metals, tended to appear in the WD plane in the SZ, but is not observed in the near-center region of the SZ, suggesting a complicated metal flow during the processing.

ACKNOWLEDGEMENT

The present work was financially supported by Grant-in-Aid for scientific research (B: project No. 13450295) and Grant-in-Aid for Cooperative Research Project of Nationwide Joint-Use Research Institutes on Development Base of Joining Technology for New Metallic Glasses and Inorganic Materials from The Ministry of Education, Culture, Sports, Science and Technology, Japan.

REFERENCES

[1] C.J. Dawes: Causing a stir in welding circles, Sheet Metal Industries, **72** [9] (1995), 4pp.

[2] R.S. Mishra and Z.Y. Ma: Friction stir welding and processing, Mater. Sci. Eng., R50(2005), 1-78.

[3] A. Schneider and A.C. Nunes, Jr.: Characterization of Plastic Flow and Resulting Microtextures in a Friction Stir Weld, Metall. Mater.Trans. **35B**(2004), 777-783.

[4] D.P. Field, T.W. Nelson, Y. Hovanski and K.V. Jata, Heterogeneity of crystallographic texture in friction stir welds of aluminum, Metall. Mater. Trans. **32A**(2001), 2869-2877.

WELDABILITY AND MECHANICAL PROPERTY OF $Ni_{53}Nb_{20}Ti_{10}Zr_{8}Co_{6}Cu_{3}$ METALLIC GLASS FOIL BY LASER WELDING

Takuya Tsumura, Katsunori Kobayashi*, Kazuhiro Nakata
Joining and Welding Research Institute, Osaka University
Ibaraki, Osaka 567-0047, Japan

Natsuki Yoneyama
Research Laboratory, Ishikawajima-Harima Heavy Industries Co. Ltd.
Isogo-ku, Yokohama, Kanagawa 235-8501, Japan

Taichi Murakami
Institute of Multidisciplinary Research for Advanced Materials, Tohoku University
Aoba-ku, Sendai, Miyagi 980-8579, Japan

Hisamichi Kimura, Akihisa Inoue
Institute for Materials Research, Tohoku University
Aoba-ku, Sendai, Miyagi 980-8577, Japan

ABSTRACT

In order to evaluate the weldability of a melt-spun Ni-based metallic glass foil with a thickness of 25 μm, bead-on-plate laser welding was carried out at different welding speeds from 2 to 12 m/min using a diode laser beam with a constant laser power of 26 W and wavelength of 808 nm. A sound weld bead, while maintaining the amorphous state, was successfully obtained at a high welding speed of 12 m/min; under these conditions, the weld bead consisted of an amorphous matrix with a very small amount of nanocrystals with a mean diameter of 14 nm. The hardness was almost the same as that of the base metal (740 HV), and the mean tensile strength of the weld bead was 1312 MPa, which is 56% of that of the base metal. At a lower welding speed of less than 6 m/min, a fusion zone appeared in the central zone of the weld bead consisting of the amorphous matrix with comparably large crystals with mean diameter of 170 nm. Further, the heat-affected zones adjacent to the fusion zone consisted of a completely crystallized structure with a high hardness of around 1200 HV, which caused a low joint strength or weld cracking due to the brittleness of crystallization or structural relaxation.

INTRODUCTION

Since the late 1980s, a number of amorphous alloys were found that exhibited a large supercooled liquid region through a distinct glass transition without any crystallization[1–3]; they were called bulk amorphous alloy, bulk glassy alloy, or metallic glass. The weldability of Zr-based metallic glass which is well known as a bulk metallic glass system with a large supercooled liquid region of about 100 K, has already been studied by spark welding[4], friction welding[5,6], and electron beam welding[7,8]. Ni-Nb-Ti-Zr-based bulk metallic glasses, which have been recently developed, have a high tensile strength, high corrosion resistance, and high crystallization temperature[9–11]. These Ni-based metallic glass alloys are expected to be applied as separator plates in solid polymer electrolyte fuel cells[11,12]. However, there has been no research conducted on the weldability of thin metallic glass sheet with a thickness lesser than 1 mm. If the welding of metallic glass with a thin sheet can be achieved by maintaining the amorphous phase, it can be expected to have wide applications in various industrial fields. The features of laser welding are local heating, high cooling rate due to the high energy density, and the process is carried out in atmosphere. Therefore, it is expected to be an effective process for welding thin metallic glasses. In this study, the diode laser welding process was applied to a ribbon-shaped Ni-Nb-Ti-Zr-based metallic glass with a thickness of 25 μm, and the weldability was investigated as a first attempt at the laser welding of thin metallic glasses.

EXPERIMENTAL PROCEDURE

$Ni_{53}Nb_{20}Ti_{10}Zr_8Co_6Cu_3$ (nominal atomic %) metallic glass specimen was fabricated in a ribbon shape with a width of 50 mm and average thickness of 25 μm by melt spinning[9]. The thermal stability associated with the glass transition, supercooled liquid region, and crystallization was examined by differential scanning calorimetry at a heating rate of 0.67 K/s, and the melting temperature was measured by differential thermal analysis[9]. The glass transition temperature (T_g), crystallization temperature (T_x), and melting point (T_m) of this alloy are 846 K, 897 K, and 1265 K, respectively[9].

Figure 1 shows the schematic views of the setup and Table I shows the welding conditions. The bead-on-plate welding of the specimen was conducted using a diode laser unit (KLY-HP300α). The diode laser beam with a wavelength of 808 nm was delivered using a glass fiber with a diameter of 0.3 mm, and it was focused on the specimen surface with a spot diameter of 0.3 mm at an angle of 80°. The specimen was fixed on a copper backing plate with a groove of 2 mm width by a copper cover plate of 2 mm thickness with a slit of 2 mm width. Argon was used as a shielding gas with a flow rate of 10 *l*/min for the central shield and 5 *l*/min for the backing shield. The welding speed was varied from 2 to 12 m/min at a laser power of 26 W.

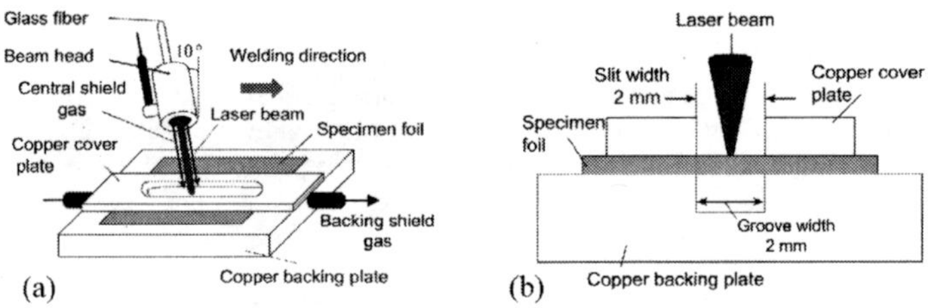

Figure 1 Schematic view of specimens during laser welding in the Ni-based metallic glass (a) setup of laser head and specimen holder and (b) cross-sectional view of specimen holder.

Table I Welding conditions used.

Welding type: Bead-on-plate	Welding speed: 2~12 m/min
Laser power: 26 W	Head angle: 10° advanced
Wavelength: 808 nm	Center shield gas: Ar, 10 l/min
Fiber diameter: 0.3 mm	Backing shield gas: Ar, 5 l/min
Spot diameter: 0.3 mm	Slit width: 2 mm
Focus: On the specimen surface	Groove width: 2mm

The surface of the welded specimen was examined using an optical microscope (OM) and by scanning electron microscopy (SEM). Micro-area X-ray diffraction (XRD) analysis, with a beam diameter of 50 μm (CuKα), tube voltage of 30 kV, and tube current of 50 mA was carried out at the center of the weld bead as well as the base metal to identify the amorphous phase. For the comparison of the XRD pattern of the crystallized phase, an annealed specimen that was heated at 973 K for 60 min and cooled in a furnace for recrystallization was used. The cross section of the welded specimen was observed using an OM; further, the microhardness profile at the center of thickness was measured at intervals of 30 μm with 0.49 N for 15 s. Moreover, the microscale crystalline phase was minutely examined using a transmission electron microscope (TEM) with an acceleration voltage of 200 kV and magnification range of 50 000 to 100 000 for light field observations. The tensile specimen with a gauge section of 25 μm × 12.5 mm was cut out of the specimen before welding for the base metal and after welding for the welded specimen respectively, using a spark-erosion wire cutting machine. The tensile test was performed on the specimen at room temperature at a crosshead speed of 0.5 mm/min.

RESULT AND DISCUSSION

Figure 2 shows the bead appearance of the welded specimen by OM and SEM at welding speeds of 2, 4, 6, and 12 m/min. The width of the weld bead decreases with an increase in the welding speed, in other words with a lower heat input per unit length. The crack along the welding direction occurred in the weld bead at the welding speed of 2 m/min, and the transverse crack occurred in the weld bead at the welding speed of 4 m/min. However, at welding speeds above 6 m/min, the crack was not observed. It is thought that the residual stress under the cooling process as well as the existence of the brittle crystalline phase may possibly be the main causes of these cracks at welding speeds below 4 m/min.

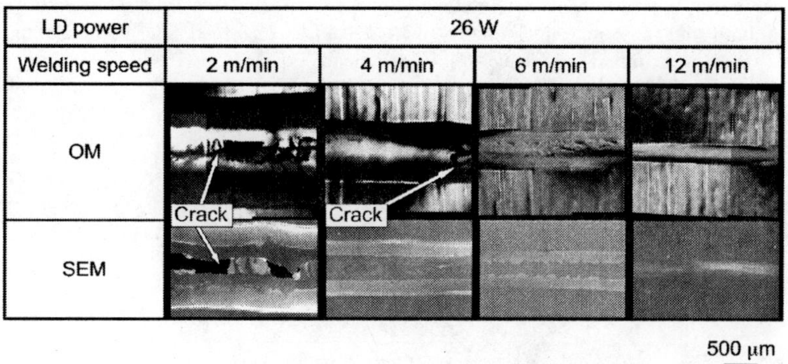

Figure 2 Bead appearance in $Ni_{53}Nb_{20}Ti_{10}Zr_8Co_6Cu_3$ laser welded specimens using OM and SEM at the welding speeds of 2, 4, 6, and 12 m/min with a constant laser power of 26 W.

Figure 3 shows the XRD patterns for the center of the weld bead surface with a constant laser power of 26 W in comparison with those for the base metal and the annealed specimen. The patterns of the base metal consist only of broad peaks showing the particular patterns obtained from the amorphous phase. On the other hand, diffraction peaks due to a crystalline phase are observed in the annealed specimen. In the weld bead, the patterns observed at the welding speeds below 4 m/min indicated the existence of the crystalline phase, but the pattern observed at welding speed of 12 m/min indicated the existence of the amorphous phase. At the welding speed of 6 m/min, the patterns seem to indicate the coexistence of the crystalline and amorphous phase. These results thus indicate that a sound weld bead could be obtained at a welding speed greater than 12 m/min while maintaining the amorphous state.

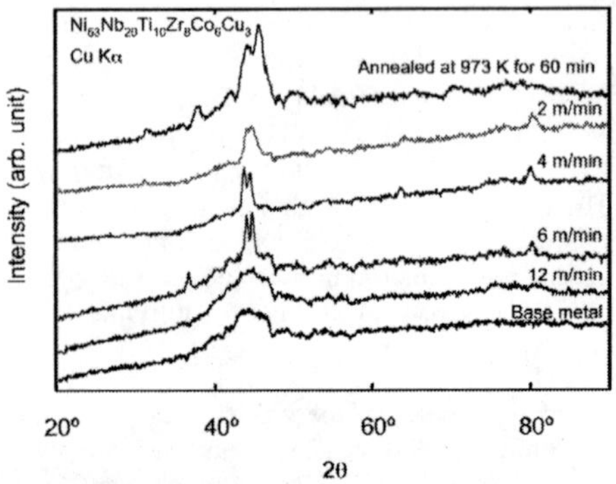

Figure 3 XRD patterns of $Ni_{53}Nb_{20}Ti_{10}Zr_8Co_6Cu_3$ laser welded specimens for the center of the weld bead at welding speeds of 2, 4, 6, and 12 m/min with a constant laser power of 26 W, base metal, and annealed specimen at 973 K for 60 min.

From the bead appearance determined by OM and SEM the welding speed of 6 m/min, as shown in Fig. 2, the weld bead consists of three zones, which are schematically shown in Fig. 4. These are zone I, with a rather jagged boundary at the weld bead center; zone II, with a smooth surface at both sides of zone I; and zone III, with a different surface state as compared to zone II and a very narrow strip that is adjacent to the base metal (B.M.). Zone I is the melted zone based on the presence of a ripple surface in it. Therefore, zones II and III are the heat-affected zones. Here, we denote zones II and III as HAZ-1 and HAZ-2, respectively. At the welding speed of 12 m/min, however, the weld bead did not melt because the ripple surface could not be observed, as shown in Fig. 2. Therefore, the weld bead may consist of HAZ-1 and/or HAZ-2

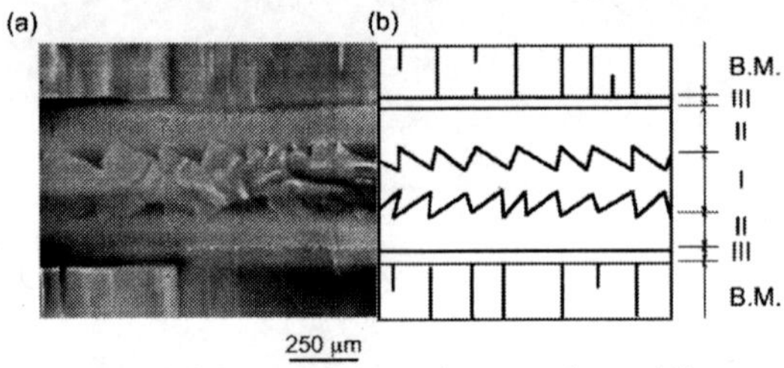

Figure 4 (a) Bead appearance of the welded specimen at the welding speed of 6 m/min, (b) schematic illustration of different zones in the weld bead.

The cross-sectional views of the microstructure and hardness profiles along with the center of the thickness at a welding speed of 6 m/min and 12 m/min are shown in Fig. 5. As shown in Fig. 5 (a), the hardness of the HAZ-2 zone was almost the same as that of the B.M. However, the hardness of the HAZ-1 zone in which the etch pit appeared, increased up to 1200 HV, and the hardness decreased to the same level as that of the base metal at the melted zone. Therefore, a close relationship is revealed between the weld structure and the hardness. On the contrary, as shown in Fig. 5 (b), the hardness is the same as that of the base metal and the etch pit cannot be seen in the weld bead for the welding speed of 12 m/min. Therefore, in this case, almost the entire welded bead consists of the HAZ-2 zone.

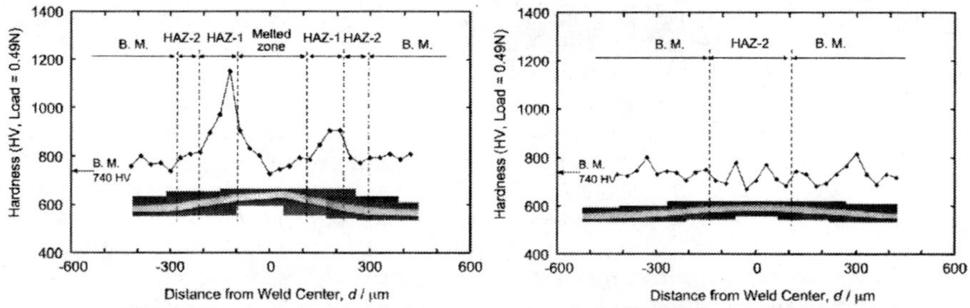

(a) Welding speed of 6 m/min (b) Welding speed of 12 m/min

Figure 5 Relation between microstructure and hardness profile on the welded specimen.

Figure 6 shows the TEM micrographs of the B.M. and the weld bead at the welding speed of 6 m/min and 12 m/min. As shown in Figs. 6 (a2) and (b2), the B.M. is a typical uniform amorphous structure and harrow ring is observed by selected-area electron diffraction. As shown in Fig. 6 (a4), the crystals with a mean diameter of 170 nm exist in the amorphous matrix as well as in the diffraction pattern for the combination of the harrow ring with the diffraction pattern by crystal at the melted zone. In slight contrast, as shown in Fig. 6 (a3), the HAZ-1 zone crystallized completely with a mean diameter of 400 nm into a clearer diffraction pattern than that of the melted zone. On the other hand, the weld bead consists of the amorphous matrix with a very small amount of nanocrystals with a mean diameter of 14 nm at the welding speed of 12 m/min, as shown in Fig. 6 (b3); this is quite different from the HAZ-1 zone images shown in Fig. 6 (a3).

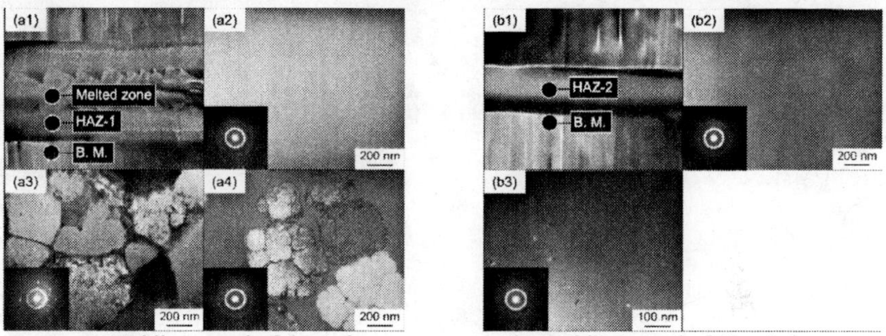

(a) Welding speed of 6 m/min (b) Welding speed of 12 m/min

Figure 6 TEM observed positions and micrographs of the weld bead; (a1) observed position at the welding speed of 6 m/min, TEM micrographs (a2) in B.M., (a3) in HAZ-1, (a4) in melted zone; and (b1) observed position at the welding speed of 12 m/min, TEM micrographs (b2) in B.M., (b3) in HAZ-2.

It was reported that the heat-affected zone crystallized more easily than the melted zone because the nose time of the TTT diagram of the metallic glass was much smaller than that of the melt[4]. In this study, the same results were concluded from the comparison of the degree of crystallization of the melted zone with that of HAZ-1, as shown in Figs. 5 (a), 6 (a3), and 6 (a4). Therefore, as shown in Fig. 4 (b), the weld bead shows three different zones (melted zone, HAZ-1, and HAZ-2) due to this feature.

Figure 7 shows the tensile strength of the base metal (B.M.) and the welded specimen at the welding speeds of 6 m/min and 12 m/min. The bars in Fig. 9 indicate the average strength. In the B.M. specimen, several fracture positions were observed. However, the fracture positions of the welded specimen were mainly in the weld metal. As an explanation for the varied strengths, it is theorized that the specimen edge by spark-erosion wire cutting is markedly uneven and/or the thickness of the specimen along its width is non-uniform. The average strength of the B.M. was 2334 MPa. Since the welded specimen at the welding speed of 6 m/min exhibits a crystalline structure, the average strength decreased to 771 MPa, which was about 32% of that of the B.M. The average strength of the specimen at the welding speed of 12 m/min was 1312 MPa, which was about 56% of that of the B.M. despite the absence of the crystalline structure. The reduction in the strength could be attributed to the concentration of stress at wrinkled surfaces, cracks starting from the nanocrystals in the weld bead, and the structural relaxation[1] caused by thermal history.

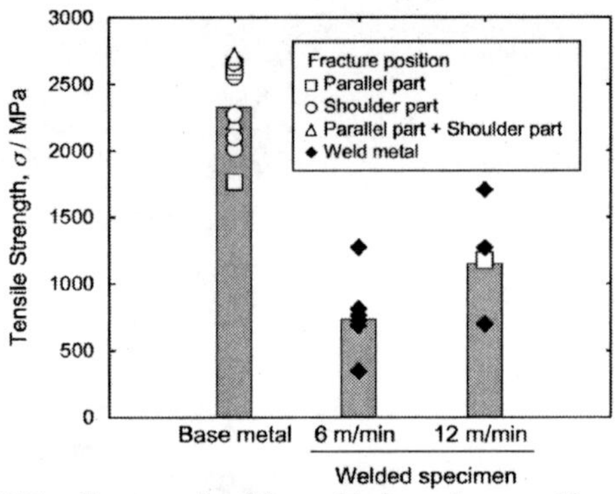

Figure 7 Tensile strength of the welded specimen and base metal.

CONCLUSIONS

The bead-on-plate welding of a melt-spun specimen of $Ni_{53}Nb_{20}Ti_{10}Zr_8Co_6Cu_3$ metallic glass was performed and the resulting microstructure of the laser weld bead and its mechanical properties were investigated. The following conclusive remarks were derived from this study.

(1) A sound weld bead can be obtained by diode laser welding at the welding speeds of 6 m/min and 12 m/min with a laser power of 26 W. The XRD pattern at the welding speed of 12 m/min exhibited the maintenance of the amorphous phase.

(2) At welding speeds lower than 6 m/min, the fusion zone of the weld bead consisted of the amorphous matrix with comparatively large crystals with a mean diameter of 170 nm, and the HAZ-1 consisted of a completely crystallized structure with a high hardness of around 1200 HV. At the welding speed of 12 m/min, the weld bead consisted of the amorphous matrix with a very small amount of nanocrystals with a mean diameter of 14 nm. The hardness was almost the same as that of the B.M. (740 HV).

(3) The average tensile strengths of the welded specimen at the welding speeds of 6 m/min and 12 m/min were 771 MPa and 1312 MPa, respectively. The joint efficiencies at 6 m/min and 12 m/min were 33% and 56% of the B.M. (2334 MPa), respectively.

ACKNOWLEDGEMENT

This work was supported by a Grant-in-Aid for Scientific Research on Priority Areas 428 KAKENHI (16039211) and Grant-in-Aid for Cooperative Research Project of Nationwide Joint-Use Research Institutes on Development Basis of Joining Technology for Metallic Glasses and Inorganic Materials from The Ministry of Education, Cultures, Sports, Science and Technology, Japan.

FOOTNOTES

*Graduate Student, Graduate School of Engineering, Osaka University, Suita 565-0871, Japan
Present address: Toyota Motor Corp., Toyota City 471-8571, Japan

REFERENCES

[1]A. Inoue, Bulk Amorphous Alloys, *Preparation and Fundamental Characteristics*, (Trans Tech Publications, 1998).

[2]A. Inoue, "Stabilization of Metallic Supercooled Liquid and Bulk Amorphous Alloys," *Acta Mater.*, **48**, 279–306 (2000).

[3]A. Inoue, "Bulk amorphous and nanocrystalline alloys with high functional properties," *Mater. Sci. Eng. A*, **304–306**, 1–10 (2001).

[4]Y. Kawamura and Y. Ohno, "Spark Welding of $Zr_{55}Al_{10}Ni_5Cu_{30}$ Bulk Metallic Glasses," *Scr. Mater.*, **45**, 127–132 (2001).

[5]T. Shoji, Y. Kawamura, and Y. Ohno, "Joining of $Zr_{41}Be_{23}Ti_{14}Cu_{12}Ni_{10}$ Bulk Metallic Glasses by a Friction Welding Method," *J. Japan Inst. Metals*, **66**, 1055–1062 (2002).

[6]C.H. Wong and C.H. Shek, "Friction Welding of $Zr_{41}Ti_{14}Cu_{12.5}Ni_{10}Be_{22.5}$ Bulk Metallic Glass," *Scr. Mater.*, **49**, 393–397 (2003).

[7]Y. Kawamura and Y. Ohno, "Successful Electron-Beam Welding of Bulk Metallic Glass," *Mater. Trans.*, **42**, 2476–2478 (2001).

[8]Y. Yokoyama, N. Abe, K. Fukaura, T. Shinohara, and A. Inoue, "Electron-Beam Welding of $Zr_{50}Cu_{30}Ni_{10}Al_{10}$ Bulk Glassy Alloys," *Mater. Sci. Eng. A*, **375–377**, 422–426 (2004).

[9]T. Zhang and A. Inoue, "New Bulk Glassy Ni-Based Alloys with High Strength of 3000 MPa," Mater. Trans., **43**, 708–711 (2002).

[10]A. Inoue, W. Zhang, and T. Zhang, "Thermal Stability and Mechanical Strength of Bulk Glassy Ni-Nb-Ti-Zr Alloys," Mater. Trans., **43**, 1952–1956 (2002).

[11]C. Qin, W. Zhang, H. Nakata, H. Kimura, K. Asami, and A. Inoue, "Effect of Tantalum on Corrosion Resistance of Ni-Nb(-Ta)-Ti-Zr Glassy Alloys at High Temperature," Mater. Trans., **46**, 858–862 (2005).

[12]S. Yamaura, H. Kimura, and A. Inoue, "The Properties of Bulk Metallic Glasses Related to Hydrogen," *KINZOKU (Mater. Sci. & Tech.)*, **75**, 48–53 (2005).

GLASS TRANSITION PHENOMENA AND HEAT CAPACITY OF $Zr_{0.55}Al_{0.10}Ni_{0.05}Cu_{0.30}$

Yosuke Moriya, Takahiro Yoshida, Hitoshi Kawaji, and Tooru Atake
Materials and Structures Laboratory, Tokyo Institute of Technology
4259 Nagatsuta-cho, Midori-ku, Yokohama, 226-8503 Japan

Mikio Fukuhara, Hisamichi Kimura, and Akihisa Inoue
Institute for Materials Research, Tohoku University
2-1-1 Katahira, Aoba-ku, Sendai, 980-8577 Japan

ABSTRACT

Molar heat capacity, C_p, of $Zr_{0.55}Al_{0.10}Ni_{0.05}Cu_{0.30}$ glassy alloy was measured for rod and ribbon samples by a thermal relaxation technique between 1.8 and 310 K. The C_p of the rod sample was also measured by differential scanning calorimetry (DSC) with the enthalpy method between 300 and 873 K; the two method are in good agreement within their overlap between 300 and 310 K. The glass transition was observed around 692 K, followed by subsequent crystallization around 725 K. From the low temperature C_p data, the electronic heat capacity coefficient and the Debye characteristic temperature were determined for both samples and compared with those of the component elements. The ribbon sample shows larger C_p than the rod sample. The difference in C_p is discussed in connection with the preparation methods.

INTRODUCTION

It is known that glassy alloys have favorable physical and chemical properties, such as high strength, high corrosion resistance and soft magnetic properties, which are superior to the corresponding crystalline alloys. Among many kinds of glassy alloys, Zr-based glassy alloys have extremely high glass-forming ability and good mechanical properties. In order to effectively weld and join these glassy alloys with different materials, fundamental thermodynamic information of each material is required. In the present study, we made precise heat capacity measurements on the $Zr_{0.55}Al_{0.10}Ni_{0.05}Cu_{0.30}$ (ZANC) glassy alloy in two different forms; rod and ribbon samples.

EXPERIMENTAL

An ingot of ZANC was prepared from a mixture of Zr, Al, Ni and Cu (99.99 mass % purity) by arc-melting in an argon atmosphere. From this ingot, bulk glassy rods with a diameter of 2 mm and a length of 45 mm were prepared by a copper mould casting technique. The ribbon sample with a width of 1 mm and a thickness of 20 μm was prepared by a single roller-type rapid quenching method.

The molar heat capacity, C_p, of both samples was measured by a thermal relaxation technique[1] using a PPMS (Physical Property Measuring System, Quantum Design, Inc., U.S.A.)[2] between 1.8 and 310 K; the sample platform area was 3×3 mm^2, and samples were attached on the platform with a very small amount of Apiezon N grease (ANG) to ensure a good thermal contact. A small piece (26.858 mg) was out from the rod and used for the C_p measurement. On the other hand, since one piece of the ribbon (3×1 mm^2) was too thin for accurate C_p measurement, these pieces were piled up to form a block with a size of $3 \times 3 \times 1$ mm^3 (11.737 mg). Each layer was adhered to one another with additional ANG (1.295 mg). The C_p of the sample was obtained by subtracting the heat capacity, C, of the additional ANG from the measured total C.

To gain information about glass transition and subsequent crystallization, differential scanning calorimetry (DSC) was performed on the rod sample (24.370 mg) by the continuous scanning method (heating rate of 20 $K \cdot min^{-1}$) using a Pyris 1 DSC (Perkin-Elmer, Inc., U.S.A.) between 300 and 873 K. The DSC temperature was calibrated with six standard materials at different heating rates. After that, further C_p measurement was made by the enthalpy method (step scanning method)[3] between 300 and 873 K. SRM720 (synthetic sapphire; NIST, U.S.A.)[4] was used for the enthalpy calibration. The sample was placed on the platform without using a sample container. Three stepwise DSC runs of blank (nothing), standard (SRM720, 12.921 mg), and sample (ZANC, rod, 23.785 mg) were made in 5K increments at 5K/min with 5 min isothermal holds. N_2 gas-flow was maintained at 20 $ml \cdot min^{-1}$ to avoid oxidation of the sample at high temperatures.

RESULTS AND DISCUSSIONS

The measured C_p of the ZANC rod sample is shown in Figure 1. The inset shows the result of the continuous scanning DSC experiment above room temperature, where the glass transition and subsequent crystallization are observed at $T_g \sim 698$ K (sigmoid inflection point) and as an exothermic peak at $T_c \sim 764$ K (peak onset), respectively. The value of the exothermic heat due to the crystallization, $\Delta_c H$, was estimated to be ~3.7 $kJ \cdot mol^{-1}$ from the area of the exothermic peak. By the step scanning method, the glass transition and crystallization are observed as a jump at $T_g \sim 692$ K ($\Delta C_p \sim 17$ $J \cdot K \cdot mol^{-1}$) and a sudden drop at $T_c \sim 725$ K, respectively. Above T_g, the sample is in a supercooled liquid phase, where the C_p is ~65 % larger than the C_p of the glassy state. On further heating, the sample suddenly crystallized at T_c and transformed into a thermodynamically stable crystalline phase, where the C_p is roughly equivalent to the C_p of the glassy state.

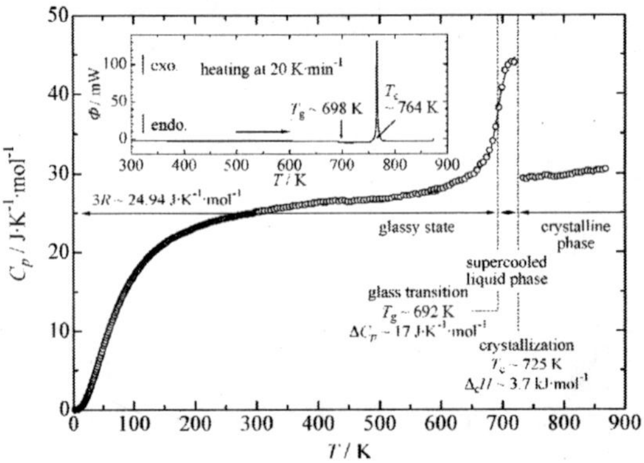

Figure 1. Measured molar heat capacity, C_p, of rod sample of $Zr_{0.55}Al_{0.10}Ni_{0.05}Cu_{0.30}$ (ZANC) glassy alloy by step scanning method. The solid line in the glass transition region shows a sigmoid fit. The inset shows the result obtained by the continuous scanning method.

Kanomata et al.[5] have measured C_p of ribbon samples of some Zr-Cu-based glassy alloys by AC calorimetry[6] and reported the following points: (1) they exhibit a broad glass transition compared to metalloid-containing glassy alloys, which show a discontinuous jump in C_p at T_g, (2) T_g does not depend on the scanning rate, while T_c increases with increasing scanning rate, and (3) C_p shows a broad peak at a certain temperature below T_g and then decreases with increasing temperature up to T_g. The glass transition observed in this study ranges over 100 K including the precursory region, in accordance with the feature (1). They explained that the broad glass transition in Zr-Cu-based glassy alloys resulted from a high space-filling rate and consequent low mobility of the component atoms with various sizes. Since the enthalpy method of DSC is categorized as not continuous but periodic heating method, the C_p was virtually measured at zero scanning rate. Comparing the two results at scanning rates of 0 and 20 $K \cdot min^{-1}$ in this study, T_g changes little (from 692 to 698 K), while T_c drastically increases (from 725 to 764 K), and these results also seem to agree with the feature (2). Concerning the feature (3), no remarkable C_p decrease was observed below T_g in the present measurement. Actually, we cannot simply compare our results with the previous results because of the difference in the methods of measurement and data analysis. More detailed examination is strongly required.

It is well known that at very low temperatures, the heat capacity of metallic materials is approximately represented by the equation:

$$C = C_{lat} + C_e = \beta \cdot T^3 + \gamma \cdot T, \qquad (1)$$

where C_{lat} and C_e are lattice and electronic contributions, and β and γ are lattice and electronic heat capacity coefficients, respectively. From the coefficient β, the Debye characteristic temperature Θ_D is calculated, which is considered as an index of lattice hardness. On the other hand, the heat capacity of glassy materials is of the same form as the equation (1) at very low temperatures[7]. Therefore, the C_p of glassy metals should have two different T-linear terms, i.e., electronic and glassy contributions (C_e and C_g). To see the data in the lowest temperature region, a $C_p \cdot T^{-3}$ vs. logT plot is given in Figure 2, where C_{lat} shows a constant value, i.e., β, in the lowest temperature region. The inset shows a $C_p \cdot T^{-1}$ vs. T^2 plot below 5.5 K, where the data are on a straight line; the slope and intercept give β and γ values, respectively. The estimated values of γ and Θ_D are given in Table I, together with those of the component elements. It should be noted that the γ value of ZANC should include both electronic and glassy contributions as mentioned above, but the former is probably dominant. The γ value of ZANC is close to that of the dominant component Zr and does not show any significant peculiarity. On the other hand, the Θ_D value of ZANC is significantly small compared to those of the component elements, which indicates the lattice vibrations should be "soft" in the glassy state. In general, a glassy state shows larger C_p than the corresponding crystalline phase especially at low temperatures because of the looser packing of atoms. Since equation (1) is only valid in the lowest temperature region (below 5.5 K in the present case), the deviation of the fit from the measured C_p becomes larger with increasing temperature as seen in Figure 2. Especially around 10 K, a large difference is observed, which indicates the existence of another low-energy excitation. In fact, many glassy materials show a large peak in $C_p \cdot T^{-3}$ at about 10 K[7], which is attributed to very high phonon density of states around 1.5 THz[8]. Nowadays, this high phonon density of states is called boson peak and it is considered characteristic of a glassy state. Actually, a peak in $C_p \cdot T^{-3}$ is commonly observed in a variety of materials, but glassy materials show an extremely large peak compared to the corresponding crystalline materials.

Table I. Electronic heat capacity coefficients, γ, and Debye temperatures, Θ_D, of ZANC and the component elements.

	γ / mJ·K^{-2}·mol^{-1}	Θ_D / K
ZANC (rod)	3.55	257
ZANC (ribbon)	3.84	250
Zr	3.03	310
Al	1.36	426
Ni	7.28	440
Cu	0.70	348

The C_p measurement of the ribbon sample was not easy compared to the rod sample because a considerable amount of ANG was used as an adhesive agent in the sample block, and the ANG shows two heat capacity anomalies above 200 K. The full details of the measurement will be published elsewhere[9], and only the data below 200 K is considered here. C_p(ribbon) is shown in Figure 3 together with C_p(rod) as a $C_p \cdot T^{-3}$ vs. logT plot. The inset shows the difference between C_p(ribbon) and C_p(rod) in percent. The ribbon sample has larger C_p than the rod sample below 200 K. This difference should be caused by the difference in lattice vibration. Especially a remarkable increase in C_p(ribbon) around 10 K implies a larger phonon density of states around1.5 THz (so-called boson beak), which signifies a higher degree of vitrification in the ribbon sample. This is also supported by the estimated values of γ and Θ_D (see Table I); the increase in γ and the decrease in Θ_D are indicative of a higher degree of vitrification, provided that the electronic states remain unchanged. This is comprehensible, considering the difference in the preparation methods; the ribbon sample was prepared by rapid quenching at a cooling rate faster than 10^4 K·s^{-1}, while the rod sample was prepared by a conventional casting technique at a cooling rate slower than 10^2 K·s^{-1}. It is conceivable that the rod sample had more time to evolve

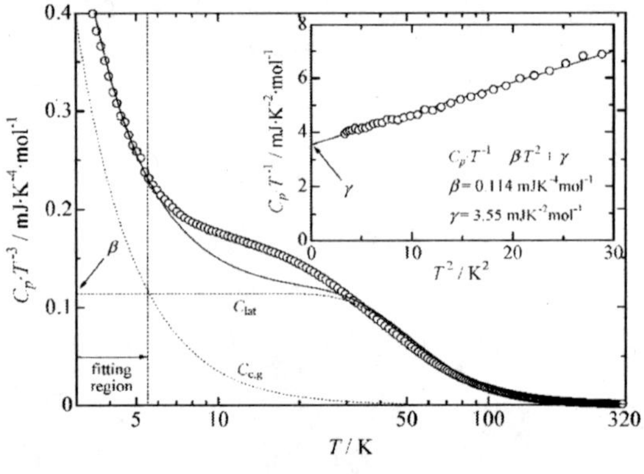

Figure 2. Plots of $C_p \cdot T^{-3}$ vs. logT of rod sample of ZANC. The inset shows $C_p \cdot T^{-1}$ vs. T^2 below 5.5 K. Solid, broken, and dotted lines show the result of fitting to the data below 5.5 K, i.e., total C_p, lattice contribution C_{lat} (Debye C_p with Θ_D = 257 K), electronic and glassy contribution $C_{e,g}$, respectively.

into a more stable state under a kinetically active condition, and is in a thermodynamically more stable glassy state than the ribbon sample. According to this idea, it is likely that in the ribbon sample, an apparent decrease in C_p measured by DSC occurs below T_g due to an exothermic effect of the structural relaxation, which was not observed in the rod sample. If no structural relaxation occurs, the glass transition should be less broad because of the higher mobility of component atoms.

CONCLUSION

The molar heat capacity, C_p, of the rod and ribbon samples of $Zr_{0.55}Al_{0.10}Ni_{0.05}Cu_{0.30}$ (ZANC) glassy alloy was measured between 1.8 and 310 K by thermal relaxation calorimetry. For the rod sample, further C_p measurement was performed by DSC, and the glass transition was observed at $T_g \sim 692$ K, which was subsequently followed by the crystallization at $T_c \sim 725$ K. The estimated electronic heat capacity coefficient is not peculiar compared to that of the component elements, but includes both electronic and glassy contributions, which should be separated for more detailed discussion. The lower Debye characteristic temperature is indicative of looser packing of atoms in the glassy state than in a crystalline state. Also the shoulder of $C_p \cdot T^{-3}$ at about 10 K implies existence of the boson peak, characteristic of glassy materials. To elucidate whether these features are of the glassy state of ZANC, it is necessary to measure the low-temperature C_p of the crystallized ZANC, which is now underway in our laboratory. The ribbon sample shows larger C_p than the rod sample, which suggests that the rod sample is in a thermodynamically more stable glassy state. Accordingly, some difference should be also observed in the high-temperature C_p and the glass transition behavior, and thus, high-temperature C_p measurement on the ribbon sample is planned. The C_p of the ribbon sample was not measured accurately above 200 K because of the existence of complex transitions in Apiezon N grease used as an adhesive agent. The most accurate C_p measurements will be carried out by adiabatic calorimetry[10], which requires no addenda in addition to a vessel but a much larger amount of sample.

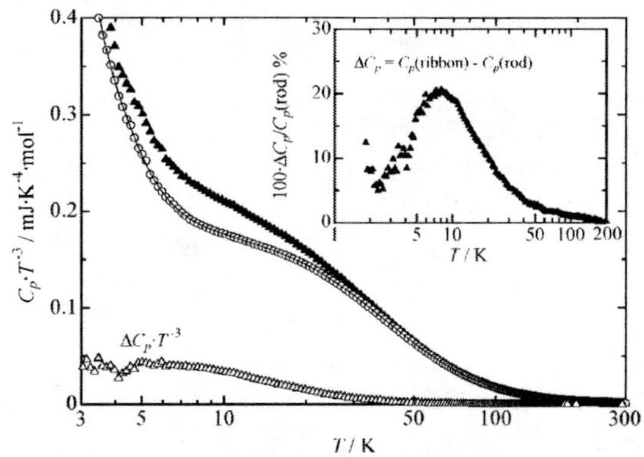

Figure 3. Plots of $C_p \cdot T^{-3}$ vs. logT of ZANC; the rod (open circles) and the ribbon (closed triangles) samples. The open triangles denote the heat capacity difference $\Delta C_p = C_p(\text{ribbon}) - C_p(\text{rod})$ divided by T^3. The inset shows ΔC_p in percent.

REFERENCES

[1]J. S. Hwang, K. J. Lin, and C. Tien, "Measurement of Heat Capacity by Fitting the Whole Temperature Response of a Heat-Pulse Calorimeter", Rev. Sci. Instrum. **68**, 94-101 (1997).

[2]J. C. Lashley, M. F. Hundley, A. Migliori, J. L. Sarrao, P. G. Pagliuso, T. W. Darling, M. Jaime, J. C. Cooley, W. L. Hults, L. Morales, D. J. Thoma, J. L. Smith, J. Boerio-Goates, B. F. Woodfield, G. R. Stewart, R. A. Fisher, and N. E. Phillips, "Critical Examination of Heat Capacity Measurements Made on a Quantum Design Physical Property Measurement System", Cryogenics **43**, 369-78 (2003).

[3]S. C. Mraw and D. F. Naas, "The Measurement of Accurate Heat Capacities by Differential Scanning Calorimetry. Conparison of D.S.C. Results on Pyrite (100 to 800 K) with Literature Values from Precision Adiabatic Calorimetry", J. Chem. Thermodyn. **11**, 567-84 (1979).

[4]D. A. Ditmars, S. Ishihara, S. S. Chang, G. Bernstein, and E. D. West, "Enthalpy and Heat-Capacity Standard Reference Material: Synthetic Sapphire (α-Al_2O_3) from 10 to 2250 K", J. Res. Natl. Bur. Stand. **87**, 159-63 (1982).

[5]T. Kanomata, Y. Sato, Y. Sugawara, S. Aburatani, H. Kimura, T. Kaneko, A. Inoue, and T. Masumoto, "Heat Capacity of Pd-Si, Ni-Si-B and Zr-Based Metallic Glasses", Sci. Rep. RITU **A43**, 89-95 (1997).

[6]I. Hatta and A. J. Ikushima, "Studies on Phase Transitions by AC Calorimetry", Jpn. J. Appl. Phys. **20**, 1995-2011 (1981).

[7]R. C. Zeller and R. O. Pohl, "Thermal Conductivity and Specific Heat of Noncrystalline Solids", Phys. Rev. **B4**, 2029-41 (1971).

[8]U. Buchenau, N. Nücker, and A. J. Dianoux, "Neutron Scattering Study of the Low-Frequency Vibrations in Vitreous Silica", Phys. Rev. Lett. **53**, 2316-9 (1984).

[9]Y. Moriya, H. Kawaji, T. Atake, M. Fukuhara, H. Kimura, and A. Inoue, "Heat Capacity Measurement on Thin Ribbon Sample of $Zr_{0.55}Al_{0.10}Ni_{0.05}Cu_{0.30}$ Glassy Alloy Using a Quantum Design PPMS", in preparation.

[10]Y. Moriya, T. Atake, T. Yoshida, H. Kawaji, M. Fukuhara, H. Kimura, and A. Inoue, "Thermodynamic Studies of $Zr_{0.55}Al_{0.10}Ni_{0.05}Cu_{0.30}$ Glassy Alloy", in preparation.

DEVELOPMENT OF FIBER LASER AIDED SPOT HEATING SYSTEM AND ITS APPLICATION TO CONTROL OF MICROSTRUCTURE IN PURE ALUMINUM

Toshiya Shibayanagi, Masahiro Tsukamoto and Nobuyuki Abe
Joining and Welding Research Institute, Osaka University
11-1,Mihogaoka, Ibaraki, Osaka, Japan, 567-0047

Nobuyuki Matsuda, Yukihiro Soga and Takamasa Matsumoto
Graduate student of Osaka University
Joining and Welding Research Institute, Osaka University
11-1,Mihogaoka, Ibaraki, Osaka, Japan, 567-0047

ABSTRACT

Spot heating of pure aluminum plate was performed by means of a fiber laser system newly developed in the present work. The present system can focus the laser beam on the surface of specimen with 30 microns in diameter and a maximum power of 100W. Pure aluminum with 1mm thick was prepared by cold rolling followed by mechanical polishing and chemical etching for the sake of appropriate surface condition.

Laser heating with a power of 30W for 20s brought about a small hole with a diameter of 10 microns, suggesting a local melting in the surface area of sample. Electron Back-Scatter Pattern analysis for this specimen was then carried out after polishing the surface mechanically, revealing the preferential recrystallization and grain growth phenomenon around the melted position. This result clearly shows that the present local heating method can generate peculiar grain structure at any positions of specimen, leading to controls of recrystallization and grain growth process that has not been realized by utilizing conventional heat treatment techniques.

INTRODUCTION

Polycrystalline materials are described by some parameters such as kind of phase, shape and size of grains, grain orientation and grain boundary structure. Spatial distributions of these microstructure parameters are known to be heterogeneous that were revealed by Electron Back-scatter Diffraction pattern (EBSD) analysis [1].

Conventional heat treatment utilizing homogeneous temperature field that has been commonly performed in industry changes these parameters in whole portion of materials inevitably. Thus some important local distributions of microstructure parameters may have some possibility of disappearing or become passive and disable to contribute to the evolution of microstructure such as texture development. For example, during the course of cube texture development, some cube-oriented nuclei in deformed matrix or small cube-oriented grains cannot grow if they are surrounded by preferentially growing grains with the other orientation.

Meanwhile, taking into account the heterogeneity of spatial distribution of each microstructure parameters, the spot heating method would become a candidate way to control evolution process of microstructure besides homogeneous heat treatment since only some selected areas are heated without giving affect to the other region.

In the present study a fiber laser-aided spot heating system has been developed, and spot heating of cold rolled pure aluminum plates has been tried to investigate microstructure change by the local heating.

EXPERIMENTAL PROCEDURE

Figure 1 schematically illustrates the ray diagram of laser heating system that has been newly developed in the present study. The system utilizes a fiber laser generator (YLR-100SM) provided by IPG Photonics Co.Ltd., with a wave-length and maximum power of 1100nm and 100W, respectively. A well-designed adjustment of optics realizes to focus the beam into less than 30μm on the focal plane. Beam profile of the laser is shown in Fig.2, representing a Gaussian distribution that would be satisfactory for the spot heating experiments in the present study. In the present paper this fiber laser-aided spot heating system will be named as FLASH.

Microstructure of the irradiated specimen was evaluated by means of optical microscopy and scanning electron microscopy (SEM) and SEM-EBSP methods. The EBSP measurement was performed on laser-irradiated specimens after grinding by about 100 μm using a suspension called OP-U that contains oxide particles.

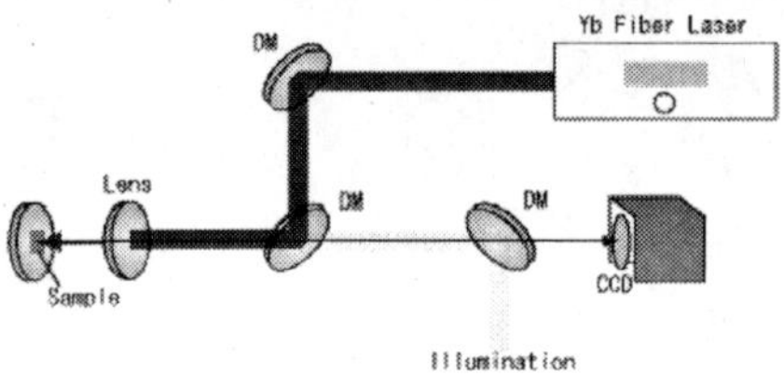

Fig.1 Schematic illustration of laser spot heating system.

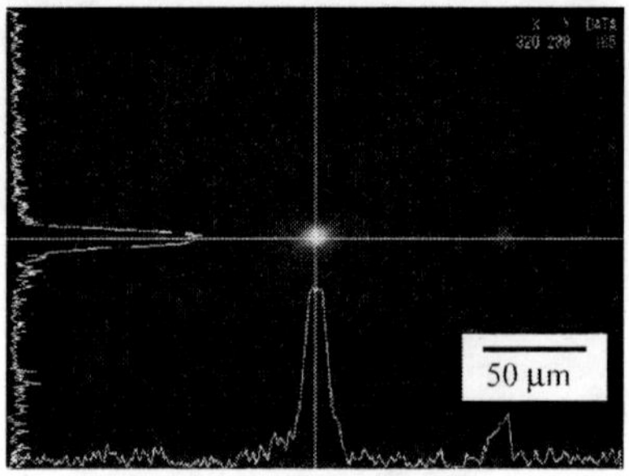

Fig.2 Beam profile.

RESULTS AND DISCUSSION

Figure 3 shows SEM micrographs of specimen surface before and after the irradiation. Comparing (a) and (b), the present spot heating system works well to focus the beam to any points. Two small holes, as indicated by arrows, appear to exist on the surface of specimen by two times of shots. The higher magnification shown in (c) clearly indicates the irradiated area with a small hole with a diameter of around 10μm. The time and power for the irradiation were

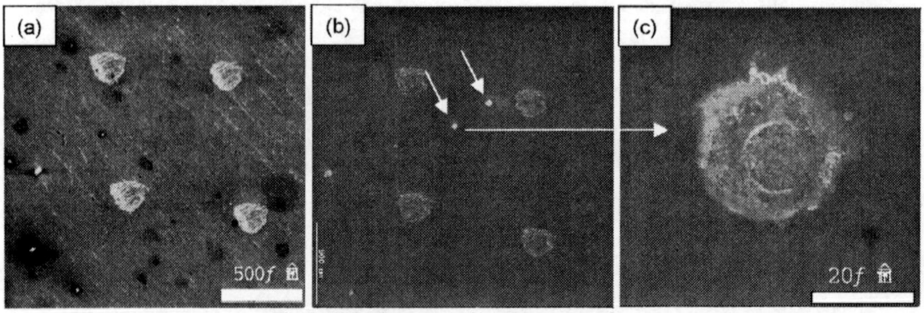

Fig.3 SEM images of specimen surface.
(a) before irradiation, (b) after irradiation, (c) higher magnification

10 s. and 30W, respectively. In this case the incident beam melted very small area of surface region resulting in an eruption of molten metal and sticking around the hole eventually. Taking into account that aluminum is a high thermal conducting material, heat-affected zone with much larger diameter should exist around the hole. The present study did not achieve suitable condition for laser irradiation that makes no molten zone and/or hole.

Figure 4 represents an orientation map based on color coding of the inverse pole figure(called inverse pole figure map) of an irradiated area together with its surrounding area of specimen. The irradiated position is located in the center of picture and grain structure is quite unclear because of the rough surface with little dimples that remained after polishing. This dimpled area is used as a marker to find the irradiated position instantly. Around this center position some large grains appear, indicating that these grains have been recrystallized in the vicinity area of the molten zone and grew towards the outside along the temperature gradient. Since the specimen was cold-rolled prior to the irradiation, no significant grain structure indicating deformed structure exists around the irradiated area.

Distribution of small angle boundaries in these large grains suggests important features of recrystallization and preferential grain growth in these areas. Grains A, B and C in Fig.4 have different orientations. Grain A and B contains few small angle boundaries, suggesting discontinuous recrystallization took place in the deformed structure and these two grains should have satisfied some growth conditions. However grain C contains much more amounts of small angle boundaries, so this area seems to have experienced a continuous recrystallization, but the reason for the changes in recrystallization mode is unclear in the present study.

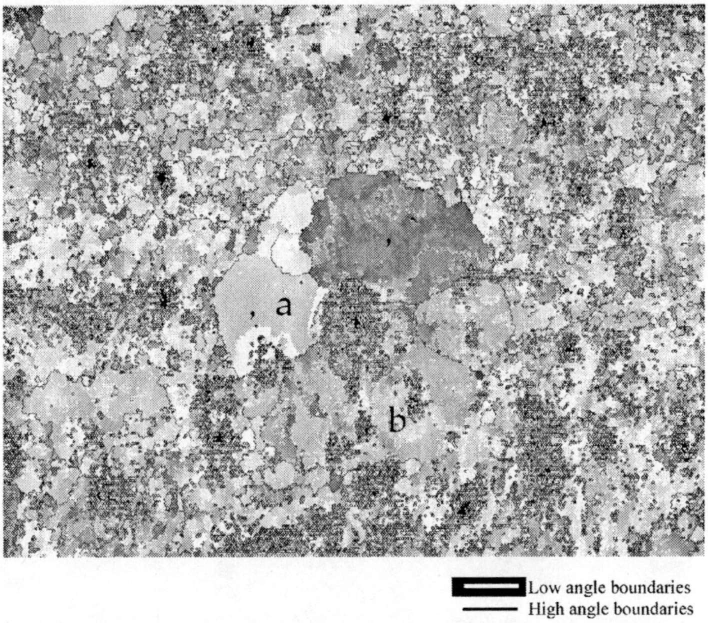

Fig.4 Inverse pole figure map of a laser spot-heated specimen.

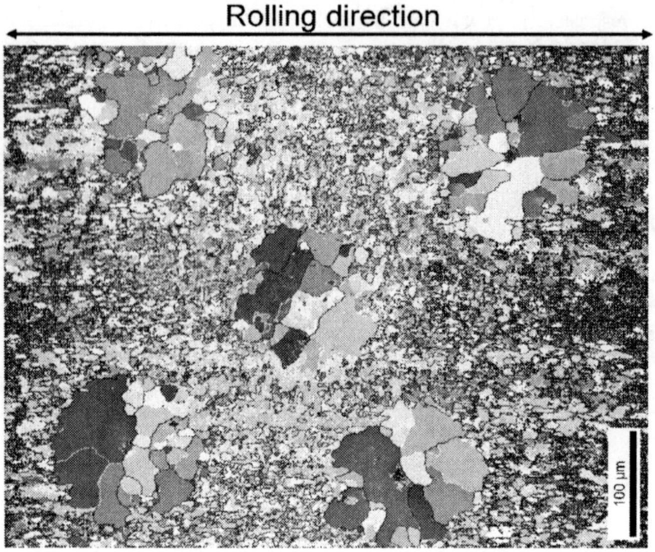

Fig.5 Inverse pole figure map of five-spot-heated specimen.

The FLASH system enables local heating at any critical position on the surface of specimen which is set on a X-Y stage. An example of five-spot heating on pure aluminum sheet is shown in Fig.5. The irradiation condition is 30W for 15s. Several grains seem to be originated from the spot-heated position or its surrounding heat affected area of each spot, that is similar to the microstructure shown in Fig.4. The size of each cluster of grains is almost the same as around 120μm. Many grains grew by the spot heating tend to belong to orientation group close to (112)//ND, but a few grains have the orientation close to (100)//ND which is indicated by an arrow in the figure. This grain is recrystallized in the deformed matrix having the orientation close to (110)//ND. Since the cube oriented grains are important for films used in the electric-condenser, the present spot heating method would contribute for the production of cube oriented pure aluminum sheet.

Size and number of grains generated in one cluster is strongly affected by surface condition of specimen as well as other physical properties of materials such as thermal conductivity. For example, specimens treated by electro-polishing did not show the recrystallized mirostructure because of quite high reflectivity for the present wave length of laser beam. Thus mechanical polishing or chemical etching is required for the sake of appropriate surface condition for the laser radiation. The present experimental technique cannot always produce the same quality of surface roughness by means of mechano-chemical polishing. Thus every specimen shows some different appearance of heat affected microstructure even under the same radiation condition.

Grains with same orientation are obtained by chance depending on heating condition and deformation structure of material. An example is shown in Fig.6. There are three grains, one

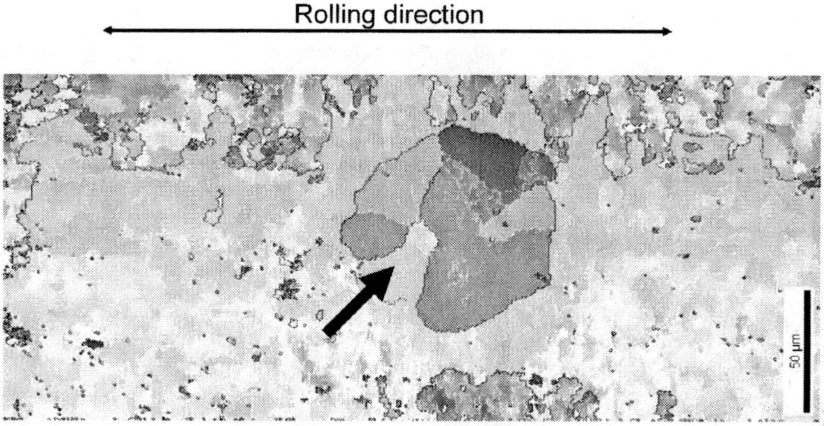

Fig.6 Inverse pole figure map of a spot-heated specimen with 30W for 15s.

large grain and two smaller grains, divided by large angle boundaries nucleated around the spot heated position. The size of this grain cluster is about 70μm, which is 42% smaller than those in the previous two figures. The minimum size of cluster or the heat affected zone will be reduced by choosing optimum laser power and surface treatment of pure aluminum. The difference of cluster size should be attributed to the surface condition of specimen. Also some other materials of lower thermal conductivity would show much smaller heat affected area yielding smaller size of grain cluster or single grain structure in the deformed matrix.

Two of the clustered grains are belonging to the orientation group of (112)//ND, but the last one grain indicated by an arrow is close to the orientation of the surrounding band-like matrix that has an orientation close to (110)//ND. This grain seems to have recrystallized continuously and grew in the deformed matrix. Orientation group of the other two grains is quite different from the matrix, indicating a discontinuous recrystallization has occurred during the spot heating. In addition a strongly developed textured microstructure has been realized in the area, suggesting that the present heating technique can produce a peculiar textured material.

Recrystallization starts in the heavily deformed structure which is consist of small regions of different orientations divided by dislocation walls or tangled dislocations. Preferential growth of the regions in deformed matrix can occur when satisfy some microstructural condition such as grain boundary structures, grain size difference [2]. Local heating has a potential to enhance the heterogeneity of growth behavior of such regions if the size of heated area is reduced to that equivalent to the size of the region in the deformed matrix. The present spot heating method would be fruitful for realizing a "pin-point heating" of the regions to develop preferentially.

Meanwhile, since the size advantage of grain leads to a preferential growth during homogeneous annealing, the FLASH technique can also produce peculiar textured structure during grain growth, that will be the forthcoming mission of the present work.

SUMMARY

Fiber-laser aided spot heating system (FLASH) has been developed in order to modify microstructure of pure aluminum by controlling recrystallization, grain growth process and texture development. Recrystallization and grain growth take place in the small area with a diameter of 120μm when spot-heated with 30W for 20s, yielding a cluster of grains. The cluster has an equi-axed shape and its size changes depending on some parameters such as surface condition and radiation condition. Depending on the radiation condition and deformation structure the recrystallized grains possess almost the same orientation, suggesting a strongly developed texture was fabricated in the limited area. The present system can introduce heat energy at any critical point on the specimen surface. Therefore we can generate a peculiar microstructure such as heterogeneous grain size distribution in any critical small area.

ACKNOWLEDGEMENT

The present work was financially supported by Grant-in-Aid for scientific research (B: project No. 13450295) and Grant-in-Aid for Cooperative Research Project of Nationwide Joint-Use Research Institutes on Development Base of Joining Technology for New Metallic Glasses and Inorganic Materials from The Ministry of Education, Culture, Sports, Science and Technology, Japan.

REFERENCES

[1]V.Randle, "Microtexture Determination and its application", *The Institute of Materials*, London, 1992.

[2]K.Matsumoto, T.Shibayanagi and Y.Umakoshi:"On the role of grain boundary character distribution in grain growth process of Al-Mg alloy", *Acta Metall.*, **45**(1997), 439-451.

PREPARATION OF AN MTES HYBRID BIOACTIVE COATING ON METAL SURFACES USING THE SOL-GEL METHOD

Yasuto Hoshikawa, Eiichi Yasuda and Takamasa Onoki
Materials and Structures Laboratory, Tokyo Institute of Technology
R3-21, 4258, Nagatsuta, Midori-ku, Yokohama 226-8503, Japan

Masaru Akao
Institute of Biomaterials and Bioengineering, Tokyo Medical and Dental University
2-3-10, Surugadai, Kanda, Chiyoda-ku, Tokyo 101-0062, Japan

Yasuhiro Tanabe
Departments of Chemical Engineering, Graduate School of Engineering, Nagoya University
Furo-cho, Chikusa-ku, Nagoya 464-8603, Japan

ABSTRACT

Sol-gel derived organically modified silicates containing calcium ions showed the ability to form hydroxyapatite (HA) in a simulated body fluid (SBF). Using the sol-gel method, we prepared a double layered coating on α-titanium (Ti) and Stainless steel 304 (SUS) plates. The double layered coating consisted of a methyltriethoxysilane (MTES) derived organic silicate layer, with a MTES-Ca-Nb-Ta layer on top which, prepared from MTES, calcium nitrate tetrahydrate, pentaethoxy-niobium and pentaethoxy-tantalum. The MTES coatings on the Ti and SUS had adhesive strengths at 5.2 and 4.6 MPa, respectively. HA formed on the surfaces of the double layered coatings on the Ti substrates after immersion in SBF at 3 days.

INTRODUCTION

Stainless steel and titanium and its alloy have sufficient mechanical properties to be utilized as artificial hip joints. Some of the recent biomaterials used under load-bearing conditions are composed of the metal substrate and bioactive ceramic coating layer including hydroxyapatite (HA), both of which has biocompatibility and mechanical properties. Many different bioactive coating methods have been reported so far, such as, plasma spraying[1], sputtering, electrolytic deposition, sol-gel coating[2], thermal decomposition coating[3], hydrothermal deposition coating[4], and immersion coating in a simulated body fluid (SBF)[5] and so on.

Bioactive ceramics exhibit the formation of bone-like HA on the surface in a physiological environment, which then it makes possible to bind with living bone[5]. Silicate glasses and glass ceramics have bioactivity[6] and are clinically used as bone repairing materials. Several organically modified silicates containing calcium ions prepared by the sol-gel processes showed the formation of bone-like HA at their surfaces during bioactivity tests using SBF[7]. An organic-inorganic hybrid gel composed of poly dimethylsiloxane and tetraethoxysilane with calcium ions also formed bone-like HA on the surface in SBF at 1 day[8]. An organic-inorganic hybrid was prepared from methyltriethoxysilane (MTES), calcium nitrate tetrahydrate, pentaethoxy-niobium and pentaethoxy-tantalum by sol-gel method, and the hybrid gel (MTES-Ca-Nb-Ta) deposited bone-like HA at 9 hours after soaking in SBF[9]. However, a sol-gel coating of a bioactive hybrid onto metal surface has not yet been reported.

In the present study, we prepared a double layered coating on SUS 304 and titanium plates that includes an initial MTES layer, with a MTES-Ca-Nb-Ta layer on the top. Bioactivity of the coatings was tested in SBF, and adhesive strength between the coatings and the metal substrates was evaluated according to Japanese industrial standards (JIS) A 6909.

EXPERIMENTAL

MTES and MTES-Ca-Nb-Ta hybrid solution

Methyltriethoxysilane (MTES, $CH_3Si(OC_2H_5)_3$) was used as a common source of polymethylsilsesquoxane precursor. Metal alkoxide components, pentaethoxy-niobium ($Nb(OC_2H_5)_5$) and pentaethoxy-tantalum ($Ta(OC_2H_5)_5$) (High Purity Chemical Laboratory), were modified with acetylacetone (AcAc) to control hydrolysis rate. The coating solution was composed of MTES, 1N hydrochloric acid, AcAc modified $Nb(OC_2H_5)_5$ and $Ta(OC_2H_5)_5$, and distilled water. This solution was mixed in ethanol and stirred for 1 hour for the initial MTES coating. For the bioactive MTES-Ca-Nb-Ta coating on the top the solution was stirred for 30 minutes, after which we added calcium nitrate tetrahydrate solution in ethanol and continued to stir for 30 minutes at room temperature. Each solution placed in an oven at 60°C for 6 hours. The reagent molar ratios are shown in Table 1.

Coating on Ti and SUS 304 plates

The size of the Ti plate used for bioactivity test was $10\times10\times0.1$ mm^3, and the SUS 304 and Ti plates for adhesion test were $10\times40\times3$ mm^3. The surfaces of the plates were polished with 2000 grit SiC grind paper. The plates were dipped into the MTES coating sol for 10 seconds. After pulling the plates from the sol, they were placed on a watch glass and dried at 80°C for 12 hours. The bioactive hybrid sol was applied to the MTES coated plate at a rate 0.3 μL/mm^2 and dried at 80°C for 3 hours to produce a double layered MTES-Ca-Nb-Ta/MTES coating.

SBF test and characterization

Bioactivity was evaluated by examining HA formation on a sample in SBF with a pH 7.40 similar to that of human blood plasma[5]. The specimen was soaked in 30mL SBF at 37°C for 3 days. After soaking in SBF, the specimens were rinsed with distilled water and dried at 40°C for 24 hours. The surfaces of both the single layered MTES and double layered specimens were examined by X-ray diffractometry (XRD; RINT-2000, Rigaku) and field emission scanning electron microscopy (SEM; S-4500, Hitachi).

MTES coating adhesive strength test

The adhesive strength between the MTES coatings and the Ti or SUS 304 substrates was evaluated using a JIS A 6909 tear-off test. Figure 1 shows a tear-off attachment and adhesion test procedure. A steel rod (8 mm diameter) was glued to the coated plate with an epoxy resin and fixed for 24 hours at room temperature. The tensile test was performed at a crosshead speed of 0.1 mm/min on a universal testing machine (Auto Graph AK-100kG, Shimads).

Table 1. Compositions for sol-gel hybrid coatings on metal substrates.

	Composition (mol)							
	MTES	$Nb(OC_2H_5)_5$	$Ta(OC_2H_5)_5$	AcAc	HCl	H_2O	C_2H_5OH	$Ca(NO_3)_2.4H_2O$
MTES	1	0	0	0	0.005	2	3	0
MTES-Ca-Nb-Ta	1	0.025	0.025	0.1	0.005	2	6	0.5

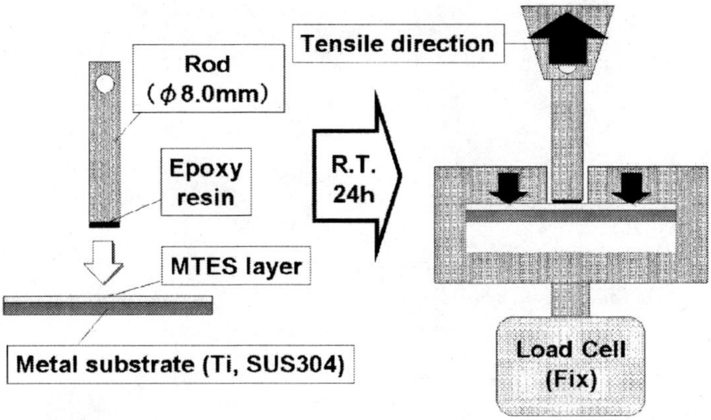

Figure 1. Setup measuring adhesive strength between MTES layer and metal substrate, tear-off testing based on JIS A 6909.

RESULT AND DISCUSSION

The single-layered MTES coating on the Ti plate and double layered MTES-Ca-Nb-Ta/MTES coating are shown in Figure 2(a) and 2(b), respectively. The single layered coating films were colorless, transparent and crack-free. On the other hand, the double layered coatings were opaque white films and seemed to be crack-free. The top layer of MTES-Ca-Nb-Ta was well attached to the underlying MTES layer. After soaking in SBF, the single layered coating of MTES-Ca-Nb-Ta came off the substrate, but the double layered coatings were quite stable.

Figure 3 shows X-ray diffraction (XRD) patterns of the single layered MTES and the double layered MTES-Ca-Nb-Ta/MTES on Ti plates after soaking in SBF for 3 days. All the XRD patterns showed broad peaks assigned to amorphous SiO_2, and intense peaks from α-Ti. The double layered coating exhibited formation of HA, and thus the coating was bioactive.

Figure 4 shows SEM images of the surfaces of the double layered coating. The surface of the coating was completely covered with spherical aggregates composed of flake-like crystals. The morphology of the crystals is similar to bone-like HA formed on common bioactive materials[8,10]. Tsuru et al. reported bioactivity of vinyltrimethoxysilane and calcium acetate derived coating on nylon 6 using an SBF immersion test. However, the coating film did not deposit bone-like HA on the surface. In contrast, a film abraded with emery paper deposited bone-like HA at 7 days[11]. This suggests that a dense silicate film may prevent HA deposition. There are two possible reasons for this: 1) HA nucleation sites provided by the silanol groups are covered with the dense silicate film, and 2) a dense silicate surface inhibits dissolution of calcium ions into SBF[11]. That is, HA forming ability enhanced optimal pore size on surface for HA nucleation site, and incorporated calcium ions to accelerate HA deposition[8].

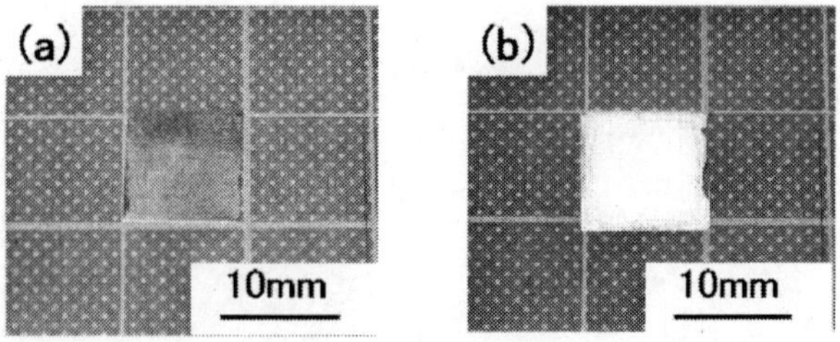

Figure 2. Photographs of the coatings (square plate located in the center) on Ti substrates, single layered MTES in (a) and double layered MTES-Ca-Nb-Ta/MTES in (b).

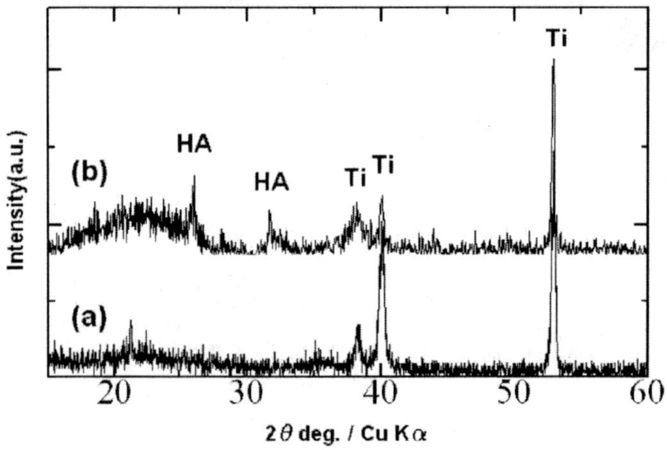

Figure 3. XRD patterns of the coatings on Ti substrate after immersion in SBF at 3 days, single layered of MTES in (a) and double layered of MTES-Ca-Nb-Ta/MTES in (b).

Table 2. Adhesive strengths between the coatings and the metal substrates, and fracture results.

Sample	Adhesive strength (MPa)	Crack path
Ti-MTES	5.2	At epoxy-coating interface
SUS-MTES	4.6	

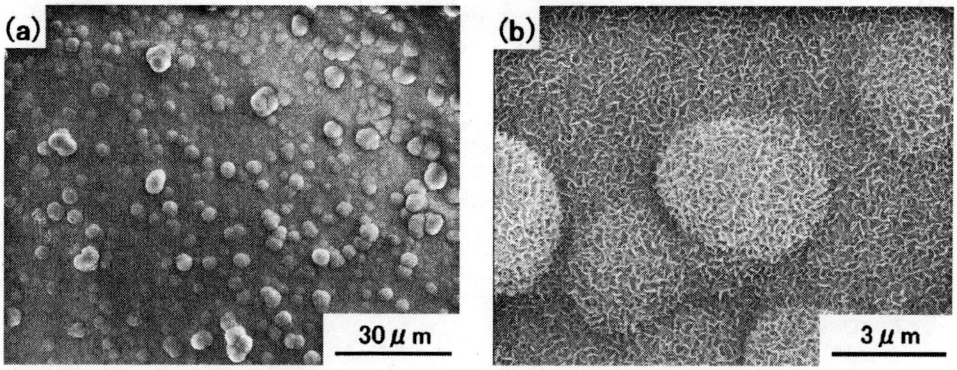

Figure 4. SEM photographs of the double layered MTES-Ca-Nb-Ta/MTES hybrid coating on Ti substrate after soaking in SBF for 3 days, magnification 1,000× (a), 10,000× (b).

The adhesive strength, and the failure mode for the coatings on Ti and SUS plates are summarized in Table 2. The single layered MTES coatings on Ti and SUS showed adhesive strengths at 5.2 and 4.6MPa, respectively. All the samples came off at the MTES-epoxy resin interfaces, because adhesion of MTES-epoxy was weaker than of MTES-metal. Hasegawa[12] measured bonding strengths between femoral bones and HA coated Ti implants prepared by plasma spraying and thermal decomposition, ranging from 3.7 to 6.6 MPa. Considering their results, the adhesion of the MTES-metals should be sufficient for bone implants; at least the underlying coating may not peel off metal surface of the bone implant.

CONCLUSIONS

We prepared MTES derived sol-gel coating on Ti and SUS substrates. The adhesive strength at the metal-MTES interface for Ti and SUS were 5.2 and 4.6MPa respectively, which may be adequate for bone implant materials. In order to add bioactivity to the metal substrates, we formed a double layered coating of MTES-Ca-Nb-Ta/MTES on the metal surfaces. The coatings showed no delamination after soaking in SBF, and showed bioactivity to form bone-like HA.

ACKNOWLEDGMENTS

This work was supported by Grant-in-Aid for Cooperative Research Project of Nationwide Joint-Use Research Institutes on Development Base of Joining Technology for New Metallic Glasses and Inorganic Materials and was partially supported by Grant-in-Aid for Young Scientists (B), 18760516, 2006 from The Ministry of Education, Culture, Sports, Science and Technology, Japan. Authors appreciated Dr. Manabu Fukushima at the National Institute of Advanced Industrial Science and Technology for his technical advises about the sol-gel techniques and Mr. Hajime Ishii at the section of technical staffs of MSL for his helps about the setup utilized by Collaborative Research Project of Materials and Structures Laboratory, Tokyo Institute of Technology.

REFERENCES

[1]S. R. Radin, P. Ducheyne, "Plasma spraying induced changes of calcium phosphate ceramic characteristics and the effect on in vitro stability," *J. Mater. Sci. Med.*, **3**, 33-42 (1992).

[2]A. Montenero, G. Gnappi, F. Ferrari, M. Cesari, E. Salvioli, S. Kaciulis and M. Fini, "Sol-gel derived hydroxyapatite coatings on titanium substrate," *J. Mater. Sci.*, **35**, 2791-97 (2000).

[3]P. Zhou and M. Akao, "Preparation and characterization of double layered coating composed of hydroxyapatite and perovskite by thermal decomposition," *Bio-Med. Mater. Eng.*, **7**, 67-81 (1997).

[4]Y. Huzushiro, M. Nishino, A. Sugimori, A. Okuwaki, T. Sato, "Coating of hydroxyapatite on various substrates via hydrothermal reactions of $Ca(edta)^{2-}$ and phosphate, " *J Mater Sci: Mater. Med.*," **12**, 333-37 (2001).

[5]H. –M. Kim, "Ceramic bioactivity and related biomimetic strategy," *Current Opinion in Solid State and Mater. Sci.*, **7**, 289-99 (2003).

[6]P. Li, and K. D. Groot, "Better Ceramics through Sol-Gel Process, " *J. Sol-Gel Sci. Tech.*, **2**, 797-801 (1994).

[7]K. Tsuru, C. Ohtsuki, A. Osaka, T. Iwamoto and J. D. Mackenzie, "Bioactivity of Sol-gel derived organically modified silicates, " *J. Mater. Sci. Med.*, **8**, 157-61 (1997).

[8]K. Tsuru, Y. Aburatani, T. Yabuta, S. Hayakawa, C. Otuki and A. Osaka, "Synthesis and In Vitro Behavior of Organically Modified Silicate Containing Ca Ions, " *J. Sol-Gel Sci. Tech.*, **21**, 89-96 (2001).

[9]M. Fukushima, "The preparation and low environmental impact process of organic inorganic hybrid materials from methyltriethoxysilane," *Doctoral dissertation of Tokyo Institute of Technology* (2004)

[10]S.B. Cho, K. Nakanishi, T. Kokubo, N. Soga, C. Ohtsuki, T. Nakamura, T. Kitsugi and T. Yamamuro, "Dependence of Apatite Formation on Silica Gel on its Structure: Effect of Heat Treatment," *J. Am. Ceram. Soc.*, **78**, 1769-74 (1995).

[11]K. Tsuru, A. Hayakawa, C. Ohtsuki and A. Osaka, "Bioactive Gel Coatings Derived from Vinyltrinethoxysilane, " *J. Sol-Gel Sci. Tech.*, **13**, 237-40 (1998).

[12]S. Hasegawa, "Characterization and Canine Bone Tissue Reaction of Hydroxyapatite-Coated Titanium Using Thermal Decomposition Method," *Bull. Tokyo Med. Dent. Univ.* **43** 25-44 (1995).

SYNTHESIS OF SiO_2-COATED MAGNETITE NANOPARTICLES AND IMMOBILIZATION OF PROTEINS ON THEM

Yoshio Kobayashi*, Mayumi Yoshida, Daisuke Nagao
Department of Chemical Engineering, Graduate School of Engineering, Tohoku University
6-6-07 Aoba, Aramaki-aza, Aoba-ku
Sendai, Japan, 980-8579

Yasuo Ando, Terunobu Miyazaki
Department of Applied Physics, Graduate School of Engineering, Tohoku University
6-6-05 Aoba, Aramaki-aza, Aoba-ku
Sendai, Japan, 980-8579

Mikio Konno
Department of Chemical Engineering, Graduate School of Engineering, Tohoku University
6-6-07 Aoba, Aramaki-aza, Aoba-ku
Sendai, Japan, 980-8579

ABSTRACT

This paper describes a method for silica-coating of magnetite nanoparticles and immobilization of proteins on the silica-coated particles. The magnetite nanoparticles were prepared according to the Massart method. The silica-coating was performed in aqueous/ethanolic solution with tetraethoxyorthosilicate and ammonia aqueous solution in the presence of the magnetite nanoparticles. TEM observation revealed that the nanoparticles were coated with silica shell. Saturation magnetization of the SiO_2-coated particles was 52.4 emu/g-Fe_3O_4, which indicated that the magnetite nanoparticles were chemically undamaged during the silica-coating. The protein immobilization was performed in aqueous solution with bovine serum albumin in the presence of the silica-coated particles surface-modified with 3-aminopropyltrimethoxysilane and glutaraldehyde. UV-VIS absorption spectroscopy revealed that the BSA was adsorbed on the surface-modified particles.

INTRODUCTION

Nanoparticles of magnetic materials, in which magnetite (Fe_3O_4) is representative, have been extensively studied because of their wide applications such as high-density storage, magnetic resonance imaging (MRI), drug delivery, and separation of biochemical products. Magnetic particles themselves can not be used in practical applications, since they tend to form large aggregates, which may result in the alteration of magnetic properties, and they can undergo rapid biodegradation when they are directly exposed to biological system.

Coating of the particles with an inert shell is a candidate technique for preventing such limitations. The role of the shell is two-fold, since it not only provides a greatly enhanced colloidal stability in water, but also can be used to keep the magnetic particles from contact with the biological system.

From this view point, there have been extensive studies on silica-coated iron oxide nanoparticles[1-6]. Iijima et al. prepared silica-coated iron oxide particles by a sol-gel method, and found that removal of ions in suspension of iron oxide particles before silica-coating was required to prevent aggregation of particles that often takes place in silica-coating [6]. However, most works on silica-coated magnetite particles have not discussed long-time chemical stability of the particles, though the stability should be very important for practical use. According to a preliminary experiment, elaborate washing of magnetite nanoparticles was an important step for preparing chemically-stable silica-coated magnetite nanoparticles. In a present work, we propose a method for preparing stable silica-coated magnetite nanoparticles, and study on effect of washing the magnetite on chemical stability of the silica-coated particles. This work also studies the immobilization of proteins on the silica-coated magnetite (SCM) particles mainly toward the drug delivery and the separation of biochemical products.

EXPERIMENTAL

A colloid suspension of magnetite nanoparticles was prepared according to the Massart method[7]. An aqueous solution of 5 ml that contained HCl and $FeCl_2$ at an equal concentration of 2 M was mixed with 20 ml of $FeCl_3$ aqueous solution at 1 M. This mixture was added to 250 ml of NH_3 aqueous solution at 0.7 M, and stirred for 1 h to complete the reaction. Figure 1 shows X-ray diffraction (XRD) pattern and transmission electron microscopy (TEM) image of the nanoparticles. A sample for the XRD was obtained by placing a magnet under the vessel of the solution and removing supernatant liquid (unwashed magnetite). The particles had an average size of 15 nm. All peaks in Fig. 1 could be indexed to magnetite (JCPDS card No. 19-629). Figure 2 shows the room temperature magnetization for the magnetite nanoparticles. No coercivity was measured because of superparamagnetism based on nano-size of the magnetite particles. The magnetite particles were washed by repeated centrifugation at 20000 rpm for 30 min, removal of supernatant, addition of water and sonication several times. The particles separated by removal of supernatant were suspended in 30 ml of 1 M tetramethylammonium hydroxide (TMA) as a dispersion stabilizer, and then diluted with the deionized water to a total volume of 250 ml.

Stöber method with tetraethylorthosilicate (TEOS) was applied to silica-coat the magnetite particles. To 0.8 ml of the magnetite colloid was added water, ethanol, NH_3 aqueous

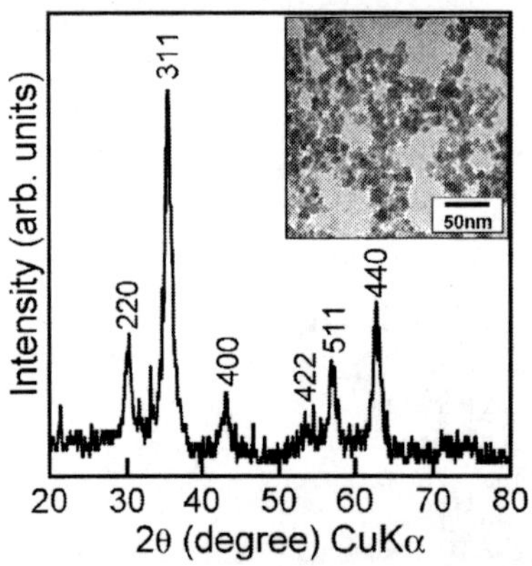

Figure 1 XRD pattern and TEM image of magnetite nanoparticles prepared by the Massart method.

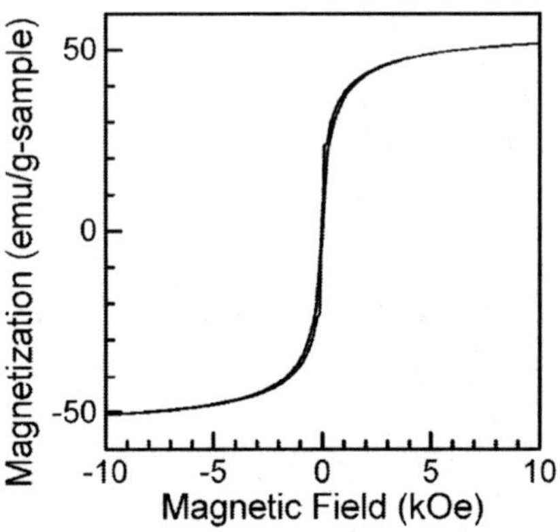

Figure 2 Room temperature magnetization vs. field data for magnetite nanoparticles.

solution and successively TEOS to adjust the total volume of the suspension to 100 ml. The concentrations of water and ammonia were 16.5 and 0.05 M, respectively. In a preliminary experiment, precipitation was produced at a TEOS concentration below 0.0045 M. Probably, this low TEOS concentration resulted in slow formation of silica shell, which provided agglomeration of the SCM particles prior to the shell formation. Accordingly, The TEOS concentration was varied in a range from 0.0067-0.015 M.

For efficient protein-immobilization, surface-modification of the SCM particles was performed with the use of APS and GA. The pH of the colloid of the SCM particles was adjusted to 6.2 by addition of aqueous NaOH solution. To 100 ml of the SCM was added 1 ml of aqueous APS solution under stirring at 100 rpm at room temperature (APS-SCM particles). After 24 h, the APS-SCM particles were washed three times by repeating centrifugation at 12000 rpm for 1 h, removal of supernatant, addition of water and sonication. After the final removal of supernatant, the particles were dispersed in 40 ml of 5 % GA/phosphate buffer solution (pH 7.4) under sonication, and aged at room temperature for 24 h (GA-APS-SCM particles).

Protein-immobilization was performed for the surface-modified particles. The GA-APS-SCM particles were three times washed by repeated centrifugation at 12000 rpm for 1 h, removal of supernatant, addition of water and sonication. After the final removal of supernatant, the particles were dispersed in 40 ml of phosphate buffer solution (pH 5.4) with sonication. Then, 10 ml of 0.12 g/l bovine serum albumin (BSA) solution was added to the 40 ml of the colloid, and stirred at 50 rpm at room temperature for 4 h.

The samples were characterized by TEM, XRD and vibrating sample magnetometry (VSM), electrophoretic light scattering (ELS) spectroscopy and ultra violet (UV) spectroscopy. TEM observation was performed (Zeiss LEO 912 and LEO 922) at accelerating voltages of 100 and 120 kV, respectively. The samples for TEM were prepared by dropping and evaporating the obtained colloid on top of a collodion-coated copper grid. XRD measurements were made with a Rigaku RU-200A diffractometer at 40 kV and 30 mA with CuKα radiation. VSM measurements of powder were made with a Toei VSM-5-15 magnetometer. ζ-potential of particles was measured with an Ohtsuka ELS-8000 electrophretic light scattering spectrophotometer. Aqueous HCl solution or aqueous NaOH solution was added to solution to vary pH of solution for ELS. To confirm BSA-immobilization on the particles, UV-VIS absorbance for supernatant of the colloid of the BSA-immobilized particles was measured with a Hitachi UV-3010 spectrophotometer. The supernatant was obtained by placing a magnet under the vessel of the solution.

RESULTS AND DISCUSSION

SCM particles

Figure 3 (a) shows TEM image of SCM particles prepared with the use of the unwashed magnetite particles. Darker and lighter parts of particles were magnetite particles and silica, respectively. Some particles were connected forming a neck-like structure. Probably, this neck-like structure was formed during preparation of the TEM samples, though precise reason for the formation of the structure is still unclear. Multiple cores were observed in the SCM particles formed. Figure 3 (b) shows TEM image of SCM particles that were twice washed by repeated centrifugation, removal of supernatant, addition of water and sonication. No silica shell was observed, which indicated that the silica shell was dissolved during the washing. In a preliminary experiment, silica particles prepared by Stöber method were dissolved in an aqueous ammonia solution. Ammonium ions could be present between the magnetite particles and the

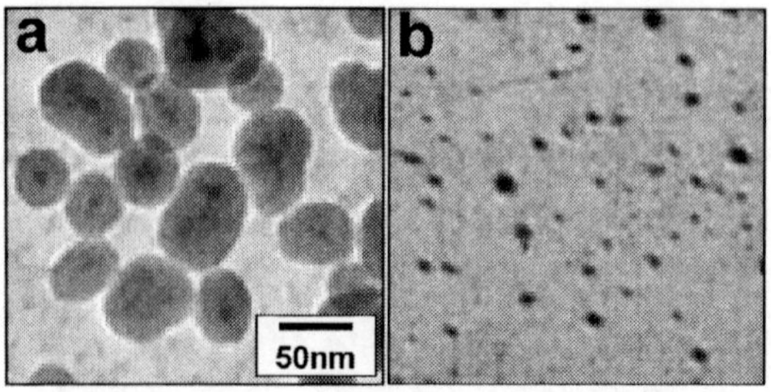

Figure 3 TEM images of SCM particles that were prepared with the use of the unwashed magnetite particles by silica-coating at 16.5 M H_2O, 0.05 M NH_3 and 0.0067 M TEOS. Samples (a) and (b) were Fe_3O_4@SiO_2 particles prior to and after washing with water, respectively.

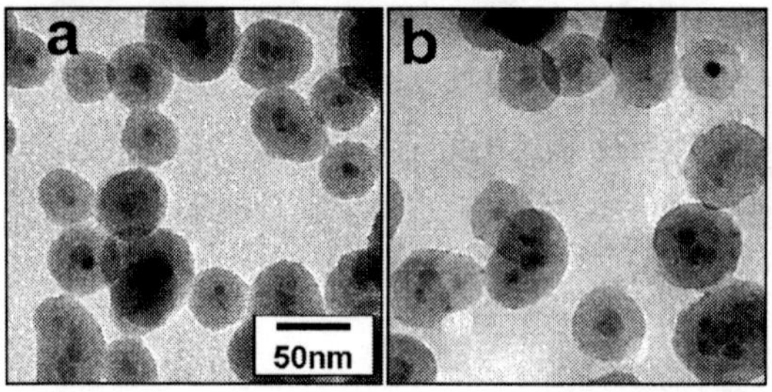

Figure 4 TEM images of SCM particles that were prepared with the use of the washed magnetite particles by silica-coating at 16.5 M H_2O, 0.05 M NH_3 and 0.0067 M TEOS. Samples (a) and (b) were magnetite@SiO_2 particles prior to and after washing with water, respectively.

silica shell in the SCM particles, since aqueous ammonia was used for producing the magnetite particles. The ammonium ions probably dissolved the silica shell from its inside because of the basicity of ammonia.

Figure 4 (a) shows TEM image of SCM particles prepared with the use of the washed magnetite particles. The SCM particles prepared similarly to the Figure 3 (a) had an average shell thickness of 12 nm. Figure 4 (b) shows TEM image of SCM particles after the same washing procedure as in the Figure 3 (b). No dissolution of the silica shell was observed in the present work, which indicated that ammonia ions that were adsorbed on the magnetite particle surface were removed by the washing process.

Figure 5 shows XRD pattern of SCM particles. A broad peak observed at ca. 23 degree was attributed to amorphous silica, and other sharp peaks were indexed to magnetite as well as the magnetite particles in Figure 1. Figure 6 shows room temperature magnetization as a function of magnetic field for the SCM particles. The saturation magnetization was around 5 emu/g-sample. Since the saturation magnetization was obtained for a unit weight of the sample that contained the magnetite particles and the silica shell, it was lower than the saturation magnetization of the magnetite particles. The measured saturation magnetization of 5 emu/g-sample was reasonable for the magnetite particle size of 15 nm and the silica shell thickness of 12 nm. These results indicated that the silica-coating process did not change the crystal structure or magnetic property of the magnetite in the particles.

Accordingly, the washed Fe_3O_4 particles were used for succeeding experiments.

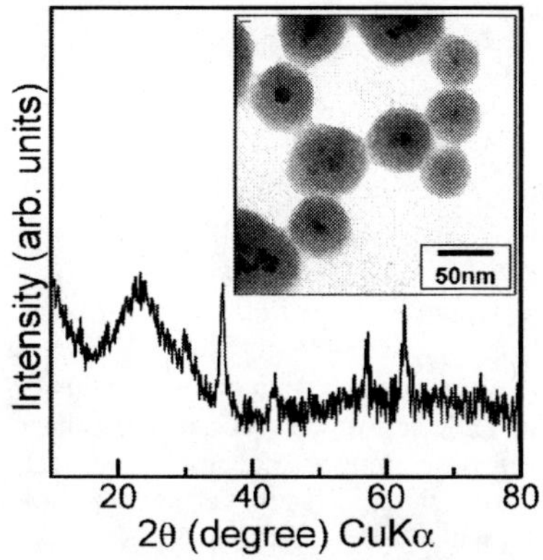

Figure 5 XRD pattern and TEM image of SCM particles prepared with the use of the washed magnetite particles. Silica-coating was performed at 16.5 M H_2O, 0.05 M NH_3 and 0.015 M TEOS.

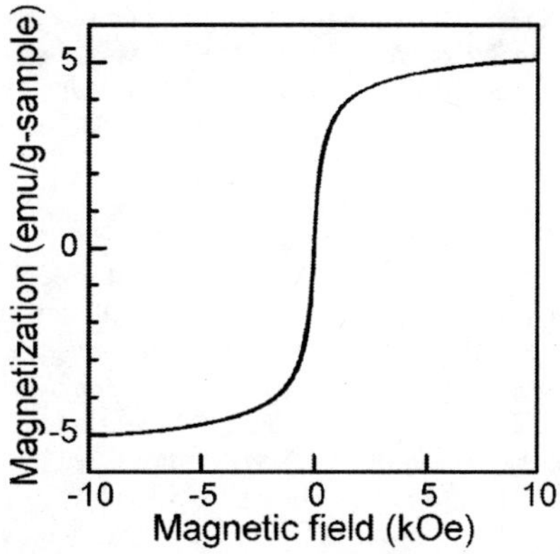

Figure 6 Room temperature magnetization as a function of magnetic field for SCM particles. Sample was the same as in Figure 5.

Surface-modification with APS

For the APS-SCM particles, no precipitation was observed by the naked eye. Figure 7 shows a TEM image of APS-SCM particles. Sticking among the particles probably took place because of strong irradiation of electron beam at high magnification. The APS-SCM particles appeared similar to the SCM particles shown in the inset of Figure 5, which indicated that the particles were dispersively stable and mechanically undamaged during the surface-modification with APS.

Figure 8 shows ζ-potentials as a function of pH for the SCM particles and the APS-SCM

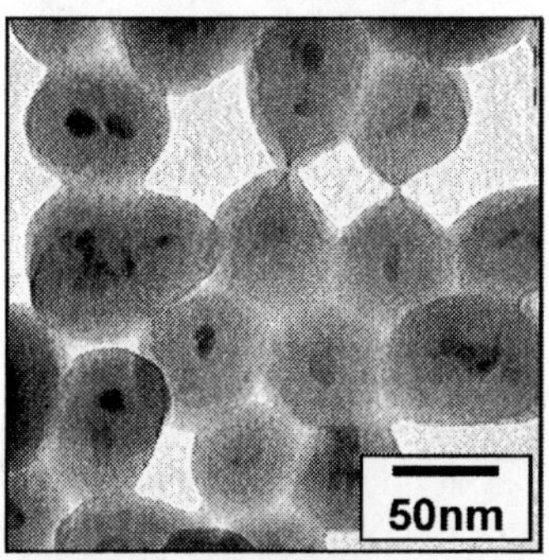

Figure 7 TEM images of APS-SCM particles. SCM particles in Figure 5 were used for the surface-modification.

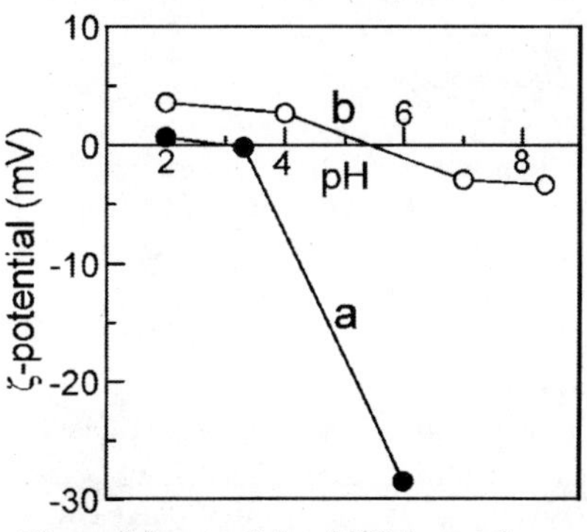

Figure 8 ζ-potentials of SCM particles as a function of pH. Samples (a) and (b) were the SCM particles and the APS-SCM particles, respectively.

particles. The SCM particles revealed iso-electric points of ca. 3.2, as a result of silica surface. The isoelectric point of the APS-SCM particles was ca. 5.5, which was probably derived from amino group of APS on surface of the particles. Accordingly, this result supported that the particle surface was successfully modified with APS.

Protein-immobilization

For the BSA-immobilized particles, no precipitation was observed by the naked eye. Figure 9 shows absorption spectrum of supernatant of colloid solution of BSA-immobilized particles. Spectra for aqueous BSA solution and supernatant of SCM particle colloid mixed with BSA are also shown. A peak observed at ca. 280 nm was due to BSA. Absorbance of the BSA peak for the SCM particle colloid mixed with BSA was lower than that of the aqueous BSA solution, which indicated that BSA was immobilized on the SCM particle surface. Absorbance of the BSA peak for the BSA-immobilized particles was low, in comparison with the one of the supernatant of the SCM particle colloid mixed with BSA, which indicated that efficient protein-immobilization was achieved by the surface-modification with the use of APS and GA. Estimation with absorbance difference revealed that ca. 30 % of initial amount of BSA was immobilized in the examined conditions.

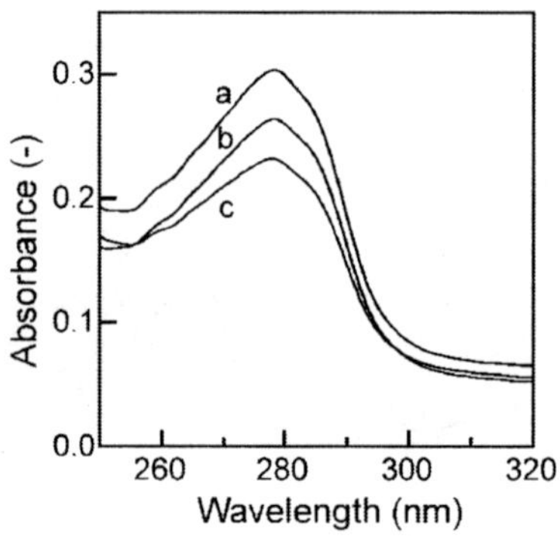

Figure 9 Absorption spectra of (a) aqueous BSA solution, (b) supernatant of SCM particle colloid mixed with BSA and (c) supernatant of colloid solution of BSA-immobilized particles.

The XRD pattern and TEM image of the BSA-immobilized particles were almost similar to the SCM particles shown in Figures 5 and 7. The saturation magnetization was around 5 emu/g-sample, which was almost the same as that of the SCM particles shown in Figure 6. These results indicated that the particles were dispersively stable and mechanically undamaged, and their crystal structure of magnetite was maintained even after the surface-modification and the protein-immobilization.

CONCLUSION

In the present work, we prepared colloidally and chemically stabilized magnetite nanoparticles surrounded by a silica shell by intensive removal of ammonia from the magnetite particle surface. The SCM particles were colloidally stable and chemically undamaged during surface-modification with APS and GA and protein-immobilization with BSA. Measurement of optical absorption showed that the SCM particles were successfully immobilized with BSA.

ACKNOWLEDGEMENTS

This research was partially supported by a Grant-in-Aid for Science Research (No. 15656163) and by the COE project, Giant Molecules and Complex Systems from the Ministry of Education, Culture, Sports, Science and Technology of Japan. The authors express their thanks to Mr. Y. Shibata in Institute of Multidisciplinary Research for Advanced Materials from Tohoku University for his assistance during TEM observations.

* Present address: Department of Biomolecular Functional Engineering, College of Engineering, Ibaraki University, 4-12-1 Naka-narusawa-cho, Hitachi, Ibaraki, Japan, 316-8511

REFERENCES

[1]X. Liu, J. Xing, Y. Guan, G. Shan, and H. Liu, "Synthesis of Amino-Silane Modified Superparamagnetic Silica Supports and Their Use for Protein Immobilization," *Colloids Surf. A*, **238**, 127-31 (2004).

[2]Y. Sun, L. Duan, Z. Guo, Y. DuanMu, M. Ma, L. Xu, Y. Zhang, and N. Gu, "An Improved Way to Prepare Superparamagnetic Magnetite-Silica Core-Shell Nanoparticles for Possible Biological Application," *J. Magn. Magn. Mater.*, **285**, 65-70 (2005).

[3]M. A. Correa-Duarte, M. Giersig, N. A. Kotov, and L. M. Liz-Marzán, "Control of Packing Order of Self-Assembled Monolayers of Magnetite Nanoparticles with and without SiO_2 Coating by Microwave Irradiation," *Langmuir*, **14**, 6430-5 (1998).

[4]Y. Lu, Y. Yin, B. T. Mayers, and Y. Xia, "Modifying the Surface Properties of Superparamagnetic Iron Oxide Nanoparticles through A Sol-Gel Approach," *Nano Lett.*, **2**, 183-6 (2002).

[5]Y.-H. Deng, C.-C. Wang, J.-H. Hu, W.-L. Yang, and S.-K. Fu, "Investigation of Formation of Silica-Coated Magnetite Nanoparticles via Sol-Gel Approach," *Colloids Surf. A*, **262**, 87-93 (2005).

[6]M. Iijima, Y. Yonemochi, M. Kimata, M. Hasegawa, M. Tsukada, and H. Kamiya, "Preparation of Agglomeration-Free Hematite Particles Coated with Silica and Their Reduction Behavior in Hydrogen," *J. Colloid Interface Sci.*, **287**, 526-33 (2005).

[7]R. Massart, "Preparation of Aqueous Magnetic Liquids in Alkaline and Acidic Media," *IEEE Trans. Magn.*, **MAG-17**, 1247-8 (1981).

NANO-PARTICLES AND POWDERS

FABRICATION OF TEXTURED ALUMINA-MULLITE-SILICON CARBIDE NANO-COMPOSITES

Yoshio Sakka, Sho Saito, Atsushi Honda and Tohru S. Suzuki
Nano Ceramics Center, National Institute for Materials Science,
1-2-1 Sengen, Tsukuba, Ibaraki 305-0047, JAPAN

ABSTRACT

The controlled development of microstructure such as nanocomposite or texture is a way for effectively improving physical and mechanical properties. Al_2O_3-mullite-SiC nanocomposites were prepared by partial oxidation of the Al_2O_3-SiC green body followed by reaction sintering. Recently, we reported that the crystal orientation of alumina can be controlled in a high magnetic field due to its crystal magnetic anisotropy. By the combination of both techniques we have demonstrated that textured Al_2O_3-mullite-SiC nanocomposites can be fabricated by slip casting of mixed suspensions of Al_2O_3 and SiC powders in a high magnetic field followed by partial oxidation and reaction sintering of the Al_2O_3 and SiC. The sintered density depended on the degree of dispersion and oxidation of SiC. Dense nanocomposites were prepared by suitable oxidation and reaction sintering at 1600 without pressure. The three-point bend strength was observed approximately 800 MPa for the nanocomposites. The strength depended on the textured plane and amount of mullite phase, and increased using the suspension treated by beads-milling due to the de-agglomeration of SiC powders.

INTRODUCTION

The controlled development of microstructure such as nanocomposite or texture is a way for effectively improving physical and mechanical properties. Nanocomposites, in which nanometer-sized second-phase particles are dispersed in a micron or sub-micron sized ceramic matrix, are known to show enhanced mechanical properties [1]. The controlled development of a textured microstructure is also known to improve the mechanical properties [2,3].

Recently, we reported that the crystal orientation of alumina can be controlled in a high magnetic field due to its crystal magnetic anisotropy [4,5]. Although the magnetic susceptibility of feeble magnetic materials is very small, there is the possibility to control the crystal orientation using the high magnetic field of a superconducting magnet due to the magnetic anisotropy. The principle of the process is that a crystal with an anisotropic magnetic susceptibility will rotate to an angle minimizing the system energy when placed in a magnetic field. The magnetic torque T attributed to the interaction between the anisotropic susceptibility and a magnetic field is estimated by Eq. (1) [5].

$$T = -\Delta\chi V B^2 \sin 2\theta / 2\mu_0 \quad (1)$$

where $\Delta\chi$, $(= |\chi_{a,b} - \chi_c|)$ is the anisotropy of the magnetic susceptibilities which are measured for the a,b-axis ($\chi_{a,b}$) and c-axis (χ_c), V is the volume of the materials, B is the applied magnetic field, θ is the angle between an easy magnetization axis in a crystal and imposed magnetic field direction, and μ_0 is the magnetic permeability in a vacuum. This magnetic torque is the driving force for the magnetic alignment. However, since feeble magnetic materials in an applied high magnetic field have a low magnetic torque, to align each particle it is necessary to disperse the powder in the suspension. Therefore, preparing a well-dispersed suspension is very important from the standpoint of a low viscosity and non-agglomerated particles for rotation. It is also very important to obtain excellent mechanical properties by decreasing the large pores due to agglomeration.

Alumina-based ceramics are known to show good mechanical properties, but their strength is low at high temperatures [6,7]. Mullite is considered to be a candidate ceramic for high temperature structural applications because of its excellent high temperature properties [8,9]. We have reported that Al_2O_3-mullite-SiC nanocomposites can be prepared by partial oxidation–reaction sintering of the Al_2O_3-SiC green body [10].

Our objective is to prepare nanocomposites that have better mechanical properties than the mullite single phase, and a flexural strength at high temperature better than Al_2O_3 and SiC. In this study, a well-dispersed suspension of SiC and Al_2O_3 was prepared, and the fabrication of textured Al_2O_3-mullite-SiC nanocomposites was fabricated by slip casting in a high magnetic field.

EXPERIMENTAL

High purity α-Al_2O_3 powder (Taimei Chemical: TM-DAR) with a mean particle diameter of 0.15μm and β-SiC powder (Ibiden Co., Ltd.) with a mean particle diameter of 0.27μm were used as the starting powders. Colloidal aqueous suspensions with a solid content of 30vol% were prepared. [10] The SiC powder content of each colloidal suspension was fixed at 5, 10 and 15vol%, and each one after consolidation was labeled T-5SiC, T-10SiC and T-15SiC, respectively. The powder mixtures were dispersed with an appropriate amount of polyelectrolyte (Polyammoniumacrylate, A-6114, Toagosei Co.). After ultrasonic vibration was applied to facilitate dispersion of the powder agglomerates[11], slip casting was done with and without a high magnetic field (12 T). In some cases, bead-milling using zirconia beads of 0.05 mm in diameter was conducted to deagglomerate the SiC fine powders. [12]

Figure 1 shows a schematic of partial oxidation and reaction sintering. The partial oxidation treatment was conducted in an alumina crucible in a stream of O_2 at 1200 and 1300 for 1h. Reaction sintering was then carried out in a stream of Ar at 1600 for 2h. The partial oxidation treatment and reaction sintering were continuously conducted. The partial oxidation condition was confirmed in advance by thermogravimetric analysis.

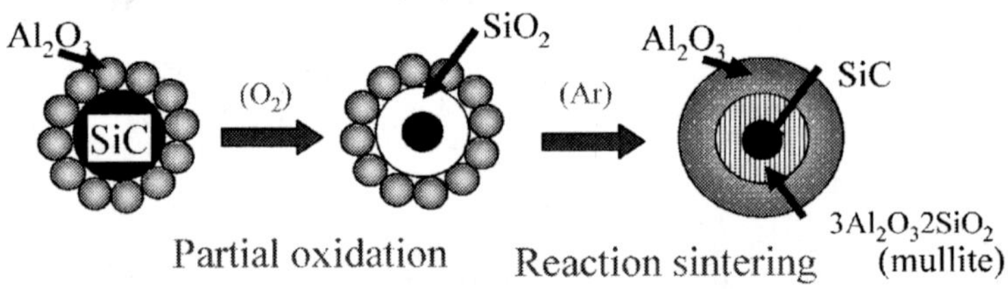

$$SiC + 2O_2 \rightarrow SiO_2 + CO_2$$

$$3Al_2O_3 + 2SiO_2 \rightarrow 3Al_2 \cdot 2SiO_2$$

Figure 1 Schematic diagram for partial oxidation and reaction sintering.

The phases and texture analyses were done by analyzing the X-ray diffraction patterns (XRD). The resulting microstructures were observed by scanning electron microscopy (SEM) and transmission electron microscopy (TEM). The samples were cut and ground using a diamond wheel into 4 mm×3 mm× 20 mm bars for the three-point bending test. The mechanical properties of the samples were evaluated by the three-point bending with a span 16.1 mm.

RESULTS AND DISCUSSION

Figure 2 shows X-ray diffraction patterns of 15 vol% SiC after reaction sintering, where the oxidation conditions at 1300 for 1, 3 and 5 h. It is seen that SiO_2 and Al_2O_3 react to form mullite and a small amount of SiC remain . By TEM observation, the existence of dispersed nano-sized SiC particles was confirmed in the matrix. These results indicate that a dense Al_2O_3-mllite-SiC nanocomposite without SiO_2 was fabricated by this method.

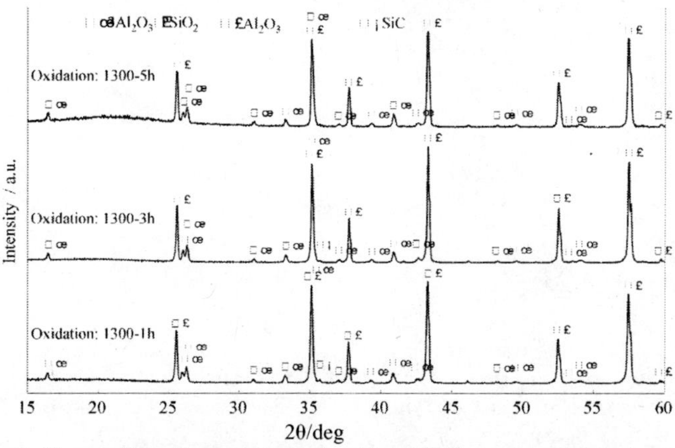

Figure 2 X-ray diffraction patterns for 15vol%SiC system sintered at 1600 (Oxidation conditions at 1300 for 1, 3 and 5h).

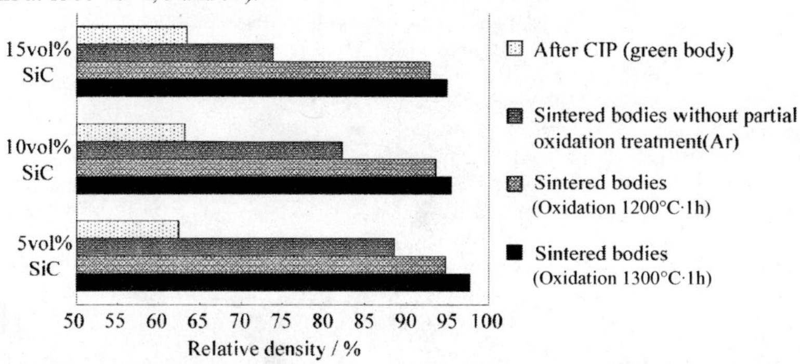

Figure 3 Relative densities of green bodies after CIP and sintered bodies with oxidation treatment and without oxidation treatment.

Figure 3 shows the relative densities of the green bodies after CIP for the sintered bodies with and without oxidation treatment, followed by sintering in an Ar atmosphere at 1600 for 2h. It was confirmed that the relative density of the sintered bodies significantly increased by the oxidation treatment. The sintered density increased with a high oxidation temperature and with a small SiC content. When a large quantity of SiO_2 was generated at a high oxidation temperature, densification was enhanced by the viscous sintering of the surface SiO_2 of SiC and Al_2O_3 [10].

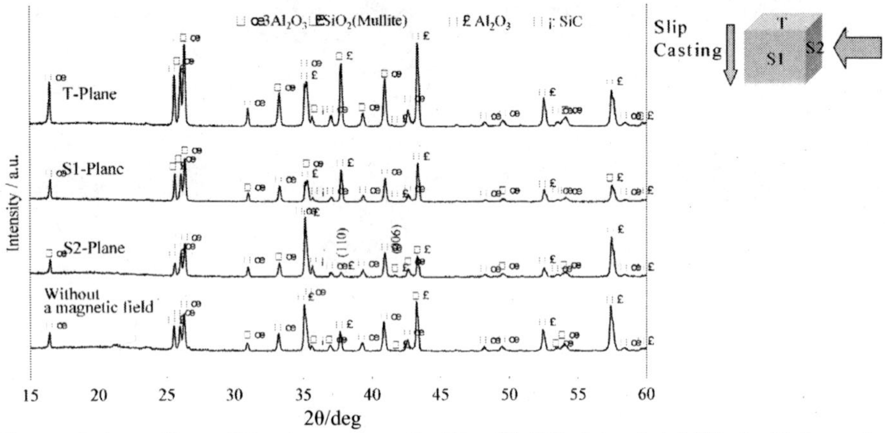

Figures 4 shows X-ray diffraction patterns for 15 vol% SiC sintered at 1600 for 2h in an Ar atmosphere after oxidation treatment at 1300 for 1h in an O_2 atmosphere, where slip casting was conducted without magnetic field and in 12 T.

Figures 4 shows X-ray diffraction patterns for 15 vol% SiC sintered at 1600 for 2h after oxidation treatment. In the T-plane and S1-plane which are parallel to the direction of magnetic field, the intensities of (110) are high while the intensities of (006) are very low. In contrast, in the S2 plane perpendicular to the high magnetic field direction, the intensity of (110) is low and the intensity of (006) is higher than that of T and S1. These results indicate that the c-axis of the alumina crystal was aligned in the high magnetic field direction. The degree of crystalline texture was determined from intensity ratio of (006) and (110) as follows.

$$P = I_{006} / (I_{006} + I_{110}) \quad (2).$$

Figure 5 illustrates the degree of crystalline texture for the specimens prepared by slip casting at 12T after reaction sintering under different oxidation

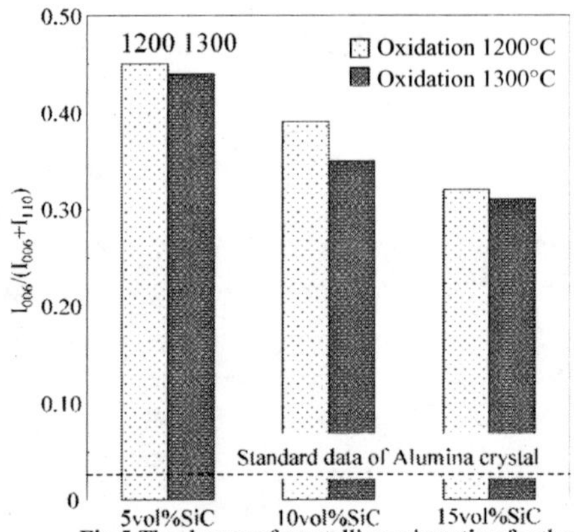

Fig.5 The degree of crystalline orientation for the specimens prepared by slip casting at 12T followed by reaction sintering.

conditions. The degree of crystalline texture significantly increased when the SiC content was low and the oxidation treatment temperature was 1200 . It is seen that the degree of crystalline texture depended on the Al_2O_3 content but not as high in comparison with the single phase of Al_2O_3[4,5].

Figure 6 shows the bend strength for the textured nanocomposite with different directions. From the SEM observations, it is seen that inter-granular fracture was observed in the fracture form but that no significant difference between the fracture surfaces was observed for the different directions of the applied magnetic field. The effect of the alumina orientation on the fracture strength was not clearly observed. This is because the temperature of the reaction sintering was 1600 and the grain growth of alumina was not significant in this experiment. The fracture strengths decreased with increased SiC (i.e., mullite) contents. The decrease in the fracture strength was due to the content of mullite in sintered bodies because the fracture strength of mullite is not very high at room temperature (300-400MPa). However, the absolute value of the bend strength was low. To clarify the reason, EDS mapping images were observed for Si and Al. Agglomeration of Si was observed, which indicates that ultrasonic irradiation was not effective for the de-agglomeration of SiC powders.

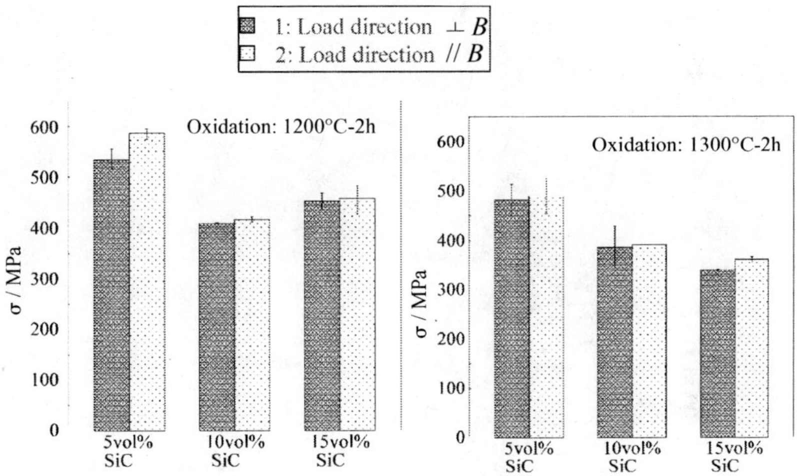

Figure 6 Three-point bend strength for bodies sintered of nanocomposites.

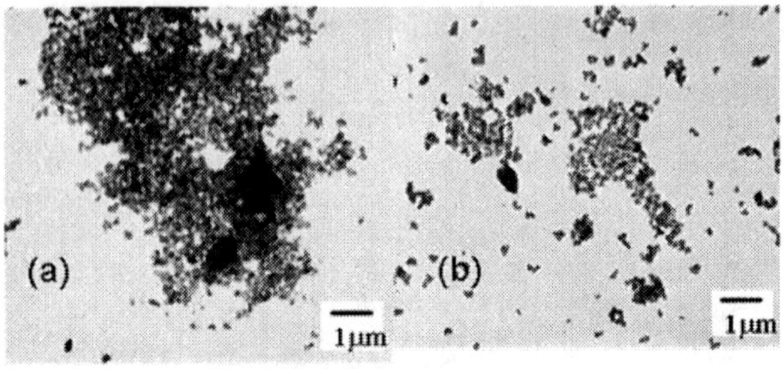

Figure 7 TEM images of 10 vol% SiC system before (a) and after (b) bead milling.

Recently it is clarified that beads-milling using small zirconia beads of 0.05 mm in diameter is effective for the deagglomeration of fine powders [12]. Figure 7 shows that the beads milling is effective for the deagglomeration of the present suspension system. Using the beads-milling treated suspensions, the fracture strength increased to above 800 MPa for 10 vol% SiC systems as is shown in Figure 8. The bending strength of the plane where load direction parallel to applied magnetic direction is larger than that load direction is perpendicular to magnetic field direction. Similar trend was also observed for undoped alumina ceramics.

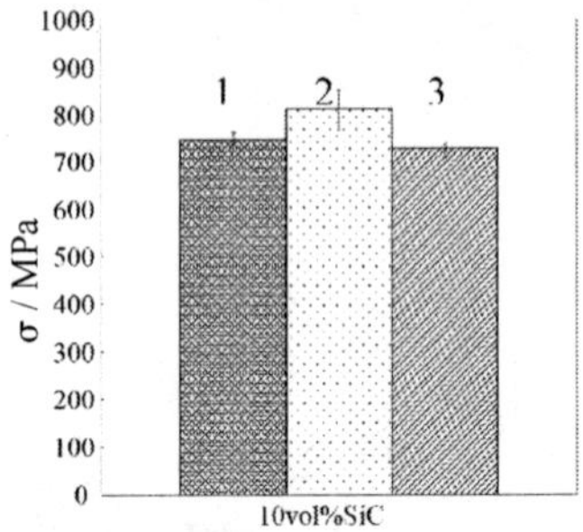

Figure 8 Three-bending strength of reaction-sintered 10 vol% SiC systems after milling treatment. (1: Load direction ⊥ *B*. 2: Load direction // *B*, 3: Without magnetic field)

ACKNOWLEDGMENTS

This work was supported in part by the Budget for Nuclear Research and by grant-in-Aid for Scientific Research from the Ministry of Education, Culture, Sports, Science and Technology.

REFERENCES

[1]K. Niihara, New-design concept of structural ceramics-ceramic nanocomposites-, *J. Ceram. Soc. Jpn.*, **99**, 974-982 (1991).

[2]Y. Yoshizawa, K. Hirao, and S. Kanzaki, Mechanical properties of textured alumina made by high-temperature deformation, *J. Am. Ceram. Soc.*, **87**, 2147-2149 (2004).

[3]T. S. Suzuki and Y. Sakka, Preparation of oriented bulk 5 wt% Y_2O_3-AlN ceramics by slip casting in a high magnetic field and sintering, *Scripta Mater.*, **52**, 583-586 (2005).

[4]T.S. Suzuki, Y. Sakka and K. Kitazawa, Orientation amplification of alumina by colloidal filtration in a strong magnetic field and sintering, *Adv. Eng. Mater.*,**3,** 490-492 (2001).

[5]Y. Sakka and T. S. Suzuki, Textured development of feeble magnetic ceramics by colloidal processing under high magnetic field, *J. Ceram. Soc. Jpn.*, **113**, 26-36 (2005).

[6]G. A. Dogotsi, Fracture toughness of ceramics and ceramic composites, *Ceram. Inter.*, **29** 777-784(2003).

[7]Y. Yoshizawa, K. Hirao, S. Kanzaki, Fabrication of low cost fine-grained alumina powders by seeding for high performance sintered bodies, J. Eur. Ceram. Soc.,**24**, 325-330 (2004).
[8]M. I. Osendi and C. Baudin, Mechanical properties of mullite materials, *J. Europ. Ceram. Soc.*, **16**, 217-224 (1996).
[9]H. Ohira, M. G. M. U. Ismail, Y. Yamamoto, T. Akiba and S. Somiya, Mechanical properties of high purity mullite at elevated temperatures, *J. Europ. Ceram. Soc.*, **16**, 225-229 (1996).
[10]Y. Sakka, D. D. Bidinger and I. A. Aksay, Processing of silicon carbide-mullite-alumina nanocomposite, *J. Am. Ceram. Soc.*, **78**, 479-486 (1995).
[11]T. S. Suzuki, Y. Sakka, K. Nakano and K. Hiraga, Effect of ultrasonication on the microstructure and tensile elongation of zirconia-dispersed alumina ceramics prepared by colloidal process, *J. Am. Ceram. Soc.*, **84**, 2132-2134 (2001).
[12]M. Inkyo and T. Tahira, Dispersion of Agglomerated Nanoparticles by Fine Beads Mill, *J. Soc. Powder Powder Technol. Japan*, **41**, 578-585 (2004). (in Japanese)

COLLOIDAL CONSOLIDATION OF MIXED POWDERS OF THE ALUMINA / INDIUM TIN OXIDE SYSTEM

Naoki Matsunaga[1], Shinji Ueno[1], Yosuke Tanaka[2] and Yoshihiro Hirata[2]
[1]Department of Applied Chemistry and Chemical Engineering, Kagoshima University, 1-21-40 Korimoto, Kagoshima 890-0065, Japan

[2]Department of Advanced Nanostructured Materials Science and Technology, Kagoshima University, 1-21-40 Korimoto, Kagoshima 890-0065, Japan

ABSTRACT

The applied pressure and suspension height during consolidation of an aqueous suspension containing submicrometer-sized alumina and ITO (10 mass% tin oxide-90 mass% indium oxide) with or without polyacrylic ammonium (dispersant, PAA) were continuously recorded using a pressure filtration apparatus. The isoelectric points of Al_2O_3 and ITO powders without PAA were at pH 7.8 and 3.2, respectively, and the addition of PAA shifted the both zeta potentials to negative values at a similar pH. The packing density of 70 vol% Al_2O_3-30 vol% ITO system without PAA after the consolidation at 19 MPa of applied pressure was in the range from 52 to 60 % and decreased with increasing suspension pH. On the other hand, the packing density of the mixed particles with PAA of 0.47 mg/m^2 (powder) was independent of the suspension pH and 52-55 % in the pH range from 3 to 10. The consolidation energy of the colloidal particles was measured to correlate to the dispersion characteristics.

INTRODUCTION

The interaction between particles in a suspension depends on particle size, concentration of particles and surface charge of particles. Recently the wet process has been applied to two-component powder systems such as Al_2O_3-ZrO_2[1] or Si_3N_4-SiC[2] to achieve high mechanical or electrical properties. When the sedimentation rates of the mixed particles are significantly different, segregation of particles occurs during the consolidation. This results in the inhomogeneous microstructure. In our previous studies, the segregation problem was not observed by adding a suitable polyelectrolyte dispersant (polyacrylic ammonium, PAA) to the suspension of Al_2O_3-ITO system[3] or by controlling the electrostatic interaction between colloidal SiC and Al_2O_3 particles[4]. Moreover, a new pressure filtration apparatus, where the membrane filter is attached to the bottom of piston in a closed cylinder and the dehydrated solution is drained toward the upper direction, was developed by our group to eliminate the gravity effect during the consolidation. This apparatus was applied to form powder compacts of SiC[5], Al_2O_3[6] and hydroxyapatite[7]. In this paper, the interaction between particles and consolidation behavior of the 70 vol% Al_2O_3-30 vol% ITO system with and without PAA were studied using the developed pressure filtration apparatus.

EXPERIMENTAL

A high purity α-Al_2O_3 powder of median size 150 nm (Al_2O_3 > 99.99 mass%, specific surface area 10.5 m^2/g, density 3.99 g/cm^3, Sumitomo Chemical Industry Co., Japan) and indium tin oxide powder of median size 200 nm (ITO, 90 mass% In_2O_3 - 10 mass% SnO_2, specific surface area 4.42 m^2/g, density 7.15 g/cm^3, Sumitomo Chemical Industry Co., Japan) were used as starting materials. The as-received ITO powder was milled by 3 mol% Y_2O_3-stabilized ZrO_2 ball of 3 mm diameter for 24 h to break the particle aggregate. Polyacrylic ammonium (PAA, molecular weight 10000, Dai-ichi Kogyo Seiyaku Co., Ltd., Japan) was also used as a polyelectrolyte dispersant. Al_2O_3 and ITO powders were mixed at pH

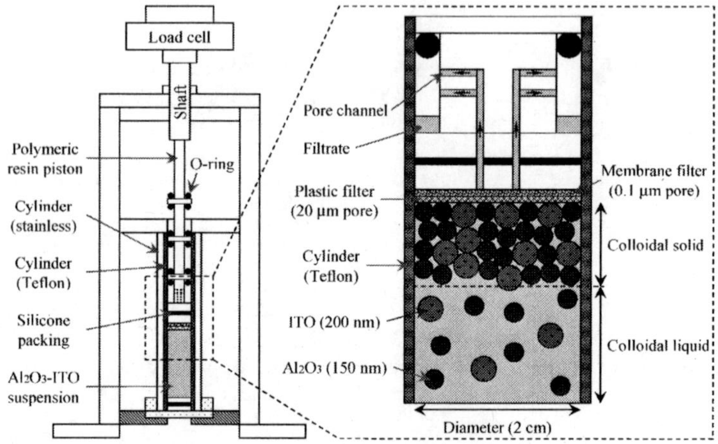

Figure 1. Schematic illustration of pressure filtration apparatus recoding the applied load and suspension height. The filtrate overflows in the open space between the piston and cylinder through the plastic and membrane filters attached to the bottom of the piston.

3.0 (near the isoelectric point of ITO), pH 6.0 (heterocoagulation region between Al_2O_3 and ITO), pH 8.0 (near the isoelectric point of Al_2O_3) and pH 10.0 (co-dispersion region of Al_2O_3 and ITO)[3, 4]. The dissociation of PAA starts at about pH 3.0 and reaches 100 % above pH 9.0[8]. The added PAA is adsorbed on the particle surfaces as a neutral polymer at a low pH and negatively charged polymer at a high pH. The adsorption sites for dissociated PAA are positively charged $AlOH_2^+$ or $InOH_2^+$ sites. A 7 vol% solid suspension of the 70 vol% Al_2O_3-30 vol% ITO system with or without PAA (0.47 mg/m^2-powder surface) was prepared. The apparent viscosity of the mixed powder suspension was measured by cone- and plate-type viscometer (Model EMD type, Tokimec Inc., Japan). The applied pressure and suspension height during consolidation of the mixed powder suspension were continuously recorded using a developed pressure filtration apparatus shown in Fig. 1[5-7]. A plastic filter of pore diameter 20 μm and two sheets of membrane filter of 0.1 μm pore diameter were attached at the bottom of the polymeric resin piston. The piston was loaded at the crosshead speed of 0.2 mm/min and the filtrate was drained to the upper direction. Thus the obtained green compact was dried at 100 °C for 24 h and then calcined at 700 °C for 1 h in air. The density of calcined compact was measured by the Archimedes method with distilled water.

RESUTLS AND DISCUSSION

Interaction energy between Al_2O_3 and ITO particles

Figure 2 shows the zeta potentials of as-received Al_2O_3 and milled ITO powders with and without PAA[3]. This figure clearly shows that the Al_2O_3 and ITO particles without PAA had the isoelectric points at pH 7.8 and 3.2, respectively. The addition of PAA (0.47 mg/m^2) shifted the zeta potentials of both the powders to negative values and increased the similarity of the surface characteristics. To understand the interaction between colloidal particles in the suspension, van der Waals attraction energy (E_a) was calculated using Eq. (1)[9],

$$E_a = -\frac{Aa_1a_2}{6(a_1+a_2)H} \qquad (1)$$

where A is the Hamaker constant, a_1 and a_2 the radius of particles 1 and 2, respectively, and H the distance between particles 1 and 2. The Hamaker constant for the interaction between Al_2O_3 and Al_2O_3 particles is reported to be 3.67 × 10^{-20} J (A_{awa}, a: Al_2O_3, w: water)[10]. The Hamaker constant for the interaction between ITO and ITO particles is 2.26 × 10^{-20} J (A_{iwi}, i: ITO)[11]. The Hamaker constant for the interaction between Al_2O_3 and ITO particles was calculated to be $A_{awi} = (A_{awa}\ A_{iwi})^{1/2} = 2.88 \times 10^{-20}$ J[11]. On the other hand, Equation (2) corresponds to the repulsion energy (E_r) by electric double layer between charged particles 1 and 2[9],

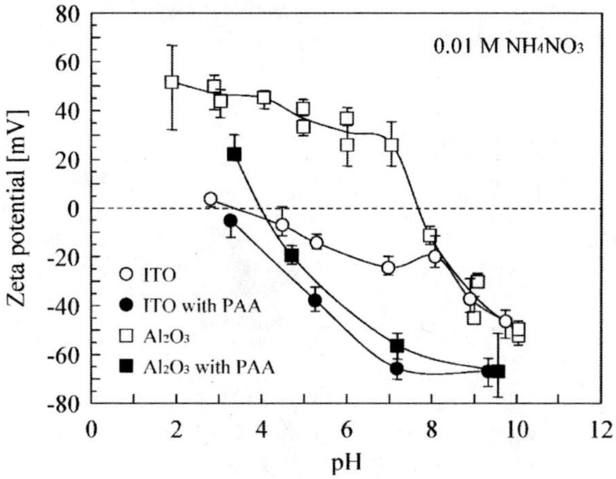

Figure 2. Zeta potentials of ITO and Al_2O_3 particles with and without PAA as a function of pH[3].

$$E_r = \frac{\pi\varepsilon\varepsilon_0 a_1 a_2}{a_1 + a_2}(\varphi_1^2 + \varphi_2^2)\left[\frac{2\varphi_1\varphi_2}{\varphi_1^2 + \varphi_2^2}\ln\left\{\frac{1+\exp(-\kappa H)}{1-\exp(-\kappa H)}\right\} + \ln\{1+\exp(-2\kappa H)\}\right] \tag{2}$$

where ε is the relative dielectric constant (78.3 for H_2O at 25 °C), ε_0 the permittivity of vacuum (8.8542 × 10^{-12} F/m), φ_1 and φ_2 the surface potential (approximated by the zeta potential) of charged particles and $1/\kappa$ the double layer thickness. The double layer thickness was calculated by Eq. (3)[9],

$$\frac{1}{\kappa} = \left(\frac{2e^2 N_A B Z^2}{\varepsilon\varepsilon_0 kT}\right)^{-\frac{1}{2}} \tag{3}$$

where e is the charge on electron (1.602 × 10^{-19} C), N_A the Avogadro number (6.022 × 10^{23} / mol), B the concentration of electrolyte, Z the charge number of electrolyte (assumed to be +1), k the Boltzmann constant (1.381 × 10^{-23} J/K) and T the temperature (assumed to be 298 K). The double layer thickness was calculated to be as follows: 9.6 nm for B = 10^{-3} M, 30 nm for B = 10^{-4} M, 96 nm for B = 10^{-5} M and 304 nm for B = 10^{-6} M. Since the suspension was prepared in the pH range from 3 to 10, the B value for H^+ or OH^- ion based on the ionic product of water ($K_w = [H^+][OH^-] = 10^{-14}$ M^2) is approximated to be 10^{-3} M for pH 3, 10^{-6} M for pH 6 and pH 8, and 10^{-4} M for pH 10.

Figure 3 shows the interaction energy, $E_a + E_r$, for the 70 vol% Al_2O_3-30 vol% ITO suspension. In the suspension at pH 3, a small repulsive energy was calculated between Al_2O_3 and Al_2O_3 particles. ITO particles make particle clusters with ITO or Al_2O_3 particles because of the attractive interaction. In the suspension at pH 6, Al_2O_3-ITO particle clusters are formed by the electrostatic attraction. In the suspension at pH 8, a small repulsive energy was calculated for ITO-ITO interaction. Most of the particles have a tendency to form particle agglomerates. On the other hand, relatively high repulsive energy was calculated for the Al_2O_3-Al_2O_3, Al_2O_3-ITO and ITO-ITO interaction at pH 10, suggesting a high dispersion state of colloidal particles.

Figure 4 shows the apparent viscosity of a 7 vol% aqueous suspension of 70 vol%

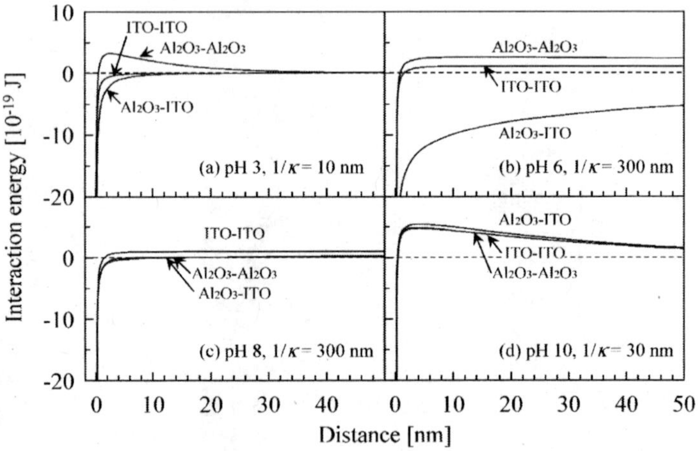

Figure 3. Calculated interaction energy for 70 vol% Al_2O_3-30 vol% ITO suspension. Electric double layer (1/κ) was assumed at 10 nm for pH 3, 300 nm for pH 6 and 8, and 30 nm for pH 10, respectively.

Al_2O_3-30 vol% ITO system with and without PAA. The apparent viscosity of the Al_2O_3-ITO system without PAA was lower for the acidic suspensions at pH 3 and 6 than for the basic suspensions at pH 8 and 10[3, 6]. A similar tendency was also observed in the alumina suspensions. The above comparison indicates that the interaction between Al_2O_3 and Al_2O_3 particles dominates the rheology of the Al_2O_3-ITO system. The decrease of repulsive energy between Al_2O_3 and Al_2O_3 at pH 8 accelerates the formation of particle clusters and increases the viscosity. However, it is difficult to explain the high viscosity at pH 10 from the corresponding interaction energy shown in Fig. 3. A possible explanation for this result is the formation of particle agglomerates through dehydration and condensation between hydrated

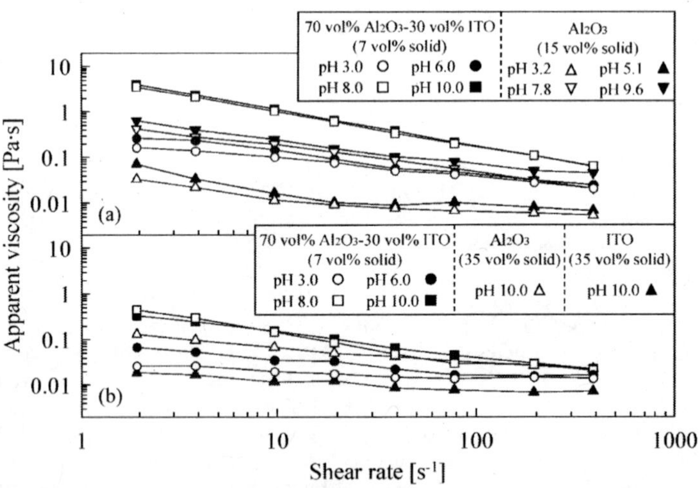

Figure 4. Apparent viscosity of a 7 vol% aqueous suspension of 70 vol% Al_2O_3 - 30 vol% ITO system (a) without and (b) with PAA as a function of shear rate.

surfaces[12]: $Al(OH)_3 + Al(OH)_3 \rightarrow Al(OH)_2 - O - Al(OH)_2 + H_2O$. The remaining hydroxyl reacts with NH_4OH solution to form the negatively charged AlO^- sites. The apparent viscosity of the suspension with PAA (Fig. 4b) at pH 3-10 decreased to 1/10 of the viscosity without PAA. This result reflects the increased dispersion of alumina and ITO particles due to the steric stabilization effect at a low pH or electrosteric stabilization effect at a high pH by the adsorbed PAA.

Pressure filtration of Al_2O_3 - ITO suspension

Figure 5 summarizes (a) the density of powder compact after calcination and (b) consolidation energy (E_c) of 1 cm^3 particles at 19 MPa of applied pressure. The area under the applied pressure-filtrate volume curve corresponds to the energy for the consolidation of the colloidal suspensions. The density of the Al_2O_3-ITO compact without PAA became higher at a lower pH of the suspension. This result was in accordance with the pH dependence of the density of 70 vol% Al_2O_3-30 vol% ITO compact consolidated through gypsum mold. The density of monolithic alumina compact consolidated using the same pressure filtration apparatus showed a similar pH dependence of the suspension. The high packing density of the Al_2O_3 or 70 vol% Al_2O_3-30 vol% ITO compact at a low pH of the suspension is due to the high dispersion of colloidal particles. The E_c value of Al_2O_3-ITO or monolithic Al_2O_3 compact showed a tendency to decrease as the pH of the suspension increased. The elimination of the solution through the porous cake at a high pH needs no significant energy. On the other hand, the packing density of the Al_2O_3-ITO compact with PAA was almost independent of the suspension pH. The E_c value increased as the suspension pH became higher. As shown in Fig. 4 (b), the viscosity of the suspension with PAA was lower for the lower pH of the suspension. In the acidic suspensions, the adsorption of neutral PAA on the particles reduces the packing density, leading to the decrease of E_c because of the formation of porous cake. On the other hand, the adsorption of negatively charged PAA on the particle surface enhanced the dispersion of the particles and the packing density increased from 49 % for no PAA addition at pH 10 to 58 % for PAA addition when the particles were consolidated through gypsum mold[3]. The results shown in Fig.5 indicate the small influence of applied pressure on the packing density for the basic suspensions with PAA. However, the packing density for the basic suspensions without PAA was significantly increased by the application of stress of 19 MPa. As a result, no

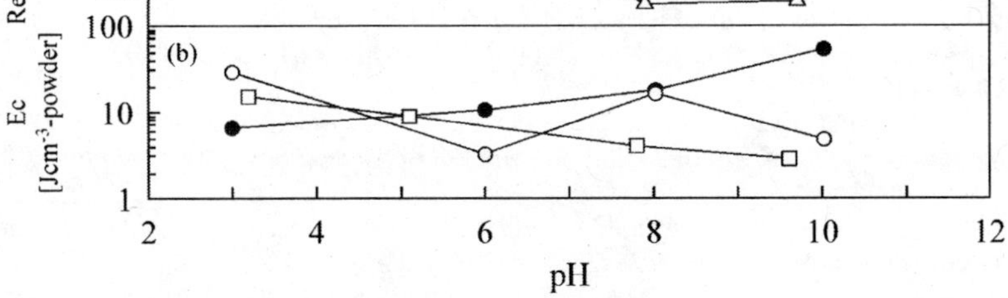

Figure 5. (a) Packing density and (b) consolidation energy (Ec) for 1 cm^3 of 70 vol% Al_2O_3 - 30 vol% ITO particles as a function of pH.

difference in the packing density was measured for the pressure filtration of basic suspensions with and without PAA in Fig.5. The adsorption of the charged PAA increased the consolidation energy of the particles during the filtration.

CONCLUSIONS

The interaction between Al_2O_3 and Al_2O_3 particles dominated the rheology of the 70 vol% Al_2O_3-30 vol% ITO system. The packing density of the Al_2O_3-ITO compact formed through gypsum mold or using a pressure filtration apparatus became higher at a lower pH of the suspension without PAA. The influence of applied pressure on the packing density was small for the dispersed suspension at a low pH and significant for the flocculated suspension at a high pH. In the acidic suspensions, the adsorption of neutral PAA on the particle surfaces reduced the packing density, leading to the decrease of the consolidation energy. The adsorption of negatively charged PAA on the particle surface in a basic suspension enhanced the dispersion of the particles but little increase of the packing density was measured between the suspensions with and without PAA at 19 MPa of applied pressure. However, the consolidation energy of the basic suspension increased by the adsorption of PAA.

REFERENCES

[1]C. Kaya, E. G. Butler and M. H. Lewis, "Co-extrusion of Al_2O_3/ZrO_2 Bi-phase High Temperature Ceramics with Fine Scale Aligned Microstructures", *J. Eur. Ceram. Soc.*, **23**, 6, 935-942 (2003).

[2]D. S. Cheong, K. T. Hwang and C. S. Kim, "Fabrication, Mechanical Properties and Microstructure Analysis of Si_3N_4/SiC Nanocomposite" *Comp. Part A*, **30**, 4, 425-427 (1999).

[3]Y. Koreeda, Y. Hirata and S Sameshima, "Analysis of Microstructure of Two-component Powder Compact by Electrical Conductivity Measurement", *J. Ceram. Soc. Jpn.*, **112**, 5, S280-S285 (2004).

[4]Y. Hirata, K. Hidaka, H. Matsumura, Y. Fukushige and S Sameshima, "Colloidal Processing and Mechanical Properties of Silicon Carbide with Alumina", *J. Mater. Res.*, **12**, 11, 3146-3157 (1997).

[5]K. Kishigawa and Y. Hirata, "Forming of Aqueous SiC Suspension by Pressure Filtration", *J. Ceram. Soc. Jpn.*, **114**, 3, 259-264 (2006).

[6]Y. Hirata, Y. Tanaka and Y. Sakamoto, "Packing Density and Consolidation Energy of Colloidal Particle thorough Pressure Filtration", *Advances in Science and Technology*, **45**, 471-479 (2006).

[7]Y. Hirata, M. Nakamura, M. Miyamoto, Y. Tanaka and X. H. Wang, "Colloidal Consolidation of Ceramic Nanoparticles by Pressure Filtration", *J. Am. Ceram. Soc.*, **89**, 6, 1883-1889 (2006).

[8]Y. Hirata, J. Kamikakimoto, A. Nishimoto and Y. Ishihara, "Interaction between Alpha-Alumina Surface and Polyacrylic-Acid", *J. Ceram. Soc. Jpn.*, **100**, 1, 7-12 (1992).

[9]F. Kitahara, "Powder - Theory and Application -", Edited by T. Kubo, G. Jinbo, E. Suito, H. Takahashi, S. Hayakawa, *Maruzen*, Japan, 1979, p.150, pp.162-163.

[10]L. Bergström, "Hamaker Constants of Inorganic Materials", *Adv. Colloid Interface Sci.*, **70** , 18, 125-169 (1997).

[11]J. Sum, B. V. Velamakanni, W. W. Gerberich and L. F. Francis, "Aqueous Latex/Ceramic Nanoparticle Dispersions: Colloidal Stability and Coating Properties", *J. Colloid Interface Sci.*, **280**, 2, 387-399 (2004).

[12]Y. Hirata, X. H. Wang, Y. Hatate and K. Ijichi, "Electrical Properties, Rheology and Packing Density of Colloidal Alumina Particles", *J. Ceram. Soc. Jpn.*, **111**, 4, 232-237 (2003).

IMPROVEMENT OF SILICA PARTICLE DISPERSABILITY IN XYLENE USING SURFACE MODIFICATION

Chika Takai, Masayoshi Fuji and Minoru Takahashi
Nagoya Institute of Technology, Ceramics Research Laboratory
10-6-29, Asahigaoka, Tajimi, Japan, 507-0071

ABSTRACT

Silica/polyisocyanate composites are kind of materials which can be applied to hard coatings and paintings owing to their hardness and excellent weather resistance. In this study, in order to obtain high particle dispersability in organic medium, surface modification using silane coupling reagent (IPTS) having similar chemical structure was performed. Degree of surface modification was confirmed by FT-IR spectrometry. From TG/DTA analysis, it was clear that half amount of surface free silanol groups was reacted at a IPTS concentration of 1 ml/g. In addition, aggregation size decreased rapidly with increase in IPTS concentration.

INTRODUCTION

Non-aqueous slurries dispersing particles have wide applications including ink jet, paint, and paste. Improvement of physical and chemical properties can be expected by filling particles. It is well known that the functionality of these products is strongly affected by particle dispersion. In order to obtain a well-dispersed suspension, it is necessary to produce stability suspension between medium and dispersing phase. Therefore, affinity at the interface between them should be considered.

Due to the development of nanotechnology, particles have been controlled in a nano-scale. Organic slurry can also have unique properties by filling nanosized particle. Owing to have higher specific surface area with decreasing particle size,[1-3] aggregation force between nanosized particles would be stronger than dispersion. That is why it is difficult to obtain uniform materials. In addition, electrostatic repulsive force among particles in organic slurry hardly works owing to absence of ions. General strategies to disperse particles into organic solvent uniformly are mechanical mixing and surface chemical modification.[4-9]

In order to improve affinity between particle surfaces and the medium, properties of both materials such as polarity should be similar. In this study, at the aim of mixing with polyisocyanate matrix, IPTS having isocyanate group was used as a surface modification reagent. Silica particles were used as dispersing particulate which has numerous hydroxyl groups on the surface. A silane coupling reagent reacts with the surface free silanol groups. The degree of surface modification with different modifier amount was confirmed using FT-IR and TG/DTA analysis. Improvement in the affinity between particle surface and medium was investigated by a dispersion evaluation in xylene which is as solvent for synthesis of polyisocyanate. Aggregation size in xylene was also measured.

EXPERIMENTAL SECTION

Fine silica powder (OX50, specific surface area = 50 m^2/g) and 3-isocyanate propyl triethoxysilane (IPTS) were obtained from Nippon Aerosil Co., Shin-Etsu Chemical Co., Ltd., respectively. Xylene was obtained from Aldrich. Hexane was purchased from Wako Pure Chemical Industries Ltd., Osaka, Japan.

Chemical surface modification is the reaction of IPTS with the hydroxyl groups of the silica substrate. This reaction was carried out using the autoclave method [10,11] at 235 ℃ under 30 atm for 1 h by varying the content of IPTS in the range from 0.1 to 1.0 ml/g as shown in Figure 1. The degree of surface modification was controlled by the modifier content for 3 -OH/nm^2 surface density of the hydroxyl groups determined by the Grignard reagent method.[12] Before the measurement, the sample was degassed at 160 ℃ for 4 h.

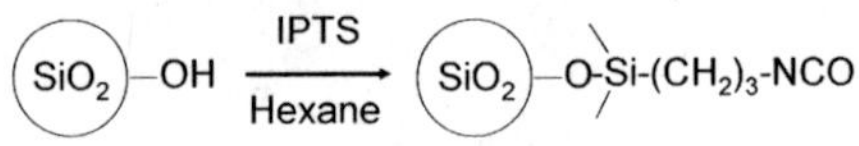

Figure 1. Surface modification with IPTS using autoclave method.

The surface density of the modifier was estimated from the number of modifiers and the specific surface area. The number of isocyanatepropyl groups on the surface was determined using a TG/DTA (Thermo Plus TG8120, Rigaku Co., Japan). [13] The measurement was carried out under oxygen at a flow rate of 250 ml/min. The surface isocyanate group density, d_A, was calculated using equation (1),

$$d_A[/\text{nm}^2] = \frac{(\Delta w_{CH} - \Delta w_{OH})N_A}{M_w S_{N2}} \times 10^{-18} \qquad (1)$$

where Δw_{CH} and Δw_{OH} are the ratio of weight loss for the modified and unmodified samples measured by thermogravimetry, M_w is the molar weight of the isocyanate group, S_{N2} is the BET specific surface area measured from the nitrogen adsorption isotherm and N_A is Avogadro's number.

The FT-IR study was carried out by FT-IR 6200 (JASCO Co., Japan) to characterize the modified particle surface.[14] Improvement of affinity between particle surface and organic medium was confirmed by visual observation and particle size measurement. Visual observation was a simple method to evaluate dispersion of the particles in xylene for 1 minute ultrasonic treatment and slurries were statically equilibrated for 24 hours.

RESULTS AND DISCUSSION

The surface modification of silica powder with IPTS was successfully completed. Figure 2 shows the FT-IR spectra of unmodified and modified with 0.1-1.0 ml/g IPTS. Absorbance of free silanol groups at 3747 cm^{-1} on unmodified silica surface disappeared on the modified particle. A broad peak between 2915 and 1985 cm^{-1} caused by the C-H stretching vibration and sharp peak at 2277 cm^{-1} caused by N=C=O appeared and increased gradually with IPTS amount. Figure 3 shows plot of the relative absorbance of the N=C=O groups obtained from FT-IR study. Rapid increase was observed with modifier concentration, which indicates the attachment of IPTS to silica surfaces.

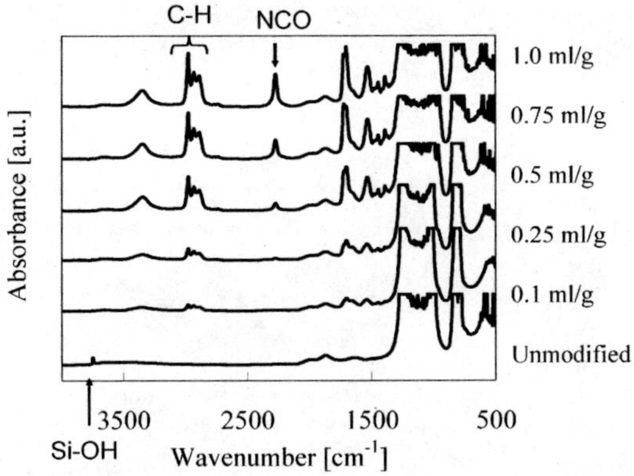

Figure 2. FT-IR spectra of unmodified and modified with different modifier amount.

The relationship between IPTS amount and surface density of the modifier groups are shown in Figure 4. The modification ratio is the fraction of covered hydroxyl groups in all amount of hydroxyl group. From this graph, the surface density of the modifier groups changed significantly depending on IPTS concentration. It increased gradually with a maximum value of 1.34 $/nm^2$ at 1.0 ml/g modifier amount. This indicates that about half of the all surface free hydroxyl groups which is 3 OH/nm^2 obtained from Grignard method reacted with IPTS. NCO group which is the end group of IPTS has a slightly rigid formation because of its two double bonds. Steric hindrance between the NCO groups might occur on silica surface. Dense surface modification would be prevented by the steric hindrance as well as lower reactivity of IPTS.

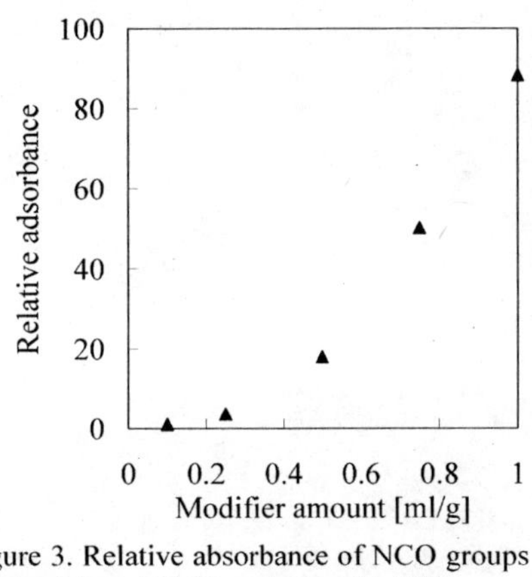

Figure 3. Relative absorbance of NCO groups obtained from FT-IR analysis with IPTS.

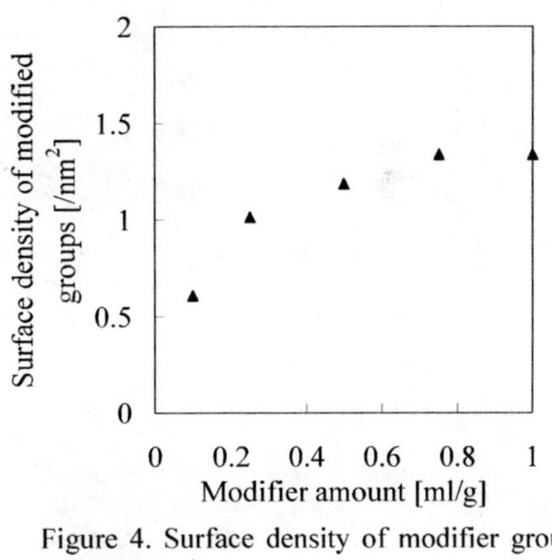

Figure 4. Surface density of modifier groups with IPTS.

The dispersion evaluation by visual observation was carried out with dispersing particles in xylene. Figure 5 shows the appearance of these slurries left for 10 minutes after ultrasonic treatment. Unmodified particles rapidly settled immediately after the treatment. Although modified particles with 0.1-0.75 ml/g settled slowly, there is no sedimentation with 1.0 ml/g modified particles.

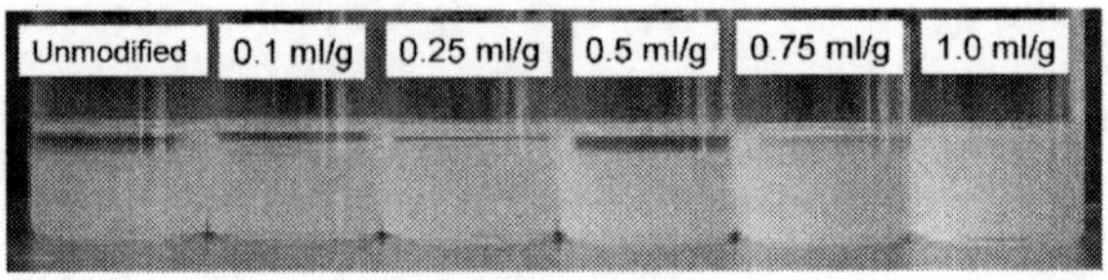

Figure 5. Visual observations of slurries dispersing unmodified and modified particle with 0.1-1.0 ml/g after ultrasonic treatment.

Figure 6 shows the change of aggregation size in xylene as a function of IPTS concentration measured using photon correlation spectroscopy which can be applied to particle size less than 3 μm. Three slurries consisting Unmodified and modified with 0.01-0.2 ml/g particles showed extremely large aggregation size which was above 3 μm. In the range from 0.5-1.0 ml/g IPTS, the size decreased significantly with increasing in IPTS. The minimum value was about 628 nm at 1.0 ml/g IPTS. Terminal velocity at the aggregation size of 628 nm can be estimated as about 0.8×10^{-5} cm/s, therefore, the particle should settle completely within 24 hours. No sedimentation observed with 1.0 ml/g IPTS indicates improvement of particle dispersability in xylene.

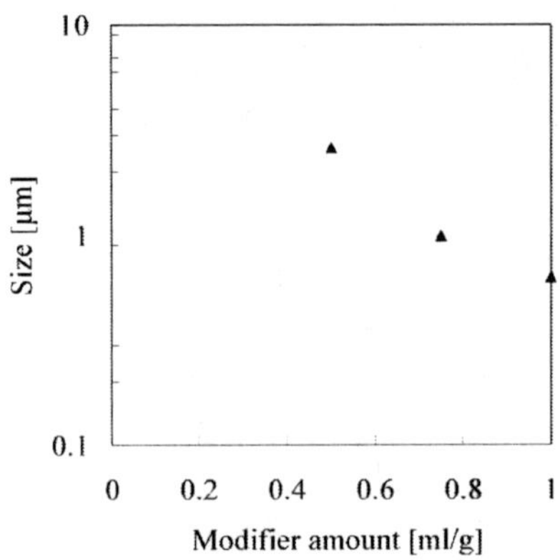

Figure 6. Change of aggregation size in xylene as a function of IPTS concentration.

Schematic drawing of estimated particle dispersion in xylene is shown in Figure 7. Fig.7 (a) and (b) are described as the unmodified and modified with 1.0 ml/g, respectively, which is estimated from the above results It should be more difficult to obtain a uniform modified surface with higher density, because steric hindrance and condensation reaction are created by the surface modifiers. Progress of the hydrolysis reaction of modifier can be changed by the amount of water in the reaction on adsorbed on particle surfaces. It is expected that further modification can be obtained as shown in Fig. 7(c) by control of these factors.

Figure 7. Schematic drawing of particle structure in xylene: (a) and (b) are the illustration of unmodified and modified with 1.0 ml/g, respectively. As shown in (c), further modification can be expected

CONCLUSION

Surface modification considering affinity between silica nano-particle and polyisocyanate was performed using IPTS which has NCO group. The degree of modification can be controlled by IPTS concentration.

No sedimentation of modified particles with 1.0 ml/g in xylene indicates improvement of particle dispersability. This approach has potential for further improvement in particle dispersability.

ACKNOWLEDGEMENTS

This research was supported in part by a grant from the NITECH 21st Century COE Program, "World Ceramics Center for Environmental Harmony", Hori Information Science Promotion Foundation and the METI (Ministry of Economy Trade and Industry) as a part of the Industrial Science and Technology Research and Development "Research and development of the ultra low dielectric insulation film and anticorrosive film using hollow nanoparticles"(2005-2007).

REFERENCES

[1]B. K. Larson and L. T. Drazal, "Interphase Development in a Liquid Composite Molding Environment," *Proceeding of the American Society of Composites, Technical Conference*, 7th, 286-293, (1992).

[2]K. W. Thomson and L. J. Broutman, "The effect of water on the fracture surface energy of fiber-reinforced composite materials," *Polymer Composites*, **3**, 113-117, (1982).

[3]K. Matsumoto, T. Yoshida, and K. Okuyama, "Dispersion of surface modified silica nanoparticles into a resin by using a twin screw extruder," *Funtai Kogaku Kaishi*, **40**, 489-496, (2003).

[4]K. Matsumoto, T. Yoshida, K. Okuyama, "Dispersion of surface modified silica nanoparticles into a resin by using a twin screw extruder," *J. Soc. Powder Technol. Jpn.*, **40**, 489–496, (2003).

SELF-DISPERSIBLE SILICA NANOPARTICLES MODIFIED WITH AMINOALKYLSILANE

Toshio Kakui, Mitsuru Ishii, Sayaka Sato and Masao Ishiguro
Chemicals Division, Lion Corporation
3-7, Honjo 1-Chome, Sumida-Ku, Tokyo, 130-8644, Japan

Koji Hisanaga
Analytical Research Center, Research & Technology Headquarters, Lion Corporation
7-13-12, Hirai, Edogawa-Ku, Tokyo, 132-0035, Japan

ABSTRACT

Self-dispersible silica nanoparticles, i.e., colloidal silica, in water were produced by a surface modification using an aminoalkylsilane coupling agent, i.e., N-2 (aminoethyl) 3-aminopropyl-trimethoxysilane (ATS). The surface of the silica nanoparticle was treated with the ATS coupling agent in an acidic aqueous solution. The Si-O-Si bond, which was formed by the reaction between the silanol group on the silica surface and the methoxysilane of the ATS coupling agent, was determined by 29Si-NMR. The silica nanoparticle modified with the coupling agent displayed a positive charge in acid and alkaline aqueous solutions, although the original silica nanoparticles had a negative charge. The positive ζ potential of the modified silica nanoparticle increased with an increase in the treated amount of the ATS coupling agent. When the amount of ATS was over 5 wt. %, the positive ζ potential reached a plateau. The average diameter of the silica nanoparticles treated with 10 wt. % of the ATS increased from 31.6 nm to 36.9 nm. Interestingly, the silica powder, which was obtained by drying the colloidal silica treated with the ATS, easily dispersed as nanoparticles in water, and produced a silica suspension with a 50 wt. % solid concentration. However, the silica powder from the untreated original silica did not disperse. The 50 wt. % silica suspension was transparent as well as the 20 wt. % original colloidal silica, and its particle diameter was in the nano-size range which was similar to that of the original silica.

INTORODUCTION

Silane coupling agents are well known as a surface modification agent of particles and materials. To improve the mechanical properties of the composite materials with particles and the dispersion of particles in the composites, the surface modification of the particles was carried out by using silane-coupling agents.[1] The reaction mechanisms of the coupling agents have also been investigated by many researchers.[2, 3] In order to facilitate the effective reaction of the silane coupling agents, the relationship between the amount of the chemisorbed coupling agent on the silica surface and the hydrolysis reaction and condensation of the silanol groups on the silica surface has been examined with reference to the pH influence.[4] However, few studies have reported the effect of the coupling agents on the dispersion stability and particle interaction. On the other hand, the surface modification of the particles by the adsorbed polymer dispersant affects the control of the dispersibility of the particle in liquid media. By using a colloidal probe AFM (atomic force microscope), we investigated the polymer dispersant adsorbed on an alumina surface that facilitated a steric repulsive force between the alumina nanoparticles. This steric repulsive force reduced the viscosity of the concentrated alumina ethanol suspensions.[5, 6] However, the effect of the coupling agents, which are able to directly modify a particle surface, has not been reported on the dispersability of particles. In this study, an acid colloidal silica was used as the silica nanoparticle. The surface modification of the silica nanoparticle was carried out by treating with the aminoalkylsilane coupling agent in an acid solution.[7] To evaluate the surface state of the modified silica nanoparticle, the ζ potential and the particle size distribution were examined. The self-dispersibility of a dried silica powder to water was examined in order to investigate the dispersibility of the modified silica nanoparticle. The surface structure of the silica nanoparticle modified with the aminoalkylsilane coupling agent will be discussed on the basis of results.

EXPERIMENTAL PROCEDURE

Preparation of Silica Nanoparticles Modified with Aminoalkylsilane Coupling Agent

Acid colloidal silica (SN-125, Catalysts & Chemicals Ind. Co., Ltd.) was used as the silica nanoparticles. The average diameter of the silica nanoparticles was determined to be almost 30 nm based on an FE-SEM image analysis (Figure 1(a)). N-2 (aminoethyl) 3-aminopropyltrimethoxysilane (ATS) (KBM-603, Shin-Etsu Chemical Co., Ltd.) was used as the aminoalkyl silane-coupling agent. After the ATS coupling agent was dropped into an acid solution of the colloidal silica at 20 wt. %, the mixture was stirred at 70 - 90°C for 1 hour to complete the reaction. The colloidal silica was treated with coupling agent amounts ranging from 0.1 to 20 wt. % versus the silica solid. The particle size of the silica nanoparticle treated with 10 wt. % ATS was almost equal to the size of the original silica in the FE-SEM image analysis as shown in Figure 1(b). Although ordinary alkaline colloidal silica (SI-50, Catalysts & Chemicals Ind. Co., Ltd.) was also examined as a silica nanoparticle, the alkaline colloidal silica did not react with the ATS coupling as shown in Figure 2(b).

(X 200,000)

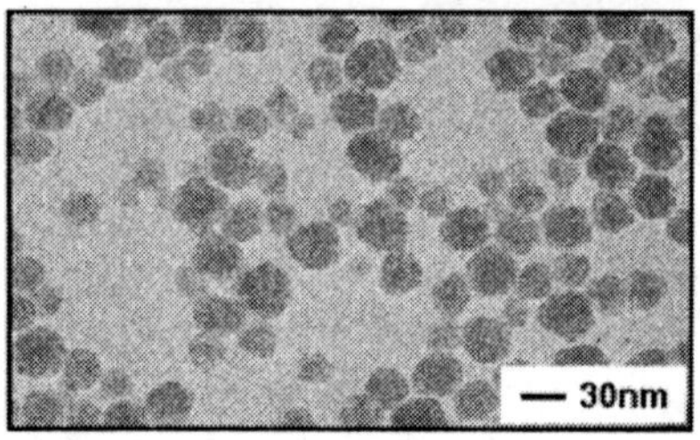

(a) Original colloidal silica

(X 200,000)

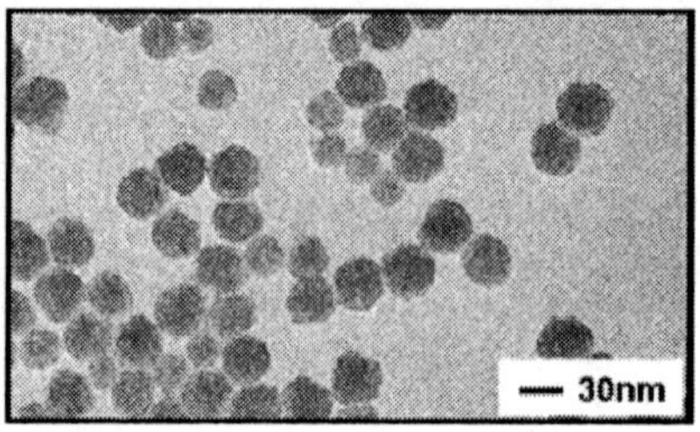

(b) Colloidal silica treated with 10 wt.% of ATS

Figure 1. FE-SEM image of the original colloidal silica and the treated colloidal silica

Determination of Reactivity of Aminoalkylsilane Coupling Agent and Colloidal Silica

To determine the reactivity of the ATS coupling agent and the silica surface, the 29Si-NMR and 13C-NMR spectra of the solids (ECX 500MHz, JEOL Ltd.) were evaluated for identification of the Si-O-Si bond, which was formed by the reaction between the silanol group of the silica surface and the methoxysilane of the ATS. The NMR signals were measured using the powder samples, which were obtained by drying the colloidal silica treated with 10 wt. % ATS and the original colloidal silica at 30 °C under vacuum.

Characterization of Silica Nanoparticles Modified with Aminoalkylsilane Coupling Agent

First, to evaluate the surface state of the silica nanoparticle, the ζ potential of the silica nanoparticles modified with the ATS coupling agent ranging from 0.1 to 20 wt.% and the original silica were measured using an Acousto Sizer (Colloidal Dynamic Ltd.). The ζ potential measurement was carried out in 20 wt. % aqueous solutions with pH ranging from 2 to 13, which were prepared using HCl or NaOH. The particle size distribution of the silica nanoparticles in the 20 wt. % aqueous solutions with pH 2 was also analyzed using dynamic light scattering. The next stage was to evaluate the self-dispersibility of the modified silica nanoparticles. The dried powder samples were mixed with water, and the mixtures were then stirred for several hours. The appearance of the obtained suspensions with solid concentrations ranging from 5 to 60 wt. % and the particle sizes were evaluated as being the self-dispersibility.

RESULTS

Determination of Reactivity of ATS and Silica Nanoparticle

The 29Si-NMR spectra of the colloidal silica before and after the reaction with 10 wt. % of the ATS coupling agent in an acid solution are shown in Figure 2(a).[8, 9] The 29Si-NMR of the alkaline colloidal silica treated with the ATS in an alkaline solution, which did not react with the ATM, is also shown in Figure 2(b). The two peaks shown at 100 and 110 ppm are caused by the silanol groups on the surface of the silica nanoparticles, which are $\underline{Si}\text{-}(OH)(O\text{-})_3$ and $\underline{Si}\text{-}(O\text{-})_4$, respectively. For the alkaline colloidal silica, the peaks before and after the reaction did not change, and no peak at in the other areas were observed. In the case of the acid colloidal silica, the two peaks that appeared at 100 and 110 ppm did not change as well as that of the alkaline state. However, the intensity of the peak shown at 90 ppm for $\underline{Si}\text{-}(OH)_2(O\text{-})_2$ decreased after the reaction, and the two new peaks shown at 57 and 67 ppm, which are attributed to $\underline{Si}\text{-}(CH_2R)(O\text{-})_3$ and $\underline{Si}\text{-}(CH_2R)(OH)(O\text{-})_2$, appeared.

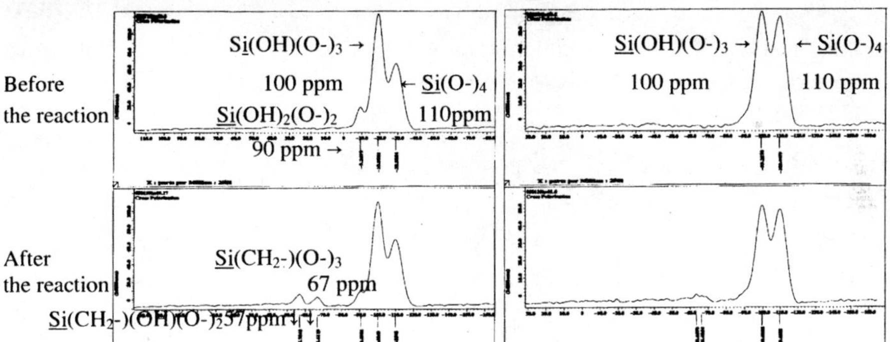

(a) Colloidal silica (b) Alkaline colloidal silica

Figure 2. 29Si-NMR of the colloidal silica before and after the reaction with the ATS coupling agent

The 13C-NMR signals of the acid colloidal silica treated with the ATS coupling agent is shown in Figure 3. The existence of carbon peaks attributed to the ATS has been confirmed, although the carbon peaks from the original silica were not observed. As a result, the silanol groups, i.e., Si-OH, on the silica surface reacted with the trimethoxysilane, i.e., Si-$(OCH_3)_3$, of the ATS coupling agent, and the surface of the silica nanoparticle was modified with the ATS coupling agent.

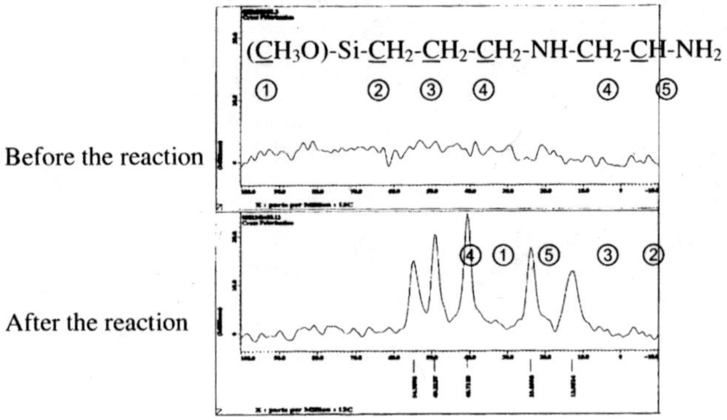

Figure 3. 13C-NMR of the colloidal silica before and after the reaction with the ATS coupling agent

Characterization of Modified Silica Nanoparticle

To evaluate the surface state of the modified silica nanoparticles, the ζ potential and the particle size distribution before and after the reaction were measured. The ζ potential of the silica nanoparticles treated with the ATS coupling agent ranged from 0.1 to 20 wt. % and the original silica in an acid solution is shown in Figure 4(a). The silica nanoparticle treated with 0.5 wt. % ATS displayed a positive charge, although the original silica nanoparticle had a negative charge. The positive ζ potential of the silica nanoparticle increased with an increase in the treated amount of the coupling agent. When the amount of ATS was over 5 wt. %, the positive ζ potential reached a plateau of about 50 mV. When the pH ranged from 2 to 13, the ζ potential of the silica nanoparticle treated with 10 wt. % ATM is plotted in Figure 4(b). The treated silica nanoparticles displayed positive charges in solutions with pH ranging from 2 to 11. The isoelectric point of the treated silica nanoparticles shifted from pH 2 to pH 12, which is that of the original silica.

Next, the particle size distribution of the modified silica nanoparticles in an aqueous solution with pH 2 was examined by dynamic light scattering. The average diameter of the silica nanoparticles treated with 10 wt. % ATS coupling agent is shown in Figure 5. The average diameter increased from 31.6 nm to 36.9 nm. The particle size also shifted to a larger size over the entire distribution.

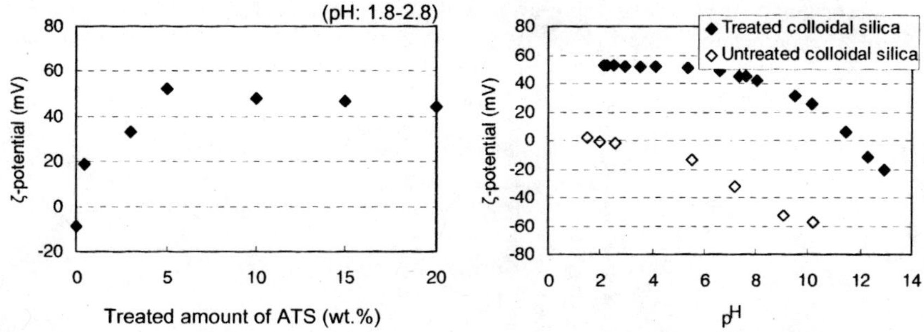

(a) Silica nanoparticles treated with the ATS ranging from 0.1 to 20 wt.%

(b) Silica nanoparticles treated with 10 wt. % ATS in solutions with pH ranging from 1 to 13

Figure 4. The ζ potential of the silica nanoparticles treated with the ATS coupling agent

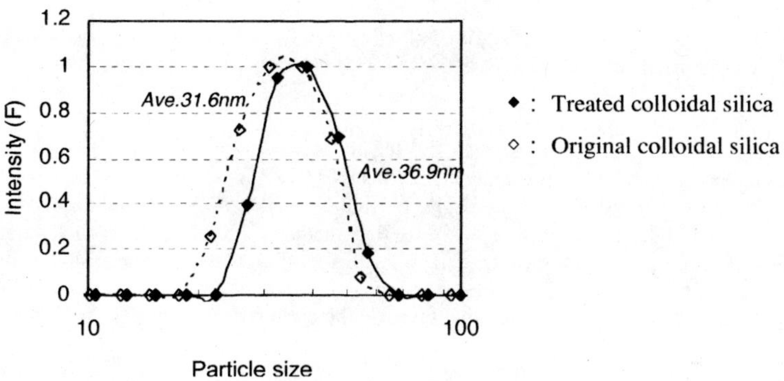

Figure 5. The particle size distribution of the colloidal silica treated with 10 wt. % ATS coupling agent

To evaluate the effect of the coupling agent on the silica surface, the self- dispersibility of the silica powder, which was obtained by drying the colloidal silica treated with the 10 wt. % ATS coupling agent, was investigated. The silica powder from the modified silica nanoparticles easily dispersed like the original silica nanoparticles. The appearance of suspensions in solid concentrations ranging from 5 to 50 wt. % obtained by self-dispersing is shown in picture 1. The suspension of 50 wt. % concentration obtained by self-dispersing was transparent as well as the

20 wt. % original colloidal silica. The average diameter of the silica nanoparticles in the 20 wt. % suspension obtained by self-dispersing the silica powder was 56.8 nm, which was larger than the 36.9 nm size of the original silica. However, the silica powder from the unmodified original silica did not disperse. The modified colloidal silica displayed a self-dispersibility.

Original colloidal silica in 20 wt.% | Silica powder dried original silica without self-dispersibility | 5 wt.% 20 wt.% 30 wt.% 40 wt.% 50 wt.% Silica powder after self-dispersing

Picture 1. The appearance of the colloidal silica obtained by self-dispersing the silica powder, which was the dried silica nanoparticle treated with 10 wt.% of the ATS coupling agent.

DISCUSSION

Based on the above-mentioned results, the surface structure of the silica nanoparticles modified with the ATS coupling agent is illustrated in Figure 6. According to 13C-NMR in Figure 3, the existence of carbon peaks attributed to the ATS on the surface of the silica nanoparticles has been confirmed. It seems that methoxy groups of trimethoxysilane, i.e., Si-$(OCH_3)_3$, of the ATS coupling agent, react with the silanol groups, i.e., Si-OH, on the silica surface. Other methoxy groups will be decomposed to silanol groups, and some methoxy groups without decomposing will be remaining, because the carbon peak of the Si-OCH_3 was observed.

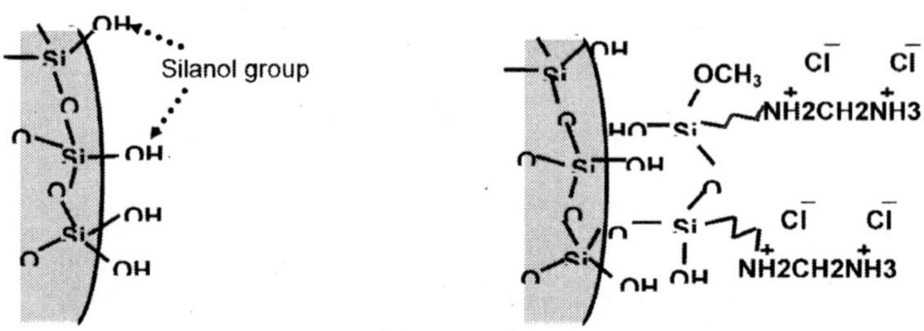

Surface structure of original colloidal silica

Surface structure of silica nanoparticle modified with the ATS coupling agent

Figure 6. Illustrations of the surface structure of the silica nanoparticle modified with the ATS coupling agent in an acid solution.

The ATS coupling agent, which reacted with the silanol group on the silica surface, will cause an increase in the particle diameter from 31.6 to 36.9 nm. The silica nanoparticle modified with the aminoalkylsilane coupling agent will have a large positive charge of about 50 mV due to the cationic ammonium chlorides. It seems that the steric and electrostatic repulsive forces between the silica nanoparticles cause the self-dispersibility of the silica powder, which was obtained by drying the modified colloidal silica solution.

CONCLUSION

The surface modification of the silica nanoparticles, i.e., colloidal silica, was carried out by treating with the aminoalkylsilane coupling agent that is N-2 (aminoethyl) 3-aminopropyl-trimethoxysilane (ATS) in an acid solution. The reaction between the silanol group of the silica surface and the methoxysilane of the ATS coupling agent was determined by 29Si-NMR and 13C-NMR. The silica nanoparticle modified with the aminoalkylsilane displayed a positive charge in solutions with pH ranging from 2 to 11, although the original silica had a negative charge. The positive ζ potential of the silica nanoparticle increased with an increase in the treated amount of the coupling agent, and reached a plateau of about 50 mV over 5 wt.%. The average diameter of the silica nanoparticle modified with the 10 wt. % ATS increased from 31.6 nm to 36.9 nm. Interestingly, the silica powder, which was obtained by drying the modified colloidal silica easily dispersed as nanoparticles. The average diameter of the self-dispersible silica nanoparticles in the 20 wt. % suspension was 56.8 nm, which was almost equal to 36.9 nm for the original silica. However, the silica powder from the unmodified original silica did not disperse. The silica nanoparticle modified with the aminoalkyl silane coupling agent displayed a self-dispersibility.

REFERENCES

[1]X. Q. Xie, S. V. Ranade, A. T. Dibenedetto, A solid state NMR study of polycarbonate oligomer grafted onto the surface of amorphous silica, *Polymer*, **40**, 6297-6306 (1999).

[2]H. Ishida, J. L. Koenig, Fourier transform infrared spectroscopic study of the silane coupling agent/porous silica interface, *J. Colloid Interface Sic.*, **64**, 565-576 (1978).

[3]M. W. Daniels, J. Sefcik, L. F. Francis, A. V. McCormick, Reactions of a Trifunctional Silane Coupling Agent in the Presence of Colloidal Silica Sols in Polar Media, *J. Colloid Interface Sci.*, **219**, 352-356(1999).

[4]M. Chaimberg, Y. Cohen, Note on the silylation of inorganic oxide supports, *J. Colloid Interface Sci.*, **134**, 576-579 (1990).

[5]T. Kakui, T. Miyauchi, H. Kamiya, Analysis of the action mechanism of polymer dispersant on dense ethanol alumina suspension using colloidal probe AFM, *J. Euro. Ceram. Soc.*, **25**, 655-661 (2005).

[6]T. Kakui, T. Miyauchi, H. Kamiya, Analysis of Action Mechanism of PEI for Dispersion of Alumina Nano-particles in Ethanol Suspension using Nano-colloidal probe AFM, *ACS, Annual meeting, Indianapolis USA*, **Apr. 22-24** (2004).

[7]A.Van Blaaderen, A. Vru, Synthesis and Characterization of Monodisperse Colloidal Organo-silica Spheres, *J. Colloid Interface Sic.*, **156**, 1-18 (1993).

[8]K.Yoshinaga, Y. Tai, Y. Tanaka, Surface modification of fine colloidal silica with copolymer silane coupling agents composed of maleic anhydride, *Colloid Polym. Sci.*, **280**, 85-89 (2002).

[9]N. Nishiyama, T. Asakubo, K. Horie, Condensation Behavior of a Silane Coupling Agent in the Presence of Colloidal Silica studied by 29Si and 13C NMR, *J. Colloid Interface Sci.*, **124**, 14-21 (1987).

LOW TEMPERATURE CO-FIRED CERAMICS (LTCC) – DESIGN AND CHARACTERIZATION OF INTERFACES

Torsten Rabe, Markus Eberstein, Wolfgang A. Schiller
Federal Institute for Materials Research and Testing
Unter den Eichen 44-46
12200 Berlin, Germany

ABSTRACT

Today's LTCC applications often require modules with highly complex structures. Examples of the variety of possible interfaces inside such LTCC multilayer are presented. Sintering reactions at the interfaces between silver conductors and ceramic dielectric, between LTCC tapes with different functions and between LTCC tape and release tape were analyzed and the results are discussed in this paper.

It is shown that silver diffuses up to 50 microns into the surrounding LTCC glassy phase during sintering at 900 °C. Hence, sintering behavior of LTCC is locally changed around silver conductors. Phase boundary layers with microstructural defects and composition notwithstanding both LTCC materials are often formed in the sintering process of mixed tape modules. It can endanger the functional capability of the module and must be controlled.

Constrained sintered LTCC modules suffer from penetration of the release tape by the glass phase of the LTCC. Due to lower viscosity by silver amounts in the glass phase around vias the penetrated layer is thicker and a stronger adhesion of release tape residues was observed.

INTRODUCTION

LTCC has enjoyed a significant market growth in recent years and possesses great potentials in several promising application areas like wireless communication, sensor technology, electronic control units and micro-systems[1]. LTCC are preferably used in hybrid microelectronics where integration of passive components (resistors, capacitors, inductors...), miniaturization (high savings in volume and mass) and high reliability are demanded. Increased demands on miniaturization and additional functionalities lead to enhanced complexity of LTCC-multilayers. Yet, to expand the application fields of LTCC, further innovations in material development and ceramic processing are necessary.

Most LTCC tapes consist of crystalline phases and considerable fractions of glass to lower the sintering temperature. The variety of glass compositions offers manifold chances to tailor the properties of LTCC. During the sintering process demixing effects and crystallization of glass melt, dissolving of the crystalline phase inside the liquid phase and solid state reactions between the liquid phase and the crystalline dispersed phase may take place[2].

Standard dielectric LTCC tapes and tapes with dissenting composition are laminated and sintered together to integrate more functions in one multilayer. Release tapes are intermittently added to reduce shrinkage variation and to prevent to a great extent sintering shrinkage in lateral direction. The successful development of complex modules requires control of reactions inside the LTCC tapes as well as reactions at the interfaces between different materials used in the module.

Figure 1 presents the schematic concept of a multi-component LTCC module consisting of two types of LTCC tapes, conductor lines of silver or silver alloys inside and on the surface of the module, electrical and thermal vias and surface resistors. To minimize lateral shrinkage, non shrinking release tapes are laminated on top and bottom of the module. A set of materials has to be developed to manufacture such a highly integrated module. Material development frequently includes the adjustment of commercially available dielectric "standard" tapes to the sintering behavior of other module components. More components as shown in figure 1 may be integrated by LTCC technology, e. g. buried resistors and capacitors, cavities, channels and optical fibers.

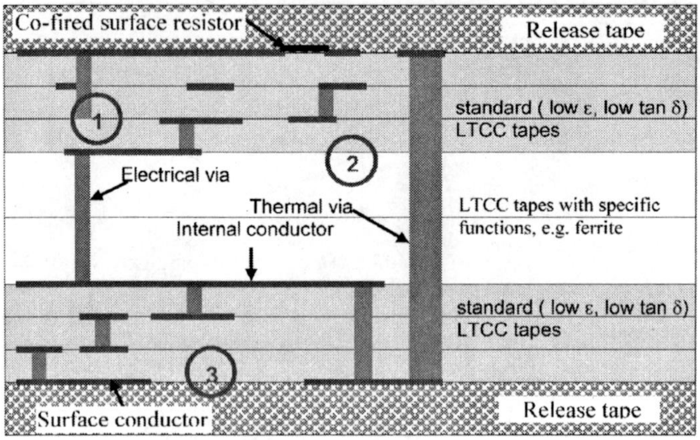

Figure 1.
Schema of a LTCC multilayer module, analyzed interfaces indicated by numbers are
1: silver via/ standard LTCC

2: standard LTCC/ ferritic LTCC

3: standard LTCC/ alumina release tape

In this paper we investigate physical and chemical processes at selected interfaces during sintering.

INTERFACE BETWEEN LTCC AND SILVER CONDUCTOR

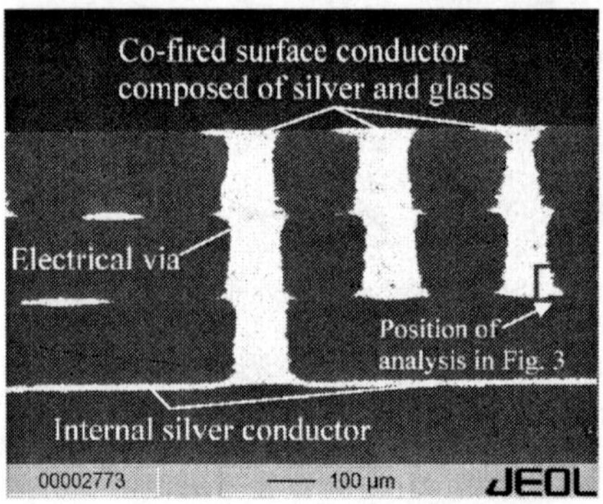

Figure 2. Scanning electron microscope (SEM) image of polished cross-section of a LTCC multilayer with silver-based metallization

Conductors in LTCC are usually based on silver because of its excellent electrical performance. Particle size distribution and composition of silver paste are tailored, e.g. by addition of glass powder to adapt the sintering behavior or by using of silver alloys mostly with palladium to improve the oxidation stability of silver during the sintering process.

Inside one LTCC module different paste compositions are frequently applied for internal conductors, surface conductors and electrical vias (Figure 2). As result of varying additives the interface reaction between silver-based conductors and LTCC can locally differ. It is well-known that silver is highly agile and moves into the glassy phase of the LTCC ceramic during the co-firing process. The mobility of silver around printed conductors and vias and the impact on the local sintering behavior as well as the microstructure were investigated in a LTCC material consisting of alumina and $CaO-SiO_2-Al_2O_3-B_2O_3$-glass.

The presence of silver in the above-mentioned glassy phase starts at about 650 °C as shown by microprobe investigations. The thickness of the silver contaminated area at this temperature amounts to less than 5 microns. Figure 3 illustrates the element distribution in the interface between silver via and LTCC (position of analysis see Figure 2) after thermal treatment at 870°C. An irregular silver distribution was detected in the LTCC. Silver exclusively enters the glassy phase. Glass areas are characterized in figure 3 by above-average concentration of silicon and calcium. Silver content of glass areas near the interface amounts up to 10 wt.%. Silver was found after sintering at 870 °C up to a distance of 20 microns from the interface. Higher sintering temperature leads to broader silver-containing areas, e. g. of up to 50 microns after sintering at 900 °C/ 30 min. The limit of silver detection by microprobe investigation is estimated by 0.3 wt.%.

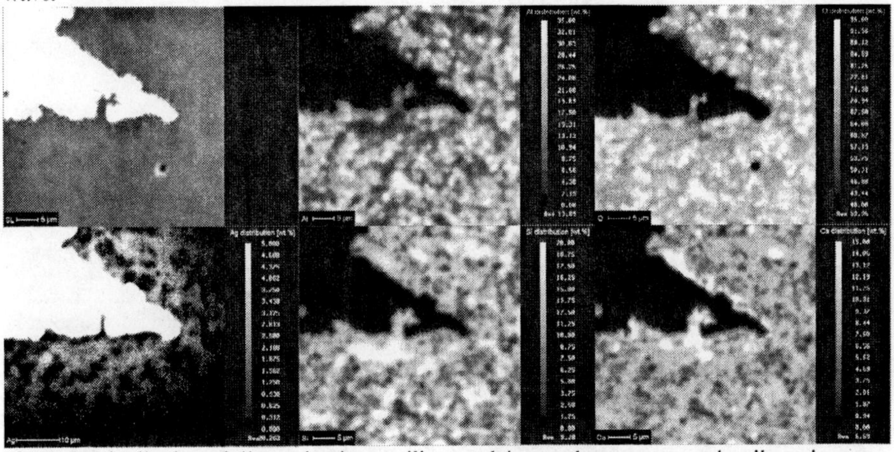

Figure 3. Distribution of silver, aluminum, silicon, calcium and oxygen around a silver via after sintering at 870 °C measured by electron probe microanalyzer (WDX).

It is to expect, that silver enters the glass melt in its oxidized state Ag_2O and takes part in the glass structure as a network modifier. In earlier investigations[2] it was shown that 5 % by mol silver content in a $CaO-B_2O_3-SiO_2$-glass decreases the viscosity of the starting glass by one order of magnitude. Mass transport during sintering densification, dissolution and crystallization is determined by diffusion of single species and can be strongly influenced by a local variation of viscosity in the reported range. Hence, in the surroundings of silver conductors local variations in the shrinkage are promoted which can lead to undesired deformations or stress in the final product. The distance between neighboring silver lines often amounts to 150 microns or less (see figure 2). Therefore detected silver transport in the glassy phases may endanger the excellent electrical isolation behavior of standard LTCC.

Besides the glass phase composition also the distribution and the grain size of alumina is shown in figure 3. Alumina crystals can be proved by above-average concentration of aluminum and oxygen.

INTERFACE BETWEEN LTCC TAPES WITH DIFFERENT COMPOSITION

Multi-functional LTCC modules require tapes with proper microstructure and tailored dielectric, electric and thermal properties. Multilayer modules consisting of tapes with different compositions have to designed, e. g. combinations of dielectric and ferritic tapes or of tapes with low and high permittivity.

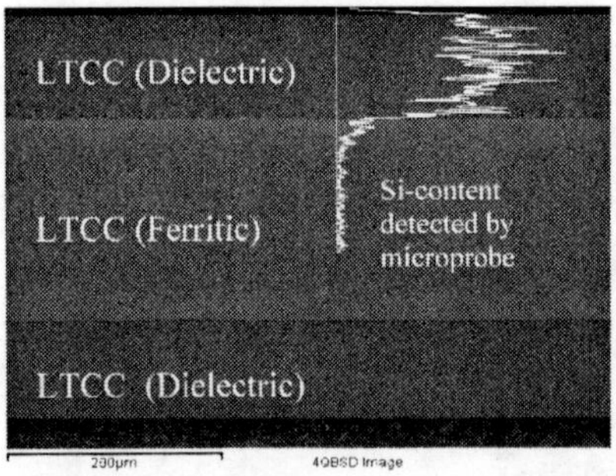

Figure 4: SEM image of LTCC multilayer with ferritic core

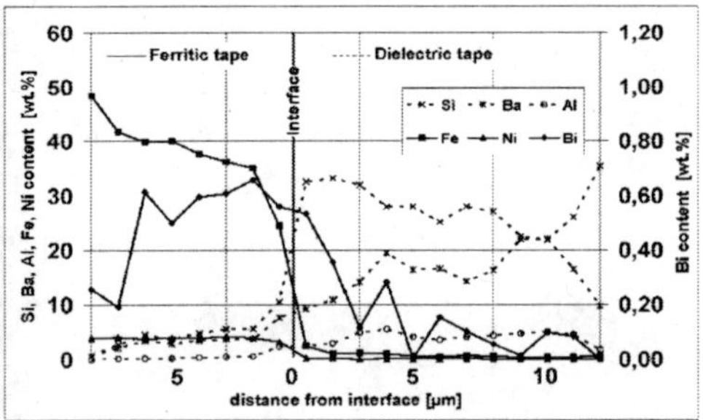

Figure 5: Microprobe line scan showing element distribution through the interface dielectric tape/ferritic tape

An interface area of some microns thickness between ferritic and dielectric tape after co-sintering at 900 °C/ 3 h is in evidence in Figure 4. During the sintering process the low-viscous glass of the dielectric tape infiltrates the ferritic tape. Silicon and barium were detected inside the ferritic tape in a distance up to 10 microns from the interface (see figure 5). Material transport in the opposite direction is taking place, too. Bismuth content is in the ferritic tape increasing with decreasing distance to the dielectric LTCC which was entered by Bismuth obviously. No iron and nickel was found in the dielectric tape after multilayer sintering,

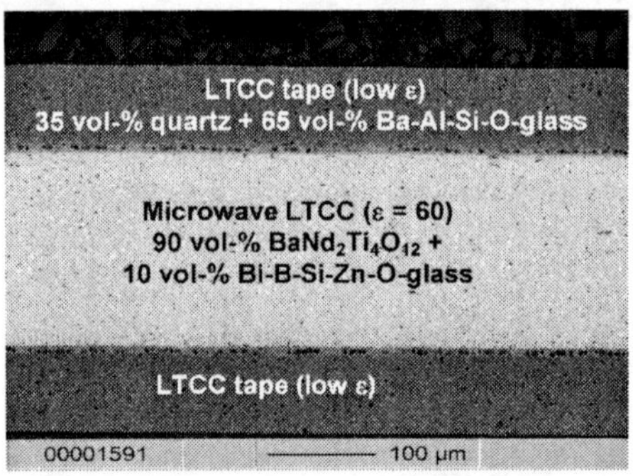

Figure 6: Microstructure of LTCC multilayer consisting of tapes with different permittivity

Figure 6 illustrates the formation of an interface layer with a thickness of about 10 microns and a composition notwithstanding both LTCC materials. Moreover pores with diameters of some microns are visible, forming a pore chain inside the interface layer. Results of microprobe investigation show complex diffusion and reaction processes between different LTCC tapes. Zinc oxide is enriched in the interface layer. The amount is higher than the initial value both in the microwave and the standard LTCC. Scanning SEM images and element distributions give evidence to the fact that the low viscous, highly reactive glass flux of the microwave tape penetrates the low ε tape and reacts with crystalline phases (quartz). Dissolving and crystallization processes, gas precipitation and pore growth are observed. The thickness of the interface layer remains small because the low ε tape is fully densified below sintering temperature range of the microwave tape. Moreover the precipitated zinc containing crystals possibly act as diffusion barrier.

INTERFACE BETWEEN LTCC AND RELEASE TAPE

Release layers are the most usual way to prevent lateral shrinkage in LTCC modules[3]. Such tapes are comprised of refractory ceramic material, typically alumina. They are co-laminated on the top and the bottom surfaces of the green LTCC multilayer. The release tapes do not shrink or sinter themselves. Important property of release layers is a sufficient porosity allowing for oxygen transport into the LTCC tape for binder burn-out and in the opposite direction the transport of gaseous debindering products.

Minimal penetration of LTCC glass into the release tape and/or limited reaction between this glass and the alumina effectively locks the x-, y- movement of the LTCC laminate during the sintering. After sintering, the constraining release tape is in the form of a porous layer in which the particles are held together by weak physicochemical forces. It can be removed by one of a number of mildly abrasive processes including sand blasting. A residue-free removal of release tape is necessary to reach sufficient surface quality (small roughness) for post-sintering processes, as surface metallization and SMD assembling.

Adhesion strength and thickness of the release tape residues can strongly differ in dependence on the LTCC tape used. Interfaces between two commercially available LTCC tapes and alumina release tapes (Ceramtape A) were investigated. Sintered LTCC tape CT 700 possesses only a thin layer of release tape residues after ultrasonic treatment (figure 7, left). However, single alumina particles strongly adhere on the surface.

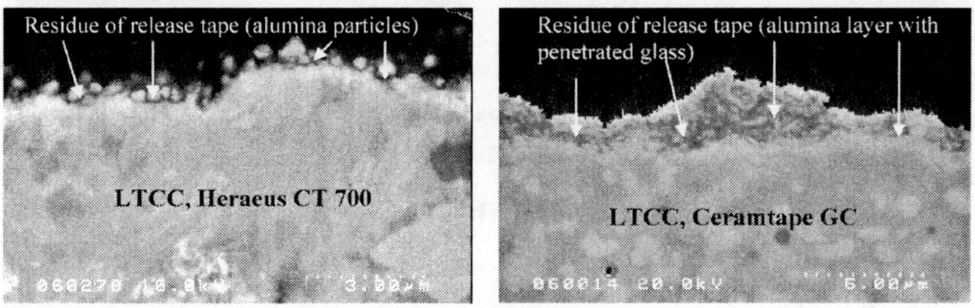

Figure 7: Scanning electron microscope images of sintered CT 700 (left) and Ceramtape GC (right) with release tape residues on the surface

Glass of Ceramtape GC penetrates some microns into the release tape. The resulting alumina/glass layers on the sintered LTCC surface can not be completely removed by mildly abrasive processes or ultrasonic treatment. The thickness of the interface layer differs between 1 and 5 microns causing high surface roughness and problems in post-sintering processes.

If we assume that the interface energies are comparable in the two tapes obviously the residual glass phase in the tape GC had a lower viscosity.

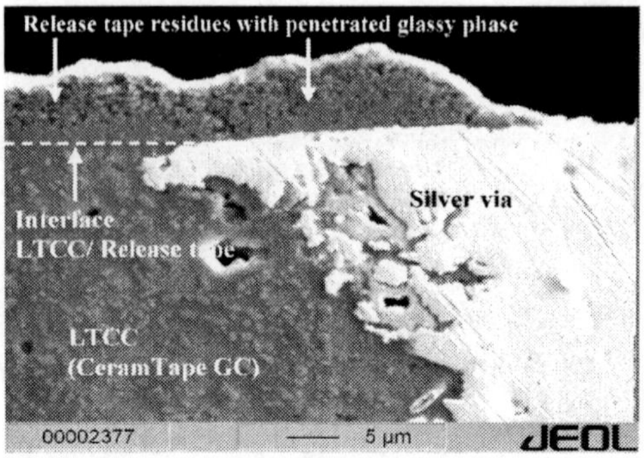

Figure 8:
Scanning electron microscope image of an LTCC module surface area with silver via and release tape residues

The strongest adhesion was observed on via edges (figure 8). As above mentioned the glass viscosity is drastically reduced in this area by silver contamination. Increased mobility of the glass leads to thicker layers of release tape residues.

CONCLUSIONS

The control of manifold high-temperature reactions inside each tape and between the module components (metallic conductors, LTCC tapes with different compositions and release tapes) is the key to a successful development of highly integrated LTCC modules and a pre-condition to a robust production.

One significant high-temperature process is the diffusion of silver into the glassy phase causing viscosity reduction of the liquid phase and a changed sintering behavior around conductor lines and vias. Silver mobility starts at about 650 °C and its intensity depends on sintering temperature, sintering atmosphere, type and amount of the glassy phase in the LTCC. It was shown, that silver moved into a surrounding CaO-SiO_2-Al_2O_3-B_2O_3-glass up to 50 microns during a standard LTCC sintering process at 900 °C.

Competing demands exist for interfaces between tapes with different compositions. On the one hand, stable interconnection between all tapes is necessary after sintering and material transport between the tapes and thickness of the interface zone should be minimized, on the other. Phase boundary layers of some microns were detected in all real mixed multilayer after sintering. Optimization of such interfaces includes the reduction of layer thickness, the prevention of material transport between the tapes and the development of a widely pore-free microstructure of the interface.

The penetration depth of glassy phase in the release tape depends on LTCC glass composition and sintering conditions. The penetrated layer is thicker and a stronger adhesion of release tape residues was observed around the vias as result of viscosity reduction after silver diffusion.

Thanks to T. Moelders for the microstructural investigation and to G. Oder for element analysis by electron probe microanalyzer.

REFERENCES

[1] D. L. Wilcox et. al, « LTCC morphing into strategic microsystems technology », Proc. Of the 2002 Ceramic Appl. for Microw. and Photonic Packaging, Providence, RI, May 2-3

[2] W. Schiller, U. Mücke, D. Linke, "Reactions between the phases in glass-alumina ceramic composites." Journal de Physique, Colloq. C 1, Suppl. 2, 47 (1986), C1/509 - C1 514

[3] T. Rabe, W.A. Schiller, T. Hochheimer, C. Modes, A. Kipka, „Zero Shrinkage of LTCC by Self-Constrained Sintering", Int. J. Appl. Ceram. Technol. 2 [5] 374-382 (2005)

CONTROLLING INTERFACIAL CHEMISTRY DURING THE PROCESSING OF MICRON SCALE SURGICAL INSTRUMENTS

Nicholas Antolino, Gregory Hayes, and James H. Adair
Particulate Materials Center, Materials Research Institute,
Department of Materials Science and Engineering
The Pennsylvania State University, University Park, PA, 16802 USA

Christopher Muhlstein
Department of Materials Science and Engineering
The Pennsylvania State University, University Park, PA, 16802 USA

Mary Frecker and Eric Mockensturm
Department of Mechanical and Nuclear Engineering
The Pennsylvania State University, University Park, PA, 16802 USA

ABSTRACT

Free standing, micron scale, yttria stabilized zirconia parts for surgical instrument applications have been fabricated using a lithography based mold forming technique. Problems and solutions inherent to molding micron scale ceramic parts, such as suspension viscosity, surface spallation, and surface roughness are addressed. Concentrated zirconia suspensions, achieved through a chemically aided attrition milling process, are gelcast into molds via a screen printing method to form parts. After sintering, near theoretical density (99.8%) is achieved with a grain size of 500 nm.

INTRODUCTION

In minimally invasive surgery (MIS), procedures are performed using small instruments (1-5 mm in diameter) inserted through small incisions. The benefits of MIS include reduced surgical trauma, shorter hospital stays, reduced postoperative use of narcotics, and more rapid return to full activity.[1, 2] There is a significant interest in applying minimally invasive techniques to a wider variety of surgical procedures and in performing existing procedures more quickly and efficiently.[1, 2, 3] There is also a need for an order of magnitude smaller and more versatile instruments in applications such as ophthalmic surgery,[4] flexible endoscopic surgery,[5] and intravascular and cardiac catheters. However, the surgical instrument industry has reached the limit of what can be done with conventional materials and manufacturing methods. There is a need for new technology that will overcome the barriers associated with the physical properties of traditional materials and existing manufacturing techniques.

Although there are numerous techniques available to fabricate micron-scale ceramic structures,[6, 7] the most economically feasible techniques are those associated with lithography due to the large number of components traditionally associated with integrated circuit manufacturing. There are two primary techniques based on lithography: soft lithography and lost-mold processing. Although soft lithography methods have been somewhat successful,[8, 9] the use of a soft mold is thought to be too flexible to form the complex, high aspect ratio shapes necessary for surgical instrument fabrication. Lost-mold processing borrows from standard UV photolithography, and uses a photo-curable resin to create a mold on top of a substrate. The mold is then filled with ceramic slurry, while ultimate mold removal is achieved via chemical dissolution[10, 11] or thermolysis.[12] This type of lost-mold technique has shown promising results, however, several challenges remain. Many of the challenges inherent to this lost-mold process relate to the mold itself, and include both ceramic part and mold cracking due to thermal expansion mismatches, substrate reflectivity, and ceramic slurry to substrate adhesion.[10]

The procedure used in this study is based on the lost-mold approach via thermolysis. This technique has the ability to create sharp edges (~1µm) that are advantageous for surgical instruments due to the rigid wall of the mold.[12] In addition, the lost-mold approach also does not require any manipulation of the weaker green ceramic part with direct infiltration and sintering on a refractory substrate. Potential disadvantages of casting directly into the lithographed mold include defects introduced during the casting, thermal demolding, and firing processes. [12] More specifically, these disadvantages include edge cracking and spallation, drying cracks, and dimensional changes between the mold and part. Processing approaches and solutions to overcome these problems will be addressed in the present study.

EXPERIMENTAL PROCEDURE

The goal of the suspension preparation was to obtain the maximum possible solid loading for mold infiltration while achieving a well-dispersed, flowable suspension. Stable suspensions of partially stabilized zirconia were prepared in an attrition mill (Union Process® SP500, Akron, OH) in order to reduce agglomeration in the as-received, granulated powder (Tosoh TZ-3Y). Suspensions were milled in batches of 200 mL at 46 vol% solid using 1 mm stabilized zirconia media at 2000 RPM.

Ammonium polyacrylate ($NH_4(PAA)$, Darvan® 821-A, R.T. Vanderbilt Company, Inc., USA) was used to electrosterically disperse the zirconia at pH 8. At this pH value, the solubility of both yttria and zirconia is minimized below 10^{-6} mol/L and 10^{-8} mol/L, respectively.[13] The amount of dispersant was determined by incrementally adding powder until the suspension approached the dilatant point; then, incremental (100µL) additions of dispersant were added to fluidize the suspension and prevent the mill from seizing. Using this technique, a 46 vol% suspension of zirconia was achieved. The degree of dispersion was evaluated using electrophoretic light scattering (ZetaPALS, Brookhaven Instruments Corp., Holtsville, NY) to determine ζ-potential and dynamic light scattering (Nano-S, Malvern Instruments, Southborough, MA) to measure particle/agglomerate size. The rheological properties of slurries with varying solids loading were determined using a parallel plate rheometer (CSL 100, Carri-Med) with a 2° cone.

Gelcasting is a ceramic forming process used to obtain complex shapes with high green strength after drying, minimal mold defects, and easy binder burnout.[14, 15] The green parts were formed using a methacrylamide/bis-acrylamide, monomer/crosslinker gelcasting system, catalyzed by TEMED (N,N,N',N'-Tetramethylethylenediamine), and initiated by ammonium peroxydisulfate.[15] The ratios of monomer to crosslinker and initiator to catalyst were 6:1 and 10:1, respectively. In preliminary formulation studies, a 50:1 ratio of initiator to monomer and crosslinker was found to gel the suspension in approximately 1 hour. A total gel-casting organic content of 5 wt% on dry powder basis was the lowest concentration at which gelation occurred; consequently, this concentration was used to minimize the amount of organic that ultimately had to be removed via thermolysis.

Molds were fabricated with SU-8 25 photoresist (MicroChem Corp., USA) on alumina substrates (Kyocera Industrial Ceramics Incorporation) using standard UV-photolithography techniques based on the SU-8 process guidelines. Mold infiltration was achieved using a process similar to screen printing except the SU-8 "screen" was attached to the substrate. The squeegee pass via the screen printing action overcomes the Bingham yield point of the suspension and promotes flow of the suspension into the mold cavities.

Following mold infiltration, the parts were gelled, dried, and sintered *in-situ* on the refractory substrate. The parts were gelled in a nitrogen atmosphere near 100 % relative humidity to prevent spallation and drying cracks. Once gelled, the parts were removed from the controlled atmosphere and dried in the ambient atmosphere at 25 °C. The zirconia parts were then fired in air with a soak to allow the SU-8 and gelled binder to combust. The binder thermolysis temperature was determined by thermogravimetric analysis (TGA, Model 2050, TA Instruments). After sintering, fully densified parts were suspended in viscous alumina slurries ($D_{50} \sim 20$ nm) and placed in a wrist action shaker (Burrell Scientific Model 75, Pittsburg, PA) for 10 minutes at maximum oscillation. This process cleaned and dressed the surface, and removed any debris that was remaining from the forming and sintering processes.

RESULTS AND DISCUSSION

Based on preliminary studies, a well-disperse, concentrated suspension is required to prevent drying cracks and non-uniform shrinkage defects in the final ceramic parts. Large capillary forces during drying are associated with the fine pore structure in the packing of well-dispersed ceramic particles.[16, 17] Attrition milling in the presence of ammonium polyacrylate reduced the particle size from approximately 100 μm granules to a suspension with an average agglomeration number of 12 particles per agglomerate.[18] The suspension had a particle size of 114 nm (D_{50} by number) via dynamic light scattering, and remained dispersed with a ζ-potential of -61 ± 3 mV at pH 8.

Using electrostatic dispersion and formulation via the milling and dilatant limit approach, a 46 vol% suspension was the highest zirconia content achievable while maintaining a yield point and viscosity low enough for the mold filling process. The suspensions have shear thinning behavior, which is ideal for the shearing action of the molding process. [12] The viscosity at high shear rate (400 - 500s^{-1} average) and Bingham yield point are plotted as a function of solid loading in Figure 1. The rapid increase in the viscosity and yield point of the suspension occurs between 40 – 45 vol%. This rapid increase, due to steric interactions among particles, [19, 20] is the limiting factor to achieve a zirconia content higher than 46 vol%.

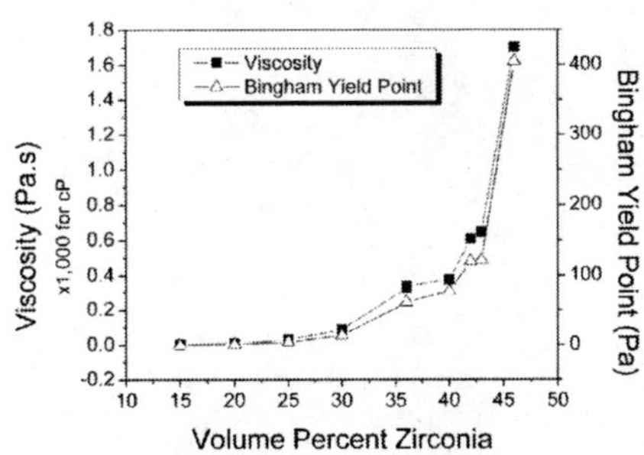

Figure 1. Viscosity and Bingham yield point as a function of zirconia content. The increase in the Bingham yield point and viscosity are attributed to steric interactions between particles, limiting the maximum solid content in suspension.

Surface defects created by spallation result in mechanical failure of the final parts at low stress, limiting part reliability. Spallation of the surface of gelled parts is common when the polymerization reaction is carried out in air due to oxygen hindering the reaction at the surface. Oxygen reacts with free the radical monomers in the gelling suspension and the forms relatively stable free radical peroxides. [21, 22] This side reaction decreases the polymerization rate, and can possibly arrest the polymerization reaction if the concentration of free radical monomers falls below a critical level. [21] A comparison of a part gelcast in N_2 with no spallation and a part gelcast in air with surface spallation is shown in Figure 2.

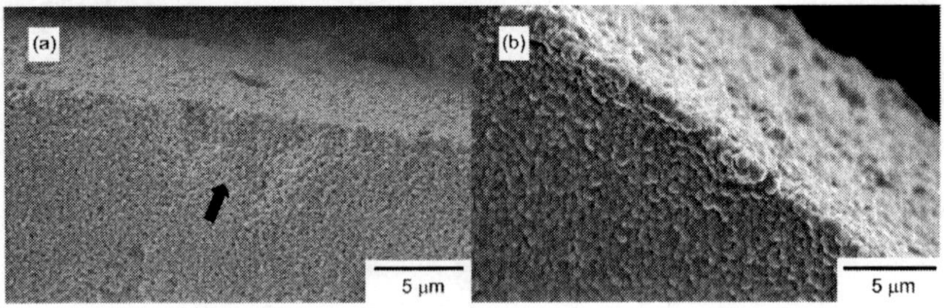

Figure 2. (a) A zirconia square gelcast in air, sintered at 1200°C for 1 hour, showing spallation along the top edge. (b) A zirconia square polymerized in N_2 and 100% relative humidity, sintered at 1300°C for 2 hours, showing no surface spallation.

Mold and polymer burnout is carried out after the gelation and part drying. Rapid mold and polymer combustion results in cracking in the part adjacent to the part/mold interface. Complete organic removal occurs below 550°C as determined by thermogravimetric analysis (TGA). Therefore, a slow ramp to 600°C was implemented, typically 1°C/min with a 2hr hold at 600°C. Following mold and polymer burnout, the parts are sintered at 1300°C to near theoretical density (99.8%). Figure 3 shows an increase in grain size as a function of temperature to a range of 300nm to 500 nm at final density.

Figure 3. Grain size increases with increasing sintering temperature. (a) Green zirconia part (b) Part surface after 1200°C for 2 hours (c) Part surface after 1250°C for 2 hours (d) Part surface after 1300°C for 2 hours.

After sintering, parts are washed with an alumina suspension to remove debris from the casting and sintering processes, and to eliminate surface defects. This step is important because surface defects are detrimental to part reliability. Shown below in Figure 4(a) are the letters P, S, and U made using our process that have not been cleaned using the nanoparticle slurry method and show large defects. Figure 4(b) shows the end of a cleaned test bar specimen showing no surface debris and minimal surface defects. Following rinsing in deionized water, the cleaned parts can then be manipulated or further processed for integration into assembled components.

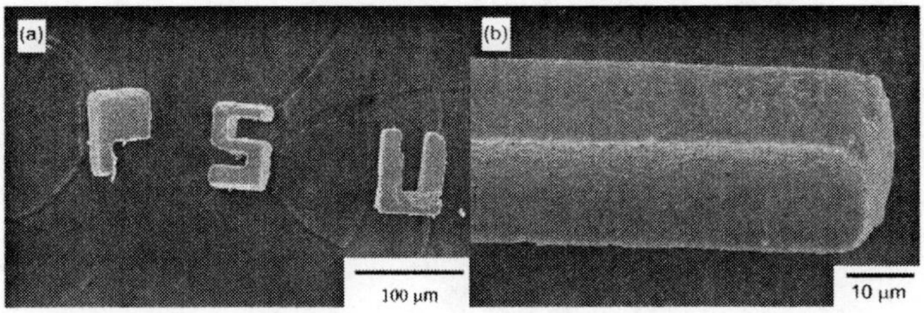

Figure 4. (a) Letters P, S, and U gelcast and fired to 1300°C for 2 hours with no surface cleaning. (b) End of a test bar specimen, gelcast and fired to 1300°C for 2 hours with surface cleaning.

CONCLUSION

Using chemically aided attrition milling, a well dispersed, high solid loading, yttria stabilized zirconia suspension is obtained. Rheological studies show that 46 vol% solid loading is suitable for the screen printing process used to fill the molds due to its shear thinning properties. Gelcasting *in situ*, while controlling the oxygen level in the ambient atmosphere, limits spallation and drying defects in the final parts. Using a controlled firing cycle, uncracked parts with 99.8 % theoretical density were sintered. Fully dense parts were cleaned and dressed by submerging them in an alumina nanoparticle slurry and applying light agitation. Future work will focus on the fabrication of arrays of parts as shown in Figure 4(b) with mechanical testing.

REFERENCES

[1]Robinson, T.N. and G.V. Stiegmann, "Minimally invasive surgery," *Endoscopy*, **36** [1], 48-51, (2004).

[2]Geis, W.P., et al., "Synergistic benefits of combined technologies in complex, minimally invasive surgical procedures - Clinical experience and educational processes," *Surgical Endoscopy-Ultrasound and Interventional Techniques*, **10** [10], 1025-1028, (1996).

[3]Mehta, N.Y., et al., "Sequence and task analysis of instrument use in common laparoscopic procedures," *Surgical Endoscopy and Other Interventional Techniques*, **16** [2], 280-285, (2002).

[4]Booth, K.C.a.D., "Utilization study of ophthalmic instruments," *Alcon Manufacturing, Ltd.,internal report*, (2004).

[5]Richards, W.O. and D.W. Rattner, "Endoluminal and transluminal surgery - No longer if, but when," *Surgical Endoscopy and Other Interventional Techniques*, **19** [4], 461-463, (2005).

[6]Tay, B.Y., J.R.G. Evans, and M.J. Edirisinghe, "Solid freeform fabrication of ceramics," *International Materials Reviews*, **48** [6], 341-370, (2003).

[7]Heule, M., S. Vuillemin, and L.J. Gauckler, "Powder-based ceramic meso- and microscale fabrication processes," *Advanced Materials*, **15** [15], 1237-1245, (2003).

[8]Xia, Y.N. and G.M. Whitesides, "Soft lithography," *Annual Review of Materials Science*, **28** 153-184, (1998).

[9]Martin, C.R. and I.A. Aksay, "Submicrometer-scale patterning of ceramic thin films," *Journal of Electroceramics*, **12** [1-2], 53-68, (2004).

[10]Schonholzer, U.P., R. Hummel, and L.J. Gauckler, "Microfabrication of ceramics by filling of photoresist molds," *Advanced Materials*, **12** [17], 1261-+, (2000).

[11]Su, B., D. Zhang, and T.W. Button, "Micropatterning of fine scale ceramic structures," *Journal of Materials Science*, **37** [15], 3123-3126, (2002).

[12]Hayes, G., et al., "Processing of sub-500 Micron Scale Ceramic Parts for Surgical Applications," *Materials Science & Technology Conference*, (2006). *In press.*

[13]Kimel, R.A. and J.H. Adair, "Aqueous degradation and chemical passivation of yttria-tetragonally-stabilized zirconia at 25 degrees C," *Journal of the American Ceramic Society*, **85** [6], 1403-1408, (2002).

[14]Young, A.C., et al., "Gelcasting of Alumina," *Journal of the American Ceramic Society*, **74** [3], 612-618, (1991).

[15]Janney, M.A., et al., "Development of low-toxicity gelcasting systems," *Journal of the American Ceramic Society*, **81** [3], 581-591, (1998).

[16]Juillerat, F., P. Bowen, and H. Hofmann, "Formation and drying of colloidal crystals using nanosized silica particles," *Langmuir*, **22** [5], 2249-2257, (2006).

[17]Scherer, G.W., "Theory of Drying," *Journal of the American Ceramic Society*, **73** [1], 3-14, (1990).

[18]Adair, J.H., "Colloidal Lessons Learned for Dispersion of Nanosize Particulate Suspensions," *Lessons learned in nanotechnology from traditional and advanced ceramics*, 244, (2005).

[19]Bergstrom, L., E. Blomberg, and H. Guldberg-Pedersen, "Interparticle forces and rheological properties of ceramic suspensions," *Novel Synthesis and Processing of Ceramics*, 119-126, (1999).

[20]Krieger, I.M. and T.J. Dougherty, "A Mechanism for Non-Newtonian Flow in Suspensions of Rigid Spheres," *Transactions of the Society of Rheology*, **3** 137-152, (1959).

[21]Ma, J.T., et al., "Elimination of surface spallation of alumina green bodies prepared by acrylamide-based gelcasting via poly(vinylpyrrolidone)," *Journal of the American Ceramic Society*, **86** [2], 266-272, (2003).

[22]Omatete, O.O., M.A. Janney, and R.A. Strehlow, "Gelcasting - a New Ceramic Forming Process," *American Ceramic Society Bulletin*, **70** [10], 1641-&, (1991).

DEVELOPMENT OF NANOENCAPSULATED CURCUMIN IN CHITOSAN FOR COSMETIC USE VIA EVAPORATION OF O/W/O EMULSION

W. Tanthapanichakoon
National Nanotechnology Center, Thailand Science Park
Klong Luang, Pathumthani 12120, Thailand
Email: wiwut@nanotec.or.th

N. Sowasod, T. Charinpanitkul
Department of Chemical Engineering, Chulalongkorn University
Pathumwan Bangkok 10330, Thailand

ABSTRACT

Chitosan is a biodegradable polymer derived from chitin. Curcumin extract comes from the rhizomes of the plant *Curcuma longa*. In this work curcumin was nanoencapsulated in chitosan cross-linked with tripolyphosphate (TPP) via the technique of evaporation of o/w/o emulsion. The resulting nanocapsules were characterized and the release characteristics of curcumin from the nanocapsules were investigated. It was found that the nanocapsules in the emulsions of interest were somewhat spherical with average particle size of 250 nm, as measured with the Dynamic Light Scattering (DLS) method. After freeze-drying, the observed nanocapsules were found by TEM analysis to be about 150 nm in size. The size discrepancy may be attributed to shrinkage during the evaporation of residual water and solvent as well as the different size analysis techniques. The yield of encapsulation ranged from 17.98 - 51.91 %w/w. FT-IR analysis confirmed that TPP did cross-link with the ammonium groups of the chitosan in the wall of the nanocapsules. Release characteristics of curcumin from the nanocapsules were measured using the USP XXIII shaking basket dissolution apparatus. The nanocapsules were found to display extended release profiles lasting up to 24 hours. Therefore the nanoencapsules display potential applications as subcutaneously injected preparations and skin-care cosmetics.

INTRODUCTION

Natural curcumin (1,7-bis(4-hydroxy-3-methoxy phenyl)-1,6-heptadione-3,5-dione) (Fig. 1) is yellowish and exhibits numerous biological activities such as anti-cancer, anti-inflammatory and anti-oxidation [1, 2]. A tough inert hydrophilic biopolymer insoluble in water [3, 4]. (Fig.1), chitosan displays biocompatibility, biodegradability, non-toxicity and biological properties attributable to the presence of amino and hydroxyl groups. To be soluble in acetic-acid, its degree of deacetylation must be 80-85 % or higher. Tripolyphosphate (TPP) is a nontoxic cross-linking agent for chitosan. It can form gel via ionic interactions between the positively charged amino groups of chitosan and the negatively charged tripolyphosphoric groups ($P_3O_{10}^{-5}$) of TPP. The objective of this study was to nanoencapsulate curcumin in chitosan cross-linked with TPP and to characterize the resulting nanocapsules.

EXPERIMENTAL

Materials

Polymer: chitosan (CS) (molecular weight ca. 227,000, degree of deacetylation 87 %) was purchased from Seafresh Chitosan (Lab), Thailand. Active agent: curcumin was purchased from Thai-China Industry, Thailand. Solvents: acetic acid and ethanol (99.7%) were obtained from Merck, England. Surfactants: sorbitan monooleate (Span 80) was obtained from Fisher Scientific, UK and polyoxyethylene sorbitan monooleate (Tween 80) and sodium tripolyphosphate were obtained from Fluka Chemical, Switzerland. All reagents were of analytical grade except commercial-grade chitosan.

Preparation of Nanocapsules

The methodology for synthesizing nanocapsules via multiple emulsion/solvent evaporation is shown schematically in Fig. 2.

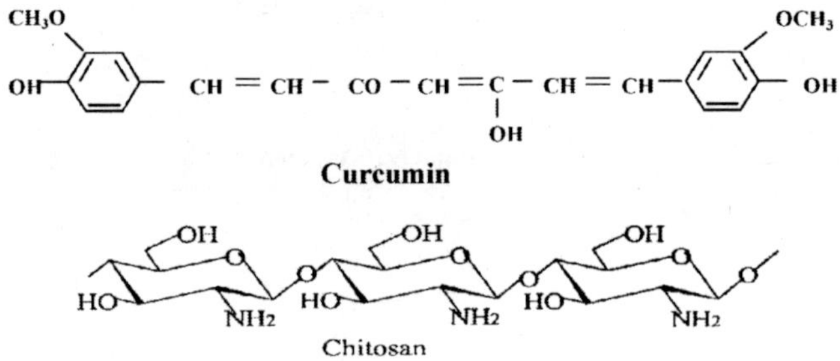

Figure 1. Structures of curcumin and chitosan.

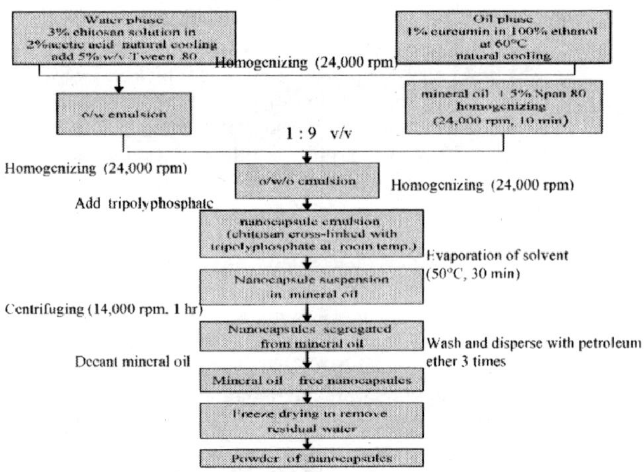

Figure 2. Preparation procedure of nanocapsules.

Characterization Techniques

Particle size: The particle size distribution of the prepared nanocapules in the o/w/o emulsion was measured in the suspension state by the Dynamic Light Scattering (DLS) method with a Nanosizer [Model Nano-ZS, Malvern Instruments Ltd., UK]

Nanocapsules morphologies: The size and morphology of the freeze - dried nanocapsules were investigated by Transmission Electron Microscopy (TEM). [JEM-2010, 200 kV]

FT-IR characterization: Fourier transform infra-red spectroscopy (FT-IR) [JASCO Corporation, Tokyo, Japan] was used to confirm linkage of the tripolyphosphoric groups of TPP with the ammonium groups of CS.

Yield of encapsulation (Y) [5]

$$Y = (L_c/T_c) \times 100 \quad (\%w/w) \qquad (1)$$

where L_c is the mass of curcumin in the loaded nanocapsules and T_c is the total mass of curcumin introduced in the preparation.

Curcumin concentration: it was analyzed by High-Performance Liquid Chromatography (HPLC) analysis [Agilent, Germany]

In vitro release studies [6]

To investigate the time release of curcumin, the nanocapsules (0.01 g) were suspended in 10-ml mixture of phosphate-buffered saline (PBS) (pH 7.4) and ethanol (50:50) in a shaking flask maintained at 37 °C, 150 cycles/min. To find out the released curcumin concentrations, samples (1 ml each) were periodically removed and the volume of each withdrawn sample was replaced by the same volume of fresh medium. The *in vitro* release studies were performed in triplicates for each of the release condition.

Preliminary Tests

- 81 formulations to find a suitable ratio of water phase to oil phase

Formation of nanocapsules were attempted by fixing the ratio of the water to oil phase at 3:1, 4:1 or 5:1 while varying the CS concentration at 2.0, 2.5 or 3%w/v, the surfactant in a blend of 5% Span 80 and 5% Tween 80 (50/50, 60/40 or 70/30 v/v), and TPP concentration of 0.7, 1.0 or 1.5%w/v. Formation of stable emulsion nanocapsules was observed for only some specific concentrations of chitosan while fixing the ratio of the water to oil phase at 4:1 (at CS 3%w/v). Though nanocapsules could be prepared when the ratio of Span 80 to Tween 80 was 70:30 v/v, the emulsion sizes were significantly bigger than the case of 50:50.

- 27 formulations to compare the effect of Span 80 and Tween 80

Formation of nanocapsules were attempted by fixing the ratio of the water to oil phase (4:1) while varying the CS concentration at 2.0, 2.5 or 3%w/v, the surfactant in a blend of 2 % Span 80 and 2 % Tween 80 (50/50, 60/40 or 70/30 v/v), and TPP concentration of 0.7, 1.0 or 1.5%w/v. At CS concentration of 2.0 %w/v, the o/w/o emulsions with 2 % Span 80 and 2 % Tween 80 contained larger particle sizes than the case of 5% Span 80 and 5% Tween 80, as analyzed by the Nanosizer

RESULTS AND DISCUSSION

The nanocapsules were synthesized by fixing the ratio of the water to oil phase at 4:1 while varying the CS concentration at 2.5 or 3%w/v, the surfactant blend (5% Span 80 and 5% Tween 80) at 50/50 or 60/40 v/v, and TPP concentration of 0.7, 1.0 or 1.5%w/v. The experimental results were summarized in Table I. The various batches of nanocapsules in the emulsion state had average size range of 253.8 - 286.4 nm. The yield of encapsulation was found to range from 17.98 - 51.91 % w/w.

Particle size of nanocapsules in o/w/o emulsion

Table I reveals that a surfactant blend of Span 80 and Tween 80 at 50/50 v/v gave smaller average particle sizes than the case of 60/40 v/v. In fact, the emulsions were stabilized by a combination of the two surfactants at 50/50 v/v because their final hydrophilic-lipophilic balance (HLB) = 10.3. Formulation B_5 was found to give the smallest average capsule size of 253.8 nm and the highest encapsulation yield of 51.91 %w/w.

Table I. Average particle size and % yield of encapsulation

CS (%w/v)	TPP (%w/v)	Span 80/ Tween80	Formulation code	Emulsion (Nanosizer) Average size (nm)	Emulsion (Nanosizer) P.I.	Freeze - dried (TEM) Average size (nm)	Freeze - dried (TEM) Yield of encapsulation
2.5	0.7	50/50	A_1	256.5	0.103	*	*
		60/40	A_2	266.5	0.174	*	*
	1	50/50	A_3	258.1	0.149	*	*
		60/40	A_4	258.1	0.151	*	*
	1.5	50/50	A_5	261.6	0.278	600	17.98
		60/40	A_6	268.9	0.270	188.9	37.64
3	0.7	50/50	B_1	276.1	0.199	*	*
		60/40	B_2	268.7	0.146	*	*
	1	50/50	B_3	258.5	0.165	*	*
		60/40	B_4	286.4	0.234	*	*
	1.5	50/50	B_5	253.8	0.262	156.8	51.91
			B_5(repeat 1)	256.9	0.531	145.5	44.55
			B_5(repeat 2)	272.7	0.218	150.8	49.11
			B_5(repeat 3)	291.2	0.189	172.7	39.11
		60/40	B_6	282.4	0.157	222.2	43.77

P.I., polydispersity index

* After freeze-dried at -80 °C for 48 h, the resulting nanocapsules agglomerated.

Particle size and shape of freeze-dried nanocapsules

The average sizes of the o/w/o emulsions and freeze-dried nanocapsules were compared in Table I. The size decrease after freeze drying is attributable to the evaporation of the residual oil in the emulsion. From the TEM images of B_5 shown in Fig. 3 (a and b) as examples, the freeze-dried nanocapsules are about 150 nm in size and somewhat spherical in shape.

Fourier transform infra-red spectroscopy (FT-IR)

Fig.4 shows the FT-IR spectra of pure chitosan and some of the nanocapsules. Compared to pure chitosan, the characteristics of the FT-IR peaks confirmed that the tripolyphosphoric groups of TPP cross-linked with the amino groups of chitosan in the nanocapsule walls, thereby causing gelation of chitosan. For example, the valley at 3434.16 cm^{-1} in Fig. 4 can be attributed to the $-NH_2$ and $-OH$ groups in the pure chitosan. In the case of B_5, a shift from 3434.2 to 3354.1 cm^{-1} was observed, which indicates enhanced hydrogen bonding. Meantime the 1641.5 and 1580.0 cm^{-1} peaks of $-NH_2$ bending vibration shifted to 1625.9 and 1572.2 cm^{-1}, respectively. The shifts are attributed to the linkages between the phosphoric and amino ions [7].

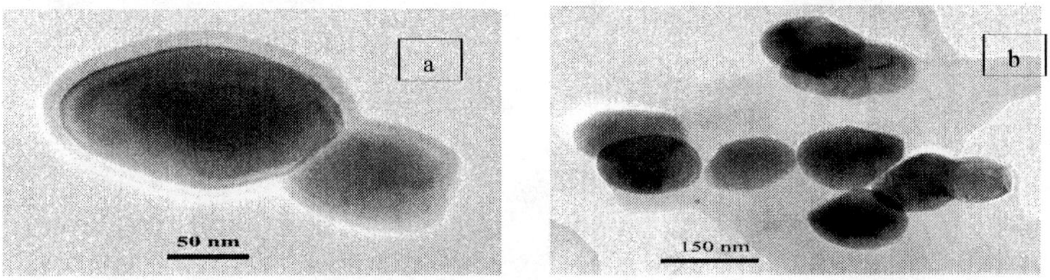

Figure 3. (a and b) TEM photographs of nanocapsules (Formulation B_5 in Table I).

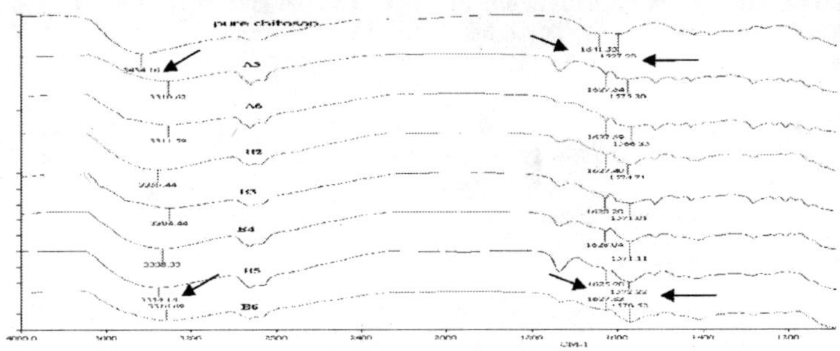

Figure 4. FT-IR spectra of pure chitosan and chitosan nanocapsules.

Yield of encapsulation

The yields of encapsulation are listed in Tables I. If the number of particles (N_p), internal volume (V_{inside}) and total surface (S_{total}) of the nanocapsules are used to predict the characteristics of ideal nanocapsules with chitosan wall thickness of 20 nm and average size of 200 nm, the maximum contents of curcumin should be 0.0359 g when the chitosan weight was

0.2418 g. However, the experimental results in the case of formula B_5 were: average size 253.8 nm, chitosan wall thickness 15.8 nm and the curcumin content 0.02 g. This indicates that the encapsulation efficiency could still be improved.

In vitro release study

Fig. 5 shows the time behaviors of curcumin released from chitosan nanocapsules prepared with TPP concentration at 1.5%w/v. In general, as the concentration of chitosan rose from 2.5% (A_5 and A_6) to 3%w/v (B_5 and B_6), the release rate was expected to decelerate. The swelling property and gel formation of chitosan in the nanocapsule walls were reported to play an important role on the sustained release behavior. In the present work, curcumin was released in PBS solution with ethanol (50:50) (pH 7.4) and the release was prolonged up to 24 hr.

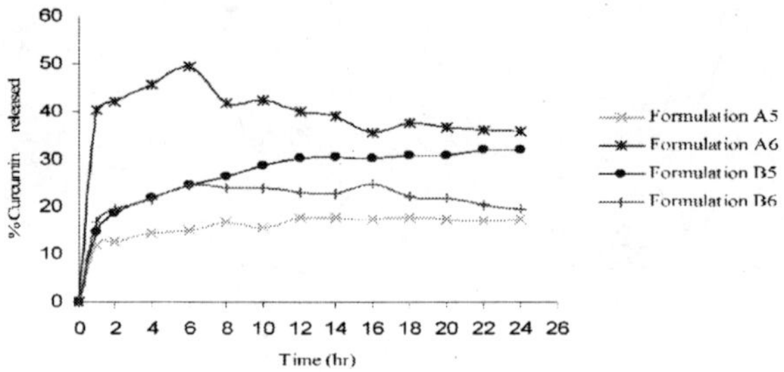

Figure 5. Release profiles of curcumin from nanocapsuls (formulation A_5, A_6, B_5, and B_6)

CONCLUSION

Under the investigated conditions, the freeze-dried nanocapsules had average size of 253.8 - 268.4 nm, and yield of encapsulation of 17.98-51.91 %w/w. Formula B5 was found to be most suitable. The average size could still be reduced and the yield improved by optimizing the synthesis conditions.

ACKNOWLEDGEMENT

The authors are grateful to Mr. Pricha Sangthirapithikul and Dr. Uracha Ruktanonchai for their valuable suggestions and assistance, and to National Nanotechnology Center as well as Center of Excellence in Particle Technology, CU, for some use of its facilities.

REFERENCES

[1]S. Dutta, S. Padhye, K. I. Priyadarsini, and C. Newton, Antioxidant and antiproliferative activity of curcumin semicarbazone, *Bioorganic & Medicinal Chemistry Letters*, **15**, 2738 – 2744 (2005).
[2]W. M. Weber, L. A. Hunsaker, S. F. Abcouwer, Lorraine M. Deck, and L. Vandert Jagt, Anti-oxidant activities of curcumin and related enones, *Bioorganic & Medicinal Chemistry*, **13**, 3811 - 3820 (2005).
[3]I. Higuera-Ciapara, L. Felix-Valenzuela, F. M. Goycoolea, and W. Arguelles-Monal, Micro-encapsulation of astaxanthin in a chitosan matrix, *Carbohydrate polymer*, **56**, 41 - 45 (2004).

[4]B. Krajewska, Membrane-based processes performed with use of chitin/chitosan materials, *Separation and Purification Technology*, **41,** 305-312 (2004).
[5]K. Bouchemal, S. Briancon, E. Perrier, H. Fessi, I. Bonnet, and N. Zydowicz, Synthesis and characterization of polyurethane and poly(ether urethane) nanocapsules using a new technique of interfacial polycondensation combined to spontaneous emulsification, *Int. J. Pharm.*, **269**, 89 - 100 (2004).
[6]J. A. Ko, H.J. Park, S.J. Hwang, J.B. Park, and J.S. Lee, Preparation and characterization for controlled drug delivery, *Int. J. Pharm.*, **249**, 165 - 174 (2002).
[7]Y. Xu, Y.Du, Effect of molecular structure of chitosan on protein delivery properties of chitosan nanoparticles, *Int. J. Pharm.*, **250**, 215-226 (2003).

STRUCTURAL AND MORPHOLOGICAL STUDY OF NANOCERAMICS PREPARED BY SPRAY PYROLYSIS

L.S. Gómez, M.E. Rabanal, J.M. Torralba
Materials Science Department, Universidad Carlos III de Madrid,
Avda de la Universidad 30, 28911, Leganés, Madrid, Spain

L. Mancic, O. Milosevic
Institute of Technical Sciences of Serbian Academy of Sciences and Arts,
K.Mihajlova 35/IV, 11000 Belgrade, Serbia

ABSTRACT

Gadolinium oxide doped with europium in different atomic proportions (0.09/0.01, 0.08/0.02, 0.07/0.03, 0.06/0.04) was produced by spray pyrolysis. The stoichiometric droplets of aerosol, generated with an ultrasonic device, produced nanoparticles after dehydration, decomposition and a reaction process in a unique step at intermediate temperatures (700ºC). The "as-prepared particles" were annealed isothermally at 800-1200^{0}C for 12 h in air in order to improve the crystallinity and functional properties. The structural and morphological characterization was done by means of x-ray diffraction (XRD), transmission electron microscopy (TEM), and scanning electron microscopy (SEM). Detailed phase analysis was carried out by combining XRD and electron diffraction. Also, an evaluation of the particle size distribution and special characteristics associated with nanoparticles including "primary nanoparticles" and "crystallite size" evolution was completed by combining TEM-HRTEM and field emission scanning electron microscopy (FESEM).

INTRODUCTION

Development of high-resolution flat panel displays (FPD) such as liquid crystal displays (LCD's), plasma displays (PD), thick or thin film electroluminescent (TFEL's) displays and field emission displays (FED) requires phosphor particles with good morphological and luminescence characteristics[1]. Europium activated gadolinium oxide has been extensively used in many different applications[1]. For good luminescent characteristics, phosphor particles have to be fine size and narrow size distribution, non-aggregated, spherical morphology, and have good crystallinity. Gadolinia exhibits three polymorphic forms classified as being hexagonal, monoclinic, and cubic,[2] and each of them particularly affects the luminescent behaviour in combination with the luminescence centre.

Spray pyrolysis has been recognized as a successful method to produce powders with high purity over a wide range of composition.[3,4] It involves passing the aerosol of a precursor solution through a temperature-graded tubular flow reactor, in which the droplets decomposed to form the oxide particles[5]. Also, a nearly uniform chemical and phase composition can be achieved, because each of the liquid droplets acts as a small reactor[6,7,8]. Several important processing parameters are able to be controlled: the carrier gas flow rate, residence time, temperature profile inside the reactor and heating and cooling rates. The aim of this work is to evaluate the structural and morphological properties of the gadolinia-doped europia phosphor particles having different concentrations of Eu^{3+} ions as a luminescence center, synthesized by spray pyrolysis.

EXPERIMENTAL

The Gd_2O_3:Eu^{3+} precursor solutions were prepared from gadolinium (III) nitrate hexahydrate and europium (III) nitrate pentahydrate (both Aldrich 99.9%). The stoichiometric amounts were dissolved in water to reach molar ratio of Gd/Eu: 0.09:0.01(nn1), 0.08:0.02 (nn2), 0.07:0.03 (nn3), 0.06:0.04 (nn4), with the solution concentration of 0.1mol/dm^3. A 2.1MHz (RBI) ultrasonic atomizer was used for aerosol generation. The droplets decomposed into a tubular flow reactor at 700°C. The carrier gas flow rate was 10 l/min and the residence time was 75 s. The as-prepared particles were post-treated from 800 to 1200°C for 12h for further crystallization. The crystal structure was studied by XRD (X`Pert Philips automatic diffractometer with CuKα radiation). A Rietveld based Fullprof program [9] was used for the lattice parameter calculation. The structure and morphology of particles were investigated using a Field Emission Scanning Electron Microscopy (JEOL 6330F, resolution 12Å) and Transmission Electron Microscopy (JEOL JEM 4000 EX, 400KV, point resolution 1.8Å, and a TEM LIBRA, 120 KV, with a corrected 90° in a column OMEGA for EELS (parallel measurement) and element specific images (elemental maps)), together with selected area electron diffraction pattern (SAEDP). EELS (electron energy loss spectra) provided the information about local coordination of the atoms.

RESULTS

Fig. 1a shows the XRDP evolution of the sample nn3 (Gd/Eu: 0.07/0.03) before and after the annealing. In the *"as-prepared"* sample, two crystalline phases are identified: the main cubic Gd_2O_3 (JCPDS: 43-1014), *Ia3* (SG=206) making up 72wt%, and the secondary one, *Fm-3m* (SG=225, similar to JCPDS: 37-1400) making up 22wt%. In annealed samples, only the *Ia3* phase is identified by XRD. The refinement of all samples, done by Rietveld method, started with the structure model of cubic-Gd_2O_3 form (JCPDS: 43-1014), *Ia3*, a= 10.813 Å and atomic positions for Gd1 in 8a, Gd2 in 24d and O in 48e. It was found that the crystallite size increase with the annealing temperature.[10, 11]. In all systems, the cell parameter is higher than the theoretical one (file card 43-1014), indicating the incorporation of Eu^{3+} into the gadolinia matrix, since the Eu^{3+} ionic radii is slightly larger than that of Gd^{3+} (Eu^{3+} - 0.095nm; Gd^{3+} - 0.094nm). Fig. 1b represents the variation of cell parameters with thermal treatment temperature for the different Gd/Eu molar ratio. The increase of the cell parameter is in the same proportion as the gadolinium-to-europium molar ratio used in the common precursor solutions.[12]

FESEM was used in order to identify the particle morphology produced by spray pyrolysis. Fig. 2 shows FESEM images taken with a secondary electron mode detector of the sample with a nominal Gd/Eu ratio 0.08/0.02 (nn2), before and after the thermal treatment in investigated range, showing the changes in morphology and size of the particles. A typical spherical particle shape is formed before annealing at 1100°C/12 (Fig. 2a-d). After this temperature, a "*sintering neck*" is observed (Fig. 2e), indicating the collision and aggregation of particles. At 1200°C/12h agglomeration is more pronounced, and a polyhedral particle shape is evident (Fig. 2f).

Fig. 3a shows a TEM image taken at intermediate magnification in bright field mode of the sample having a Gd/Eu ratio 0.08/0.02 (nn2) after a treatment at 900°C/12h,. Observations reveal the presence of primary nanoparticles ranging in size from 20-38 nm, with polyhedral to spheroidal shapes.

The EELS spectra for the samples with nominal Gd/Eu ratio 0.09/0.01 (nn1) and 0.08/0.02 (nn2) is shown in Fig.3b. The energy loss spectra of the rare earths are characterized by sharp *M* 4, 5 edges. Notice that in the range of 1185-1217 eV there are two intense peaks at~ 1194eV and ~1225 eV, corresponding to the Gd in the *M*5 and *M*4 shells, respectively, as well as the peaks with less intensity, corresponding to the Eu in the *M*5 and *M*4 shells (~1131eV-1163 eV). The comparison between the spectra intensities (nn1 and nn2) implies the increasing in the Eu content and the corresponding decreasing of Gd content for

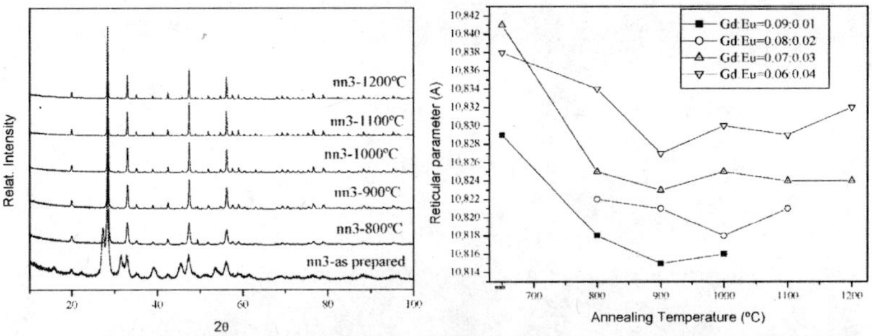

Figure 1.The temperature dependence of XRPD (nn3) (a), and the cell parameters (b).

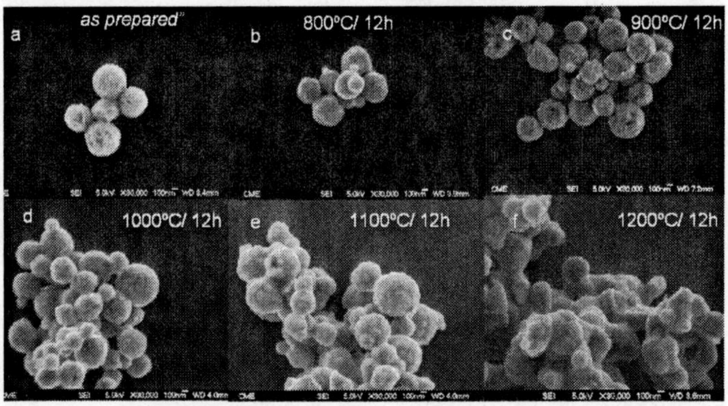

Figure 2. FESEM images in secondary electron mode of nn2 sample

Fig. 3c shows a SAEDP for a sample with molar ratio Gd/Eu, 0.08/0.02 after annealed at 900°C/12hours, showing the polycrystalline character. Based on the ring patterns, the cubic *Ia3* phase with the cell parameter a~ 10.81Å (JCPDS: 43-1014) is identified. In Fig. 3d the SAEDP reveals the presence of a monoclinic phase with the approximate unit cell parameters a~13.6 Å, b~3.56 Å and c~8.6Å, ß~100.6°, formed in the same sample after treatment at 1100°C/12hours, as presented elsewhere.[12] The formation of this metastable phase is probably associated with the short residence time as well as high heating and cooling rates during synthesis.

CONCLUSIONS

Gd_2O_3: Eu^{3+} un-agglomerated spherical particles with different doping concentrations of europium were synthesized ultrasonically from common nitrates solutions. Detailed phase and structural analyses completed using XRPD, TEM-HRTEM and FESEM. XRPD revealed the presence of two crystalline phases in the as-prepared powders: a cubic *Ia3* and a *Fm-3m* phase. After annealing, XRPD revealed only the cubic *Ia3* phase, with the cell parameters affected by Eu^{3+} doping. HR-TEM investigation revealed the appearance of a locally metastable monoclinic structure at temperatures around 1100°C.

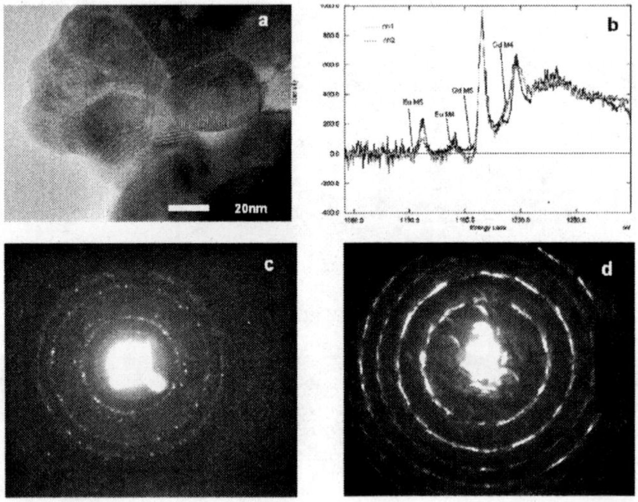

Figure 3. HR-TEM bright field image of nn2 after 900 °C/12h (a); Parallel EELS spectra for nn1 and nn2 (b); SAEDP of nn2 after 900°C/12 h, (c) SAEDP of nn2 after 1100°C/12 h (d).
nn2, in agreement with the nominal ratio Gd/Eu ratio, and indicates the same oxidation state for Gd and Eu in both samples.

ACKNOWLEDGEMENT

This research is supported though the Comunidad de Madrid (ESTRUMAT-CM, MAT/77), Ministry of Science and Environment Protection, Serbia (#142010) and COST 539 Action. Authors thank to the J. L. Baldonedo, Centro de Microscopía "Luis Bru" and Nanotechnology System Division, FEI for TEM observations. Coauthor, O.M., acknowledges the support of the Ministry for Education and Science, Spain (SAB 2004-0035).

REFERENCES

[1] Rack, P.D., Holloway, P.H, "The structure, device physics and materials properties of thin film electroluminescent displays", *Mater.Sci.Eng.*, **4,** 171-219, (1998)

[2] A.M. Pires, M.R. Davolos, C. O. Paiva-Santos, E. B. Stucchi and F. Flor, "New X-ray powder diffraction data and rietveld refinement for Gd2O3 monodispersed fine spherical particles" *J. Solid State Chem.* **171** (1-2), 420-423 (2003)

[3] G.H.Haertling, "Ferroelectric ceramic: history and technology" *J. Am. Ceram. Soc.* **82,** 797-818, (1999)

[4] K. Okuyama, I. W. Lenggoro, "Preparation of nanoparticles via spray route" *Chem. Eng. Sci.,* **58,** 537-547, (2003)

[5] P.S. Patil, "Versatility of chemical spray pyrolysis technique" *Mater. Chem. Phy.* **59** (3) 185-198, (1999)

[6] G.L. Messing, S.C. Zhang, G.V. Jayanthi, "Ceramic powder synthesis by spray pyrolysis" *J. Am. Ceram. Soc.* **76** (11) 2707-2726, (1993)

[7] A.H. Lefebvre, Atomization and Sprays, Hemisphere, New Cork, 1989

[8] G.V. Jayanthi, S.C. Zhang, G.L. Messing, "Modeling of solid particle formation during solution aerosol thermolysis-the evaporation stage" *J. Aerosol Sci. Techn.* **19** (4) 478-90, (1993)

[9] Rodríguez Carvajal "Rietveld Pattern Matching Analysis of powder pattern". ILLL Grenoble 1990

[10] O. Milosevic, L. Mancic, M.E. Rabanal, B. Yang and P.D. Townsend, "Structural and luminescence properties of $Gd_2O_3:Eu^{3+}$ and $Y_3Al_5O_{12}:Ce^{3+}$ phosphor particles synthesized *via* aerosol", *J. Electrochem. Soc.*, **152,** 707-13, (2005)

[11] M.E. Rabanal, L. Gómez, A. Khalifa, L. Mancic, O. Milosevic, J.M. Torralba, " Structural properties of europium-doped gadolinia synthesized through aerosol" *J. Europ. Ceram. Soc.,* in press, (2006)

[12] Y Wang, O Milosevic, L. Gomez, M. E. Rabanal, J.M.Torralba, B Yang, and P D Townsend, "Thermoluminescence responses from europium doped gadolinium oxide" *J. Phys: Cond. Matter.* **18** (40) 9257-9272, (2006)

INFLUENCE OF WATER CONTENT OF STARTING POWDER MIXTURE ON THE MECHANOCHEMICAL SYNTHESIS OF STRONTIUM DOPED LANTHANUM MANGANITE

Jintawat Chaichanawong, Kazuyoshi Sato, Hiroya Abe, Makio Naito
Joining and Welding Research Institute, Osaka University,
11-1 Mihogaoka, Ibaraki, Osaka, 567-0047 Japan

Tawatchai Charinpanitkul
Department of Chemical Engineering, Chulalongkorn University,
Payathai Road, Pathumwan, Bangkok, 10330 Thailand

Wiwut Tanthapanichakoon
National Nanotechnology Center,
Thailand Science Park, 111 Paholyothin Road, Klong Luang, Pathumthani 12120 Thailand

ABSTRACT

The influence of water content of starting powder mixture on the mechanochemical synthesis of $La_{0.8}Sr_{0.2}MnO_3$ (LSM) powder was investigated. Industrial grade La_2O_3, $SrCO_3$ and Mn_3O_4 powders were used as starting materials. The water content of the starting powder mixture was varied up from dry state to 2.0 mass%. With water content of 2.0 mass%, disordering of crystalline starting powders rapidly proceeded. However, single phase LSM powder was not obtained. With lower water contents, fine grinding of the mixture occurred at an early stage of the milling and single phase LSM powder was obtained after 120 min of milling. The specific surface area (SSA) of the LSM was 5.0 m^2/g and the average particle size calculated from the SSA was approximately 180 nm.

INTRODUCTION

Strontium doped lanthanum manganite (LSM) is an attractive material because of its good properties for a wide range of potential applications such as colossal magneto-resistance[1] and catalytic activities[2]. In addition, LSM is a candidate cathode material for the solid oxide fuel cell (SOFC) because of its good electrical conductivity, catalytic activity for reduction of oxygen, and chemical and thermal compatibilities with other SOFC components[3]. Mechanochemical methods have been investigated for the synthesis of various fine particles including LSM powder[4,5]. Recently, Sato et al.[6] have demonstrated that $LaMnO_{3+\delta}$ (LM) can be mechanochemically synthesized after milling for only 30 min from industrial-grade La_2O_3 and Mn_3O_4 powders. An attrition type milling apparatus was used without any media balls, and a special feature of the milling was to do it under humid atmosphere. The fine grinding sufficient to trigger the mechanochemical synthesis of LM was achieved at an early stage of the milling (10 min), mainly due to the formation of $La(OH)_3$ on the surface of La_2O_3 powder.

The influence of water or hydroxyl groups on mechanically induced reaction (soft mechanochemical reactions) has extensively been investigated and reported by Senna and co-workers[7,8]. Senna[7] reported that rapid complexation was enhanced by the presence of extraordinarily polarized hydroxyl groups due to structural imperfections of the hydroxyl supporting substrates. Mechanochemical reactions of hydroxides or hydrous compounds have also been reported[9]. Avvakumov et al.[10] reported that mechanochemical reactions in mixtures of hydrated oxide occurred faster than those in anhydrous oxide mixtures. However, it has been reported that after mechanochemical reaction, the water in Cr_2O_3-nH_2O has been transferred as free water, which works like a binder combining fine particles in agglomerates into one large particle[11]. The proper water content in a starting powder for the synthesis of fine particles without reducing the reactivity is under investigation.

In this study, mechanical milling without any media balls was conducted on industrial-grade La_2O_3, $SrCO_3$ and Mn_3O_4 powders by varying the water content of these starting powder mixtures. Phase evolution and specific surface area of the milled powders are shown, and the influence of the water content of the starting powder mixture on the formation of LSM will be discussed. The proper water content of starting powder mixtures for the synthesis of LSM powder will also be reported.

EXPERIMENTAL

The starting materials were industrial-grade La_2O_3 (purity: 99.99%, KCM Corp., Japan), $SrCO_3$ (97%, KCM Corp., Japan) and Mn_3O_4 (99%, KCM Corp., Japan) powders. Their average particle sizes as calculated from the specific surface area were about 800, 400 and 70 nm, respectively. They were mixed at the stoichiometric molar ratio of cations as La: Sr: Mn = 0.8: 0.2: 1. The water content in the starting powder mixture was varied up to 2.0 mass% by exposing the dry starting powders to humid atmosphere (RH 80%, 25 °C). The powder mixture was mechanically activated using the attrition type milling apparatus[6]. The rotating speed of the rotor was 3000 rpm and the total milling time was 120 min.

Phase evolution during the milling was examined by XRD measurement (JDX-3530M, JEOL, Japan) using Ni filtered CuKα radiation as X-ray source. The XRD patterns were collected in the range of 10-70° with a step size of 0.02°. The specific surface areas (SSA) of the starting and the milled powder mixtures were measured with a nitrogen gas adsorption instrument (Micromeritics ASAP 2010, Shimadzu, Japan) based on the BET method. The morphology of the LSM powder was observed by scanning electron microscopy (SEM, ERA-8800FE, Elionics, Japan).

RESULTS AND DISCUSSION

Figures 1 (a) and (b) show the phase evolution of the powder mixture as a function of the milling time, the moisture contents of the starting powder mixture being < 0.2 and 0.8 mass%, respectively. The phase evolutions in both cases exhibited similar behavior. Peaks of the individual starting powders, La_2O_3, $SrCO_3$ and Mn_3O_4, were clearly observed before the milling. After milling for 10 min, the peak intensities of the starting powders slightly decreased. After milling for 20 min, the peak intensities of the starting powders decreased further. In addition, the peaks corresponding to the LSM started to appear. This suggests that the mechanical forces applied by milling were utilized efficiently for the formation of LSM during the first 20 min. After that continued milling reduced the peak intensities of the starting powders. Single phase LSM was successfully obtained after milling for 120 min. In a third case in which the water content of the starting powder mixture was 2.0 mass%, although disordering of the starting crystalline powders rapidly proceeded, single phase LSM was not obtained even after the powder mixture was milled for 120 min, as shown in Figure 1. (c). It should be noted that the peaks identified as $La(OH)_3$ appeared at the start and at 10 min of milling. The formation of $La(OH)_3$

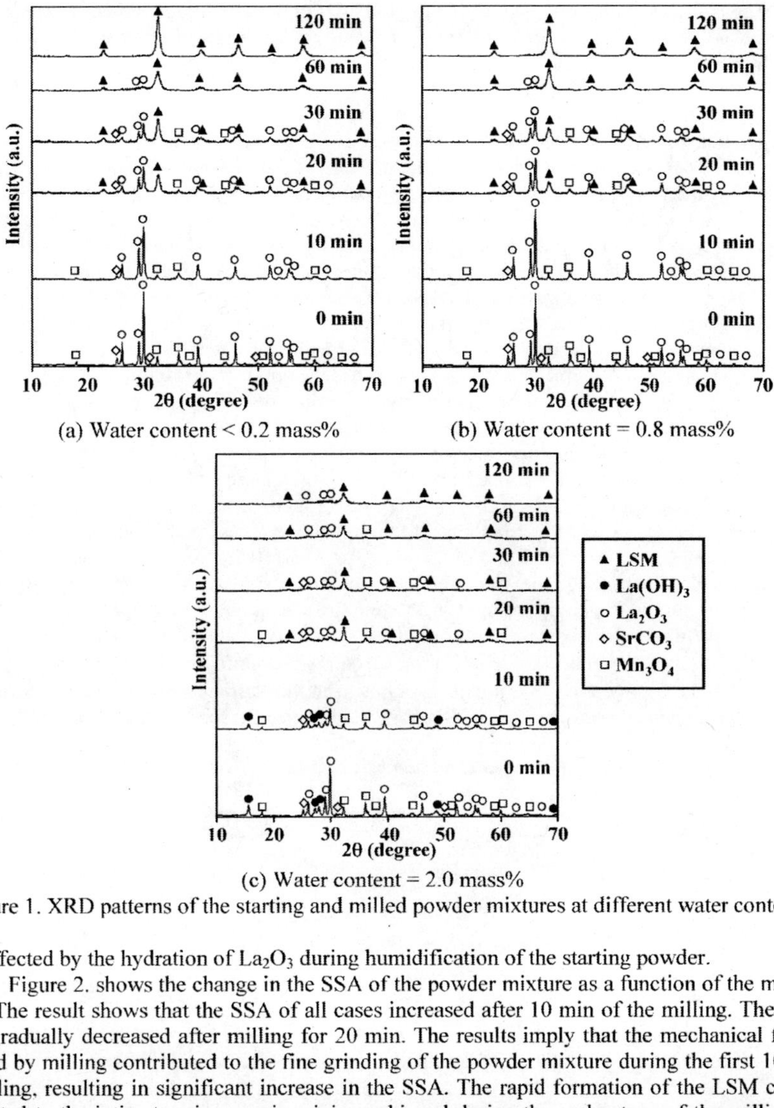

Figure 1. XRD patterns of the starting and milled powder mixtures at different water content.

was effected by the hydration of La_2O_3 during humidification of the starting powder.

Figure 2. shows the change in the SSA of the powder mixture as a function of the milling time. The result shows that the SSA of all cases increased after 10 min of the milling. Then, the SSA gradually decreased after milling for 20 min. The results imply that the mechanical forces applied by milling contributed to the fine grinding of the powder mixture during the first 10 min of milling, resulting in significant increase in the SSA. The rapid formation of the LSM can be attributed to the intimate microscopic mixing achieved during the early stage of the milling. In the case of water content lower than 0.2 mass%, the SSA again increased after milling for 30 min, then there was no significant change until 60 min. The SSA significantly increased after milling for 120 min. A similar behavior has been reported on the synthesis of LSM powder by milling constituent oxides of La_2O_3-SrO-Mn_2O_3-MnO_2 in dry condition[5]. When the initial water content was 0.8 mass%, the SSA did not significantly change after further milling for 30-120 min. On the other hand, at 2.0 mass% initial water content, the prolonged milling resulted instead in a significant decrease of the SSA.

As seen from the results, after the SSA reached a maximum (around 10 min), the peaks corresponding to the LSM on the XRD patterns appeared and those of the starting powders significantly decreased. This confirms that the increase in the SSA and disordering of crystalline structures of the raw substances were of vital importance for solid state reaction of LSM in the present milling. In this case, the presence of proper water content in the starting powder is shown to be an important factor. La_2O_3 is easily hydrated in the presence of water vapor, while Mn_3O_4 and $SrCO_3$ are not. Generally, the mechanical hardness of metal hydroxides is significantly lower than that of their corresponding oxides. Therefore, the Mn_3O_4 and $SrCO_3$ powders act as media balls and preferentially ground the $La(OH)_3$ formed on the surface of La_2O_3 powder, resulting in an increase in the SSA. Consequently, a large area of contacts between the reactant substances is achieved[6]. Such a microscopic mixing state promotes the solid state reaction of LSM. However, after mechanochemical reaction, the excess water in the starting powder worked as a binder combining the fine particles into agglomerates that merged into large particles[11]. In the present case, it is obvious that, when the water content of the starting powder increased to 2.0 mass%, the SSA decreased as the milling time increased. No single phase LSM was obtained. This result implies that strong agglomeration has taken place rather than the mechanochemical reaction.

Figure 3. shows SEM micrographs of the products synthesized by mechanical milling for 120 min. When the water contents were less than 0.8 mass% (Figure 3. (a) and (b)), the milling products seemed to be agglomerates. These agglomerates consisted of fine primary particles about several hundred nanometers in size. In contrast, strong agglomeration was observed at a water content about 2.0 mass% (Figure 3. (c)). The size of the resulting primary particles observed in the micrograph was about 1 μm. These results confirm that the formation of large particles may be ascribed to the excess water content in the starting powder mixture. At the two lower water contents, less than 0.2 and 0.8 mass%, the specific surface area (SSA) of the LSM

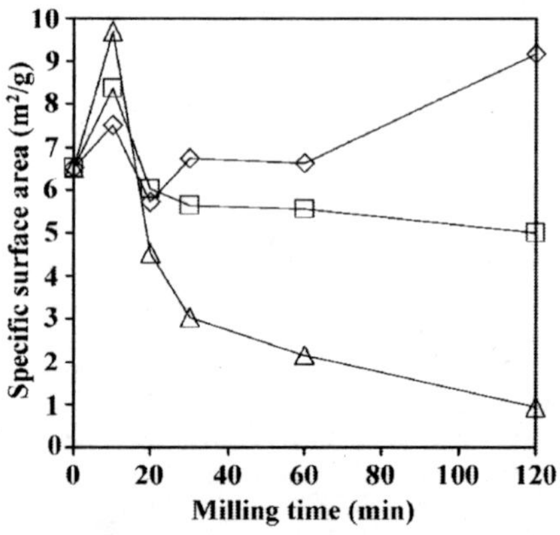

Figure 2. Specific surface area (SSA) of starting and milled powders at different water content: ◇ < 0.2 mass%; □ 0.8 mass%; △ 2.0 mass%.

(a) Water content < 0.2 mass%

(b) Water content = 0.8 mass%

(c) Water content 2.0 mass%

Figure 3. SEM micrograph of LSM powder synthesized by milling powder mixture for 120 min.

was 9.2 and 5.0 m^2/g, respectively. Their particle size calculated from the SSA was approximately 100 and 180 nm, respectively.

CONCLUSIONS

In this study, the influence of the water content up to 2.0 mass% of the starting powder mixture on the mechanochemical synthesis of $La_{0.8}Sr_{0.2}MnO_3$ (LSM) powder was investigated with attrition type milling apparatus. At the two of the lower water contents (less than 0.2 and at 0.8 mass%), fine grinding of the mixture occurred at the early stage of the milling and single phase LSM powder was obtained after 120 min of milling. The specific surface area (SSA) of the LSM was 9.2 and 5.0 m^2/g and the particle size calculated from the SSA was about 100 and 180 nm, respectively. In contrast, at a water content of 2.0 mass%, disordering of the starting crystalline powders rapidly proceeded while strong agglomeration was observed. However, no single phase LSM powder was obtained at the higher water content.

ACKNOWLEDGEMENTS

This work was partially supported by Grand-in-Aid for Cooperative Research Project of Nationwide Joint-Use Research Institute on Development Base of Joining Technology for New Metallic Glasses and Inorganic Materials from The Ministry of Education, Culture, Sport, Science and Technology, Japan. J. Chaichanawong and W. Tanthapanichakoon also received research grants from the Thailand Research Fund (TRF) [Grant No. PHD/0135/2546 of Royal Golden Jubilee Ph.D. Program].

REFERENCES

[1]D. Grossin, and J.G. Noudem, "Synthesis of Fine $La_{0.8}Sr_{0.2}MnO_3$ Powder by Different Ways," *Solid State Sci.*, **6**, 939-44 (2004).

[2]T. Hibino, K. Suzuki, K. Ushiki, Y. Kuwahara, and M. Mizuno, "Ultra-fine Grinding of $La_{0.8}Sr_{0.2}MnO_3$ Oxide by Vibration Mill," *Appl. Cat. A: General*, **145**, 297-306 (1996).

[3]N.Q. Minh, "Ceramic Fuel Cells," *J. Am. Ceram. Soc.*, **76**[3], 563-88 (1993).

[4]Q. Zhang, and F. Saito, "Mechanochemical Synthesis of $LaMnO_3$ from La_2O_3 and Mn_2O_3 Powders," *J. Alloys and Comp.*, **297**, 99-103 (2000).

[5]Q. Zhang, T. Nakagawa, and F. Saito, "Mechanochemical Synthesis of $La_{0.7}Sr_{0.3}MnO_3$ by Grinding Constituent Oxides, *J. Alloys and Comp.*, **308**, 121-5 (2000).

[6]K. Sato, J. Chaichanawong, H. Abe, and M. Naito, "Mechanical Synthesis of $LaMnO_{3+\delta}$ Fine Powder Assisted with Water Vapor," *Mater. Lett.*, **60**, 1399-402 (2006).

[7]M. Senna, "Grinding of Mixture under Mild Condition for Mechanochemical Complexation," *Int. J. Miner. Process*, **44-45**, 187-95 (1996).

[8]T. Watanabe, T. Isobe, and M. Senna, "Mechanisms of Incipient Chemical Reaction between $Ca(OH)_2$ and SiO_2 under Moderate Mechanical Stressing: I: A Solid State Acid–Base Reaction and Charge Transfer Due to Complex Formation," *J. Solid State Chem.*, **122**, 74-80 (1996).

[9]T. Temuujin, K. Okada, and K.J.D. MacKenzie, "Role of Water in the Mechanochemical Reactions of $MgO-SiO_2$ Systems," *J. Solid State Chem.*, **138**, 160-77 (1998).

[10]E.G. Avvakumov, "Soft Mechanochemical Synthesis as a Basis for New Chemical Processes," *Chem. Sustain. Dev.*, **2**, 1-15 (1992).

[11]Q. Zhang, J. Lu, and F. Saito, "Mechanochemical Synthesis of $LaCrO_3$ by Grinding Constituent Oxides," *Powder Technol.*, **122**, 145-9 (2002).

CARBON NANOTUBES AS PHOTOCATALYTIC CARRIERS

Georgios Pyrgiotakis and Wolfgang Sigmund
University of Florida, Department of Materials Science and Engineering
Gainesville, FL, 32611, USA

ABSTRACT

In the current research, the goal was to synthesize a composite consisting of a multi-wall carbon nanotube (MWNT) core, and a thin coating of anatase-titania. The purpose was to maximize the surface area, and improve the electrical properties of the composite material. The synthesis was completed via a sol-gel deposition route followed by heat treating to crystallize the coating. Raman and XPS were used to characterize the interface formed between the two materials. Results from Raman and XPS showed that the composite has to fulfill two conditions two enhance the photocatalysis; there has to be good contact between the nanotubes, and there has to be a bond between the titania coating and the underlying graphene layers.

INTRODUCTION

In the last few decades the demand for safer environmental conditions has increased dramatically. One reason is the constantly increasing biological threat that can be expressed in every aspect of the daily life, ranging from cases as simple as food bacterial contamination (*E. coli* and *salmonella*) to epidemic outbreaks (*Ebola* and *SARS*) and biological warfare (*anthrax* and *smallpox*). The need for effective and efficient disinfection is driving the industry in the development of a wide range of products.

Recently semiconductor photocatalysis has attracted a great deal of attention since it has a wide range of applications.[1,10] One of the most interesting materials is titania (TiO_2).[3,4,9] It is also the material that is used here as coating on carbon nanotubes. It is widely available since it is used as pigment in many applications and the production is fairly cheap.[4] Since 1972 when Fujishima et al. reported the photocatalytic splitting of water on TiO_2 electrodes [2] a great deal of research has been done to enhance the properties of titania and to develop them into applications. The applications range from photovoltaic cells to biological disinfection.[3,4,9]

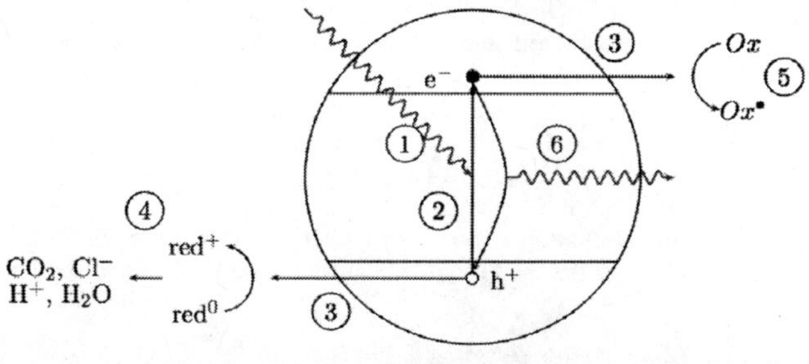

Figure 1: Schematic representation of the reactions taking place in titania. **1** Light strikes the semiconductor. **2** An electron-hole pair is formed. **3** Electrons and holes migrate to the surface. **4** The holes initiate oxidation leading to CO_2, Cl^-H^+, H_2O. **5** The conduction band electrons initiate reduction reactions. **6** Electrons and holes recombine to yield heat or light.

Two of the most popular applications are microbial sterilization and self-cleaning surfaces.[6,13-20] There are, however, certain limitations coming primarily from the electronic properties of the titania, that reduce its efficiency [4,9]. The biggest breakthrough happened in 1985 by Matsunaga et al. [14] when by mixing titania with silver particles a significant enhancement of catalysis was observed. Since that time the range of applications has increased dramatically.

BASIC PRINCIPLES

Figure 1 schematically represents the steps of photocatalysis. Initially when a photon of proper energy ($h\nu \geq E_g$) strikes the surface of the semiconductor it generates an electron hole pair (h^+-e^-). Both electrons and holes, either recombine or migrate to the surface, where, they proceed with chemical reactions. The holes generate [$OH^\bullet$] radicals and the electrons H_2O_2. A very important factor for these processes is the required time.[4] The reaction time has been measured with laser flash photolysis.[11,12] Judging by the time frame of those reactions it is obvious that the dominant reaction is the recombination rather than [$OH^\bullet$] radical generation. Some of the most widely used methods to overcome this are to pair titania with a conductor and dope it with C, N or transition metals.

SYNTHESIS OF THE COMPOSITES

In this paper titania and carbon nanotubes will be combined in the form of a nano-composite that will deliver a highly efficient photocatalyst. Several researchers already reported about the synthesis, and here an in-depth study and analysis of the properties is presented[5,7,8]. In general there are two distinct trends in combining two nanomaterials: either in the form of titania pellet with the nanotubes embedded, or in a more sophisticated approach, the titania is applied as a coating on the nanotubes. Here, the second approach will be reported since it takes full advantage of the nanotube properties, by creating a composite with a single nanotube as core.

For the synthesis of the nanocomposite earlier reported methods were followed.[5,7,8] Arc discharge nanotubes were bought in soot form, so initially the soot was ground with mortar and pestle to yield fine powder. 50 mg of this powder was mixed in 200 ml of highly concentrated HNO_3 (63% or 10N). The solution was sonicated for 3 hours to further disperse the powder and was then refluxed in an oil bath at 140°C for 10 h. Then the heat was turned of and the solution was left for an additional 3 h until the temperature dropped below 30°C. Triple washing with de-ionized water followed. After the final washing the CNT suspension (30 mg of CNTs in 300 ml of water) were placed in a three-way 300 ml flask. The flask was placed in an oil bath at 40°C and was refluxed under constant stirring speed. After the temperature was constant the pH was fixed at ~3 with 0.1N HNO_3. The precursor ($Ti_2(SO_4)_3$) was injected and the mixture was stirred while reacting for 1 hour. The coated tubes were then collected and washed three times with water and the composite was dried at 40°C for two days. The heat treatment was performed at 500°C for 3 hours with a heating rate of 10 K/min.

The TEM of the coated nanotubes revealed a coating with some variation in the thickness ranging from 3 to 10 nm with an average value of 4 nm (figure 2 (a)). It is worth to note that some anatase particles also nucleate on the surface of the nanotubes. Furthermore, the coating does not always yield complete surface coverage. Some spots can be seen where the coating may not be complete. XRD and XPS both confirmed that the coating is indeed the anatase phase of titania.

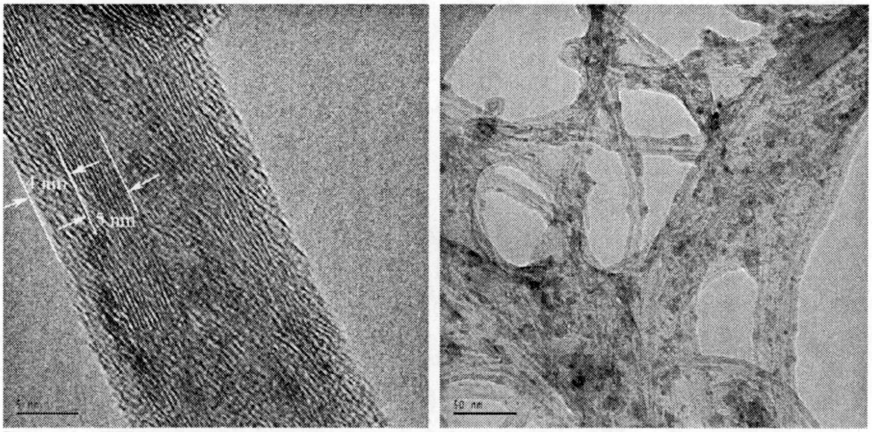

Figure 2: TEM images of coated carbon nanotubes. A uniform coating is seen in the picture on the left with an anatase coating thickness of approximately 4 nm. The scale bar is 5 nm. On the right, the scale bar is 50 nm. Here, an overview of several coated nanotubes is presented.

The TEM results are also in good agreement with the BET results which yielded a specific surface area of 172 m^2/g.

The particles that are presented here have already been tested successfully for destruction of organic hazardous compounds and spores in comparison with Degussa P25, which is reported elsewhere.[8] The first report by Lee et al. showed that the particles were more effective than Degussa P25 against *Bacillus Cerius* spores in UV light.[8] Pyrgiotakis et al. reported that the particles were more effective against organic dyes compared to Degussa P25 in UV light and even more in visible light. This paper will focus on the XPS and Raman spectroscopy of these materials.

SPECTROSCOPIC ANALYSIS

The unexpected enhancement in photocatalytic properties can be better understood when fundamental properties and structures are known for these nanocomposites. Raman spectroscopy and XPS are most suitable techniques to study the details. Raman spectroscopy is an established tool to characterize the electronic properties of carbon nanotubes as well as their structure. XPS is a surface sensitive technique that is ideally suited to investigate bond energies and electronic states of the titania coating.

Figure 3 shows the Raman spectra for the carbon nanotubes prior and after coating. The G-Band showed a very distinct peak splitting which is characteristic of conducting nanotubes. Furthermore the lower frequency, G^-, of that split (1583 cm^{-1}) was approached a lot better with the Breit-Wigner-Fano line shape that is also an exclusive characteristic of the metallic tubes. The low G^- band peak is associated with charge transfer and is shifting downwards for donors and upwards for acceptors. In this case it is indicated that the nanotubes are accepting electrons, having therefore the role of an electron sink. Direct comparison of the nanotube spectra prior and after the coating reveals a significant peak shifting towards lower wave numbers which is an indication for bonding. There are two possible bonds that can give rise to such shifts, Ti-C or Ti-O-C.

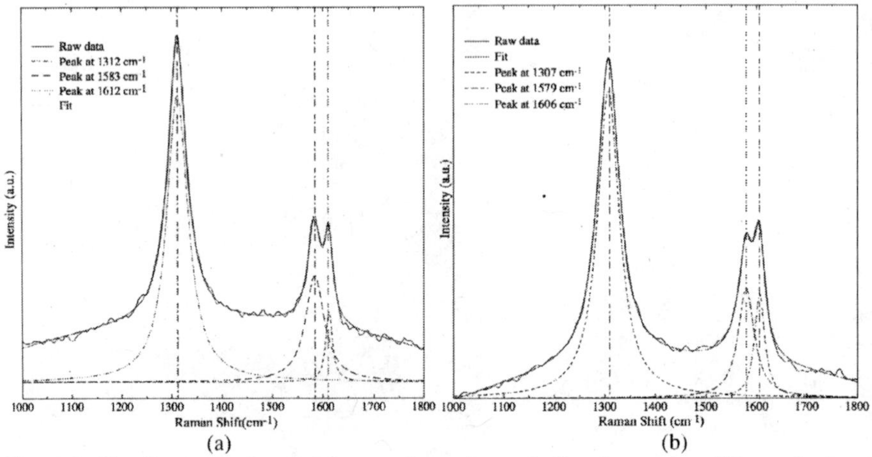

Figure 3: The Raman Peaks of the nanotubes prior and after the coating. Prior and after the coating the G-Band (~1600 cm^{-1}) shows a very distinct split, a characteristic of metallic carbon nanotubes. In addition all the peaks after the coating show a shift to lower wave numbers (higher energies) that indicates the existence of a new bond.

To further investigate the nature of this bond XPS was utilized. Figure 4 shows the C1s peak of carbon. The tubes just after the functionalization (before coating) show only the elemental carbon peak and the C-OH. After the coating a new peak appears at 289.6 eV that is being attributed to the titania coating with the CNTs. Since the bond T-C is at energies around 283 eV it is concluded that the bond is Ti-O-C. This is supported by the existence of the 285.3 eV peak which is attributed to the C=0, C-OH bond. During the synthesis of the titania coated carbon nanotubes -COOH groups were generated on the surface of the tubes, a result of the oxidizing acid treatment. Those groups act later as anchoring points for the *sol-gel* precursors. This is the fundamental reason that the bond between the carbon nanotubes and the titania coating is in the form of C-O-Ti. This type of bond will incorporate the carbon as active dopant in the structure of titania. This will result in a smaller band-gap that will make the material effective in the visible light.

CONCLUSIONS

Spectroscopic analysis verified the metallic nature of the nanotubes and the presence of C-O-Ti bonding, both of which are critical to enhance photocatalysis. The metallic nature will provide the necessary conduction to scavenge the generated electrons to retard the recombination, while the bond will activate the carbon dopant in the structure. In addition, the final composite has a very large specific surface area (172 m^2/g), which will increase the overall photocatalytic efficiency. Carbon nanotubes therefore can be successfully used as photocatalytic carriers, since they can provide electron sinking, carbon doping, and large specific surface area.

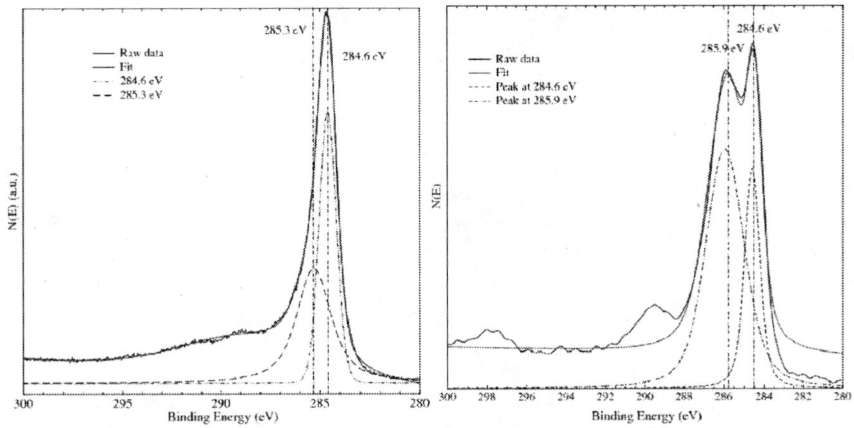

(a) (b)

Figure 4: XPS spectra along with fit curves for carbon nanotubes prior (a) and after (b) the coating. The C1s after the coating shows a new peak at 289.6 eV. This peak is attributed to the Ti-O-C bond.

REFERENCES

[1]G. Ertl. Heterogeneous catalysis on the atomic scale. *The Japan Chemical Journal Forum*, 1(1):33–45, 2001.

[2]A. Fujishima and K. Honda. Electrochemical photolysis of water at a semiconductor electrode. *Nature*, 238(5358):37–40, 1972.

[3]A. Fujishima, T. N. Rao, and D. A. Tryk. Titanium dioxide photocatalysis. *Journal of Photochemistry and Photobiology C: Photochemistry Reviews*, 1(1):1–21, 2000.

[4]M. R. Hoffmann, S. T. Martin, W. Y. Choi, and D. W. Bahnemann. Environmental applications of semiconductor photocatalysis. *Chemical Reviews*, 95(1):69–96, 1995.

[5]A. Jitianu, T. Cacciaguerra, R. Benoit, S. Delpeux, F. Beguin, and S. Bonnamy. Synthesis and characterization of carbon nanotubes - tio2 nanocomposites. *Carbon*, 42(5-6):1147–1151, 2004.

[6]K. P. Kuhn, I. F. Chaberny, K. Massholder, M. Stickler, V. W. Benz, H. G. Sonntag, and L. Erdinger. Disinfection of surfaces by photocatalytic oxidation with titanium dioxide and uva light. *Chemosphere*, 53(1):71–77, 2003.

[7]S. H. Lee and W. M. Sigmund. Inactivation of bacterial endospores by photocatalytic nanocomposites. *Abstracts of Papers of the American Chemical Society*, 227:U1275–U1276, 2004.

[8]S. W. Lee and W. M. Sigmund. Formation of anatase tio2 nanoparticles on carbon nanotubes. *Chemical Communications*, (6):780–781, 2003.

[9]A. L. Linsebigler, G. Q. Lu, and J. T. Yates. Photocatalysis on TiO_2 surfaces - principles, mechanisms, and selected results. *Chemical Reviews*, 95(3):735–758, 1995.

[10]J. Liqiang, S. Xiaojun, S. Jing, C. Weimin, X. Zili, D. Yaoguo, and F. Honggang. Review of surface photovoltage spectra of nano-sized semiconductor and its applications in heterogeneous photocatalysis. *Solar Energy Materials and Solar Cells*, 79(2):133–151, 2003.

[11]S. T. Martin, H. Herrmann, W. Y. Choi, and M. R. Hoffmann. Time-resolved microwave conductivity .1. TiO_2 photoreactivity and size quantization. *Journal of the Chemical Society-Faraday Transactions*, 90(21):3315–3322, 1994.

[12]S. T. Martin, H. Herrmann, and M. R. Hoffmann. Time-resolved microwave conductivity .2. quantum-sized TiO_2 and the effect of adsorbates and light-intensity on charge-carrier dynamics. *Journal of the Chemical Society-Faraday Transactions*, 90(21):3323–3330, 1994.
[13]T. Matsunaga, Y. Namba, and T. Nakajima. Electrochemical sterilization of microbial-cells. *Bioelectrochemistry and Bioenergetics*, 13(4-6):393–400, 1984.
[14]T. Matsunaga, R. Tomoda, T. Nakajima, and H. Wake. Photoelectrochemical sterilization of microbial-cells by semiconductor powders. *Fems Microbiology Letters*, 29(1-2):211–214, 1985.
[15]A. G. Rincon and C. Pulgarin. Photocatalytical inactivation of *E. Coli*: effect of (continuous-intermittent) light intensity and of (suspended-fixed) TiO_2 concentration. *Applied Catalysis B-Environmental*, 44(3):263–284, 2003.
[16] A. G. Rincon and C. Pulgarin. Bactericidal action of illuminated TiO_2 on pure escherichia coli and natural bacterial consortia: post-irradiation events in the dark and assessment of the effective disinfection time. *Applied Catalysis B-Environmental*, 49(2):99–112, 2004.
[17]A. G. Rincon and C. Pulgarin. Effect of pH, inorganic ions, organic matter and H_2O_2 on *E-Coli* K12 photocatalytic inactivation by TiO_2 - Implications in solar water disinfection. *Applied Catalysis B-Environmental*, 51(4):283–302, 2004.
[18]A. G. Rincon and C. Pulgarin. Field solar e-coli inactivation in the absence and presence of TiO_2: is UV solar dose an appropriate parameter for standardization of water solar disinfection? *Solar Energy*, 77(5):635–648, 2004.
[19]A. G. Rincon, C. Pulgarin, N. Adler, and P. Peringer. Interaction between *E. Coli* inactivation and DBP-precursors - dihydroxybenzene isomers - in the photocatalytic process of drinking-water disinfection with TiO_2. *Journal of Photochemistry and Photobiology a-Chemistry*, 139(2-3):233–241, 2001.
[20]T. Sato, Y. Koizumi, and M. Taya. Photocatalytic deactivation of airborne microbial cells on TiO_2-loaded plate. *Biochemical Engineering Journal*, 14(2):149–152, 2003.
[21]G. Pyrgiotakis, S.-H. Lee, W. Sigmund, Advanced phtocatalysis with titania coateted MWNTs, *Proceedings of MRS Spring Meeting 2005: Symposium R Nanoporous and Nanostructured Materials for Catalysis, Sensor, and Gas Separation Applications*, Editors: S.W. Lu, H. Hahn, J. Weissmuller, J.L. Gole, San Francisco, p. R5.7, 2005

ANISOTROPIC SINTERING SHRINKAGE AND GRAIN GROWTH FOR SPHERICAL ALUMINA POWDER COMPACTS ALIGNED IN HIGH MAGNETIC FIELD

Anze Shui[1], Lingke Zeng[1], Yanchun Liu[1], Atsushi Makiya[2] and Keizo Uematsu[2]

[1]College of Materials Science and Engineering, South China University of Technology, Guangzhou 510640, PR China

[2]Department of Chemistry, Nagaoka University of Technology, Nagaoka 940-2188, Japan

ABSTRACT:

Spherical alumina powder compacts were prepared by drying alumina slurry in a high magnetic field and cold isostatic pressing. Shrinkage and grain growth during sintering were examined for the alumina compacts in detail. The compacts prepared in 10 T showed significant anisotropies of sintering shrinkage and grain growth. The sintering shrinkage was larger in the direction parallel to the magnetic field direction (i.e., the c-axis direction of alumina crystal) than in its perpendicular direction at any sintering times (2-24 h), and the grain growth rate was faster in the plane normal to the magnetic field direction (i.e., the c-plane of alumina crystal) than in its perpendicular plane (i.e., the a-b plane of the alumina crystal). On the other hand, the compacts prepared in 0 T did not undergo these anisotropies. The experimental results indicate that sintering shrinkage of alumina powder compact depends on alumina crystal axis direction, and grain growth depends on alumina crystal plane. Origin of these anisotropies for the spherical alumina powder compacts can be attributed to particle orientation caused by the high magnetic field.

INTRODUCTION

Anisotropic shrinkage and grain growth during sintering were systematically discussed for uniaxially pressed compacts of alumina particles of elongated shape in our previous papers.[1-10] The pressed compacts of elongated alumina particles showed sintering shrinkage and grain growth anisotropies, but the pressed compacts of spherical alumina particles did not undergo the anisotropies. Origin of the anisotropies was ascribed to particle orientation in the compacts. Otherwise, the origin of the sintering shrinkage anisotropy for magnetic materials also was the particle orientation.[11]

However, the effect of the particle orientation on sintering shrinkage is not well understood. The following two theories may be considered for the sintering shrinkage mechanism of the particle orientation. The first one is the theory of anisotropic grain boundary number, which was propounded by Coble.[12,13] The second is the theory of anisotropic mass transport, which was suggested in our past papers.[3,5]

In the former theory, the grain boundary number per unit length influences sintering shrinkage, because mass transport to neck occurs along grain boundary during sintering. When alumina particles align with the long axis normal the pressing direction during uniaxial pressing, the grain boundary number per unit length becomes more in the pressing direction, and larger sintering shrinkage occurs in the pressing direction.[12,13]

In the latter theory, the mass transport rate (or diffusion coefficient) to neck along the grain boundary changes with crystal plane for materials of anisotropic crystal structure, such as alumina, because the atom packing density and surface structure vary with the crystal plane. The atom packing is denser in the c-plane than in a-b plane. The surface energy is lower in the former than in the latter at temperature above 973 K.[14-18]

According to the grain boundary number theory, spherical alumina powder compacts of oriented particles should not show sintering shrinkage anisotropy, because the grain boundary number is isotropic in the compacts. But according to the mass transport rate theory, they should show sintering shrinkage and grain growth anisotropies, because of the chemical potential difference.

However, the anisotropies of sintering shrinkage and grain growth for spherical alumina powder compacts of orientated particles have not been studied in detail. It is important to examine the anisotropies for the spherical powder compacts, in order to assess the above theories, the sintering shrinkage mechanism and the grain growth mechanism of the particle orientation.

In this study, the spherical alumina powder compacts of oriented particles are prepared by drying alumina slurry in a high magnetic field and with cold isostatic pressing. The anisotropies of shrinkage and grain growth during sintering are examined for these alumina compacts in detail.

EXPERIMENT

Spherical alumina particles were used in this study, AA05 (Sumitomo Chemical Company, Japan). It is one of high purity alumina powders (>99.99%), and alpha alumina single crystal with an average particle size 0.5μm from the manufacturer. Their average aspect ratio is 1.1.[3,5] The alumina particles and dispersant (0.6 wt%; D305, Chukyouyushi, Japan) were mixed with distilled and deionized water, and ball milled for 24 hours to make alumina slurry. The slurry was applied to a shallow plate, and dried in a high magnetic field (0-10 T) to make a pellet shaped compact. Subsequently, the compact was CIPed at 200 MPa to enhance the homogeneity in particle packing density.

The compact was sintered in an electric furnace in air at 1873 K for 2-24 h with heating rate 10 K/min. Sintering shrinkage of the compact was determined from its dimension change before and after the sintering in the direction parallel to magnetic field direction (i.e., the c-axis direction of alumina crystal, described as H direction hereafter) and its perpendicular direction (described as D direction hereafter). In SEM observation, the plane parallel to magnetic field direction (described as H plane hereafter) and its perpendicular plane (i.e., the c-plane of alumina crystal, described as D plane hereafter) of the sintered body were examined, respectively.

RESULT

Figure 1 shows sintering shrinkage change with sintering time in the H and D directions at 1873 K for the compact prepared in 10 T. The sintering shrinkage always was larger in the H direction than that in the D direction at any sintering times. The compact of particle orientation showed anisotropic sintering shrinkage at any sintering times, although the grain boundary number was same in the two directions.

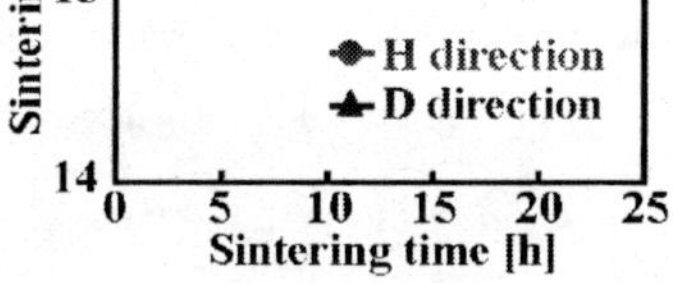

Figure 1 Sintering shrinkage change with sintering time for the compact prepared in 10 T.

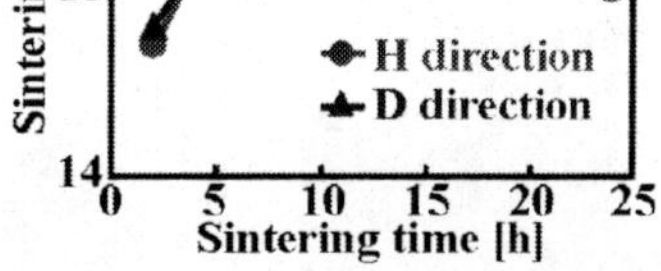

Figure 2 Sintering shrinkage change with sintering time for the compact prepared in 0 T.

Figure 2 is sintering shrinkage change with sintering time in the H and D directions at 1873 K for the compact prepared in 0 T. The sintering shrinkage essentially was same in the two directions at any sintering times. Clearly, the compact without particle orientation held isotropic sintering shrinkage.

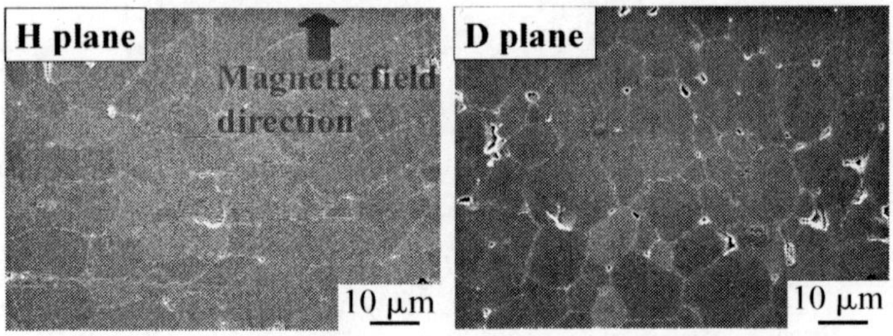

Figure 3 SEM micrographs of the sintered body prepared in 10 T

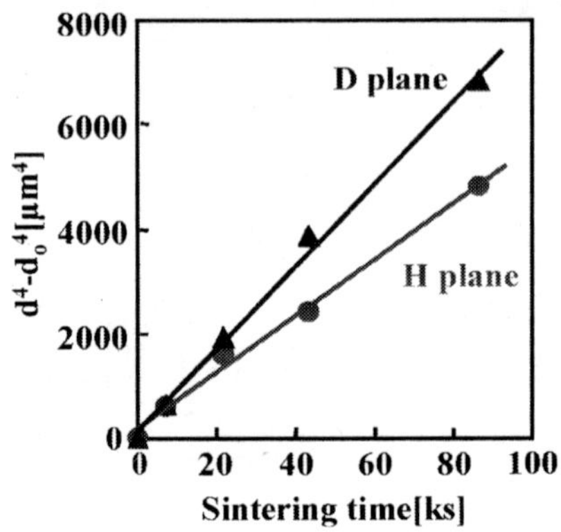

Figure 4 Grain growth during sintering at 1873 k for the compact prepared in 10 T

Figure 3 is SEM micrographs of the sintered body prepared in 10 T for the H (i.e., the plane parallel to the magnetic field direction) and D (i.e., the plane perpendicular to the magnetic field direction, the c-plane of alumina crystal) planes at 1873 K for 24 h. There was a clear difference in microstructure in the two planes. In the H plane, grains showed elongated shape, although their initial shape was spherical before sintering. Additionally, the tendency of grain alignment was observed with the short axis (i.e., the c-axis of alumina crystal) parallel to the magnetic field direction. But shape of grains in the D plane was relatively spherical comparing with in the H plane.

Figure 4 represents grain size to the fourth power as a function of sintering time at 1873 k for the compact prepared in 10 T. The fourth power law fits the grain growth data in both the H and D planes. Clearly, the grain growth rate was faster in the plane normal to the magnetic field direction (i.e., the c-plane of alumina crystal) than in its perpendicular plane (i.e., the a-b plane of the alumina crystal).

DISCUSSION

In Figs. 1 and 2, the H and D directions showed different sintering shrinkage behavior, respectively. Since their green densities (or the particle orientation structures) were different. The green density was 60.0% and 60.8%, for the former and the latter, respectively. The maximum sintered density was 99.4% and 99.7%, for the former and the latter, respectively. The sintered density dropped with increasing sintering time for both the two compacts, when the sintering time exceeded 6 h.[19)]

The compacts prepared in 10 T showed anisotropic sintering shrinkage and grain growth. There was an anisotropic microstructure in their sintered bodies. The particle orientation in the compacts was confirmed with the immersion liquid method of polarized light microscope developed by Uematsu et al.[20)] Additionally, the grain alignment in the sintered bodies was also observed with X-ray diffraction, the c-plane of alumina crystal corresponded to the D plane in Figure 3. These were same to Sakka' s results.[21)] On the other hand, the compacts prepared in 0 T held isotropic sintering shrinkage, because the particle orientation did not exist in the compacts. These experimental results are similar to the uniaxially pressed alumina powder compacts. The uniaxially pressed compacts with the particle orientation also held the sintering shrinkage and grain growth anisotropies, and the anisotropies did not also exist in the compacts without the particle orientation.[2-5)] In the same particle orientation degree condition, the elongated powder compacts have large the sintering shrinkage anisotropy comparing with the spherical powder compacts.[10)]

In the evaluation of the particle orientation structure, the X-ray diffraction is not resultful especially for ceramic compact. The immersion liquid technique of polarized light micrograph is a useful method for evaluating the particle orientation quality.[21)]

Clearly, sintering shrinkage and grain growth of spherical alumina powder compact depend on alumina crystal axis direction and crystal plane. The sintering shrinkage is larger in the c-axis direction than in its perpendicular direction. The grain growth rate is faster in the c-plane than the a-b plane. These results indicate that the mass transport rate to neck is quicker along the c-plane than that along the a-b plane, the diffusion coefficient is larger in the former than that in the latter; the surface energy is lower in the c-plane than in the a-b plane.

CONCLUSION

The compacts prepared in 10 T showed significant anisotropies of sintering shrinkage and grain growth. The sintering shrinkage was larger in the direction parallel to the magnetic field direction (i.e., the c-axis direction of alumina crystal) than in its perpendicular direction, and the grain growth rate was faster in the plane normal to the magnetic field direction (i.e., the c-plane of alumina crystal) than in its perpendicular plane (i.e., the a-b plane of the alumina crystal). On the other hand, the compacts prepared in 0 T did not undergo these anisotropies. The experimental results indicate that sintering shrinkage of alumina powder compact depends on alumina crystal axis direction, and grain growth depends on alumina crystal plane.

REFERENCES

[1]Shui, A., Zhang, Y., Uchida, N. and Uematsu, K., Origin of Shape Deformation during Sintering of Alumina Compacts, J. Ceram. Soc. Japan, Vol. 106, pp. 873-876 (1998).

[2]Shui, A., Kato, Z., Tanaka, S., Uchida, N. and Uematsu, K., Isotropic Sintering Shrinkage in Pressed Compact of Near-Spherical Alumina Particles, Am. Ceram. Soc. Bull., Vol. 80[9], pp. 29-32 (2001).

[3]Shui, A., Kato, Z., Tanaka, S., Uchida, N. and Uematsu, K., Sintering Deformation Caused by Particle Orientation in Uniaxially and Isostatically Pressed Alumina Compacts, J. Euro. Ceram. Soc., Vol. 22[3], pp. 311-316 (2002).

[4]Shui, A., Saito, M., Uchida, N. and Uematsu, K., Development of Anisotropic Microstructure in Uniaxially Pressed Alumina Compacts, J. Euro. Ceram. Soc., Vol. 22[8], pp. 1217-1223 (2002).

[5]Shui, A., Uchida, N. and Uematsu, K., Origin of Shrinkage Anisotropy during Sintering for Uniaxially Pressed Alumina Compacts, Powder Technology, Vol. 127[1], pp. 9-18 (2002).

[6]Shui, A., Makiya, A., Tanaka, S., Uchida, N. and Uematsu, K., Effect of Cold Isostatic Pressing on Microstructure and Shrinkage Anisotropy during Sintering of Uniaxially Pressed Alumina Compacts, J. Ceram. Soc. Japan, Vol. 110[4], pp. 264-269 (2002).

[7]Shui, A., Saito, M., Tanaka, S., Uchida, N. and Uematsu, K., Anisotropic Grain Growth in Uniaxially Pressed Alumina Compacts, Key Engi. Mater., Vol. 206, pp. 249-252 (2002).

[8]Shui, A., Tanaka, S., Uchida, N. and Uematsu, K., Deformation during Sintering in Alumina Ceramic Made Using Uniaxial Pressing, Key Engi. Mater., Vol. 247, pp. 61-66 (2003).

[9]Shui, A., Makiya, A., Tanaka, S., Uchida, N. and Uematsu, K., Effect of Particle Orientation on The

Anisotropic Sintering of Alumina Ceramics, Key Engi. Mater., Vol. 253, pp. 293-302 (2003).

[10]Shui, A., Zeng, L. and Uematsu, K., Relationship between Sintering Shrinkage Anisotropy and Particle Orientation for Alumina Powder Compacts, Scripta Materialia, Vol. 55[11], pp. 831-834 (2006).

[11]Zavaliangos, A., Missiaen, J.M., Bouvard, D. and Carry, C.P., Crystallographic Texture as Potential Origin of Anisotropy in Sintering of Magnetic Materials, Sintering 03, The Pennsylvania State University, State College, PA, pp. 793-99 (2003).

[12]Coble, R., L., Sintering Crystalline Solids 2: Experimental Test of Diffusion Models in Powder Compacts, J. Appl. Phy., Vol. 32[5], pp. 793-99 (1961).

[13]Coble, R., L., Sintering Crystalline Solids 1: Intermediate and Final State Diffusion Models, J. Appl. Phy., Vol. 32[5], pp. 787-92 (1961).

[14]Song, H., and Coble, R., L., Morphology of Platelike Abnormal Grains in Liquid-Phase-Sintered Alumina, J. Am. Ceram. Soc., Vol. 73[7], pp. 2086-90 (1990).

[15]Choi, J., H., Kim, D., Y., Hockey, B., J., Wiederhorn, S., M., Handwerker, C., A., Blendell, J., E., Carter, W., C. and Roosen, A., R., Equilibrium Shape of Internal Cavities in Sapphire, J. Am. Ceram. Soc., Vol. 80[1], pp. 62-68 (1997).

[16]Kitayama, M., Powers, J., D., Kulinsky, L. and Glaeser A., M., Surface and Interface Properties of Alumina via Model Studies of Microdesigned Interfaces, J. Euro. Ceram. Soc., Vol. 19[10], pp. 2191-209 (1999).

[17]Suzuki, H., Matsubara, H., Kishino, J. and Kondoh, T., Simulation of Surface and Grain Boundary Properties of Alumina by Molecular Dynamics Method, J. Ceram. Soc. Japan, Vol. 106[12], pp. 1215-222 (1998).

[18]Blonski, S. and Garofalini, S., H., Molecular-Dynamics Simulations of Alpha-Alumina and Gamma-Alumina Surfaces, Surf. Sci., Vol. 295, pp. 263-74 (1993).

[19]Shui, A., Zeng, L., Liu, P., Cheng, X. and Uematsu, K., Anisotropic Sintering Shrinkage of Spherical Alumina Particle Compact Aligned in High Magnetic Field, J. Ceram. Soc. Japan, Vol. 114[3], pp. 290-292 (2006).

[20]Uematsu, K., Ito, H., Zhang, Y. and Uchida, N., Novel Characterization Method for the Processing of Ceramics by Polarized Light Microscope with Liquid Immersion Technique, Ceram. Trans., Am. Ceram. Soc., Vol. 54, pp. 83-89 (1995).

[21]Sakka, Y. and Suzuki, T. S., Textured Development of Feeble Magnetic Ceramics by Colloidal Processing under High Magnetic Field, J. Ceram. Soc. Japan, Vol. 113, pp. 26-36 (2005).

EFFECT OF Mo POWDER SURFACE CONDITION ON FABRICATION OF $Mo-SiO_2$ FUNCTIONALLY GRADED MATERIALS WITH SLIPCASTING METHOD

Ayumu UMEMOTO and Koichi HAYASHI
TOTO Ltd., Advanced Ceramic Division
2-1-1 Nakashima, Kokurakita-ku, Kitakyushu 802-8601, Japan

Kyoko HAYANO, Noritaka SAITO, Kenji KANEKO and Kunihiko NAKASHIMA
Department of Materials Science and Engineering, Kyushu University
744, Motooka Nishi-ku, Fukuoka 819-0395, Japan

ABSTRACT

The conventional high intensity discharge lamps are made of Mo foil-sealing method, in which the Mo-metal of 30 μm thick is sandwiched by the fused silica envelope. The conventional high intensity discharge lamps therefore requires cooling system to reduce the thermal stress between Mo and fused silica during the switching operation and a long sealing zone is usually required to complete hermeticity, especially in high power devices, so that the production of compact high intensity discharge lamps has been a Herculean task.

We show the invention of the new type of hermetically-sealed high intensity discharge lamps, made of $Mo-SiO_2$ functionally graded material (FGM) as an electrode and a sealing component. In the case of high intensity discharge lamps with $Mo-SiO_2$ FGM (FGM-Lamp), the thermal expansion coefficient between Mo and SiO_2 is functionally graded so that it tolerates a large numbers of heating cycles, with no cooling system are required. Furthermore lamps survive without breakage. The W electrode is totally separated from the lamp envelope by the FGM, so that no leakage of the luminous elements or gases takes place, when a large gas pressure exists inside a lamp.

In this paper, effects of Mo powder surface conditions on the fabrication of $Mo-SiO_2$ FGM have been investigated. $Mo-SiO_2$ FGM with as-received Mo powder revealed that the Mo particles were homogenously distributed, which resulted in no microscopic compositional gradient. In contrast, $Mo-SiO_2$ FGM with pre-treated Mo powder provided a uniform gradient from the conductive part (Mo-rich phase) to the sealing part (amorphous silica).

INTRODUCTION

The concept of the functionally graded material (FGM) has been conceived as a material in which the chemical composition and/or microstructural parameters gradually changes over the definable geometric distances, resulting in a corresponding variation in the properties of the material. Potential applications of FGMs are; high efficiency engine components, ceramic turbine components, direct metal tools for industrial use, biomaterials used in artificial human implants, drug delivery devices with controlled release rates, armor and armament components for defense and others. For these applications, material composition control is necessary for improving a variety of properties such as thermo-mechanical performance, toughness and strength, and for reducing interfacial stresses between dissimilar materials [1-5].

In general, two powders with different compositions are layered to prepare bulk FGMs [6-8], which results in macroscopically graded compositions. However, they are not perfectly graded microscopically, since there are always interfaces present between each layer. In late years, $Mo-SiO_2$ FGM for high intensity quartz envelopes discharge lamps was successfully achieved using a combined sedimentation and pressurized slip casting method [9-11] with uniform grading from the conductive part (Mo-rich phase) to the sealing part (amorphous silica). Furthermore, we have shown the invention of the new type of hermetically-sealed high intensity discharge lamps, made of Mo-SiO_2 functionally graded material (FGM) as an electrode and a sealing component [12]. However, in these earlier studies, it has not clarified how the surface oxidizing condition of Mo raw powder affected on the FGM fabrication. In this study, we intend to investigate the effects of Mo powder surface condition on the fabrication of Mo-SiO_2 FGM with slipcasting method.

EXPERIMENTAL

Powders of Mo with 99.9% purity and SiO_2 with 99.999% purity were used in the fabrication of Mo-SiO_2 FGMs. A part of Mo raw powder was rinsed with hydrochloric acid, namely pre-treated Mo powder. The microstructural observation of Mo raw powders was conducted using scanning electron microscopy (SEM, Keyence, VE-8800) and transmission electron microscopy (TEM, Phillips, TECNAI-20). For TEM observation, thin specimens were prepared using a ultramicrotome apparatus (Leica, ULTRACUT C). The FGMs were fabricated using a Mo-SiO_2 mixed water suspension with 50 wt% content of solids, prepared by an ultrasonic stirring. The Mo-SiO_2 mixed slurry was placed in a vacuum desiccator to exclude air bubbles from the slurry. The FGM was prepared using a combined sedimentation and pressurized slip casting method [9-12]. The Mo-SiO_2 mixed slurry was cast in a plastic mold on a porous ceramic block, under a controlled pressure (cf. Figure 1). Cylindrical green compacts were cast in this manner. After drying the samples at room temperature in vacuum for 24 hours, the samples were sintered in vacuum at the temperature just below the melting temperature of SiO_2 for fixed time.

The microstructure of the Mo-SiO_2 FGMs was observed using reflected light optical

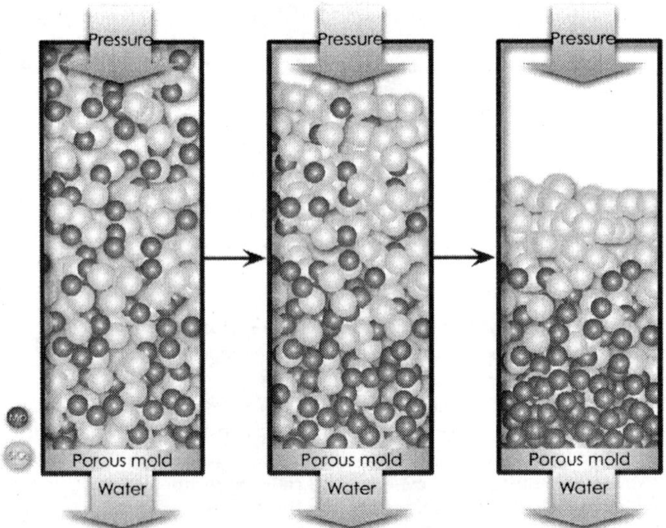

Figure 1 Schematic illustration of fabrication of Mo-SiO_2 FGM green compact with slip casting method.

microscopy and by scanning electron microscopy (SEM). To analyze the element of the FGM along the gradient, thin slices (about 1mm thick) were cut perpendicular to it. The average Mo content of each slice was measured using calibrated EPMA of both faces. Knoop hardness of FGMs was measured by applying a load of 0.3 kgf for 15 s to the FGMs whose surface had been polished to a 1 μm finish. Twelve measurements of Knoop hardness were made on each sample, and the average of 10 values without maximum and minimum values was provided as the Knoop hardness. Further microstructural analysis of the FGMs was conducted using transmission electron microscopy (TEM) with energy dispersive spectroscopy (EDS). For TEM observation, thin specimens were prepared using a focused ion beam apparatus (Hitachi, FB-2000).

RESULTS AND DISCUSSION

Figure 2 shows the SEM images (upper row) and the cross-sectional TEM images (bottom row) of the as-received and the pre-treated Mo powders, respectively. There was no significant difference in surface physical appearance in SEM images, but in the case of the pre-treated Mo powder, small particles (< 1 μm) were disappeared. As can be seen from TEM images, the Mo particles were covered with oxide layer. XPS spectra revealed that the oxide layers of both powders consisted mainly of MoO_3. The thickness of these oxide layers which covered the as-received and the pre-treated Mo powders were about 10 nm and 5 nm, respectively. It is evident that the difference in the thickness of oxide layers in Fig. 2 bottom raw are due to chemically

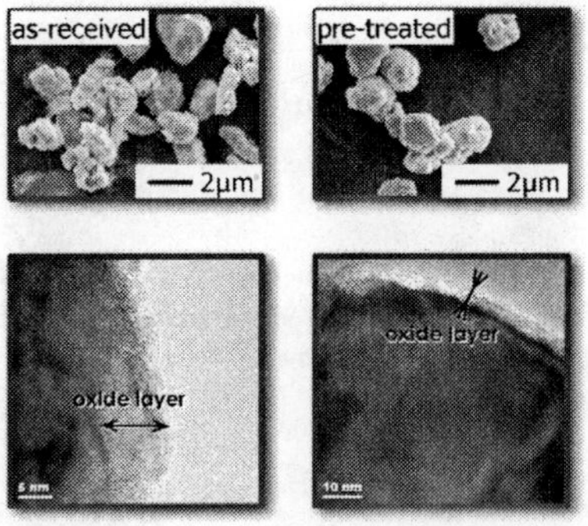

Figure 2 SEM(upper row) and TEM BF(bottom row) images of as-received and pre-treated Mo particles.

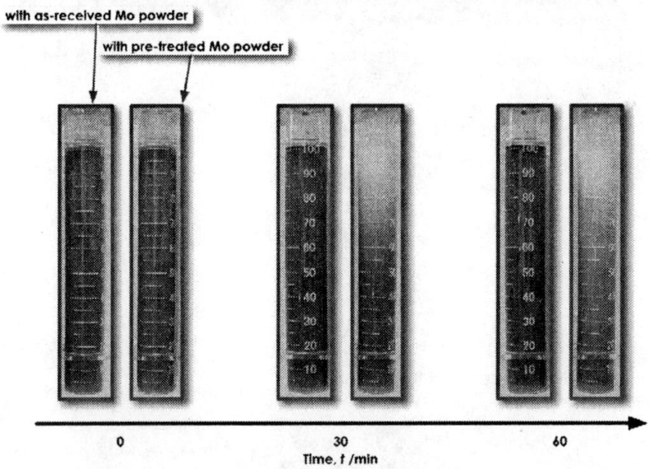

Figure 3 Time changes of the appearance of Mo-SiO_2 water suspension with different surface condition Mo powders.

dissolution with hydrochloric acid treatment.

Figure 3 reveals the time changes of the appearance of Mo-SiO_2 water suspension with the as-received and the pre-treated Mo powders. As can be seen here, the sedimentation behavior of Mo-SiO_2 water suspension with the pre-treated Mo powder was obviously different with the other suspension. Specifically, the Mo-SiO_2 water suspension with the pre-treated Mo powder clearly revealed the sedimentation of Mo particles with time, accordingly, Mo content in the water suspension varied gradually in appearance after 60 min observation. In contrast, the Mo-SiO_2 water suspension with the as-received Mo powder indicated no sedimentation of Mo particles after 60 min observation. Consequently, as can be seen from Figure 4, the green compact of Mo-SiO_2 FGM with the as-received Mo powder had no compositional gradient, which made a remarkable contrast with the other uniform green compact grading from the Mo-rich phase to the silica. This difference in the sedimentation behavior is probably attributed to a ζ-potential difference with the thickness of the oxide layer on the Mo particle surface. However, for detail description, further investigation will be required such as a ζ-potential measurement for Mo particle in Mo-SiO_2 binary water suspension.

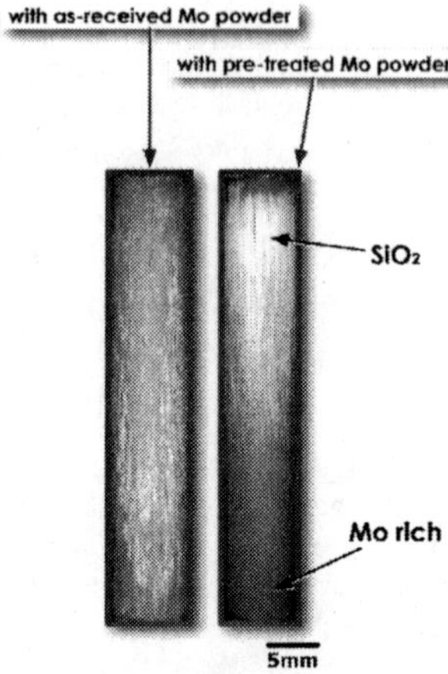

Figure 4 Images of green compacts of Mo-SiO2 FGM with different surface condition Mo powder.

Figure 5 and 6 indicates the microscopic characterizations of Mo-SiO_2 FGM with the as-received and the pre-treated Mo powders, respectively. All SEM micrographs represent the typical images of different sections taken from the FGM sample (Fig.5-a, 6-a). The appearance of the FGM sample taken with a optical microscope (Fig.5-b, 6-b) and the compositional gradients with Knoop hardness of the FGM (Fig.5-c, 6-c) are shown, where the gray lines (SC-glass) indicates the Knoop hardness of the silica glass fabricated with slipcasting method and sintering [13]. As mentioned above, Mo-SiO_2 FGM with the as-received Mo powder had no compositional gradient as shown in Figure 5-a, -b and -c. Knoop hardness of Mo-SiO_2 FGM with the as-received Mo

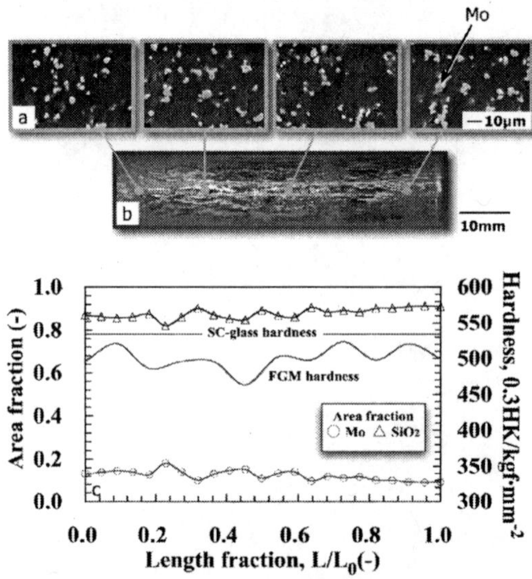

Figure 5 Characterization of Mo-SiO2 FGM with as-received Mo powder. a) SEM images of different regions of FGM, b) Appearance of FGM and c) Compositional gradient and Knoop hardness change with length fraction of FGM.

powder indicates no significant change with length fraction, either. As can be seen from Fig.6-a, on the other hand, in the case of the Mo-SiO_2 FGM with the pre-treated Mo powder, Mo particles are homogeneously distributed and their distributions and sizes are gradually varied, which resulted in a microscopic compositional gradient as shown in Fig.6-c. The maximum Mo content is almost 30vol% at the front (L/L_0 nearly equal to 0) of the total FGM length, and one-fourth of the sample is practically pure silica. In addition, Knoop hardness of the Mo-SiO_2 FGM with the pre-treated Mo powder gradually increased with increasing SiO_2 content and indicated the similar value with SC-glass at the rear (L/L_0 equal to 1.0).

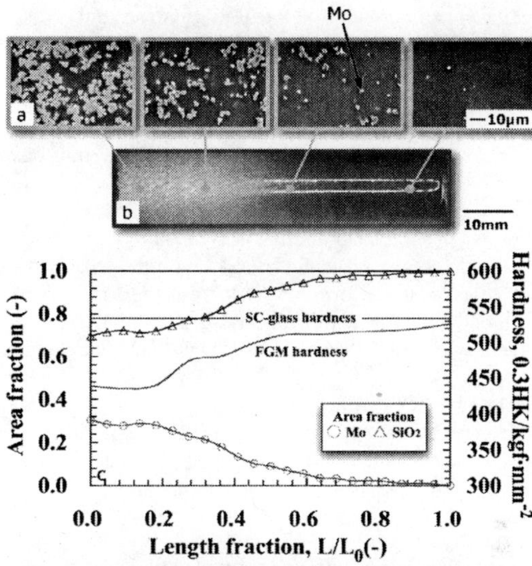

Figure 6 Characterization of Mo-SiO_2 FGM with pre-treated Mo powder. a) SEM images of different regions of FGM, b) Appearance of FGM and c) Compositional gradient and Knoop hardness change with length fraction of FGM.

Figure 7 shows the TEM-BF micrograph and STEM-EDS line profile of Mo/SiO_2 interface in Mo-SiO_2 FGM with pre-treated Mo

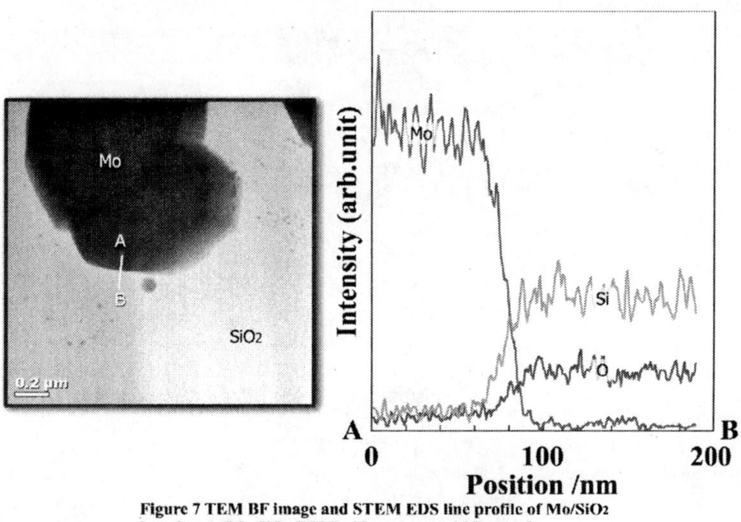

Figure 7 TEM BF image and STEM EDS line profile of Mo/SiO_2 interface in Mo-SiO_2 FGM with pre-treated Mo powder.

powder. It can be seen that the TEM analyses did not detect any reaction component between Mo particle (dark contrast) and SiO_2 matrix (bright contrast) after sintering or the presence of intergranular phase between Mo particles.

CONCLUSION

We have described the effects of Mo raw powder had the different surface oxidizing conditions on the fabrication of Mo-SiO_2 FGM with slipcasting method. Mo-SiO_2 FGM with as-received Mo powder covered with ~10 nm thick oxide layer revealed that the Mo particles were homogenously distributed, which resulted in no microscopic compositional gradient. In remarkable contrast, Mo-SiO_2 FGM with pre-treated Mo powder covered with ~5 nm thick oxide layer revealed a uniform gradient from the Mo-rich phase to the amorphous silica. It would be concluded that the compositional gradient of Mo-SiO_2 FGM can easily be controlled by the surface condition of Mo raw powder with pressurized slip casting method.

ACKNOWLEDGMENT

This work was supported by a Grant-in-aid for "Special Coordination Fund for Promoting Science and Technology" from the Ministry of Education, Culture, Sports, Science and Technology of Japan, which is acknowledged.

REFERENCES

1. B. H. Rabin and I. Shiota, "Functionally gradient materials", *MRS Bull.*, **20**, 14-18 (1995).
2. M. Koizumi and M. Niino, "Overview of FGM research in Japan", *MRS Bull.*, **20**, 19–21 (1995).
3. S. Suresh, A. Mortensen, Fundamentals of functionally graded materials: processing and thermomechanical behaviour of graded metals and metal-ceramic composites, IOM Communications, London, (1998).
4. L. D. Teng, F. M. Wang and W.C. Li "Thermodynamics and microstructure of Ti–ZrO_2 metal–ceramic functionally graded materials", *Mater. Sci. and Eng.* **A293** 130-136 (2000).
5. R. R. Kumer and S. Maruno "Functionally graded coating of HA-G-Ti composites and their in vivo studies", *Mater. Sci. and Eng.* **A334** 156-162 (2002).
6. A. J. Sanchez-Herencia, R. Moreno, J. R. Jurado, "Electrical transport properties in zirconia/alumina functionally graded materials", *J. Euro. Ceram. Soc.*, **20**, 1611-20 (2000).
7. L. D. Teng, F. M. Wang and W.C. Li, "Thermodynamics and microstructure of Ti–ZrO_2 metal–ceramic functionally graded materials", *Mater. Sci. and Eng.*, **A293**, 130-36 (2000).
8. C. S. Lee, X. F. Zhang and G. Thmas, "Novel joining of dissimilar ceramics in the Si_3N_4–Al_2O_3 system using polytypoid functional gradients", *Acta Mater.*, **49**, 3775-80 (2001).
9. J. Chu, H. Ishibashi, K. Hayashi, H. Takebe and K. Morinaga, "Slip casting of continuous functionally gradient material", *J. Ceram. Soc. Jpn.*, **101**, 841-44 (1993).
10. A. P. Tomsia, E. Saiz, H. Ishibashi, M. Diaz, J. Requena and S. Moya, "Powder processing of mullite/Mo functionally graded materials", *J. Euro. Ceram. Soc.* **18**, 1365-71 (1998).
11. H. Ishibashi, H. Tobimatsu, T. Matsumoto, K. Hayashi, A.P. Tomsia and E. Saiz, "Characterization of Mo-SiO_2 functionally graded materials", *Metallurgical and Materials Transactions A*, **31A**, 299-308 (2000).
12. A. Umemoto, K. Hayashi, K. Nakashima, N. Saito, K. Kaneko, "High-Intensity Discharge Lamp with Mo-SiO_2 Functionally Graded Material", *J. Am. Ceram. Soc.*, **89,** 1133-35 (2006).
13. T. Kajiwara, A. Umemoto, K. Hayashi, N. Saito and K. Nakashima, unpublished work.

A MECHANICALLY SYNTHESIZED $La_{0.8}Sr_{0.2}MnO_3$ FINE POWDER FOR THE CATHODE MATERIAL OF AN INTERMEDIATE TEMPERATURE SOLID OXIDE FUEL CELL (IT-SOFC)

Kazuyoshi Sato, Akira Kondo, Hiroya Abe and Makio Naito
Joining and Welding Research Institute,
Osaka University,
11-1 Mihogaoka, Ibaraki,
Osaka, 567-0047 Japan

Jintawat Chaichanawong
Department of Chemical Engineering,
Chulalongkorn University
Payathai Road, Pathumwan,
Bangkok, 10330 Thailand

ABSTRACT

Strontium-doped lanthanum manganite ($La_{0.8}Sr_{0.2}MnO_3$, or LSM) synthesized through mechanical activation was used for the cathode of intermediate temperature solid oxide fuel cell (IT-SOFC). The average particle size of the LSM powder was 0.28 μm, and the size distribution was narrow. In the resultant cathode, a homogeneous porous structure with fine grains (<0.5 μm) was observed. The cathodic polarization was 240 and 90 mV under the current density of 0.5 $A \cdot cm^{-2}$ at 700 and 800 °C, respectively. The mA aximum power density of the anode supported cell with the cathode was 0.34 and 0.75 $W \cdot cm^{-2}$ at 700 and 800 °C, respectively.

INTRODUCTION

The solid oxide fuel cell (SOFC) is one of the most important power generation technologies of the future. During the last decade, a great deal of effort went into the development of an intermediate temperature SOFC (IT-SOFC). [1-3] In the intermediate temperature range (700-800°C), a metallic interconnector can be used instead of an expensive $LaCrO_3$ based ceramic interconnector. Additionally, long term stability of the cell and the auxiliary components can also be expected. These advantages lead to decreased operating cost of the cell. The issue to be solved was the increased internal resistances related to the ionic conductivity of the electrolyte and electrochemical reactions at the electrodes. An anode supported cell which consisted of thin electrolyte (thickness < 20 μm) supported by thick porous anode was developed as an alternative design of IT-SOFC to achieve high power output even at intermediate temperature due to reduced ohmic losses of the thin electrolyte and cathode, as well as the low overpotential at the highly porous and conductive anode. [1] A major part of the internal resistance of the cell comes from the cathode side. [4] Thus, decreasing cathodic polarization is necessary to further increase the cell performance.

Strontium doped lanthanum manganite ($La_{1-X}Sr_XMnO_3$), or LSM has been widely used as promising cathode material of SOFC because of its reasonably high electrochemical activity, and chemical compatibility as well as close thermal expansion coefficient with yttria stabilized ZrO_2 (YSZ) electrolyte. Sasaki *et al.* reported the effect of size and shape of starting LSM powder on microstructure and polarization of the cathode. They demonstrated that fine and non-agglomerated powder leads to homogeneous porous microstructure, resulting in lower cathodic polarization than large and/or agglomerated one. [5]

A variety of methods such as spray pyrolysis [5], co-precipitation [6] and gel combustion [7] have been developed for synthesis of fine LSM powder. Recently, the authors demonstrated that single phase LSM nano particles can be synthesized by mechanical activation of the raw powder mixtures followed by heating at considerably lower temperature (<900°C) compared with conventional solid state reaction. [8] In the present study, microstructure-property relationship of the cathode made from the LSM powder was investigated and applicability of the LSM for IT-SOFC was discussed.

EXPERIMENTAL PROCEDUR

Stoichiometric $La_{0.8}Sr_{0.2}MnO_3$ (LSM) powder was synthesized by mechanical milling of La_2O_3, $SrCO_3$ and Mn_3O_4 powder mixture followed by heating at 900 °C for 6 h. Details of the powder synthesis were described elsewhere. [8] Fig. 1 shows X-ray diffraction pattern of the LSM powder. All diffraction peaks were identified as LSM. The powder was ball milled in ethanol for 24 h to remove large agglomerates. After the drying, the powder was mixed with polyethylene glycol (molecular weight: 400) and ethanol by ball milling for 1 h. Screen printing paste was obtained by evaporating ethanol at 80 °C.

Two configurations, electrolyte supported [9] and anode supported cells [10] were employed to evaluate cathodic polarization and power generation property of the single cell, respectively. For the former, YSZ electrolyte with the thickness about 300 μm was made by tape casting and subsequent firing at 1450 °C for 2 h. Then the NiO-YSZ anode layer was screen printed onto the electrolyte and then fired at 1350 °C for 2 h. For the latter, YSZ thin film was made by tape

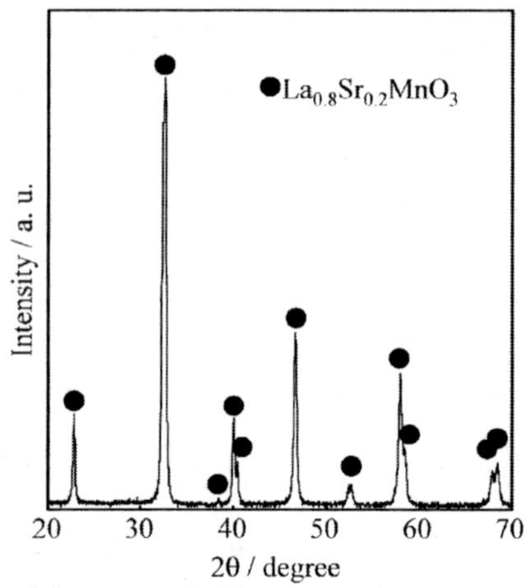

Fig. 1 X-ray diffraction pattern of the LSM powder

casting, and then NiO-YSZ slurry was applied onto the YSZ film also by tape casting. The NiO-YSZ/YSZ bilayer was co-fired at 1350 °C for 2 h. The thickness of YSZ and NiO-YSZ layers after the co-firing were about 15 and 500 μm, respectively. LSM layer was screen printed onto the YSZ for both cells and then fired at 1100 °C for 2 h. Area of the cathode electrode was 0.28 cm^2. Fracture cross-section of the cathode was observed with scanning electron microscopy (SEM).

Cathodic polarization and power generation property of the cells were measured at 700 and 800 °C. Air and H_2-3%H_2O was supplied for the cathode and the anode, respectively. For evaluation of cathodic polarization, Pt wire was attached to circumference of the electrolyte as reference electrode. The cathodic polarization was measured with current interruption technique.

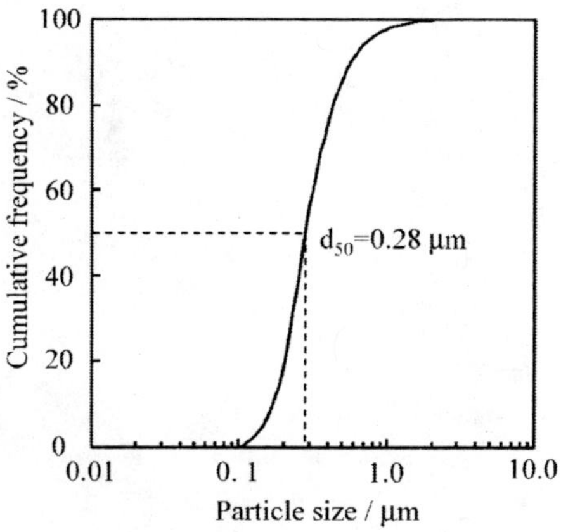

Fig. 2 Particle size distribution of the LSM powder

RESULTS AND DISCUSSION

Specific surface area (SSA) of the powder was 12.4 m^2g^{-1} after the ball milling. Corresponding primary particle size was about 75 nm. Figure 2 shows particle size distribution of the LSM powder after the ball milling measured by light scattering method. Average particle size was about 0.28 μm. Particle size distribution was narrow and no large agglomerate was observed.

Fig. 3 Cross-sectional image of the LSM cathode

Figure 3 shows the cross-sectional image of the LSM cathode. The grains were small (almost < 0.5 μm) even after the firing at 1100 °C for 2 h. No abnormally large grain was observed. This can be attributed to fine as well as narrow size distribution of the starting powder as shown in Fig. 2. The fine particles also would contribute to good sinterability of the LSM onto YSZ electrolyte.

Figure 4 shows cathodic polarization

measured at 700 and 800 °C. The values were 240 and 90 mV under the current density of 0.5 $A \cdot cm^{-2}$ measured at 700 and 800 °C, respectively. Recently, Jiang *et al.* reported the polarization of A-site deficient $La_{0.72}Sr_{0.18}MnO_3$ on YSZ electrolyte. [6] The values were about 1050 and 750 mV under the current density of 0.5 $A \cdot cm^{-2}$ at 700 and 800 °C, respectively. Grain size of their LSM cathode was in the range of 0.7-1.2 μm. Contrarily, Sasaki *et al.* reported that $La_{0.85}Sr_{0.15}MnO_3$ cathode with the grain size about 0.5 μm exhibited cathodic polarization lower than 200mV at 800 °C under the same current density. [5] Our LSM cathode consists of smaller grains (see Fig. 3) demonstrated lower cathodic polarization compared with these studies. These results suggested that cathodic polarization strongly depends on the grain size rather than A-site deficiency and/or slight difference of Sr content. The narrow size distribution of the starting powder leads to fine grains with uniform porous structure *i.e.* longer triple phase boundary at the cathode/electrolyte interface, resulting in lower cathodic polarization.

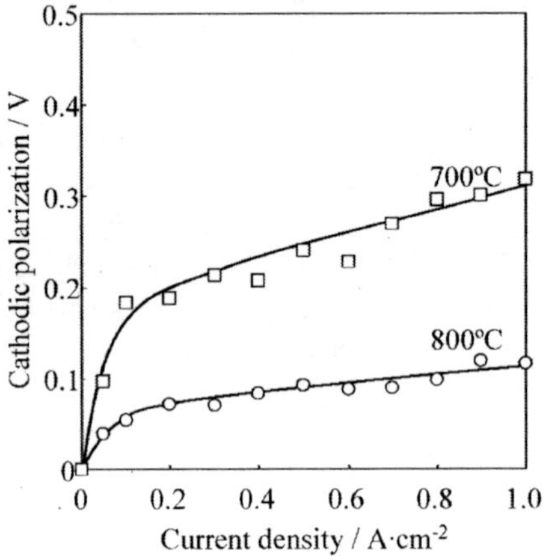

Fig. 4 Cathodic polarization of the electrolyte supported cell measured at 700 and 800 °C

Figure 5 shows the power generation property of the anode supported cell measured at 700 and 800 °C. Open circuit voltage was about 1.1 V and was agreed with that calculated by Nernst equation. This agreement indicated that thin electrolyte film was fully dense. Maximum power density was 0.34 and 0.75 $W \cdot cm^{-2}$ at 700 and 800 °C, respectively. The low cathodic polarization as shown above would contribute to good cell performance. These results indicated that the LSM powder synthesized by mechanical

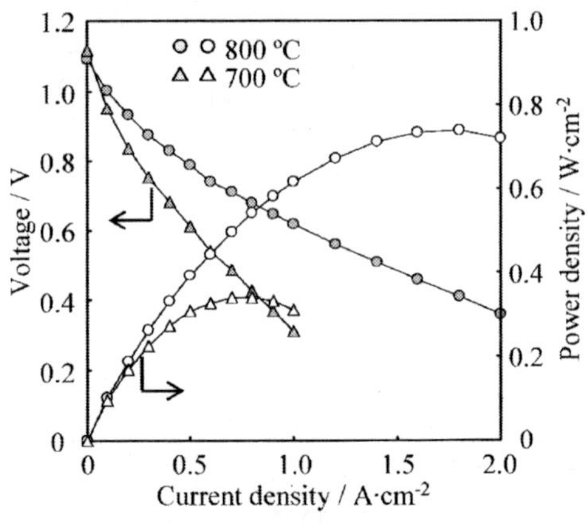

Fig. 5 Power generation property of the anode supported cell measured at 700 and 800 °C

activation can be applied for IT-SOFC. Further refinement of cathode microstructure, including application of the composite cathode consisting of the LSM and electrolyte materials such as YSZ and doped ceria, will increase cell performance significantly. [1, 2, 6] This will also allow replacement of mixed electric and ionic conducting materials such as $La_{0.6}Sr_{0.4}Co_{0.2}Fe_{0.8}O_3$ (LSCF) by LSM, which will lead to enhanced durability of the cell. Application of the composite cathode will be reported in the future work.

CONCLUSION

Fine LSM powder synthesized by mechanical activation was used for the cathode electrode of an IT-SOFC. The powder produced a cathode with a homogeneous porous structure consisting of fine grains (< 0.5 μm) after firing at 1100°C, and the cathode showed low cathodic polarization. It can be concluded that the LSM powder was beneficial for an IT-SOFC.

ACKNOWLEDGEMENT

This study was partially supported by Industrial Technology Research Grant Program from New Energy and Industrial Technology Development Organization (NEDO) of Japan.

This study was also partially supported by Grant-in-Aid for Cooperative Research Project of Nationwide Joint-Use Research Institutes on Development Base of Joining Technology for New Metallic Glasses and Inorganic Materials from Ministry of Education, Science, Sports and Culture, Japan.

REFERENCES

[1]S. D. Souza, S. J. Visco and L. C. D. Jonghe, "Reduced-Temperature Solid Oxide Fuel Cell Based on YSZ Thin-Film Electrolyte," *J. Electrochem. Soc.*, **144**, L35-L37 (1997)

[2]P. V. Dollen and S. Barnett, "A Study of Screen Printed Yttria-Stabilized Zirconia Layers for Solid Oxide Fuel Cells," *J. Am. Ceram. Soc.*, **88**, 3361-3368 (2005)

[3]S. P. Simner, M. D. Anderson and J. W. Stevenson, "La(Sr9FeO3 SOFC Cathodes with Marginal Copper Doping," *J. Am. Ceram. Soc.*, **87**, 1471-1476 (2004)

[4]T. Ishihara, T. Kudo, H. Matsuda and Y. Takita, "Doped perovskite Oxide, $PrMnO_3$, as a new Cathode for Solid-Oxide Fuel Cells that Decreases the Operation Temperature," *J. Am. Ceram Soc.*, **77**, 1682-1684 (1994)

[5]K. Sasaki, J.-P. Wurth, R. Gschwend, M. Gödickemeier and L. J. Gauckler, "Microstructure-Property Relations of Solid Oxide Fuel Cell Cathodes and Current Collectors," *J. Electrochem. Soc.*, **143**, 530-543 (1996)

[6]S. P. Jiang and W. Wang, "Novel structured mixed ionic and electronic conducting cathodes of solid oxide fuel cells," *Solid State Ionics*, **176**, 1351-1357 (2005)

[7]H. Taguchi, D. Masuda, M. Nagao, K. Tanihata, Y. Miyamoto, "Synthesis of Perovskite-Type $(La_{1-x}Sr_x)MnO_3$ ($0 \leq x \leq 0.3$) at Low Temperature," *J. Am. Ceram. Soc.*, **75**, 201-202 (1992)

[8]J. Chaichanawong, K. Sato, H. Abe, K. Murata, T. Fukui, T. Charinpanitkul, W. Tanthapanichakoon and M. Naito, "Formation of strontium doped lanthanum manganite ($La_{0.8}Sr_{0.2}MnO_3$) by mechanical milling without media balls," *Adv. Powder Technol.*, in press

[9]T. Fukui, K. Murata, S. Ohara, H. Abe, m. Naito and K. Nogi, "Morphology control of Ni-YSZ cermet anode for lower temperature operation of SOFCs," *J. Power Sources*, **125**, 17-21 (2004)

[10]H. Abe, K. Murata, T. Fukui, W. -J. Moon, k. Kaneko and M. Naito, "Morphology control of Ni-YSZ cermet anode for planer thin-film solid oxide fuel cells," *Thin Solid Films*, **496**, 49-52 (2006)

HYDROTHERMAL SYNTHESIS OF NANOSTRUCTURED Bi_2Te_3 POWDER

H. Kaga, Y. Kinemuchi and K. Watari

Advanced Manufacturing Research Institute
National Institute of Advanced Industrial Science and Technology (AIST)
2266-98 Anagahora, Shimo-shidami, Moriyama-ku, Nagoya 463-8560, Japan

ABSTRACT

Nanostructured thermoelectric Bi_2Te_3 powders were hydrothermally synthesized using various conditions to investigate the mechanism of powder formation. A homogeneous mixture in solution during the hydrothermal synthesis is critical to obtain a single phase of Bi_2Te_3. The growth of Bi_2Te_3 nanostructures can be categorized three stages: The synthesized powders are nanoparticles when the reaction temperature was below 100°C, regardless of the reaction time. As the reaction temperature is increased to 150°C, the Bi_2Te_3 products obtained show different morphologies depending upon the reaction time. Rod-like nanostructures with lower and higher aspect ratios are directly formed after 6 h and 12 h of reaction time, respectively. Further increasing the reaction time to 24 h at the same synthesis temperature results in the decomposition of the 1D nanostructures to nanoparticles. As a result, the anisotropic nanostructure is mainly controlled by the reaction temperature and time.

INTRODUCTION

Nanostructured materials have been the subject of intensive research for their potential applications in a wide variety of areas since a carbon nanotube was found in early 90's.[1] Recent theoretical and experimental results in thermoelectric materials have shown that quantum size-effects may enable much more efficient thermoelectric energy conversion.[2-4] Low-dimensional structures such as nanodots, nanotubes, nanowires and superlattice thin film effectively decrease the thermal conductivity due to the phonon-blocking effect of the nanostructures.[4-7]

For example, the thermal conductivity of Bi_2Te_3 nanowire along the longitude direction has been reported at least one order of magnitude lower than that of its bulk.[8] In addition to this, the state-of-the-art material of doped Bi_2Te_3 is currently the most efficient thermoelectric materials at room temperature, therefore, the nanostructured Bi_2Te_3 is particularly interesting and several methods have been introduced to synthesize a low dimensional nanostructured Bi_2Te_3. The structure of Bi_2Te_3 belongs to the space group *R3m*, and consists of atomic layers stacked in the order of $Te^{(1)}$-Bi-$Te^{(2)}$-Bi-$Te^{(1)}$, here the superscripts of Te denote the two different coordinative types of Te atoms, along the *c*-axis that so called quintets. Inside each quintet the strong covalent bonds prevail, while the quintets are bound by weak Van der Waals bonds.[9] This structure gives Bi_2Te_3 inherent helical chirality, as well as a tendency toward *c*-axis growth.

Physical templates such as porous alumina template are usually adapted to generate nanostructures and a large collection of nanowires can be obtained.[10-11] However, the physical templates are often limited in terms of production scale, thus different approaches are needed to be introduced. Chemical routes could alter a more promising and practical methods for nanostructured materials. Solvothermal and/or hydrothermal methods as a simple chemical method without need of costly device have been utilized to synthesize Bi_2Te_3 powders and their morphologies, such as nanotubes, nanowires, nanosheets and nanoparticles with a different synthetic approach, have been reported.[12-15] However, it is very difficult to produce nanostructured materials with well-controlled morphologies. Here we report a hydrothermal synthesis of nanostructured Bi_2Te_3 with various morphologies of powders.

EXPERIMENTAL

In a typical process, the reagent grade of $BiCl_3$ and pure tellurium (Kojundo Chemical, Saitama, Japan) were weighed to have a molar ratio of 2 : 3 and the mixture was put into a Teflon-lined autoclave. The autoclave was then filled with distilled water up to 70% of its capacity (100ml). Then, appropriate amount of ethylenediaminetetraaetic acid disodium salt (EDTA) and $NaBH_4$ was added to the solution. NaOH was then added to control of the pH 13 because acid solution is critical to synthesis the nanostructured Bi_2Te_3.[14] The autoclave was sealed and heated at 100°C to 150°C for 3 to 24 h. During the hydrothermal reaction, the solution was stirred by a stirrer with a rotational speed of 0 and 200 rpm. After a reaction time, the autoclave was gradually cooled down to room temperature and allowed to age the solution. The black color synthesized powders were obtained and these powders were clean out with distilled water, ethanol and acetone several times. Washed powder was then dried at vacuum oven at 100°C for 6 h. The phase composition was characterized by X-ray diffraction (XRD, RINT2000, Rigaku, Tokyo, Japan) with CuK_α radiation (λ=0.154056 nm). The powder morphology was determined by SEM (S-4300, Hitachi, Tokyo, Japan).

RESULTS AND DISCUSSION

Figure 1a shows the XRD patterns of hydrothermally synthesized powders at 150°C for 12 h with and without stirring the solution. In the expanded patterns from the local part of 2θ = 37 - 43° in Fig. 1a are also shown in Fig. 1b because the XRD peaks of Bi_2Te_3 are very close to those of Te phase.

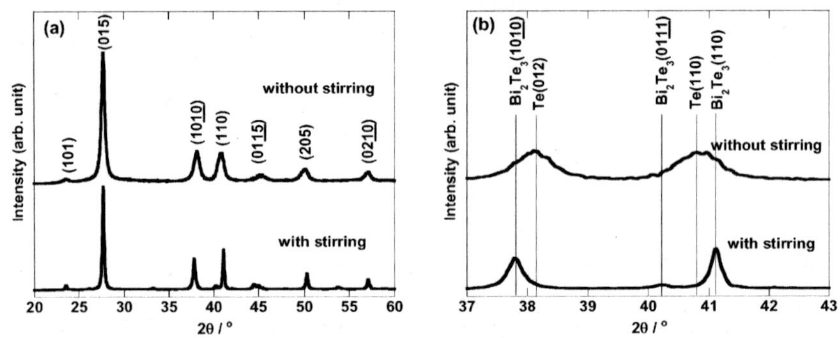

Figure 1 (a) XRD patterns of hydrothermally synthesized Bi2Te3 powers at 150oC for 12 h with and without stirring the solution. (b) Expanded patterns from the local part of 2θ = 37° to 42° in XRD pattern.

Clearly, the Te peaks remained in the XRD pattern of the sample without stirring while the XRD pattern with stirring had only Bi_2Te_3 phase and corresponds to the reflection of rhombohedral phase with lattice constants of a=4.38 Å and c=30.49 Å, which are in good agreement with the reported values.[15] The reason for unreacted Te remained in the final product is due to inhomogenous mixture of the solution so that stirring the solution is crucial to obtain a single phase of Bi_2Te_3 powders. Note that the XRD patterns of other samples with stirring showed only Bi_2Te_3 peaks regardless of reaction time and temperature except for reaction time of 3 h.

After finishing the reaction processes for 3 h at 100°C and 150°C with stirring the solution of 200 rpm, the synthesized products showed composite phases of Te and Bi_2Te_3. However, the formation of unreacted and/or unreduced Te is completely different in morphology. The purple solution was observed when reacted at 100°C, while the dark color powders were precipitated when reacted at 150°C. The purple color solution was due to the formation of amorphous Te colloids.[16] After aging the solution at room temperature for one day, it was found that solid black colored powders were obtained. The XRD observation revealed that the precipitated powders were well crystallized hexagonal Te. On the other hand, reacted at 150°C for 3 h, partially reduced Te powders were observed as seen in Fig. 2.

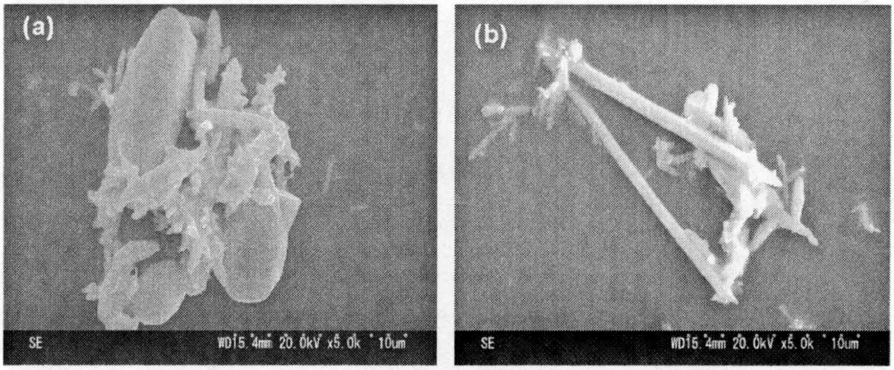

Figure 2 SEM images of hydrothermally synthesized powders at 150°C for 3 h. (a)Initial reduction of Te to form rod-like powders (b) further reduction was proceeded to form a higher aspect ratio of needle-like Te powders.

A lower aspect ratio of rod-like Te powders was clearly observed in Fig. 2a while a higher aspect ratio of needle-like Te powders was also observed with a diameter of 500 nm and length of several tens micrometer, and cross section was hexagonal shape as seen in Fig. 2b. The Te powder was initially reduced to rod-like powders and then further reduction was proceeded to form needle-like powders. These results suggest that Te powder was tendency to reduce 1D structure at 150°C. However, such large Te powders could not react with Bi ions to form Bi_2Te_3 so that reaction time was not sufficient to reduce the Te completely. The difference of the formation of Te powders between the reaction temperatures is yet not to be completely understood. It was likely that the higher reaction temperature is more stable to form needle-like powder rather than forming the amorphous phase. Note that reacted at 100°C for longer reaction time, a single phase of Bi_2Te_3 nanoparticle with a mean diameter of 50 nm was obtained.

As increasing the reaction time to 6 h, no purple color solution was observed after the reaction. The lower aspect ratio of Bi_2Te_3 nanorods was obtained with length up to 500nm and average width of about 200 nm as shown in Fig. 3a. As a rough estimation from the SEM observation, about one thirds of the powder is nanorods and the other two third is nanoparticles.

The reduction of Te was further proceeded as reaction time was increased to 12 h. At this time, the morphology of Te had a higher aspect ratio and caused to form narrower Bi_2Te_3 as shown in Fig. 3b. With increasing reaction time to 24 h, once formed nanorods were decomposed to nanoparticles. The particles are nearly spherical and sizes are 50 to 100 nm as seen in Fig. 3c. As a result, a longer reaction time caused additional reduction of Te to form the nanoparticles. The average particle size synthesized at 150°C for 24 h was bigger than that synthesized at 100°C for 6h to 24h because of increasing reaction time and temperature.

Figure 3 SEM images of hydrothermally synthesized powders at 150°C for (a)6 h, (b)12 h and (c)24 h. The aspect ratio of Bi_2Te_3 nanorod was increased as reaction time was increased from 6 h to 12 h.

CONCLUSION

Nanostructured Bi_2Te_3 powders were hydrothermal synthesized using various conditions. Single phase nanostructured Bi_2Te_3 was obtained using $BiCl_3$ and Te as the starting sources, and a homogeneous solution mixture during the reaction was critical to form a single phase. The morphologies of the synthesized powders were nanoparticle and nanorods with different aspect ratios depending upon the reaction time and temperature. At lower and higher reaction temperatures, the Te powder has a tendency for 0D and 1D growth, respectively. The nanorods having a diameter of 100 nm, with a length up to 1 μm were synthesized at 150°C for 12 h. However, 1D nanostructured Bi_2Te_3 decomposes as reaction time increased to form nanoparticles. The nanostructured Bi_2Te_3 may be incorporated into bulk structures to enhance properties, in particular, to decrease the thermal conductivity of the original parent materials.

REFERENCES

[1] S. Iijima, "Herical microtubules of graphitic carbon", *Nature*, **354**, 56-57 (1991).
[2] L. D. Hicks and M. S. Dresselhaus, "Effect of Quantum well structures on the thermoelectric figure of Merit", *Phys. Rev. B*, **47**, 12727-12732 (1993).
[3] R. Venkatasubramanian, E. Silvola, T. Colpitts, and B. O'Quinn, "Thin-film thermoelectric devices with high room-temperature figures of merit" *Nature*, **413**, 597-602 (2001).
[4] K. F. Hsu, S. Loo, F. Guo, W. Chen, J. S. Dyck, C. Uher, T. Hogan, E. K. Polychroniadis and

M. G. Kanatzidis, "Cubic $AgPb_{(m)}SbTe_{(2+m)}$ bulk thermoelectric materials with high figure of merit", *Science*, **303**, 818-821 (2004).
[5] R. Venkatasubramanian, "Lattice thermal conductivity reduction and phonon localizationlike behavior in superlattice structures", *Phys. Rev. B*, **61**, 3091-3097 (2000).
[6] L. D. Hicks and M. S. Dresselhaus, "Thermoelectric figure of merit of a one-dimensional conductor", Phys. Rev. *B*, **47**, 16631-16634 (1993).
[7] T. C. Harman, P. J. Taylor, M. P. Walsh, and B. E. LaForge, "Quantum dot superlattice thermoelectric materials and devices", *Science*, **297**, 2229-2232 (2002).
[8] T. Borca-Tasciuc, G. Chen, A. Prieto, M. S. Martin-Gonzalez, A. Stacy, T. Sands, M. A. Ryan and J. P. Fleurial, "Thermal properties of elctrodeposited bismuth telluride nanowires embedded in amorphous alumina", *Appl. Phys. Lett.*, **85**, 6001-6003 (2004).
[9] D. M. Rowe,"CRC Handbook of Thermoelectrics", CRC Press, Boca Raton, ch. 19, pp.212 (1994).
[10] A. Prieto, M. Sander, M. S. Martin-Gonzalez, R. Gronsky, T. Sands and A. Stacy, "Electrodeposition of orders Bi2Te3 nanowire arrays", *J. Am. Chem. Soc.*, **123**, 7160-7161 (2001).
[11] W. Wang, F. Jia, Q. Huang and J. Zhang, "A new type of low power thermoelectric micro-generator fabricated by nanowire array thermoelectric materials", *Microelectronic Engr.*, **77**, 223-229 (2005).
[12] X. B. Zhao, X. H. Ji, Y. H. Zhang, T. J. Zhu, J. P. Tu and X. B. Zhang, "Bismuth telluride nanotubes and the effect on the thermoelectric properties of nanotube-containing nanocomposites", *Appl. Phys. Lett.*, **86**, 062111 (2005).
[13] Y. Deng, X. S. Zhou, G. D. Wei, J. Liu, C. W. Nan, S. J. Zhao, "Solvothermal preparation and characterization of nanocrystalline Bi_2Te_3 powder with different morphology", *J. Phys. Chem. Solid.*, **63**, 2119-2121 (2002).
[14] X. B. Zhao, T. Sun, T. J. Zhu and J. P. Tu, "In-situ investigation and effect of additives on low temperature aqueous chemical synthesis of Bi2Te3 nanocapsules", *J. Mater. Chem.*, **15**, 1621-1625 (2005).
[15] JCPDS card No. 15-0863
[16] B. Mayers and Y. Xia, "One-dimensional nanostructures of trigonal tellurium with various morphologies can be synthesized using a solution-phase approach", *J. Mater. Chem.*, **12**, 1875-1881 (2002).

INFLUENCE OF NANOPOROUS STRUCTURE ON SILANE COUPLING SURFACE MODIFICATION BEHAVIOR AND ADHESION PROPERTIES OF SPHERICAL SILICA PARTICLES

Toshiyuki Kani[1, 2]*, Maki Tamonoki[1], Takahiro Suzuki[2], Mayumi Tsukada[1], Hidehiro Kamiya[1]

[1] *Institute of Symbiotic Science and Technology*
Tokyo University of Agriculture and Technology
2-24-16 Naka-cho, Koganei, Tokyo 184-8588, Japan

[2] *Nihon L'Oreal K. K.*
KSP R&D-B1113, 2-1, Sakado 3-chome, Takatsu-ku, Kawasaki-shi, Kanagawa-ken 213-0012, Japan

Corresponding author: T.KANI@rd.loreal.com

ABSTRACT

In cosmetic powder, there are several kinds of powder materials widely used, such as mica and porous silica on which nanoparticles adhered on the surface of particle. However, it is not widely known that nanoparticles are adhered on the surface of particle. In order to analyze the effect of a nanoporous structure on the surface modification behavior by using a silane coupling agent and the surface interaction between silica and mica, four types of silane coupling agents with different molecular structures were used for the surface treatment of spherical silica with and without a nanoporous structure. The microscopic adhesion force between each surface-modified silica particle and mica surface was characterized by using colloid probe atomic force microscopy (AFM). The adhesion force between the nonporous silica particles and mica surface was increased by the surface modification using almost all silane coupling agents. In contrast, the adhesion force between nanoporous silica particles and mica was decreased by the surface modification. Since the amount of reacted silane coupling agent per unit surface area of nanoporous silica particles was much less than that of nonporous silica particles.

1. INTRODUCTION

Spherical silica powder with a nanoporous structure and mica powder are added as a base for cosmetic powders to absorb sweat and sebum.[1], [2] However, the effect of these nanopores on various properties including the dispersibility and extensibility of cosmetics has not been well understood. Silane coupling treatment is applied in various fields such as ceramics, electronic materials, and catalysts.[3] In the cosmetic industry, the silane coupling treatment has been applied for surface modifications such as pigment.

In order to control the dispersibility of cosmetic powder and to analyze the effect of nanoporous structure on surface modification behavior by using silane coupling agent and surface interaction between silica particles and mica, in this study, four types of silane coupling agents with different molecular structures were used for the surface treatment on spherical silica particles with and without a nanoporous structure. The adhesive force between the spherical silica particles before and after the surface modification and mica was determined by a colloid probe AFM method.[4] Silane coupling agents with different molecular structures are applied in cosmetic powder; however, the effect of the nanoporous structure on surface modification properties and powder behavior has not been investigated.

2. EXPERIMENTAL PROCEDURE

Two types of commercially available silica—porous silica particles (CCIC, mean diameter: 5 μm, silica microbeads P-1500) and nonporous fused silica particles (Denki Kagaku Kogyo Co., Ltd., mean diameter: 4.5 μm, FB-6S)—were used in this study. Each silica particle was cleaned in a water bath with an ultrasonic disperser (SND Co., Ltd, US-3, 38 kHz) using ultrapure water. The number of repeated ultrasonic cleaning required to obtain a clean surface without nanoparticle adhesion was determined to be six in a previous paper.[4] Nanoporous silica particles before and after the surface treatments were observed using field emission scanning electron microscopy (FE-SEM) (JEOL. Ltd., FE-SEM JSM-6335F). Four types of commercial silane coupling agents with different molecular structure weight and the minimum covering area calculated from Si-O structure of each coupling agents produced by Shin-Etsu Chemical Co., Ltd. were used, as shown in **Table 1**.

Table 1 Four various silane coupling agents.

	Products	Chemical structure	molecular weight	minimum covering area (/g)
N1	LS-3120	$CH_3(CH_2)_2Si(OC_2H_5)_3$	206.4	379
N2	LS-2920	$CH_2{=}CHCH_2Si(OC_2H_5)_3$	204.3	382
P1	KBE-903	$NH_2(CH_2)_3Si(OC_2H_5)_3$	221.4	353
P2	KBE-603	$NH_2(CH_2)_2NH(CH_2)_3Si(OC_2H_5)_3$	264.5	295

The specific surface area and pore size distribution of each silica particle was measured by the Bunauer-Emmett-Teller (BET) method before and after the silane coupling. Pore diameter distribution and pore volume of porous silica were measured by the nitrogen adsorption and desorption method (Quantasorb Jr., Quantachrome). The sample was pre-treated at 100□ and 1.0 □10^{-4} Pa for 6 hours, and was measured on -196.15□. The amount of coupling agent that reacted with the silica particles on the surface was estimated by the increase in the carbon and nitrogen content after the surface treatment characterized using CHN Analyzer (J-Science Lab Co., Ltd., with the silica particles on the surface was estimated by the increase in the carbon and nitrogen content after the surface treatment characterized using CHN Analyzer (J-Science Lab Co., Ltd., MICRO CORDER JM10). The surface functional groups on the silica particle on the surface before and after cleaning were characterized by Fourier-transform infrared (FT-IR) (Nicolet Co., Ltd., AVATAR 360) spectroscopy.

The pH of ultrapure water was adjusted to 4.0 by the addition of HCl aqueous solution. Subsequently, a silane coupling agent was added to the aqueous solution and this solution was stirred for 30 min to accelerate the hydrolysis of the silane coupling agent. 7.0 g of each silica particle was mixed with this solution. NH_4OH aqueous solution was added to the silica suspension in drops in order to increase the PH to 12.0. The suspension was then placed for 24 h in an oven. The silica suspension was filtered by suction filtration and dried at 120°C for 24 h in order to promote the condensation of the reaction of the silane coupling agent on the silanol group of silica. The dried silica particles were then washed in methanol to remove the unreacted coupling agent. Washing and drying with methanol was repeated twice for each silica particle.

The amount of each silane coupling agent was determined by using the minimum covering area of each silane coupling agent and the specific surface area of each silica particle before surface treatment. The values are 147 m^2/g and 0.259 m^2/g for porous and nonporous silica, respectively; further, 12.8 mol/□ of each silane coupling agent was added on a unit surface area.

In the colloidal probe AFM method, a single particle is bonded to the tip of a commercially available AFM system to measure its interaction with a substrate such as mica. To measure surface interaction, mica (Yamaguchi Mica Co., Ltd.) was used as a substrate in the colloid probe AFM method. The adhesion force was measured using an AFM system in the contact mode with atmosphere control (Molecular Imaging Inc., Pico SPM). A humidity and temperature measurement device (IC Co., Ltd., Thermo 1022) was also placed in the chamber. This method was described in our previous paper.[4]

3. RESULTS AND DISCUSSION

3-1. Surface reaction behavior of silane coupling agents

The surface functional groups on nanoporous silica particles before and after the silane coupling treatment were characterized by FT-IR. (**Figure 1**)

The absorption of free silanol groups at 3740 cm^{-1} [5], [6] decreased after silane coupling reactions. The absorption of double bonded carbon ν(C=C) at 3050cm^{-1} and 3072cm^{-1} [7] for N2, and the C-H stretching vibration of methyl and methylene groups at 2962 cm^{-1}, 2930 cm^{-1}, and 2860 cm^{-1} [8] of each coupling agent was also detected after surface treatment. Based on these results, each coupling agent reacted with the surface silanol group on the silica particle. The amount of carbon and nitrogen content (C_D) per unit surface area (A) was calculated to discuss the surface coverage ratio of silane coupling agents by using the CHN analyzer and is shown in **Figure 2**. The amount of carbon and/or nitrogen per unit surface area on nonporous silica particles was much higher than that on porous silica particles for all silane coupling agents.

The specific surface area and pore size distribution before and after surface treatment (N2, P2) on porous silica particles were determined by N_2 adsorption desorption method and is shown in **Table 2**. Based on this result, the specific surface area, pore volume, and mean pore size were decreased by the surface modification.

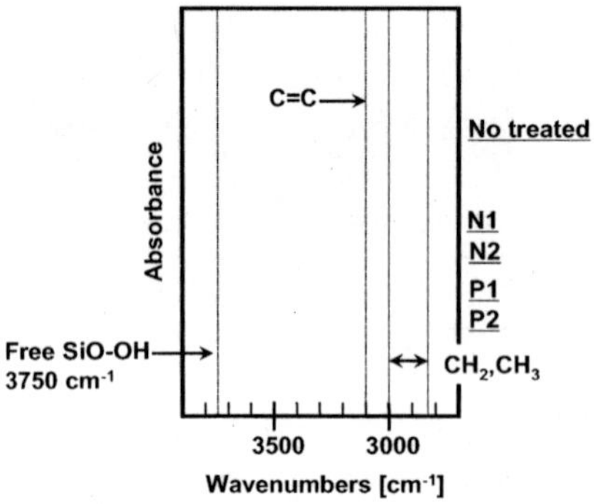

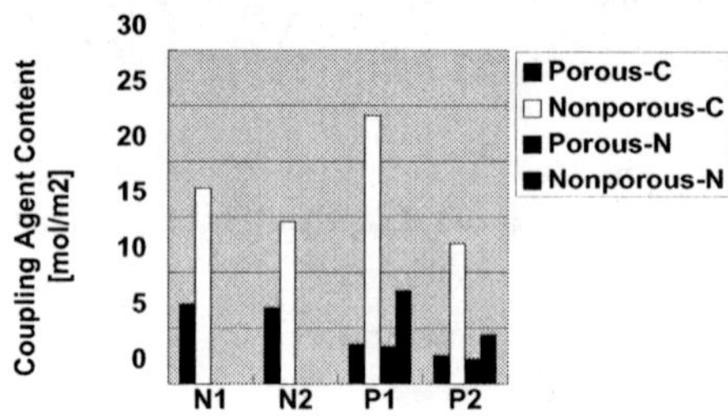

Figure 1 Characterization of surface functional groups on porous silica particle by FT-IR before and after silane coupling treatment of nanoparticles on their surface FT-IR spectra. (The left figure)
Figure 2 The result of CHN analysis. (The right figure)

Table 2 Specific surface area, pore volume, and mean pore size before and after surface modification of porous silica particles.

	BET surface area [m^2/g]	Pore volume [ml/g]	Mean pore size [nm]
Original porous silica	147.3	0.346	6.59
N2 (C=C-C-)	133.6	0.253	5.16
P2 (N-C-C-N-C-C-C-)	115.7	0.299	6.17

3-3. Adhesion force between surface-treated silica particles and mica

The effect of surface modification with silane coupling agents on the adhesion force of porous and nonporous silica particles was investigated by the colloid probe AFM method at different humidity conditions of 30% and 80% RH to discuss the effect of liquid bridge on adhesion force; this is shown in **Figure 3**. The mean adhesion forces of porous silica particles were decreased by surface modification at both humidity conditions except in the case of the N1 agent. On the other hand, at both humidity conditions of 30% and 80% RH, the mean adhesion forces of nonporous silica particles was increased by surface modification except in the case of the N1 agent.

From the CHN analysis, the amount of reacted silane coupling agent per unit surface area of the nanoporous silica particle was considerably less than that of the nonporous silica particle. Furthermore, since the pore volume and mean pore size were decreased by the silane coupling treatment with both N2 and P2 agents (Table 2), it is inferred that silane coupling agent reacted at the periphery of the nanopore on the surface of the porous silica particle. To discuss the decreasing mechanism of nanoporous silica after surface treatment, TEM observation of nanoporous silica before and after treatment was shown in **Figure 4**. In this study, pH in solution was controlled from 4 to 12 after mixing silica particles to promote the condensation reaction between silane coupling agent and surface silanol groups. At the high pH condition, since small amount of silica component on nanoporous silica was dissolved in the solution, nanoparticles was nucleated and adhered on the surface of nanoporous silica particle. It seems that adhered nanoparticles reduced the adhesion force between nanoporous silica and mica substrate. On the contrary, since nanoparticles were not produced on nonporous silica particles after surface modification at same pH condition, the adhesion force was increased by surface treatment.

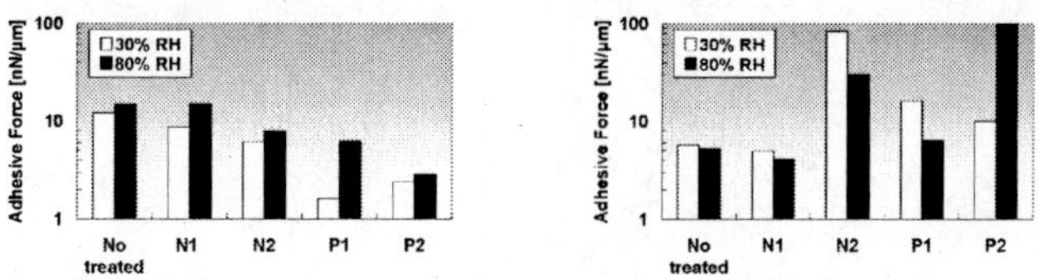

(a) (b)

Figure 3 Effect of humidity on the mean adhesion force between mica and (a) porous silica particle (P-1500) (b) nonporous silica (FB-6S) with and without treatment with each silane coupling agent.

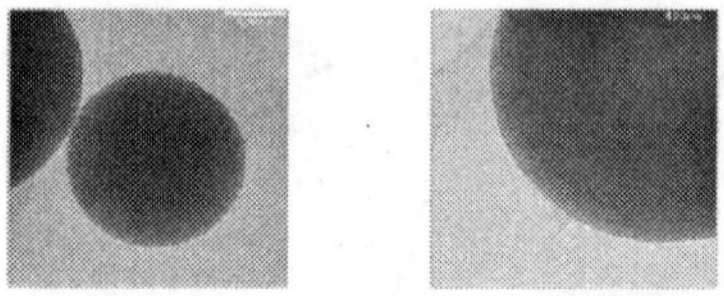

(a) (b)

Figure 4 TEM observation of (a) before and (b) after surface treatment with P2 agent.

4. CONCLUSION

The adhesion force between nonporous silica particles and mica surface was increased by a surface modification except in the case of N1 agent; this was observed by using colloid probe AFM. In contrast, the adhesion force between nanoporous silica particles and mica was decreased by surface modification except in the case of the N1 agent.

REFERENCES

[1] T. Mitsui, New Cosmetology, First edition, 385–388 (1993)

[2] Y. Kohara, Japanese published unexamined patent No. 54-55739

[3] T. Sato and R. Takahashi, Japanese published unexamined patent No.2003-54915

[4] T. Kani, M. Tamonoki, T. Suzuki, M. Tsukada, and H. Kamiya, "Influence of Surface- Adhered Nanoparticles and Nanoporous Structure on Particle-Particle Interaction of Silica Powder", Powder Technol. submitted.

[5] T. Ishikawa, M. Matsuda, A. Yasukawa, K. Kandori, S. Inagaki, T. Fukushima, and S. Kondo, "Surface Silanol Groups of Mesoporous Silica FSM-16", J. Chem. Soc., Faraday Trans., 92(11), 1985 (1996)

[6] X. S. Zhao, G. Q. Lu, A. K. Whittaker, G. J. Millar, and H. Y. Zhu, "Comprehensive Study of Surface Chemistry of MCM-41 Using ^{29}Si CP/MAS NMR, FTIR, Pyridine-TPD, and TGA", J. Phys. Chem. B, 101, 6525 (1997)

[7] R. Anwander, I. Nagl, M. Widenmeyer, G. Engelhardt, O. Groeger, C. Plam, and T. Röser, "Surface Characterization and Functionalization of MCM-41 Silica via Silazane Silylation", J. Phys. Chem. B, 104, 3532 (2000)

[8] A. Matsumoto, K. Schumacher, and K. K. Unger, "Surface Functionalization and Stabilization of Mesoporous Silica Spheres by Silanization and Their Adsorption Characteristics", Langmuir, 18, 4014 (2002)

EFFECT OF MAGNETIC FIELD ON ORIENTATION OF DIAMAGNETIC CERAMIC PARTICLES DISPERSED IN SLURRY

Satoshi Tanaka, Atsushi Makiya and Keizo Uematsu,
Department of Material Science and Technology, Nagaoka University of Technology
1603-1 Kamitomioka Nagaoka Niigata 940-2188 Japan,
Yutaka Doshida
Taiyo Yuden Co., Ltd, 6-16-20 Ueno, Taito-ku, Tokyo 110-0005, Japan

ABSTRACT

The influence of magnetic field was studied on the orientation structure during shaping using a high magnetic field (0-10 Tesla). The relationship between strength of magnetic field and orientation in an applied field was described by using Langevin's theory with Boltzman average of the function. Energy of rotation depended on the difference of anisotropy energy, which was increased with a square of magnetic field. The theoretical degree of orientation was calculated by an order parameter. Experimental degree of orientation was evaluated by an optical method and XRD using bismuth titanate particles. The dilute slurries prepared were set and dried perfectly in a magnetic field. The calculated orientation degrees of bismuth titanate were increased with magnetic field from 0 to 1 as Langevin's function. Experimental degree of orientation also similarly increased from 0 to 0.7. The results were explained with theory derived from Langevin's theory, and experimental result was smaller than theoretical data for all samples. The key of orientation was dispersion and crystal structure in each particle.

INTRODUCTION

High anisotropy in property indicates that a special processing method is required to achieve useful properties in the bulk electric material. The orientation of crystals in microstructures in ceramics is a very effective method for drastically improving their properties [1-8]. Among the various fabrication processes that are available, the use of a magnetic field has a high potential for aligning crystal grains when processing materials and does not require any physical contact [9-12]. Regular powders with nearly equi-axed shape can be oriented comparing with conventional methods such as hot forging, uniaxial pressing, and doctor blade forming, in which needle-like or plate-like shape particles are needed [13-15]. We and some researchers have already reported the use of this method with a high magnetic field to orient particles of bismuth titanate, alumina, and zinc oxide in forming [16-18]. In this method, we put dispersed slurry in a container in the magnetic field and dry a medium in the slurry (**Fig.1**). Particles in slurry rotate according to their crystal magnetic anisotropy.

The objective in this study is to examine the influence of magnetic field and particle size on the orientation structure during the shaping using a high magnetic field (0-10 Tesla) from theoretical and experimental view points. The relationship between strength of magnetic field and orientation in an applied field is described by using Langevin's theory with Boltzman average of the function.

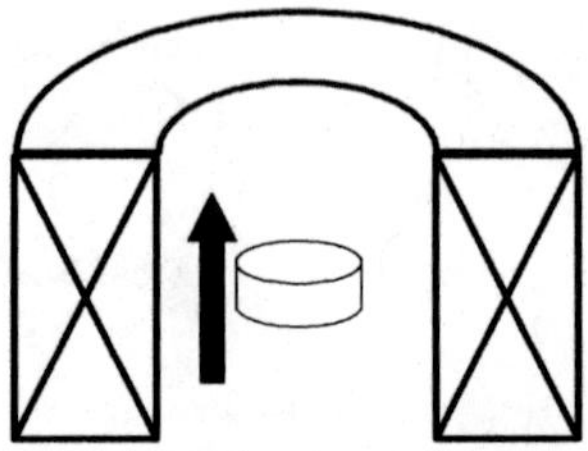

Fig.1 Mold and superconducting magnet

THEORETICAL APPROACH for PARTICLE ORIENTATION

Particle, which is a single crystal, rotate to minimize the magnetized energy in a system in high magnetic field. Magnetized energy *U* was changed for angle of crystal axis of dia- and para-magnetic materials and a magnetic field, and it is known by describing as follows,[19]

$$U(\theta) \cong -\frac{1}{4\mu_0} \cdot V \cdot B^2 \cdot (2\chi + \Delta\chi \cdot \cos 2\theta), \qquad (1)$$

where μ_0 is the magnetic permeability of vacuum, $\Delta\chi$ is the anisotropic susceptibility of particle, V is the volume of the particle and θ is the angle between the magnetic field and crystal axis as shown in **Fig.2**. Bismuth titanate is dia-magnetic material, and aligned for *a,c*-axes of particles parallel to the high magnetic field ($\chi_b < \chi_{a,c} < 0$).

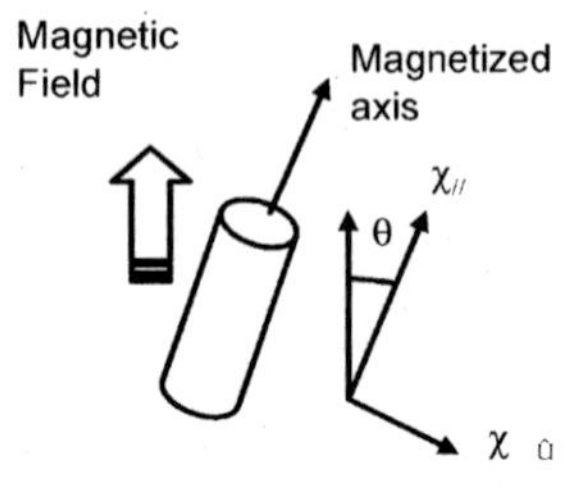

Fig.2 Particle direction in magnetic field

When differential magnetized energy ΔU is larger than kinetic energy by Brownian motions kT, some particles can rotate statistically to the stable direction to reduce the magnetized energy in the magnetic field. Here, the probability of rotated particles in the system was assumed to obey a Boltzmann distribution, the probability of particles rotation was described as follows [20],

$$P(\theta,U)\cdot \sin\theta d\theta = \frac{exp[-\Delta U(\theta,B)/kT]\sin\theta d\theta}{\int_0^{\pi} exp[-\Delta U(\theta,B)/kT]\sin\theta d\theta}, \quad (2)$$

$$\Delta U(\theta) = -\frac{1}{4\mu_0}\cdot V\cdot B^2\cdot \Delta\chi \cdot \cos 2\theta, \quad (3)$$

where k is Boltzmann constant and T is temperature. Distribution of oriented particles in slurry can be calculated by substituting eq.(3) to eq.(2). This distribution of oriented particles in a high magnetic field is corresponded to Langevin's theory[21], which is principle of magnetization behavior for para-magnetic materials.

We indicate the degree of orientation as order parameter, using equation (2),(3).

$$<m> = \frac{1}{2}\left\langle 3\cos^2\theta - 1\right\rangle = \frac{\frac{1}{2}\int_0^{\pi}\left\langle 3\cos^2\theta - 1\right\rangle exp[-U(\theta,B)/kT]\sin\theta d\theta}{\int_0^{\pi} exp[-U(\theta,B)/kT]\sin\theta d\theta} \quad (4)$$

The order parameter is 1 at perfect orientation, and -1/2 at random orientation.

Figure 3 shows the order parameter of bismuth titanate as function of θ at 10T. Particle diameter is 1μm. Anisotropic magnetic susceptibility $\Delta\chi$ is assumed to about 10^{-8}. Order parameter increased with magnetic field in similar to Langevin function and saturated in high magnetic field.

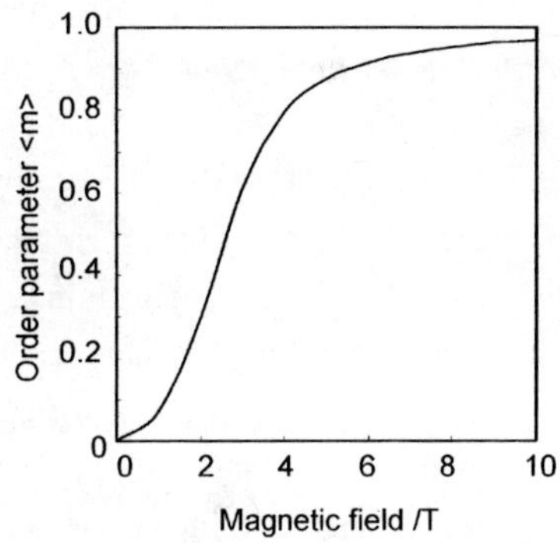

Fig.3 Order parameter

EXPERIMENTAL

Bismuth titanate powders were synthesized from some raw materials, bismuth oxide and titanium oxide [17]. Particle size was about 1μm. The synthesized powder was placed in a ball mill with distilled and de-ionized water, dispersant, and alumina balls, and mixed for 2h to prepare dilute slurry with a solids loading of 5vol%. The slurry (1 x10^{-6} m^3) was placed in a shallow container made of Teflon in **Fig.1**. The container was placed in a 10T magnetic field in a super conducting magnet (TM-10VH10, Toshiba, Japan), and left until drying at the room temperature. The magnetic field was applied horizontally to the container 1-10T. The resultant green compact was heated at 5°C/min to 1200°C and held for 1h, followed by cooling to the room temperature.

The crystal orientation of surface was examined by x-ray diffraction analysis with the Lotgering method in green and sintered body. The degree of orientation (hkl) is indicated as Lotgering factor f [22],

$$f = \frac{p - p_0}{1 - p_0}, \qquad (5)$$

where, $p = \Sigma I(h0l) / \Sigma I(hkl)$, $p_0 = p$ for randomly oriented ceramic samples, and $\Sigma I(h0l)$ is sum of the XRD peak intensities.

RESULTS and DISCUSSION

Figure 4 shows SEM micrograph of the bismuth titanate powder synthesized in this study. The particles have a nearly equi-axed shape, but the larger particles tend to have a platelet shape. The major face of the large plate-like particles should be the b face of bismuth titanate. In a single crystal, the b axis of the bismuth titanate lies perpendicular to the major face of plate-like particles.

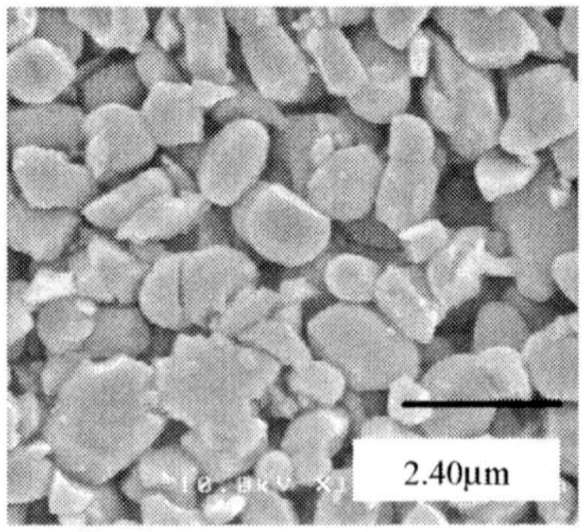

Fig.4 BiT powders

Figure 5 shows the x-ray diffraction patterns taken for the surface of the green specimens; the plane of analysis is in the direction parallel to the applied magnetic field for the specimen prepared in the magnet. In the specimen prepared in the high magnetic field, the strong, diffraction peaks are those associated with the a-c planes of the crystal, such as (200) and (002). The strongest peak, (171), in the raw powder, is almost absent from specimen prepared in the magnetic field. X-ray results and the anisotropic microstructure suggest that the b axes of particles are aligned normal to the applied magnetic field. Bismuth titanate has a tendency to form a plate shape, with the b plane being the largest face of the grain.

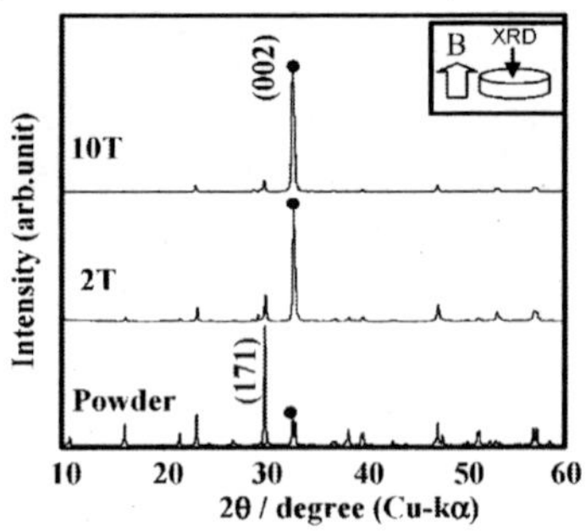

Fig.5 x-ray diffraction patterns taken for the surface of the green specimens

Figure 6 shows Lotgering factor, that is, the orientation degree, of the green samples. The Lotgering factor of the sample increased with magnetic field from 0 to about 0.7. The Lotgering factor shows the same tendency with theoretical order parameter <m> as shown in **Fig.2**. In all samples, the Lotgering factor of green body has lower value than theoretical estimation. The contribution to oriented structure by small particles was lower than that of large particles. The dependence of particle diameter in green body is understandable from differential of rotation energy in the high magnetic field, which is proportional to volume of particles. The agglomerates also reduce to make orientation structure. The dispersion of slurry is a key of orientation, and crystal structure in each particle is also important factor for orientation.

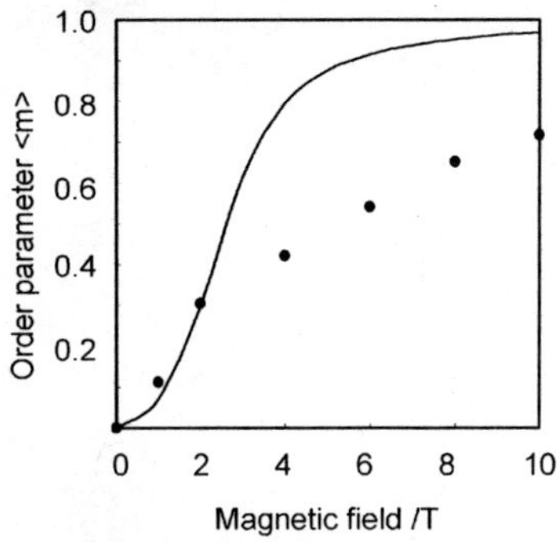

Fig.6 Order parameter (line) and Lotgering factor (dots)

CONCLUSION

The influence of magnetic field was studied on the orientation structure. (1) The relationship between strength of magnetic field and orientation in an applied field was described by using Langevin's theory with Boltzman average of the function. (2) The calculated orientation degrees of bismuth titanate were increased with magnetic field from 0 to 1 as Langevin's function. Experimental degree of orientation also similarly increased from 0 to 0.7. The tendency of orientation can be explained with theory derived from Langevin's theory, but data were smaller than theoretical data.

ACKNOWLEDGEMENT

This study was supported by the Japanese Ministry of Education, Culture, Sports, Science and Technology (17760541) and the Asahi Glass Foundation.

REFERENCES

[1] T.Takenaka, "Grain Orientation Effects on Electrical Properties of Bismuth Layer-structured Ferroelectric Ceramics", J. Ceram. Soc. Japan, 110, 215-224 (2002).

[2] S.Ikegami and I.Ueda, "Piezoelectricity in Ceramics of Ferroelectric Bismuth Compound with Layer Structure," Jpn. J. Appl. Phys., 13, 1572-77(1974).

[3] T.Takenaka and K.Sakata, "Grain Orientation and Electrical Properties of Hot-forged $Bi_4Ti_3O_{12}$ ceramics," ibid, 19, 31-39(1980).

[4] I.-S.Yi and M.Miyayama,"Electrical Properties of Layer-Strkuctured $Pb_2Bi_4Ti_5O_{18}$ Single Crystals," J. Ceram. Soc., Japan, 106, 285-89(1998).

[5] S.E.Cummins and L.E.Cross, "Electrical and Optical Properties of Ferroelectric $Bi_4Ti_3O_{12}$ Single Crystals," J. Appl. Phys. 39, 2268-74(1968).

[6] S.C.Peterson and M.J.Cima, "Magnetic Inducement of Texture in $Ba_2YCu_3O_{6.9}$ Particle

Assemblies under Cryogenic Conditions, "J. Am. Ceram. Soc., 71, C458-C459 (1988).

[7] G.L.Messing, S.T.McKinstry, E.M.Sabolsky, C.Duran, S.Kwon, B.Brahmaroutu, P.Park, P.W.Rehrig, K.B.Eitel, E.Suvaci, M.Seabaugh, and K.S.Oh, Solid State and Materials Science, 29, 45-96 (2004)

[8] T.Kimura, Application of Texture Engineering to Piezoelectric Ceramics, J.Ceram.Soc.Japan, 114, 15-25 (2006)

[9] T.S.Suzuki, Y.Sakka, and K.Kitazawa,"Preferred Orinetation of the Texture in the SiC Whisker-dispersed Al_2O_3 Ceramics by Slip Casting in a High Magnetic Field," J.Ceram.Soc.Japan, 109, 886-890 (2001).

[10] Y.Sakka and T.S.Suzuki, Textured Development of Feeble Magnetic Ceramics by Colloidal Processing under High Magnetic Field, ibid, 113, 26-36 (2005)

[11] K.Uematsu, D.Shoji, T.Ishikawa, T.Kimura and K.Kitazawa, Patent P2002-53367A.

[12] A.Makiya, D.Shoji, S.Tanaka, N.Uchida, K.Uematsu, T.Kimura, "Grain Oriented Microstructure Made in High Magnetic Field", Key Eng.Mater. 206-13, 445-448 (2002)

[13] Y.Inoue, T.Kimura and T.Yamaguchi, "Sintering of Plate-Like $Bi_4Ti_3O_{12}$ Powders," Am. Ceram. Soc. Bull., 62, 704-707(1983).

[14] H.Watanabe, T.Kimura. and T.Yamaguchi, "Sintering of Platelike Bismuth Titanate Powder Compacts with Preferred Orientation", J. Am. Ceram. Soc., 74, 139-47(1991).

[15] M.M.Seabaugh, I.H.Kerscht and G.L.Messing, "Texture Development by Templated Grain Growth in Liquid-Phase-Sintered α-Alumina", J.Am.Ceram.Soc., 80, 1181-88(1997).

[16] A.Makiya, D.Kusano, S.Tanaka, N.Uchida, K.Uematsu, T.Kimura and Y.Doshida, "Particle oriented bismuth titanate ceramics made in high magnetic field", J.Ceram Soc. Japan, 111, 702-705 (2003).

[17] Y.Doshida, K.Tsuzuku, H.Kishi, A.Makiya, S.Tanaka, K.Uematsu and T.Kimura, "Crystal-Oriented $Bi_4Ti_3O_{12}$ Ceramics Fabricated by High-Magnetic-Field Method", Japan.J.Appl.Phys., 43 9B, 6645-6648 (2004)

[18] S.Tanaka, A.Makiya, Z.Kato, N.Uchida, T.Kimura and K.Uematsu, "Fabrication of c-axis orientated polycrystalline ZnO using a rotating magnetic field and following sintering", J.Material Res., 21, 703-707 (2006)

[19] A.Yamagishi, E.Nagao and M.Date, "High Field Diamagnetic Susceptibility and the Curie-Weiss Law in Organic Liquid", J. Phys. Soc. Japan, 53, 928-931 (1984)

[20] A.Yamagishi, T.Takeuchi, T.Higashi and M.Date, "Diamagnetic Orientation of Polymerized Molecule under High Magnetic Field", ibid, 58, 2280-2283 (1989)

[21] P.Langevin, "Sur Les Birefringences Electrique et Magnetique", Le Radium 7, 249-260

(1910)

[22] F.K.Lotgering, Topotactical reactions with ferrimagnetic oxides having hexagonal crystal structures I, J.Inorg.Nucl.Chem., 9, 113-123 (1959)

MICRO-PATTERNING OF TIN OXIDE BY MICRO-MOLDING IN CAPILLARIES

Junko Imasu, *Hiroshi Fudouzi, Yoshio Sakka
Nano Ceramics Center, National Institute for Materials Science,
1-2-1 Sengen, Tsukuba, Ibaraki 305-0047, JAPAN
*Optronic Materials Center, National Institute for Materials Science,
1-2-1 Sengen, Tsukuba, Ibaraki 305-0047, JAPAN

ABSTRACT

This study describes the fabricating of ceramic micro-scale patterns over a wide area in the Micromolding in Capillaries (MIMIC) method using ceramic colloidal suspensions of a few vol% solids loading. Since the filling mechanism of the MIMIC process is capillary action, there is a problem that the patterned areas were small when using high-solids loading suspensions, because of their high viscosities. We applied tin oxide suspensions of low solids loading to MIMIC process for the purpose of patterning over a wide area. We demonstrated to fabricate striped patterns over a mm-scale area and found that the structures were dense and the thicknesses were almost 1/3 of the depth of the mold channel. The generation of thick and dense structures from only a few vol% suspensions is considered to be due to an additional infiltration when evaporation of the inside suspension occurred after capillary action, which was reported on other colloidal suspensions. Furthermore, the fabricated structures were strengthened by heat treatment. We expect this MIMIC process of low-solids loading suspensions to widely expand to other kinds of ceramic suspensions.

INTRODUCTION

Recently, the miniaturization of electro-devices has been promoted, with that the developments of micro-scale ceramic patterning methods are required. MIMIC is one of the soft-lithographies introduced by Whitesides et al. which uses elastomeric molds for patterning[1, 2]. In the MIMIC method, the capillary action is used to fill mold channels[2, 3]. Several studies are reported about applying ceramic colloidal suspensions to MIMIC. Heule's group patterned SnO_2 [4], while Moon's group applied Al_2O_3 and NiO to the improved MIMIC process by adding a vacuum-assisted step [5]. Most of the suspensions they used were of high solids loadings of more than 20 vol %. When using the suspensions of high solids loadings, the generated structures have high particle densities and the sizes are just the mold size. However, the suspensions of high solids loading have high viscosities and act as plastic flow. Therefore, the suspensions are difficult to infiltrate voluntary and the patterned area was small [4]. In this reason, some assists were needed for patterning over a millimeter-scale area [5, 6].

In the case of other colloidal suspensions, there are reports that from only a few vol% suspensions of polystyrene microspheres [7] and carbon paints [8], dense and almost mold-sized structures were generated. The generation mechanism of these structures is believed to be a result of the suspension flow in the mold channel caused by the evaporation of solvents through the exits or the walls of the elastomeric molds, after the suspensions infiltrated by capillary action. With this evaporation flow, the suspensions were supplied from the pools in front of the mold entrance, resulting in enough particles to make mold-sized structures entered into the channels.

We have studied application of this mechanism to ceramic powders, and fabricated ceramic micro-scale patterns by colloidal suspensions of low solids loading to MIMIC, because using suspensions of a lower viscosity is beneficial to patterns over a wide area. In this experiment, we used a few vol% suspensions of tin oxide nano-powder (30 nm in diameter) dispersed in ethanol, and thick, dense striped patterns were generated over a millimeter-scale area with repeatability. An additional infiltration after the capillary action is considered to lead to this result, like previous cases. However, there was a difference in the thicknesses of the products, which was only about 1/3 of the depth of the mold channel. Furthermore, the fabricated structures were strengthened by heat treatments at 300°C. In this study, we report the characteristics of the generated structures from this MIMIC process using low-solids loading suspensions.

EXPERIMENTAL

We used commercially available tin oxide powder (NanoTek, KANTO Chemical Co., Inc. Tokyo, Japan) for suspensions. The BET surface area is 23 m^2 /g, and the particle size calculated from this BET surface area is 37 nm under the assumption that the particle is a sphere. The experiment was carried out with suspensions of four kinds of solids loadings; 0.5 vol%, 1 vol%, 2 vol% and 4 vol%. The suspensions were prepared by dispersing tin oxide powder in ethanol by ultrasonication for about 10 minutes.

Figure 1 shows the schematic of the MIMIC process. The mold was made using a Poly - dimethylsiloxane (PDMS) elastomer kit (Sylgard 184, Dow Corning Inc., Midland, MI) by replicating from a master mold. The channels of this PDMS mold configured a striped pattern, and the each channel was 1.5 cm in length, 23 μm in width and 3.5 μm in depth. The PDMS mold was put on to a silicon wafer substrate (N-Type (1 0 0), coated with 1 μm-thick oxide film, Furuya Metal Co., Ltd., Tokyo, Japan). To make the surface hydrophilic, the substrate was previously plasma-treated. The suspension was placed at the entrance of the channels of the PDMS mold, at first 10 μl, and added as needed to a pool in front of the entrance. After the channels were filled and the suspension dried up, the PDMS mold was peeled away. In this experiment, three samples were fabricated from each suspension.

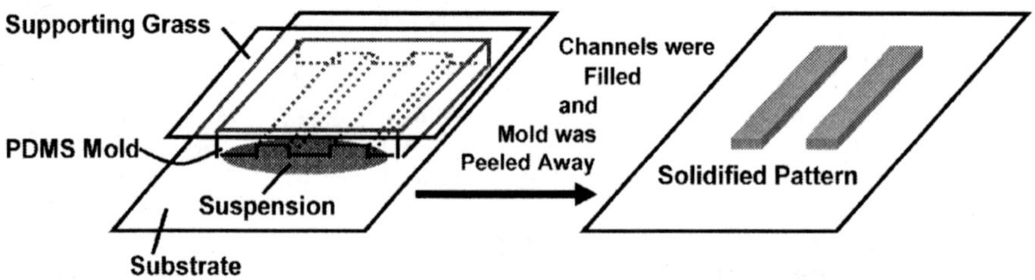

Figure 1. Schematic of MIMIC Process

The Zeta potential of suspension was measured by DT-1200 (Dispersion Technology Ltd., U.S.A) Acoustic and Electroacoustic Spectrometer, and the viscosity was determined using a viscometer, RC-500 (Toki Sangyo Co. Ltd., Japan). For the observation and analysis of the fabricated structures, FUJIX HC 300 (FUJIFILM Inc., Japan) digital camera, JSM 6500 (Jeol

Ltd., Tokyo, Japan) scanning electron microscope and SPI 3700 (Seiko Instruments Inc., Tokyo, Japan) probe station were used.

RESULTS AND DISCUSSION

The suspensions acted as almost Newtonian fluids, and the viscosities at 25°C were determined as shown in Figure 2 in the relation to the solids loading. While the zeta-potential of the 2 vol% suspension was 47mV, which indicates that the suspension is well-dispersed.

From the observation of MIMIC process, we found that the capillary action lasted only about 10 seconds, however it was more than 100 minutes before the suspension completely dried. During this period, infiltration caused by evaluation flow and aggregation went on in series.

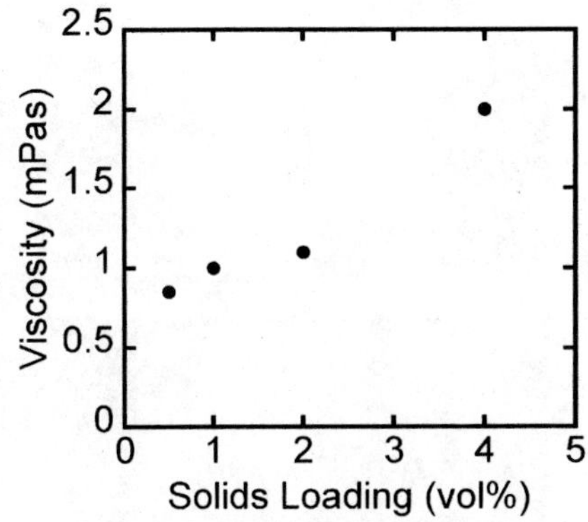

Figure 2. The viscosities of suspensions at 25°C with respect to the solids loadings.

Figure 3 shows patterned structures made from 2 vol% tin oxide suspensions dispersed in ethanol. A CCD image of the patterned area is shown in Fig. 3A. The suspension was infiltrated in a direction from the right side to left side in this photograph. The striped pattern consists of more than one hundred lines with lengths in mm scale. In the lines, the tin oxide particles are deposited at the exit side, and at the entrance side, the suspension flow remains. Fig. 3B shows a magnified SEM image that represents deposited sections. The patterned structure was a deficit -free stripe of uniform width. The front edge and the back edge of the deposited region are shown in the SEM images of Fig. 3C and Fig. 3D, respectively. Both edges stand vertically to the substrate and they are almost the same thickness. Also, from these images, the surface of the structure is smooth and flat. Furthermore, the generated structures are found a little narrower than the mold channels. There are two considerable reasons for this, one is the shrink of generated structure on drying, and another is the reduction of mold channels by swelling of the PDMS mold during the process because the PDMS mold material is known to slightly swell on exposure to ethanol. Then, we examined the inside constitution from a SEM image of a fractured cross-section of a patterned line (Fig. 3E). The structure was dense with tin oxide particles and a density contrast was not found between the upper part and lower part. At present stage we have not measured the green density of the SnO_2 film due to its small volume, but the green density consolidated by a slip casting using the well-dispersed SnO_2 suspension was approximately 30 % of theoretical density.

Then we investigated the shape of the structures in relation to the solid loading, using every three samples from the suspensions of 0.5, 1, 2, 4 vol% solids loading. Figure 4A shows the mean lengths of the infiltration and deposited section. The lengths were measured from the CCD image taken with a scale. The infiltration length means the length from the entrance of the channel to the exit edge of the fabricated structure. Infiltration length had relativity to the solids

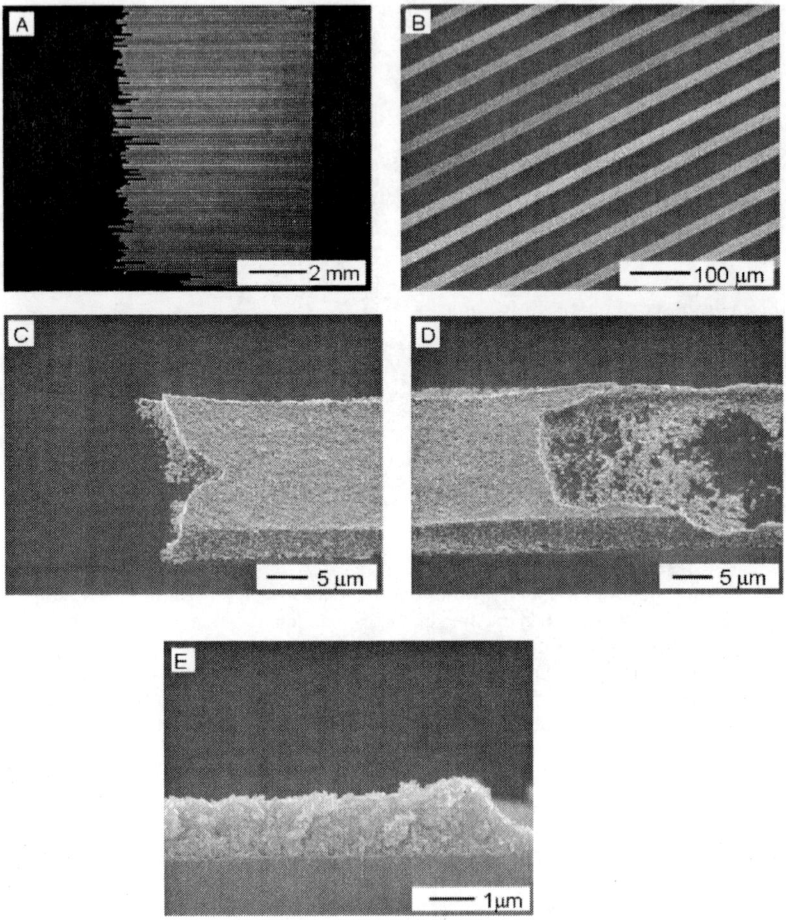

Figure 3. SEM images of SnO2 structures. (A) Entire image. (B) Typical patterned section of (A). Magnified image of a front edge (C) and a back edge (D) of the deposited region. (E) SEM image of a fractured cross-section of a patterned line.

loading of monotonically decreasing, while the deposited length had not much difference at any solids loadings. In addition, in the patterns from 4 vol% suspension, non-deposited sections were not found. As for the thickness, we chose three points randomly from the deposited section of each sample, then measured and analyzed them by SPM. In the graph of Figure 4B, the dots indicate the thicknesses obtained from nine data points of each solids loading. The thickness monotonically increased in relation to the solids loading and at 2 vol% the mean value was 1.2

μm, which corresponds to about 1/3 of the depth of the mold channels. The scatters of the data value were small at 1 vol% and 2 vol% solids loadings.

The measurements show that thick striped structures of several mm-long were fabricated with repeatability by the MIMIC process, using a few vol% tin oxide suspensions dispersed in ethanol. This result indicates additional suspension infiltrated from outside after the capillary action, as in the previous study of polystyrene microspheres and carbon paints. However the thickness of the generated structure was about 1/3 of the depth of the mold channels, which is different from the previous cases. It is inferred from the characteristic of the structures that the particles coagulated rapidly with the high condensation of the suspension. In this regard, since the evaporation of the suspension solvent contributes to the fabrication of these structures in a large way, it is considered that by changing the evaporation conditions of suspensions in the mold channels and in front of the entrance, the shape of the fabricated structures will change.

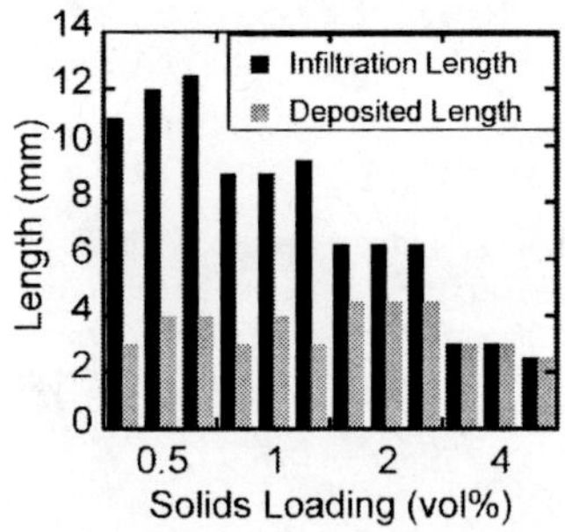

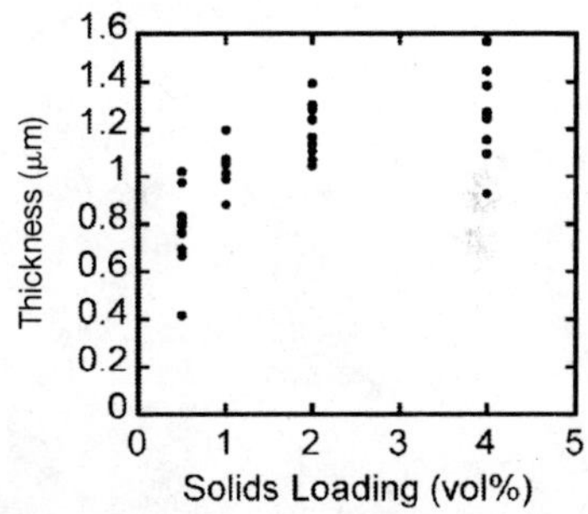

Figure 4. (A) Mean lengths of infiltration and deposited sections. (B) Thicknesses obtained from nine data points for each solids loading.

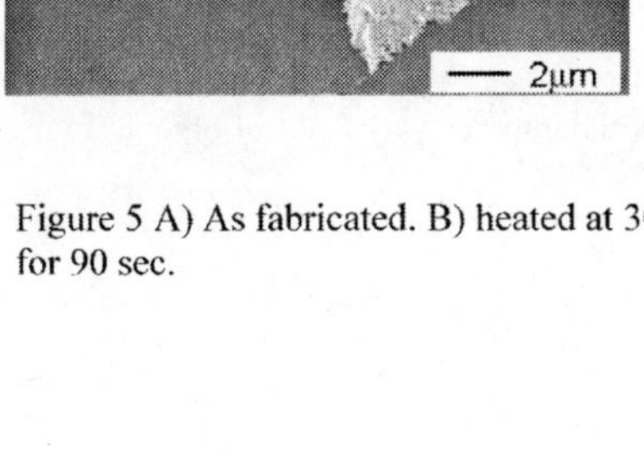

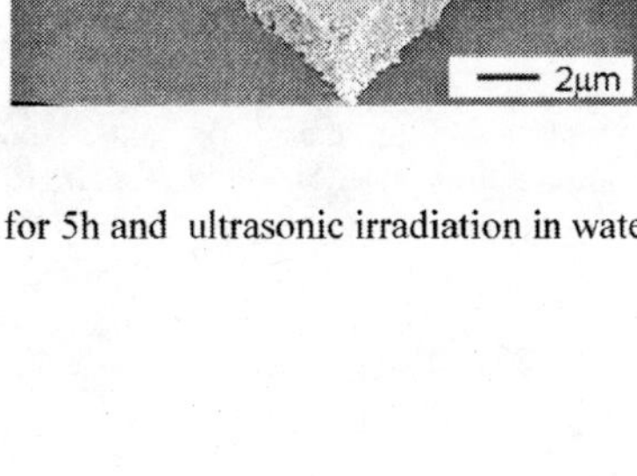

Figure 5 A) As fabricated. B) heated at 300°C for 5h and ultrasonic irradiation in water for 90 sec.

Figure 5 shows the SEM images as-fabricated and heated at 300°C for 5h. The heated film is strong enough to hold its structure even after ultrasonic irradiation in water for 90 sec. While, the volume resistivity of the film heated at 300°C was 10 kΩcm.

CONCLUSION

We demonstrated the ability to fabricate almost uniform-thickness micro-scale patterns over a mm-scale area by applying a low solids loading of Tin oxide suspensions dispersed in ethanol to MIMIC. The main developments of our study are as follows. First, we used ceramic suspensions of a few solids loadings to MIMIC. It leads the patterning over a millimeter scale area which became possible with no assistances, because the suspensions of low solids loading are of low viscosity and express Newtonian flow thus the infiltrate is widely by capillary action. Secondly, we confirmed that the additional infiltration occurs in the case of tin oxide ethanol suspension, and as a result thick and dense structures were fabricated. Next is that we found the thicknesses of the structures were almost 1/3 of the depth of the mold channels. Furthermore, the structures were demonstrated to be strengthened by heat treatment. In this study we used tin oxide as a ceramic powder, and we expect this process is applicable to other ceramic powders by making an adjustment to the suspensions.

ACKNOWLEDGMENTS

This work was supported by Hosokawa Powder Technology foundation and a Grant-in-Aid (Kiban B: # 16350117) for Scientific Research from the Ministry of Education, Science, and Culture of Japan. The master molds, from which the PDMS molds were fabricated, were made with guidance of Mr. Shinichi Hara in Nano Foundry Station of National Institute for Materials Science. We greatly appreciate his help.

REFERENCES

[1]E. Kim, Y. Xia, and G. M. Whitesides, "Polymer microstructures formed by moulding in capillaries" *Nature,* **376**, 581-84 (1995).
[2]E. Kim, Y. Xia, and G. M. Whitesides, "Micromolding in capillaries: Applications in Materials Science" *J. Am. Chem. Soc.,* **118**, 5722-31 (1996).
[3]C. R. Martin, and I. A. Aksay, "Submicrometer-Scale Patterning of Ceramic Thin Films" *J. Electroceramics,* **12**, 53-68 (2004).
[4]M. Houle, J. Schell, and L. J. Gauckler, "Powder-Based Tin Oxide Microcomponents on Silicon Substrates Fabricated by Micromolding in Capillaries" *J. am. Ceram. Soc.,* **86**, 407-12 (2003).
[5]S. J. Ahn, and J. Moon, "Vacuum-Assisted Microfluidic Lithography of Ceramic Microstructures" *J. am. Ceram. Soc.,* **88**, 1171-74 (2005).
[6]J. Imasu, Y. Sakka, and H. Fudouzi, "Micro-Scale Patterning of Ceramic Colloidal Suspension by Micro Molding in Capillaries (MIMIC) with Assistance of Highly Infiltrating Liquid" *J. Ceram. Soc. Japan,* **114**, 725-728 (2006).
[7]E. Kim, Y. Xia, and G. M. Whitesides, "Two- and Three-Dimensional Crystallization of Polymeric Microspheres by Micromolding in Capillaries" *Adv. Mater.,* **8**, 245-47 (1996).
[8]J. A. Rogers, Z. Bao, and V. R. Raju, "Nonphotolithographic fabrication of organic transistors with micron feature size" *Appl. Phys. Lett.*, **72**, 2716-18 (1998).

GELCASTING FORMULATION OF ALUMINA SLURRY OFFERING SOME ADVANTAGES IN CERAMIC SHAPING

Ruben L. Menchavez[1], Masayoshi Fuji[1], Hiroaki Takegami[1], Tomohiro Yamakawa[1], and Minoru Takahashi[1*]
[1]Ceramics Research Laboratory, Nagoya Institute of Technology
Asahigaoka 10-6-29, Tajimi City, Gifu Prefecture, 507-0071, Japan
*Email: m-takahashi@nitech.ac.jp

ABSTRACT

Gelcasting by use of traditional monomer formulations encounter several limitations, such as inhibition of gelation in the presence of oxygen, cracking during drying and sintering steps, neurotoxicity, and others. Many reports have successfully resolved each individual issue, but none addressed all limitations in single slurry formulation. In this study an attempt is made to develop an alumina gelcasting slurry containing water-soluble polymers, such as methacrylamide, a long-chain cross linker, polyacrylamide, and plasticizer, for resolving several of the preceding issues. Rheological analysis showed that the 50 vol% solid loading exhibited a strong thinning behavior, which fitted the Herschel-Bulkley model. The gelation process was promoted with the addition of an initiator and catalyst, and was held at room temperature for 3 hours in air. The freshly gelled sample was dried at normal room condition without additional control, and it was successfully debindered and sintered. Subsequently, several tests were conducted such as TGA/DTA, flexural strength, and microstructure analysis to explain the findings.

INTRODUCTION

For more than a decade, gelcasting has been recognized as a versatile process when it is compared with traditional shaping methods for fabricating advance ceramic materials. The process involves in-situ polymerization of a cast slurry, which consists of ceramic powder, water and water-soluble polymers. After casting, a macromolecular network or gel is created to locking the ceramic particles into the desired shape. Several advantages are offered by this method including controllable casting and solidification, near-net shaping of complex shapes, strong machinable green bodies, and many others.

Despite the advantages of gelcasting as mentioned above, several limitations are encountered in practice such as surface exfoliation in the presence of oxygen, cracking during drying and sintering, neurotoxicity, and others. Many reports have successfully resolved each issue individually, but none addressed all limitations in single slurry. For example, Ma et. al.(1) used a high molecular weight polymer as an additive to an acrylamide system for eliminating surface exfoliation, however, the process is likely limited to small ceramics components and the problem of high neurotoxicity remains. Bio-gelcasting addressed the problem of neurotoxicity, but it is difficult in this process to remove the agglomerates and air bubbles (2). In the fabrication of large ceramics, particularly large alumina sheets, it is desirable that the gelcasting parameters such as temperature and shaping atmosphere are kept at room conditions. Typically, plasticizers are added to conventional gelcasting formulations for room condition drying of gelcast parts (3) but the problem of surface exfoliation in air is not sufficiently addressed. Thus, there is a need to develop a gelcasting system that possesses several qualities adaptable to room conditions.

In this study, therefore, an attempt is made to develop a gelcasting alumina slurry that offers several advantages in room condition ceramic shaping. In order to achieve this goal, the gelcasting solution was prepared with soluble polymers such as methacrylamide, a long-chain cross linker, polyacrylamide, and a plasticizer. This general gelcasting formulation offers the opportunity to shape large alumina sheets at room conditions, which has significant industrial applications.

EXPERIMENTAL PROCEDURE

Materials and gelcasting formulations

The selected powder was alumina (AL 160SG-4, Showa Denko Co., Yokohama, Japan) with mean particle size (D_{50}) of 0.45 μm. The dispersant was a 40% aqueous solution of a polycarboxylate ammonium salt (Seruna D-305, Chuukyou Yushi Company, Nagoya, Japan) with an average molecular weight of 8000–10,000 g/mol. The crosslinking agents were methacrylamide (MAM, Wako Chemical, Osaka, Japan) and poly(ethylene glycol) dimethacrylate (PEGDMA, Sigma-Aldrich Inc.). A polyacrylamide (PAM, Mw = 10000 in 50% solution, Aldrich Chemical Company Inc., USA) was used for eliminating surface exfoliation. A low molecular weight polyethylene glycol (PEG 400, Kanto Chem. Company, Japan) was used as a plasticizer. The polymerization reaction was initiated using ammonium peroxodisulfate (10% solution in distilled water), and catalyzed by N,N,N`,N`-Tetramethylethylenediamine (Wako Chemical, Osaka, Japan).

Three slurry formulations were prepared for comparative evaluation, as shown in Table 1. The amount and proportion of each monomer in the slurries were based on the previous studies (1,3-6) with modification to fit the purpose of the current study. The dispersant was held constant at 0.36wt.% of powder in all slurries.

Table 1. Actual gelcasting formulations implemented in this study.

Components	**slurry A**	**slurry B**	**slurry C**
Alumina	83.07	80.11	80.14
Distilled Water	12.97	14.42	13.41
Dispersant (40%sol.)	0.75	0.72	0.72
MAM	1.84	1.88	2.05
PEGDMA	0.61	0.63	0.49
PAM	---	2.24	2.24
PEG 400	0.76	---	0.96
Solid loadings (vol.%)	55.16	50.26	50.28
Initiator/catalyst(μL/g)*	1.02/0.17	2.53 / 1.27	

*μL/g, microliter per gram of slurry

Gelcasting fabrication

Premix solutions were prepared according to the proportions indicated in Table 1. All components excluding the alumina powder were dissolved in polyethylene bottle loaded with zirconia media, and milled for 20 minutes to dissolve the polymer completely. Afterwards, the alumina powder was added to the media filled solution and milled for 24 hours to ensure breakdown of the agglomerates. After milling, the slurry was screened to remove the zirconia media. The screened slurry was de-aired under vacuum for 10 minutes, subsequently the initiator and catalyst were added to the slurry followed by an additional 2 minutes of mixing under vacuum.

This was followed by sieving (100 mesh) the slurry to remove the agglomerates built up during the addition of initiator and catalyst. The sieved slurry was poured into a Teflon mold with ten rectangular cavities (10mmx8mmx63mm) to form mechanical testing bars. To determine the feasibility of fabricating large shapes, the three slurries were gelcasted into a circular Teflon mold (D=120mm, T=15mm) in air. After gelation, the bodies were demolded, wrapped with wet paper and put onto a plastic screen, and allowed to dry for one week at room conditions. This was followed by drying to a constant weight in an oven at 80°C. The well dried bodies were debindered at 0.25°C/min to 700°C with a 2 hours hold, and naturally cooled inside the furnace. Sintering schedule was carried out separately at 1600°C for 2 hours in air. The heating rates were 1°C/min from room temperature to 1250°C, 0.5°C/min to 1600°C, and cooled at 1°C/min to 1000°C, followed by furnace cooling.

The rheological studies were carried out on the three slurries using rotational viscometer (Model: RheoStress 600, M/s Haake, Karlsruhe, Germany). The rheological behavior of the three gelcasting slurries were studied by measuring their viscosity and shear stress as a function of shear rate from 10 to 120 per seconds. The resulting rheograms were analyzed using the statistical package associated with the equipment. TGA/DTA analyses (TG 8120, Rigaku Thermo Plus, Japan) of dried samples were conducted until 800°C at a rate of 10°C/min with 100 ml/min flow of oxygen. Scanning electron microscopy (JSM-6100, JEOL) was performed on sintered samples. Bulk density was determined using Ultrapycnometer 1000 (Quantachrome Instruments, Japan). Furthermore, flexural strength test was carried out in three point mode (span=30mm, cross-head speed=0.5mm/min).

RESULTS AND DISCUSSION

Rheological properties

A rheograms of the three slurries were plotted for insight into the flow behavior, as shown in Figure 1. The shear stress versus shear rate data were fitted to existing rheological flow models and the Herschel-Buckley model fitted to all slurries with the highest R^2 correlation value, Table 2. The Herschel-Buckley model is

$$\tau = \tau_O + k(\gamma)^n \qquad (1)$$

where τ_0 is the yield stress, τ is the shear stress, γ is the shear rate, n is the power law exponent, and k is a consistency factor. Table 2 lists the values of the fitted parameters for the three slurries. Slurry A has lower yield stress compared with slurry B or C. This observation is corroborated with the plot of viscosity as a function of shear rate for the three slurries, as shown in Figure 1. The viscosity increases from sample A to C is due to the increased viscosity of the premixed solutions as a result of the polyacrylamide content, which has a high molecular weight coiled chain structure. The plasticizer effect on viscosity is less pronounced due to its low molecular weight. When the polyacrylamide and plasticizer were added together to the slurry, the viscosity increased dramatically but to a tolerable value suitable for processing and casting. This is because of the fact that both polymers do not preferentially adsorb on the particle surface due to the presence of the dispersant on alumina particles (1,5). As a result, the free polymers in solution screen the surface charges of the particles that result in a reduction of the repulsive forces (6). Nevertheless, as shown in Table 2, all slurries have values of $n < 1$ depicting a relatively strong shear thinning behavior for all slurries. This behavior is typical of concentrated stable suspension which undergoes perturbation of its structure due to shear.

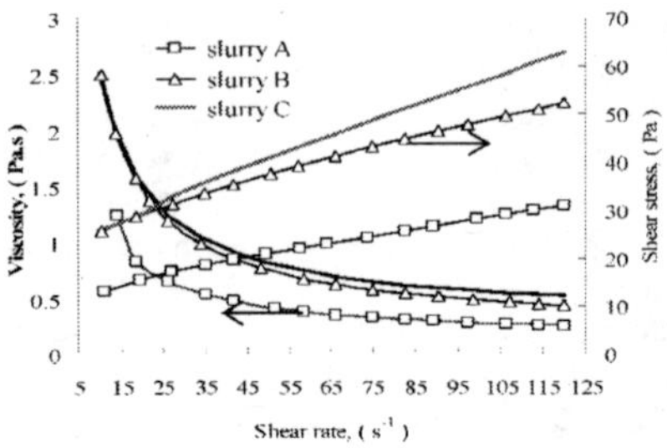

Figure 1. Rheological plot of the slurries.

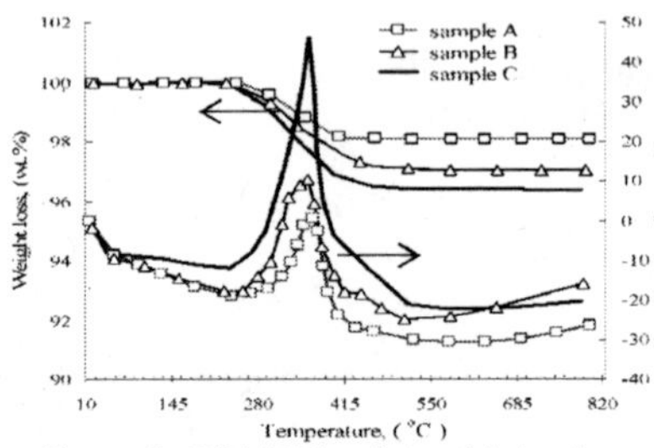

Figure 2. TGA/DTA of the dried gelcast samples.

Table 2. Herschel-Bulkley parameters of the slurries.

Slurries	Yield stress τ_0 (Pa)	Viscosity $120s^{-1}$(Pa.s)	Consistency (k)	Power law exponent (n)	Correlation (R^2)
Slurry A	9.764	0.257	0.695	0.711	0.999
Slurry B	20.300	0.436	1.033	0.716	1.000
Slurry C	20.380	0.523	0.773	0.835	1.000

b) Thermogravimetric analysis

Thermogravametric analysis of the well dried gelcast samples is shown in Figure 2. Samples B and C were debindered more easily than sample A. In order to understand the observed degradation process, the activation energy of binder burnout for each sample was determined using the Coats-Redfern equation (8) with the aid of thermograms (TGA, Fig. 2). This equation is expressed as

$$\ln\frac{g(\alpha)}{T^2}=\ln\frac{AR}{\beta E_a}-\frac{E_a}{RT} \quad (2)$$

where $g(\alpha)$ is the integral function of conversion, A pre-exponential factor, βheating rate, E_a activation energy, T the absolute temperature, and R the gas constant. In this study, the function of conversion was assumed to be first order over the 15 to 50% conversion range. The activation energy was determined from the slope of $\ln g(\alpha)/T^2$ versus $1000/T$ plot. The determined activation energies for samples A, B, and C, are 47.4, 45.4, and 41.8 J/mole.K, respectively. The reduction in activation energies may be due to the possibility that the long chain cross-linker prevents the formation of hydrogen bonding in polymerized PAM (1,7), resulting into a less rigid gel which can be easily removed upon heating. The presence of plasticizer separates the complex molecular chains of PAM causing a reduction of normal intermolecular forces. This produces a more flexible gel, which provides further reduction of the activation energy during binder degradation. Also, during heating the plasticizer volatilizes first and its gaseous phase is mobile within the polymer matrix and can diffuse easily to react with available oxygen, thereby helping the elimination of remaining binder components and their reaction products (9). Accordingly, the DTA analysis shows an increasing exothermic peak from samples A to C, as shown in Figure 2. The increasing exothermic heat flow with peak broadening, all happen at about 363°C, reveals the strong effects of PAM and plasticizer during binder burnout.

Physical characteristics and properties of dried and sintered samples

The room condition gelled samples are shown in Figure 3. Samples without the addition of polyacrylamide suffer serious surface exfoliation as shown in Figure 3a (indicated by arrows). Severe flaking is visible along the edge of the sample which exposes the dried slurry layer as evidential of oxygen-inhibited gelation, thereby creating a lack of dimensional control for this formulation. Surface exfoliation was eliminated by the addition of polyacrylamide to the conventional monomer-crosslinker system (1), but the sample is prone to cracking, as illustrated by sample B (Fig. 3b). Surface exfoliation and cracking are eliminated by the addition of both polyacrylamide and plasticizer, as shown in Figure 3c. The mechanism of elimination of surface exfoliation is a complicated process, but it could be due to the formation of a complex network structure via hydrogen bonding between the free amino and keto groups of PAM during polymerization process. This is typical for polymers containing hydroxyl groups that bind with the carbonyl groups forming a bridging effect with a high density of hydrogen bonds, causing the reduction of oxygen diffusion (10). On the downside, the polymer chains become rigid and lack of mobility that result into a crack sensitive gelcast part. The polar plasticizer (PEG) minimizes the polar connections between the polymer molecules thereby effectively reducing the interaction force between the polymer molecules (7). Thus, it provides flexibility to the gel network leading to the reduction of internal stress, but also reducing slightly the dried strength. Corresponding dried strengths in Table 3 justify the preceding statement. The strength decreases slightly from sample A to C, but still maintains enough strength for mechanical handling and polishing.

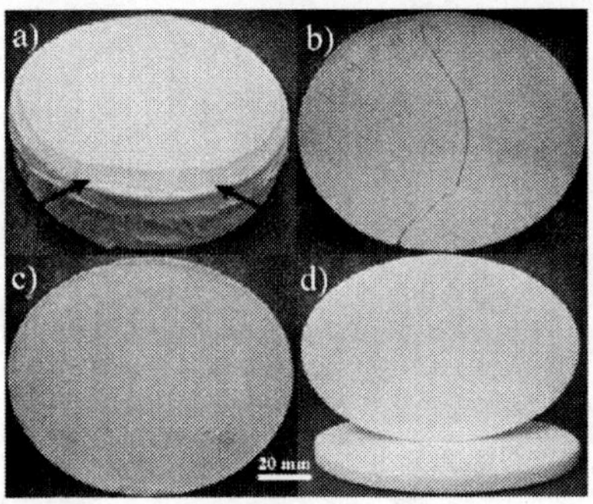

Figure 3. Photographs of well dried a) sample A, b) sample B, and c) sample C, and (d) sintered samples of A and C.

Figure 4. SEM micrograph of sample C sintered to 1600°C for 2 hours in air.

Table 3. Strengths and bulk densities of dried and sintered samples.

Samples	Dried strength (MPa)	Dried relative density (%)	Sintered strength (MPa)	Sintered relative density (%)
A	13.7	56.39	349	96.84
B	12.6	58.65	371	98.07
C	12.8	59.90	390	98.25

Large size sintered samples of gelcast A and C slurries (D=100mm, T=11mm), are shown in Figure 3d. Sintered sample B was not shown due to drying cracks. The flexural strengths of the three sintered samples were measured, Table 3. Sample C yielded the highest strength at 390 MPa, which is higher than in previous studies of 347 MPa (1,4). It should also be noted that the strength and bulk density increase from samples A to C. Typical SEM micrograph of sintered sample C in Figure 4 shows a highly uniform microstructure, typical of quality sintering with a low degree of porosity and grain growth.

CONCLUSION

This study demonstrated a promising formulation for a gelcasting slurry that addressed several gelcasting parameters at room conditions. This was achieved by adding polyacrylamide and plasticizer into the conventional gelcasting slurry. Corresponding rheological analysis showed strong shear thinning behavior which fitted the Herschel-Bulkley model, typical for a gelcasting slurry. The improved gelcasting slurry can be gelled and dried under room conditions without careful control, resulting in a part which debinds and sinters well. This formulation offers potential for the shaping of large alumina sheets under room conditions, which has important industrial applications.

ACKNOWLEDGEMENT

This research was supported by NITECH 21st Century COE program, "World Ceramics Center for Environmental Harmony, 2003-7". Also, this study was partially supported by the Ministry of Education, Culture, Sports, Science and Technology (MEXT), project on "Cooperation for Innovative Technology and Advanced Research in Evolution Area, 2006-8".

REFERENCES

[1]J. Ma, Z. Xie, H. Miao, Y. Huang, Y. Cheng, and W. Yang, "Gelcasting of alumina ceramics in the mixed acrylamide and polyacrylamide systems," *J. Euro. Ceram. Soc.*, 23, 2273-79 (2003).

[2]E. Adolfsson, "Gelcasting of Zirconia Using Agarose," *J. Am. Ceram. Soc.*, 89, 1897-902 (2006).

[3]M.A. Janney, and C.A. Walls, "Gelcasting compositions having improved drying characteristics and machinability," *Patent no. US 6,228,299* (2001).

[4]A.C. Young, O.O. Omatete, M.A. Janney, and P.A. Menchhofer, "Gelcasting of Alumina," *J. Am. Ceram. Soc.*, 74, 612-18 (1991).

[5]B.P. Singh, R.L. Menchavez, C. Takai, M. Fuji and M. Takahashi, "Stability of dispersions of colloidal alumina particles in aqueous suspensions," *J. Col. Interf. Sci.*, 291, 181-6 (2005).

[6]M. Takahashi, K. Adachi, R.L. Menchavez, and M. Fuji, "Fabrication of Semi-conductive Ceramics by Combination of Gelcasting and Reduction Sintering," *J. Maters. Sci.*, 41, 1965-72 (2006).

[7]A. Kasgoz, Z. Ozbas, H. Kasgoz, and I. Aydin, "Effects of monomer composition on the mechanical and machinability properties of gel-cast alumina green compacts," *J. Euro. Ceram. Soc.*, 25, 3547-52 (2005).

[8]S. Vyazovkin, and C.A. Wight, "Model-free and model-fitting approaches to kinetic analysis of isothermal and nonisothermal data," *Thermochimica Acta*, 340-41, 53-68 (1999).

[9]V.N. Shukla, and D.C. Hill, "Binder evolution from powder compacts: Thermal profile for injection-molded articles," *J. Am. Ceram. Soc*, 72, 1797-803 (1989).

[10]K. Cai, Y. Huang, and J. Yang, "A synergistic Low-Toxicity Gelcasting System by Using HEMA and PVP," *J. Am. Ceram. Soc.*, 88, 3332-37 (2005).

EFFECT OF SURFACTANTS ON THE FORMATION OF HOLLOW $CaCO_3$ PARTICLE BY BUBBLE TEMPLATE METHOD

Yong Sheng HAN, Li Wei LIN, Masayoshi FUJI, Takeshi ENDO, Hideo WATANABE, Minoru TAKAHASHI
Ceramics Research Laboratory, Nagoya Institute of Technology,
10-6-29 Asahigaoka, Tajimi 507-0071, Japan

ABSTRACT

Hollow $CaCO_3$ particle is a promising material with widely potential applications in the fields of papermaking, food production and drug delivery. A novel process to prepare hollow $CaCO_3$ particle with the nomination of bubble template method has been proposed in our group. To further improve the production of hollow particles, the effect of surfactants on the formation of hollow particles was investigated in this paper. Two kinds of surfactants, namely cationic surfactants and nonionic surfactants□were studied with the help of Zeta potential measurements, density measurements and microstructure characterization.

1. INTRODUCTION

The impetus to the intense research in the production of hollow particles, typically in spherical geometry, emanates from their scientific and technological interest. Hollow particles have been utilized for the encapsulation and controlled release of various substances (e.g. drugs, cosmetics, dyes, and inks) in catalysis and acoustic insulation, in the development of piezoelectric transducers and low dielectric constant materials, and for the manufacture of advanced materials [1-3]. A new and facile method, namely bubble template method [4], has been proposed in our group, which provided a new way for the production of hollow particles with the potential for industrial applications. Hollow $CaCO_3$ particles have been successfully synthesized by this new method through assembling nano-sized initial $CaCO_3$ precipitates on the surface of bubble, forming solid shells. The succeeding research was mainly focused on improving the production of hollow particles.

Even though the formation mechanism of hollow particles by bubble template method has not been finally verified, it is no doubt that the key point for the formation of hollow particles by this method is the interaction between primary particles and floating bubbles. Before the primary particles attach on the surface of bubbles and form a three phase contact, surface force govern the particle-bubble interaction [5]. The attachment of particles to the surface of bubbles is mainly determined by the electrostatic force and hydrophobic force. Electrostatic force contributes to the particle-bubble attachment only when the particles and the bubbles carry opposite signs of electrostatic potential. Since both bubbles and $CaCO_3$ particles are negatively charged in basic condition in which hollow particles were produced [6], the electrostatic force makes a negative contribution for the attachment of particles to bubbles. Moreover, the $CaCO_3$ primary particles are hydrophilic in aqueous solution, so hydrophobic force also has no positive contribution for the formation of hollow particles. The way for the formation of hollow particles by homogenous nucleation of primary precipitates in solution followed by the attachments of primary precipitates on the surface of bubbles looks not workable. The prepared hollow particles observed in our samples may be formed by the heterogeneous nucleation of precipitates directly on the surface of bubbles, which is seldom case in our experiments due to the unstable surface of floating bubbles. To improve the production of hollow particles, it is necessary to modify the surface of particles and bubbles to facilitate the precipitation of primary particles on the surface of bubbles.

In this paper, two kinds of surfactants were studied to test their influence on the formation of hollow particles. One is cationic surfactants, which was employed with the purpose of modifying surface charge of particles and bubbles and facilitating the attachment of primary particles to bubbles. Another type is nonionic surfactants, which was used to induce the hydrophobic force between particles and bubbles, improving their attachment. Each surfactant was discussed by their effects on the surface properties of particles and bubbles and their influences on the density and microstructure of final products.

2. EXPERIMENTAL SECTIONS

MATERIALS: Calcium chloride and 25% ammonia solution were supplied by Wako Pure Chemicals, Japan. Nitrogen and carbon dioxide gas were bought from Taka Gas, Japan. Two cationic surfactants, namely 1,5-Diaminopentane dihydrochloride ($ClH_3N(NH_2)_5NH_3Cl$) and Triethyl-methylammonium chloride ($CH_3N(C_2H_5)_3Cl$), were used without further purification. Two nonionic surfactants, namely Cyclohexanemethanol ($C_6H_5CH_2OH$) and Polyethyleneglycol dodecyl ether ($H(OC_2H_4)nOH$), were purchased from Aldrich.

ZETA POTENTIAL MEASUREMENTS: Zeta potential of $CaCO_3$ particles in NH_4Cl solution and in the solution containing different surfactants were measured using a Mastersizer 3000 HS (Manbrn Instruments, UK). The pH of solution was changed from 7.5 to 11.5 during measurement.

PREPARATION OF HOLLOW PARTICLES: The preparation of hollow $CaCO_3$ particles was similar as the description in reference 4. One difference is that the surfactants were added into calcium chloride solution before the introduction of CO_2/N_2 gas. After reaction, the white precipitates were collected and used for further examination.

SEM and TEM: A JSM-7100FO (JEOL, Japan) Scanning Electron Microscopy (SEM) was used to observe the morphology of samples. Transmission Electron Microscope (TEM) was performed using a JSM 2000 EXII TEM equipment (JEOL, Japan) with an accelerating voltage of 160kV. The TEM samples and cut SEM samples were prepared by solidifying prepared precipitates in polymer firstly following with cutting the solidified samples by LEICA EM UC 6 Ultramicrotome (Leica USA).

DENSITY MEASUREMENT: The densities of prepared samples were measured by a pycnometer (Ultra pycnometer 1000, Quantachrome Instruments, USA) with 40 times repeating measurement.

3. RESULTS AND DISCUSSIONS

3.1 Cationic Surfactants

Figure 1 showed the zeta potential of $CaCO_3$ in NH_4Cl solution containing different concentrations of 1,5-Diaminopentane dihydrochloride (DAPDC). With the introduction of this surfactant, the surface charge of particles changed from negatively charged (without surfactants) to positive charged, which indicated the adsorption of surfactants on the surface of particles. When the concentration of surfactant increased, the zeta potential also increased, especially a big increase arose while the concentration amounted to 0.6mass%. DAPDC is a cationic surfactant with one cationic functional group on each end. When it closes to negatively-charged $CaCO_3$ particles, there are two possible ways for their adsorption. First way is that only one functional group attach on the surface of particle while another functional group extending in the solution. Second way is that both of these two functional groups attach on the surface of particles and leave hydrophobic alkyl chain in the solution, which usually happen at low concentration of surfactants. When the concentration of surfactants is high, the first case prefers to happen. Due to the high concentration of 0.6mass%, surfactants prefer to adsorb on the surface of particles in the first way, leaving more positive charge functional group on the surface of particles, which lead to a big increase of zeta potential at 0.6mass%, as shown in Fig.1.

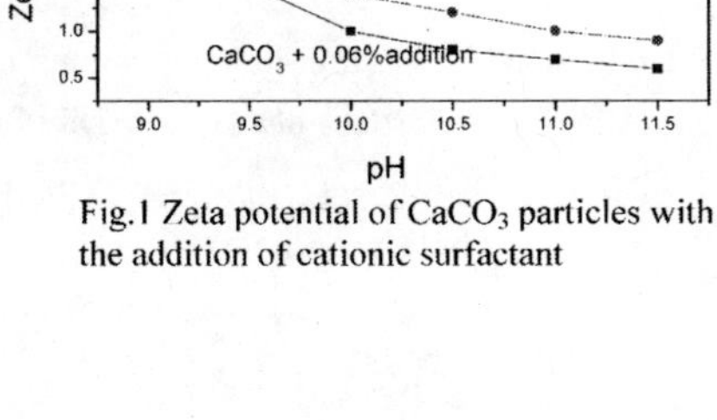

Fig.1 Zeta potential of $CaCO_3$ particles with the addition of cationic surfactant

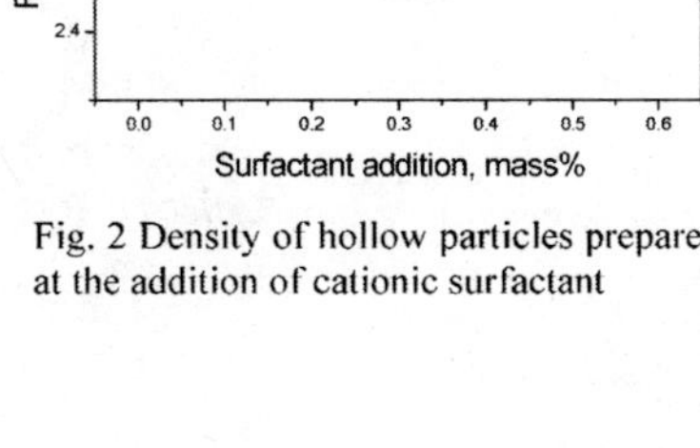

Fig. 2 Density of hollow particles prepared at the addition of cationic surfactant

This cationic surfactant has been introduced during the preparation of hollow $CaCO_3$ particle. The density of precipitated samples with different addition of surfactants has been measured, as shown in Figure 2. With the addition of surfactant, the density decreased firstly, which indicated that more hollow particles were formed as the result of surfactants addition. Therefore, this surfactant is effective in improving the formation of hollow particles. When the concentration of surfactant increased to 0.6mass%, the density trended to increase, which indicated that the less formation of hollow particles. The less formation of hollow particles may be the result of surfactant adsorbing on both bubbles and particles surface at high concentration and preventing their attachment by carrying the same positive charge. Another reason for the less formation maybe result from the loss of hydrophobic force due to the reduction of two functional groups adsorbing on the particles surface at high concentration of surfactants, as discussed above.

The morphologies of prepared samples with 0.3mass% surfactant addition were shown in Fig.3. Some broken particles with hollow structure inside were observed, as shown in Fig.3a. The inset of Fig.3a was a TEM image of unbroken particle, which revealed that some unbroken particles were also hollow inside, deducing from the more electrons dense in center of particle. Fig.3b showed the images of cut surface of samples. The bottom inset is an image of cut dense particle while the up inset is an image of cut hollow particle. A lot of cut hollow particles were observed in Fig.3b, which further prove the effectiveness of this surfactant.

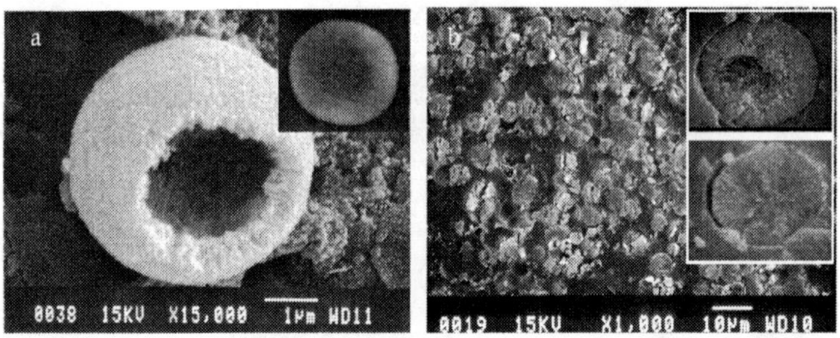

Fig. 3 Morphologies of hollow $CaCO_3$ particles prepared with the addition of cationic surfactant

Beside this surfactant, one more cationic surfactant, namely Triethyl-methylammonium chloride, was tested in our experiments. The addition of this surfactant also resulted in the decrease of density, which indicated that this surfactant with only one ammonia cationic functional group is also effective in improving the formation of hollow particle.

3.2 Nonionic Surfactants

Bubbles in a liquid are unstable. After the introduction of bubbles in solution, the bubbles start to rise (float) in solution and rapidly accelerate to its terminal velocity which is determined by the balance between the buoyant rise force and the drag force [7]. The drag force is usually determined by bubble size and bubble interface. One purpose of our study in this paper is to increase the drag force, slowing down the rise of bubbles and facilitating the attachment of particles to bubbles. When the surfactants are present in the liquid, they tend to concentrate at the rear portion of a floating bubble [8], which gives rise to a lower interfacial tension at the rear of the bubbles and thus inducing a surface tension gradient towards the front of the bubble. This surface tension gradient results in the increase of drag force and slows down the rise of bubbles. In our experiments, two nonionic surfactants, namely Cyclohexanemethanol (GHM) and Polyethyleneglycol dodecyl ether (PEG), were employed to modify the surface of bubbles since this kind of surfactants are hydrophobic and easily attach on the surface of bubbles.

When surfactants are supposed to be employed in aqueous solution, the first factor we should consider is the solubility of surfactant, which determines the efficiency of surfactants. Since PEG has a longer chain than that of GHM, the solubility of PEG in aqueous solution is limited. Besides, the longer chain of PEG results in a long time to reach the adsorption equilibrium on the surface of bubbles, which also limits the efficiency of this surfactant. So GHM was chosen as surfactant for further investigation to prepare hollow particles. The density of samples prepared with the addition of this surfactant is 2.544g/cc, which is lower than that of no addition. So the addition of GHM is also effective in improving the formation of hollow particles. Fig.4 is the SEM image of the prepared samples. Some broken hollow particles were observed in Fig.4a. The cut surface of sample, as shown in Fig.4b showed that many hollow particles were formed in this sample.

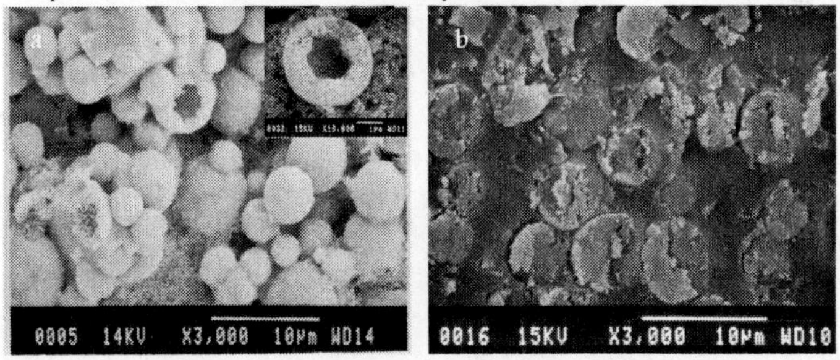

Fig. 4 SEM images of hollow $CaCO_3$ particles prepared with the addition of nonionic surfactant

4. CONCLUSIONS

Influences of surfactants on the production of hollow particles by bubble template method were investigated in this paper. It was found that the cationic surfactants, especially

$ClH_3N(NH_2)_5NH_3Cl$, is effective in modifying the surface of $CaCO_3$ particles and facilitating the formation of hollow particles. The nonionic surfactant ($C_6H_5CH_2OH$) also showed an effective impact on the formation of hollow particles. The results in this paper indicated that the addition of selected surfactants is effective for enhancing the production of hollow particles by bubble template method.

ACKNOWLEDGEMENTS

This research was supported by a grant from the NITECH 21st Century COE Program, "World Ceramics Center for Environmental Harmony". A part of this research has been financed by Japan Science and Technology Agency (JST).

REFERENCES

1. Hollow and Solid Spheres and Microspheres: Science and Technology Associated With Their Fabrication and Application, Vol 372 (Eds: D. L. Wiloox, M. Berg, T. Bernat, D. Kellerman, J. K. Cochran) Materials Research Society Proceeding, Pittsburgh, 1995, and reference therein)
2. Yong Sheng Han, Masayoshi Fuji, Minoru Takahashi, Synthesis and applications of hollow particles, (submitted).
3. Frank Caruso, Hollow capsule processing through colloidal templating and salt-assembly, Chem. Eur. J. 2000, 6, 413-419.
4. Yong Sheng Han, Gunawan Hadiko, Masayoshi Fuji, Minoru Takahashi, A novel approach to synthesize hollow calcium carbonate particles, Chem. Lett. 34 (2), 2005, 152-153.
5. M. Preuss, H. J. Butt, Int. J. Miner. Process., 56 (1999) 99.
6. Yong Sheng Han, Li Wei Lin, Hiroaki Takegami, Masayoshi Fuji, Minoru Takahashi, Bubble template method for synthesis of ceramic hollow particles, submitted for Shaping 3, in Limoges, France 2006.
7. Surface Chemistry of Froth Flotation, Second edition, Revised by S. Ramachandra Rao, First edition by Jan Leja, Kluwer Academic/Plenum Publishers, 2002
8. B. Guenot, J. Magnauet, B. Spennato, The effect of slightly soluble surfactants on the flow around a spherical bubble, J. Fluid. Mech. 339, 1997, 25-53.

A FACILE METHOD TO SYNTHESIZE ZNO TUBES BY INVOLVING AMMONIA BUBBLES

Liwei Lin, Yongsheng Han, Masayoshi Fuji, Takeshi Endo, Hideo Watanabe and Minoru Takahashi*
Ceramics Research Laboratory, Nagoya Institute of Technology
10-6-29, Asahigaoka, Tajimi, 507-0071, Japan
m-takahashi@nitech.ac.jp

ABSTRACT

A large quantity of submicrometer-sized zinc oxide tubes was directly prepared by a novel simple method through introducing ammonia bubbles into $ZnCl_2$ solution. Hexagonal-faceted ZnO tubes with length of 1~2μm, diameter of 300~500nm and shells thickness of 20~50nm were synthesized by this new method. The prepared tubes were characterized by XRD, SEM and EDS. It was found that the final pH of solution played an important role on the formation of pure zinc oxide tubes. The effect of pH on the formation of ZnO tubes was investigated in this paper.

INTRODUCTION

Zinc oxide is a distinguished material with some special properties, such as a direct band gap of 3.37eV and large exciton binding energy of 60 meV at room temperature. ZnO has received great interests in scientific research, potential utilization in optical and electronic devices. Especially, one-dimensional (1D) and quasi one-dimensional ZnO,[1,2,3,4] such as nanowires, nanosprings, nanorods, nanobelts, and so on, have been widely treated as one of the most promising material for nanoscale or microscale electronic, photochemical, catalytic, optoelectronic devices, such as gas sensor[5], varistor[6], etc.

Up to date, several methods have been developed for synthesis of submicrometer-sized zinc oxide tubes, such as thermal evaporation of mixed Zn/ZnO powder at 1300 °C[7], physical evaporation of the mixture of ZnO and graphite powders[8] and surfactants-induced template method[9].Generally speaking, these methods require additional template to build hollow architectures and the template needs to be removed later. The limitations of these methods are rigid condition, complex procedure and not easy to control. So, it is necessary to prepare zinc oxide tubes by a simple method which can product ZnO tubes efficiently and amelioratively.

A new and one step method to synthesize ZnO tube by introducing of ammonia bubbles into zinc salt solution has been proposed in our group. The new method offers some advantages, such as low temperature and easily controlled condition. It was found that the formation of the prepared ZnO tubes was strongly depended on the final pH of reaction solution. The effect of the final pH on the formation of ZnO tubes was discussed in this paper.

EXPERIMENT

ionic concentration is 0.5M, was put into a water bath where the temperature was kept 90° C. The solution was stirred at the constant rate by the Teflon-coated magnetic stirring bar. The ammonia bubbles were introduced into the bottom of the solution through a micro-sized bubble maker (Tekeno, Japan). During the reaction, the pH and temperature of solution were continuously measured by a pH meter (SG8, Mettler Toledo, Switzerland). When the pH increased to a given value, the ammonia bubbles were stopped immediately. The white precipitate was filtered, and then the products were dried at 80° C for 24 hours and used for measurement.

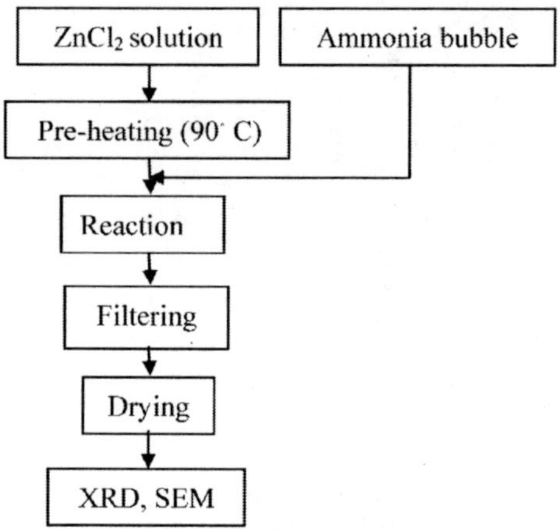

Fig.1 Schematic diagram of preparation procedure

The as-synthesized samples were characterized by the field emission scanning electron microscope (JEOL JSM-7000FO, Japan). The content of elements was examined by the energy dispersive spectroscopy (EDS). The composition of products were confirmed by X-ray diffraction (XRD, RINT1100, Rigaku, Japan) by employing a scanning rate of 0.02°/s in the 2θ range from 20° C to 80° C with Cu Kα radiation.

RESULTS AND DISCUSSION

Fig.2 shows the change of pH and temperature during reaction. The starting pH was about 4.5. After the introduction of ammonia bubbles into solution, the pH started to increase, while the temperature of solution sharply increased firstly and then decreased, which indicated that it was an exothermic reaction for the formation of zinc oxide. The increase of temperature was slowed down, when the reaction was nearly ending.

The reaction equations of the formation of ZnO single crystalline are usually accepted as followings [11]:

$$ZnCl_2 + NH_4OH + H_2O \longrightarrow xZnCl_2 \bullet yZn(OH)_2 \bullet zH_2O \downarrow + NH_4^+ \quad (1)$$

$$xZnCl_2 \bullet yZn(OH)_2 \bullet zH_2O + NH_4OH \xrightarrow{PH\sim12} aqueous\ solution \xrightarrow{PH\sim7} Zn(OH)_2 \downarrow + NH_4Cl + NH_3 \uparrow \tag{2}$$

$$Zn(OH)_2 \xrightarrow{T>55\,C} ZnO + H_2O \tag{3}$$

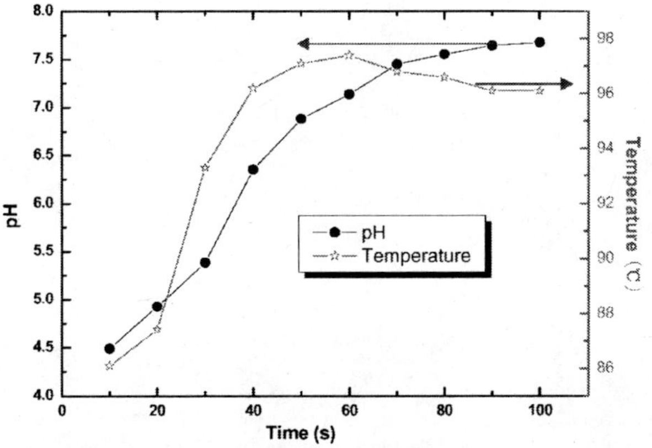

Fig.2 Dependence of pH and reaction temperature on reaction time

Fig.3 shows the XRD patterns of the samples prepared at different pH. It was found the products were mostly zinc hydroxyl salt precursor ($Zn_5(OH)_8Cl_2H_2O$), when the pH was 5 or 6. However, when the pH was about 8, the products became pure zinc oxide, which peaks can be indexed to the hexagonal phase of zinc oxide (wurtzite structure, space group P63mc). The lattice parameters were calculated to be a=3.25, c=5.21, which agreed well with the reported value (JCPDS card NO 36-1451) and there was no characteristic peaks of impurities observed, such as $Zn(OH)_2$, zinc hydroxyl salt precursor. Therefore, it can be concluded that the formation of single crystalline was strongly dependent on the final pH of reaction solution in our experiments. Only at a relative high pH was it possible for the formation of the single crystalline ZnO.

Fig.4 shows the SEM images of the samples prepared at different final pH. Fig.4 (a) and Fig.4 (b) are the images of samples prepared at pH 5 and 6, respectively. The products were layer-like irregular particles without the observation of zinc oxide tubes. However, many incomplete zinc oxide tubes were observed in Fig.4(c) at pH 7. The average diameter of the tubes was in the 300nm~500nm with the length in the 1μm~2μm and the thickness of the wall was in the 20nm~50nm. When the pH arrived 8, almost all products were rod-like particles, as shown in Fig.4 (d) (the inset image is the amplified rod-like image).

Fig.5 is the EDS analysis of rod-like particles prepared at pH 8, which demonstrates that the products consisted of only zinc and oxygen two element (except ♦–carbon paste and ●–O_sO_4 coating), of which the atomic radio was nearly 1:1, further proved these the rod-like particles were zinc oxide.

Fig.6 is the morphology of the cutting surface of the rod-like zinc oxide particle. The cuttings of samples were carried out by putting samples into the polymer, and then dried at 60°C for 12 hours to solidify polymer. The cutting slices were carried out by LEICA EM UC6. From Fig.6, we can find all of the rod-like zinc oxide were hollow, that is to say, the products were zinc oxide tubes at the final pH 8.

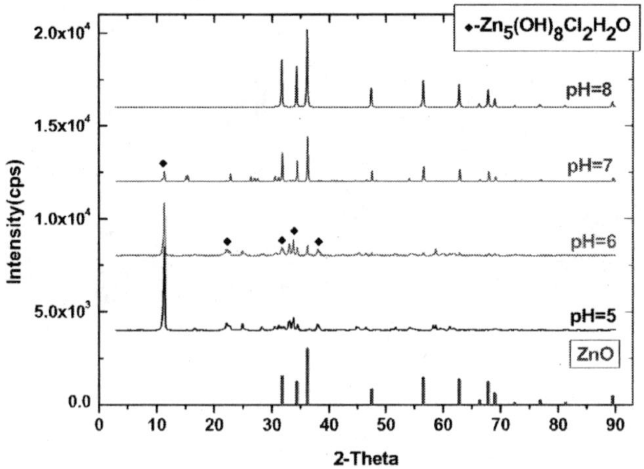

Fig.3 XRD patterns of as-prepared products at different pH

Fig.4 SEM images of samples prepared at different final pH
(a–pH 5, b–pH 6, c–pH 7 and d–pH 8)

Fig.5 EDS analysis of the rod-like particle
(♦—carbon paste and ●—O_sO_4 coating)

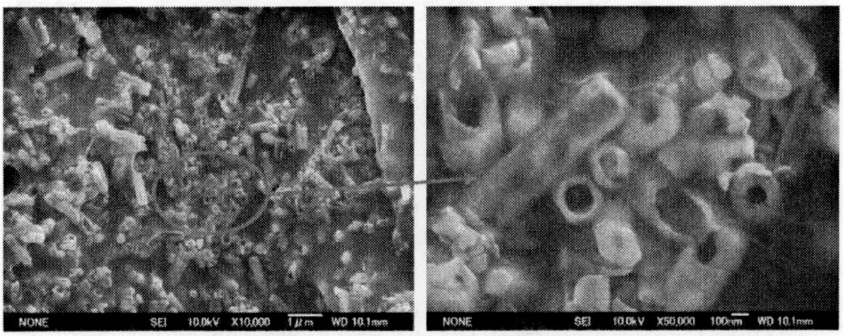

Fig.6 Cutting surface of the rod-like particle

For the formation mechanism of zinc oxide tubes, a brief explanation has been proposed in our precious works[10,12]. The formation of zinc oxide tubes maybe involves ammonia bubbles template. When the ammonia bubbles floating in the solution, the newly formed ZnO nanocrystals preferred to attach on the rear of bubbles and form bowl-like coating. In the following, the crystal growth of zinc oxide preferred to happen at edge of coating due to the rich of growth ions there carried by flowing solution.

Since this is a new method for the synthesis of zinc oxide tubes, it is necessary for us to continue this investigation in order to further elaborate the formation mechanism of ZnO tubes.

CONCLUSION

In this paper, zinc oxide tubes with a single crystalline composition have been efficiently synthesized by a novel simple method involving ammonia bubbles. The morphology and composition of zinc oxide tubes have been confirmed by SEM, XRD and EDS analysis. The tubes presented the diameter in the range of 300nm~500nm, and the wall thickness in the range of 20nm~50nm, the length of the tubes in the range of 1µm~2µm. The possible growth mechanism for the growth of zinc oxide tubes has not been elaborated.

Further investigation on the formation mechanism of zinc oxide tubes is to be reported in the future work.

ACKNOWLEDGEMENT

This research was supported by a grant from the NITECH 21st Century COE Program, "World Ceramics Center for Environmental Harmony". A part of this research has been financed by Japan Science and Technology Agency (JST).

REFERENCES

[1]Z.L. Wang, "Zinc oxide nanostructures: growth, properties and applications," *J. Phys.: Condens. Matter.*, 16, 829-858 (2004).

[2]P.X. Gao, Y. Ding, W.J. Mai, William L. Hughes, C. Lao, Z.L. Wang, " Conversion of Zinc Oxide Nanobelts into Superlattice-Structured Nanohelices," *Science*, 309,1700-04(2005).

[3]M. Yin, Y.Gu, Igor L.Kuskovsky, T. Andelman, Y. Zhu, G.F.Neumark, and Stephen OBrien, " Zinc Oxide Quantum Rods," *J.Am.Chem.Soc.*,126,6206-07(2004).

[4]Z. Wang, X. Qian, J. Yin, and Z. Zhu, " Aqueous solution fabrication of large-scale arrayed obelisk-like zinc oxide nanorods with high efficiency," *J. Solid State Chem.*, 177, 2144-49(2004).

[5]C.S. Rout, S.H. Krishna, S.R.C. Vivekchand, A. Govindaraj, C.N.R. Rao, "Hydrogen and ethanol sensors based on ZnO nanorods, nanowires and nanotubes," *Chem.Phy.Lett.*, 418, 586-90(2006).

[6]V. Tohver, S.L. Morissette, J.A. Lewis, B.A. Tuttle, J.A. Voigt, D.B. Dimos, " Direct-Write Fabrication of Zinc Oxide Varistors," *J. Am. Ceram. Soc.*, 85, 123-28(2002).

[7]Y. J. Xing, Z. H. Xi, Z. Q. Xue, X. D. Zhang, and J. H. Song, "Optical properties of the ZnO nanotubes synthesized via vapor phase growth," *Applied. Phys. Lett.* 83, 1689-91(2003).

[8]X.H Hong, X.M. Sun, X.L. Li, Y.D. Li, "Catalytic growth of ZnO nanotubes," *Mater. Chem. Phys*, 82, 997-1001(2003).

[9]J. Duan, X. Huang, E. Wang, "PEG-assisted synthesis of ZnO nanotubes," *Mater. Lett.*, 60, 1918-21(2006).

[10]Y.S. Han, L.W. Lin, H. Takegami, M. Fuji, M. Takahashi, "A Novel Approach to Synthesize Hollow ZnO Particles Using Bubbles as Templates." Proceeding of CICMT2006, Denver, USA, April 24-27, 2006.

[11]S. M. Haile, D. W. Johnson, "Aqueous Precipitation of Spherical Zinc Oxide Powders for Varistor Application", *J. Am. Ceram. Soc.*, 72(10), 2004-08(1989).

[12]Y.S. Han, L.W. Lin, M. Fuji, M. Takahashi, "A Novel One-Step Solution Approach to Fabricate Tubular ZnO Nanostructures." (Submited)

RAPID DRYING TECHNIQUE FOR SLIP CAST BODY

Takashi Shirai, Masaki Yasuoka, Yoshiaki Kinemuchi, Yuji Hotta, and Koji Watari
National Institute of Advanced Industrial Science and Technology (AIST),
2266-98 Anagahora, Shimoshidami, Moriyama, Nagoya 463-8560, Japan

ABSTRACT

We investigated the effects of rapid drying on ZnO slip cast bodies by microwave heating and compare with those dried by conventional heating. The extent of warpage of a microwave dried body is less than by conventional drying. Furthermore, the number and size of pores are smaller by microwave drying than for conventional. The selective heating by microwaves contributes to uniformities in temperature and water content in a green body during drying. Microwave heating results in small gradients of water content between in the inside and outside of the green body, because the free and bound waters with NH_4^+ salt of poly (acrylic acid) as a dispersant in the green body are selectively heated due to the higher dielectric constant compared with the ZnO powder, which results in efficient evaporation and transport of water under microwave heating. Therefore, the extent of warpage of the dried body is reduced by microwave drying compared to conventional drying, and the microstructures of sintered bodies are more uniform by microwave drying.

INTRODUCTION

Slip casting [1-4] is an essential technique for producing technical and/or advanced ceramics. A major disadvantage of slip casting is that it requires long periods to form a body of the desired thickness and further periods to dry the cast body. Typically, rapid drying should increase defects, such as warpage, cracks, and density distribution. These defects adversely affect the characteristics of sintered bodies [5].

Advances in ceramic drying technology have been increasingly stimulated over the past years, so that, in addition to the traditional convective drying with warm air, other drying methods are becoming more and more important. In this context, the possibility of applying high frequency drying in the drying process of ceramics is of great interest [6-8]. Some studies have dealt with the possibility of predicting the rate of heating with microwaves [9, 10], while other studies have been concerned with the acceleration of drying by microwave heating [11].

Previously, we examined rapid microwave drying of a ZnO ceramic slip cast body and compared the characteristics of such bodies with those of conventionally dried cast bodies [12]. This work indicated that microwave heating has merits in terms of both rapid drying and structural uniformity of the dried bodies. Microwaves not only reduced the drying time, but also improved the characteristics of sintered bodies [12].

In this report, we investigated the differences in the water evaporation behaviour of a ZnO slurry between microwave heating and conventional heating.

EXPERIMENT

A commercially available sub-micron ZnO powder (JIS 1, Hakusuitech Co., Ltd.) with an average diameter of d_{50}=0.5 μm was used for the preparation of slurries that contained 40 vol% solid phase and 0.32 mass% of a commercially available NH_4^+ salt of poly (acrylic acid) (Aron A-6114, Toagosei Co., Ltd.) as a dispersant.

After 2min of mixing with an electric mixer (AR-250, THINKY Corp.), the suspension was ball-milled for 12 h with high-grade Al_2O_3 balls (diameter 5 mm) to prepare the ZnO slurry.

The ZnO slurry was poured into a beaker ($\phi 22 \times 20$ mm) and then dried by microwave and conventional drying. In oven drying, the slurry was dried at 80±3°C (RH 40%) for at least 26 h. This drying period appeared to be sufficient to evaporate all free water and most bound water. A microwave oven (MW-Master, 2.45 GHz, Mino Ceramic Co. LTD.) was used for microwave drying. The drying temperature was controlled by a proportional integral differential (PID) controller. The green body was heated to 80±3°C at a constant heating rate of 2 K/min and then kept at this temperature for 1.6 h. An equivalent amount of free water and bound water was evaporated in the microwave furnace compared to the conventionally dried body.

RESULTS

Figure 1, reprinted from our previous study [12], shows the extent of warpage of the dried bodies examined on their bottom face using the laser confocal displacement meter. The extent of warpage of the dried body is smaller when dried by microwave than by the other conventional drying methods. Figure 2, reprinted from our previous study [12], shows a SEM image of the polished surface of the sintered bodies. The bottom part of the air-dried sample, Fig. 2-(a), shows many large pores. Local differences in water content arise during the long drying period of the air-dried green body, forming many pores on the bottom of the part. In oven drying, many large pores occurred in the upper part of the green body particularly at the surface as seen in Fig. 2-(b). In this case, the green body is heated from the outside generating a temperature difference in the green body, resulting in rapid evaporation of free water occurs from the upper surface, producing many large pores. With microwave drying, the pores are smaller in number and size than those obtained with other drying methods, moreover, the pores were distributed uniformly.

Figure 3 shows a mass loss with drying time for the ZnO slurry for the different drying techniques. In conventional drying, Fig. 3-a, a minimum of 9 h is required to bring the water content to an almost constant value. The remaining water could be mostly ascribed to the water at the meniscus regions and adsorbed water. On the other hand, microwave drying, Fig. 3-b, can eliminate the water to the same limits in only 1.6 h.

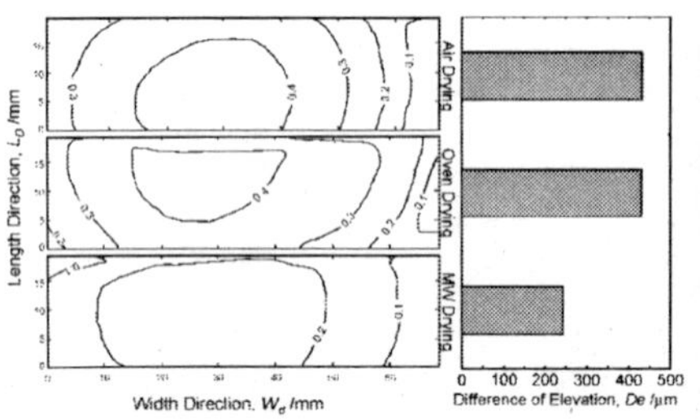

Figure 1: Extent of warpage of dried bodies different drying methods [12].

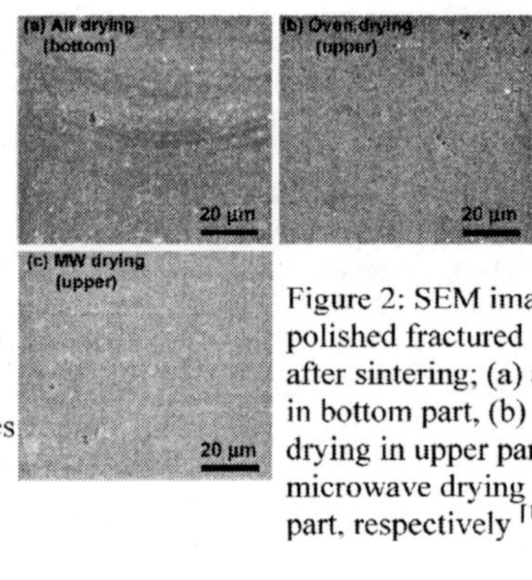

Figure 2: SEM image of polished fractured surface after sintering; (a) air drying in bottom part, (b) oven drying in upper part, and (c) microwave drying in upper part, respectively [12].

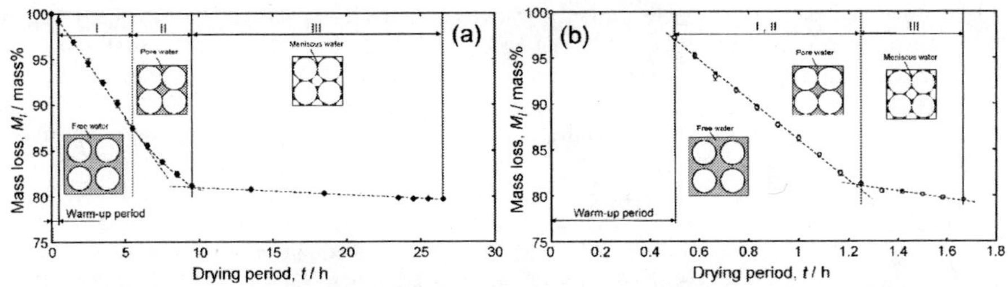

Figure 3: Comparison of conventional drying (a) and microwave drying (b) on the water evaporation behavior. In conventional drying, the weight loss curve can be divided into three steps. In comparison, by microwave drying, the weight loss curve can be separated into only two steps.

DISCUSSION

Drying of green body has recently become more important due to the use of ultrafine powder. The capillary size of the porosity in the green body reduces considerably as the particle size of the powder decreases. The transport of water through the capillaries less than 1/5 of the particle diameter exerts a strong drying stress on the green body. Therefore, the drying must be conducted either at an extremely slow rate so as to preserve the body intact. The evaporation and transport of water are the keys to drying and controlling the kinetics of drying, subsequently affecting the properties of green body [13].

In conventional drying, shown in Fig. 3-a, the drying rate was fixed at first, and it decreases gradually with drying time. The mass loss curve can be separated into three steps. Generally speaking, in the case of a slip cast body, the drying process by conventional heating follows three steps [5]. The green body consists of particles and water before drying. In the first step, the free water around the particles (as shown of the diagram A in Fig. 3-a) evaporates, and the resulting shrinkage is due to the decrease in the distance between particles. After that, the drying process shifts to the second step in which particles make contact with each other. The turn point of the curves to the second step of drying in Fig. 3 is close to a mass loss of ca. 12 mass% that is corresponded to green density after drying. In the second step, the pore water (as shown of the diagram B in Fig. 3-a) is transported to the surface reducing the total content in the green body. However, a water film will still remain in the contact region (as shown of the diagram C in Fig. 3-a) forming a meniscus shape. Third step starts as the evaporation of the water at the meniscus regions becomes dominant [5].

Drying is basically an interface control process. The drying rate is theoretically controlled by the evaporation at the outer surface. As long as the environmental conditions and the surface area of the gas/water interface remain constant, the drying rate is fixed [13]. Therefore, the first step has a constant drying rate which means the rate of water loss is constant with respect to time. However, in the second step, the drying rate is decreased by reduction of the capillary size due to shrinkage. A parabolic relation between the drying time and mass loss is observed which means this step is diffusion controlled drying, as shown in the second step of Fig. 3-a. The evaporation and transport of water are the keys to drying and controlling the kinetics of drying, subsequently affecting the properties of green body [13]. Therefore, drying must be conducted at an extremely slow rate for the body to remain intact.

In comparison, for the case of microwave heating, the drying proceeds in two steps, Fig. 3-b. The mass loss curve for microwave heating has no distinct boundary between first step and second step. These steps have a constant drying rate which means the rate of water loss is constant with respect to time.

Microwave heating is fundamentally different from conventional heating. In microwave heating, heat is generated internally within the material instead of from external heating sources [7]. A material needs to have a high dielectric loss to absorb microwaves efficiently. Figure 4, reprinted from our previous study [14], shows the temperature-time curves for the ZnO slurry, water with PAA, water, and a ZnO powder under the following conditions: f=2.45 GHz, 150 W. After MW irradiation for 6 min, the ZnO slurry and the water with PAA reached 75°C and 71.4°C, respectively. The results for the ZnO slurry and water with PAA showed similar temperature-time curves. On the other hand, the temperature of the ZnO powder only slightly increased to 27°C. From these results, the ZnO powder did not undergo an increase in temperature under microwave irradiation. In the case of the ZnO slurry, the water with PAA was heated selectively due to a higher dielectric constant compared with the ZnO powder. Comparing the water with PAA to only water, the rate of temperature increase for the water with PAA was higher than that of only water, and reached a temperature almost 20°C higher than that reached by only water after 6 min. It is believed that a small addition of PAA to water has important effects on microwave absorption. From these results, the water with PAA in slurry is heated selectively by microwaves. This means that the temperature distribution of green body is small due to selective heating by microwave, because it can be considered that water is distributed in a green body equally. Besides, in the case of conventional heating, the green body has lower internal temperature than outside temperature, because green body is heated from the outside. On the other hand, in the case of microwave heating, the green body has higher internal temperature than outside temperature, because green body is heated by itself due to selective heating. The disappearance of the diffusion control process by microwave heating in figure 3-b is the transport of the water during drying, enhanced in the stage of linear kinetics through the reduction of water surface tension due to selective heating by microwave.

Based on these results, microwave heating can greatly improve the drying kinetics due to selective heating and the drying stage can be reduced to two steps. Therefore, the drying defects such as warpage, cracks and density distribution are fewer by microwave drying than by conventional drying, and the microstructure of the sintered bodies are more uniform when dried by microwaves, as reported in our previous results [12].

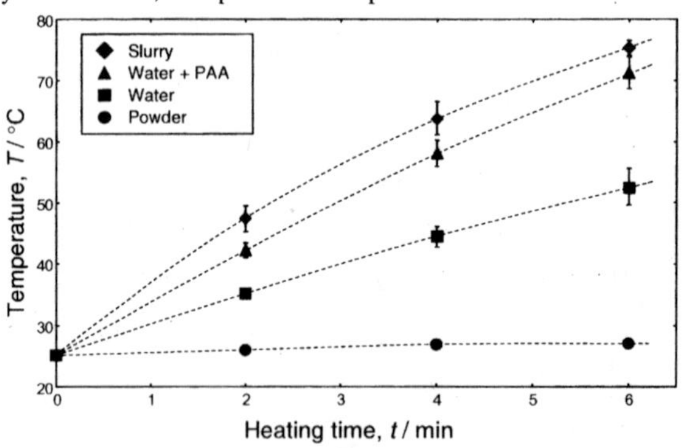

Figure 4: Temperature-time curves for the ZnO slurry, water with PAA, water, and a ZnO powder under the following condions: f=2.45 GHz, 150 W [14].

CONCLUSIONS

We observed a difference in the water evaporation behaviour in a ZnO slurry between microwave heating and conventional heating. The following conclusions were drawn.

(1) Water with PAA in the slurry is heated selectively with microwave drying. A small addition of PAA to water has important effects on microwave absorption.

(2) Microwave heating can greatly improve the drying kinetics due to selective heating and the drying stages can be reduced to two steps.

(3) The transport of the water during drying is enhanced in the linear kinetic stage through the reduction of water surface tension due to selective microwave heating.

(4) Microwave heating results in a small gradient of the water content between the inside and the outside of a green body, because the evaporation of water is more efficient. Therefore, the extent of warpage of the dried body is smaller by microwave drying than by conventional drying, and the resulting microstructure of the sintered bodies are uniform.

REFERENCES

[1]Y. Hotta, T. Tsugoshi, T. Nagaoka, M. Yasuoka, K. Nakamura, and K. Watari, "Effect of Oligosaccharide Alcohol Addition to Alumina Slurry and Translucent Alumina Produced by Slip Casting", *J. Am. Ceram. Soc.*, **86**, 755-760 (2003).

[2]K. S. Chou, and L. J. Lee, "Effect of Dispersants on the Rheological Properties and Slip Casting of Concentrated Alumina Slurry", *J. Am. Ceram. Soc.*, **72**, 1622-1627 (1989).

[3]L. B. Garrido, and E. F. Aglietti, "Pressure Filtration and Slip Casting of Mixed Alumina-Zircon Suspensions", *J. Eur. Ceram. Soc.*, **21**, 2259-2266 (2001).

[4]N. Omura, Y. Hotta, K. Sato, Y. Kinemuchi, S. Kume, and K. Watari, "Slip Casting of Al_2O_3 Slurries Prepared by Wet Jet Milling", *J. Ceram. Soc. Japan*, **113**, 495-497 (2005).

[5]Y. Shiraki, "Ceramic-Seizo-Process III", *Gihodo Shuppan*, pp. 39-43 (1980).

[6]I. W. Turner and P. G. Jolly, "Combined Microwave and Convective Drying of a Porous Material", *Drying Technol.*, **9**, 1209-1269 (1991).

[7]W. H. Sutton, "Microwave Processing of Ceramic Materials", *Am. Ceram. Soc. Bull.*, **68**, 376-386 (1989).

[8]D. Skansi, and S. Tomas, "Microwave Drying Kinetics of a Clay-Plate", *Ceram. Int.*, **21**, 207-211 (1995).

[9]R. Y. Ofoli, and V. Komolprasert, "On the Thermal Modeling of Foods in Electromagnetic Fields", *J. Food Processing and Preservation*, **12**, 219-241 (1988).

[10]P. Jolly, and I. W. Turner, "Non-Linear Field Solutions of One-Dimensional Microwave Heating", *J. Microwave Power and Electromagnetic Energy*, **25**, 3-15 (1990).

[11]P. Jolly, "Temperature Controlled Combined Microwave- Convective Drying", *J. Microwave Powder*, **23**, 65-74 (1986).

[12]T. Shirai, M.Yasuoka, Y. Hotta and K. Watari, "Rapid Microwave Drying for Slip Cast Bodies", *J. Ceram. Soc. Japan*, **114**, 217-219 (2006).

[13]Wen-Cheng J. Wei and Chang-Li Hsieh, "Drying Kinetics of Ultrafine Alumina Cake with Drying Control Cemical (DCC)", *J. Ceram. Soc. Japan*, **107**, 313-317 (1999).

[14]T. Shirai, M. Yasuoka, Y. Hotta, Y. Kinemuchi and K. Watari, "Microwave Drying for Slip Cast Body", *Adv. In Tech. of Mat. Proc. J. (ATM)*, (2006), in press.

INTERFACE CHARACTERIZATION AND CONTROL

IMPROVEMENT OF MOLTEN-METAL WETTABILITY ON CERAMICS BY MICRO-MACRO LEVEL MORPHOLOGY CONTROL

Hideki Kita, Hideki Hyuga, Katsumi Yoshida and Naoki Kondo
National Institute of Advanced Industrial Science and Technology (AIST)
2266-98 Shimo-Shidami, Moriyama-ku, Nagoya, 463-8560, Japan

ABSTRACT

For the effective use of silicon nitride ceramics in the metal casting industry, it is necessary to decrease the wettability of the ceramic by the molten metal, because wetting by the molten metals leads to adhesion of solidified metals, which causes several problems. Surface structure is known to affect wettability; therefore, silicon nitride plates with ordered hemispherical protrusions were fabricated by slip-casting, and were subjected to wettability tests using molten aluminum. A surface with protrusions exhibited a higher contact angle compared to flat surfaces. It was supposed that the contact angle depends on the fraction of solid surface, so that the experimental results were discussed using Cassie's theoretical model.

INTRODUCTION

Silicon nitride ceramics have good mechanical properties, high wear resistance, corrosion resistance, stability at elevated temperatures, low thermal conductivity, and low wettability for molten metals. Taking advantage of these properties, silicon nitride has been used for tools in aluminum casting lines such as ladles, stalks and protection tubes etc [1-3]. For the effective use of silicon nitride ceramic tools in the metal casting industry, it is necessary to lower the wettability to molten metal. The wettability of a liquid to a solid surface is generally known to be affected by adjustment of i) chemical composition, and ii) control of the surface structure.

Many studies regarding the wettability of a liquid to a solid surface have dealt with water and oil in the low-temperature range, and on joint interfaces and other smooth surfaces [4-6]. However, there have been few studies that investigate bulk materials requiring a high level of thermal shock resistance and durability, as is the case for metal casting tool applications. With this background, the present study follows a physical approach and attempts to fabricate silicon nitride plates with controlled surface structures formed by the slip-casting method. The surfaces were then subjected to wettability tests, and the influence of surface geometry on the wetting behavior was investigated.

THEORETICAL BACKGROUND AND DESIGN OF THE PLATE

Cassie's theory assumes that when a composite surface is formed, a liquid is completely lifted up by rough features on the surface (Fig.1). The apparent contact angle (CA) on the basis of Cassie's model, θ_{rc}, is given by [6],

$$\cos \theta_{rc} = f_s \cdot \{(1 + \cos \theta_b) - 1\}$$

where f_s is the contact area fraction of the solid surface.

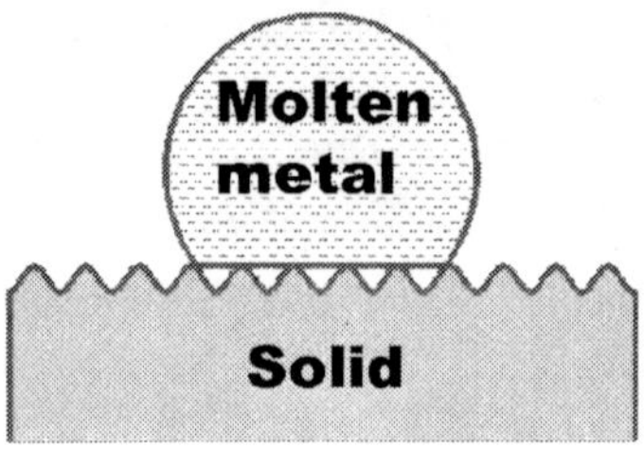

Figure 1. Cassie's contact model

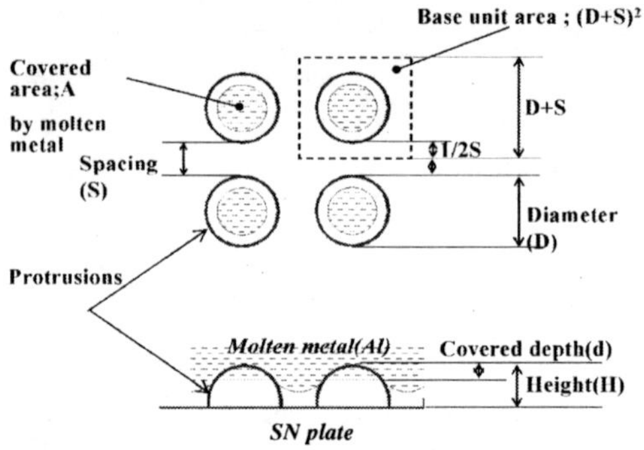

Figure 2. The top and cross-sectional schematic views of a rough surface geometry of hemispherical protrusions.

Figure 2 shows the top and cross-sectional schematic views of a rough surface geometry of hemispherical protrusions. For the wettability test, each top of the hemisphere protrusions were covered with molten aluminum. Figure 2 shows the size of the coverage area marked as A, and a regular rough surface consisting of an array of hemispherical protrusions, where the diameter, height and spacing of the protrusions are denoted as D, H, and S respectively. The covered depth (d) of molten aluminum was measured from photographic images, and A was arithmetically calculated. Therefore, using the base unit area, $(D+S)^2$, the fraction of the solid surface (f_s) is given as:

$$f_s = A/(D+S)^2$$

Therefore, the apparent CA, according to Cassie's theory, is given by the function of the geometric parameters of the surface

$$\cos \theta_{rc} = A/(D+S)^2 \{(1 + \cos \theta_b) - 1\}$$

Figure 3 shows curves for the predicted apparent CAs according to Cassie's theory as a function of the geometric parameter f_s with a given value of θ_b. According to the predictions, for example, f_s needs to be lower than 0.5 to achieve 150° at θ_b = 140°. The wettability of Al_2O_3 and SiC by molten aluminum has been investigated[7)8)]. The results obtained show significant difference of 90° to 165° at 800°C.

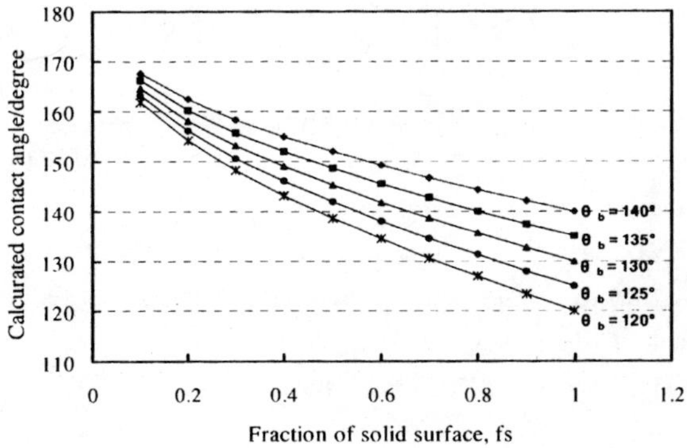

Figure 3. The apparent contact angles on a surface with protrusions, according to Cassie's model.

EXPERIMENTAL PROCEDURE

Raw silicon nitride powder with sintering additives (5 wt% Y_2O_3 and 5 wt% Al_2O_3) was mixed with water to make a slurry. The slurry was cast into a gypsum mold with ordered hemispherical dimples for consolidation. The dimples were transferred to the surface of the consolidated green body. The green body was then sintered at 1800 °C for 2 h in a 0.5 MPa N_2 atmosphere. Thus, silicon nitrides plates with ordered hemispherical protrusions were obtained. For comparison, silicon nitride samples with as-sintered or polished flat surfaces were also fabricated. The contact angles of these specimens were measured by the sessile drop method. Pure aluminum was used as the molten metal. The sample was placed near the center of a ceramic substrate having protrusions, and the sample-substrate system was inserted in the sessile drop apparatus. The temperature was increased at a rate of 10°C/min. The temperature was slowly increased; then the aluminum sample melted at 800°C, became spherical, and then stabilized. The sample was maintained at this temperature for approximately 10 min. The measurements were conducted under an Ar gas atmosphere. Real-time images were taken using a CCD camera, and the contact angles were measured from the images.

RESULTS AND DISCUSSION

Table 1 The geometric parameters of the prepared plates

(mm)

Sample code	Diameter	Height	Spacing	Array pattern
SN-A	1.69	0.84	0.41	tetragonal
SN-B	0.8	0.4	0.16	tetragonal
SN-C	0.45	0.21	0.42	tetragonal
SN-D	0.83	0.41	0	hexagonal

Table 1 shows the geometric parameters of the prepared plates. Several parameters pertaining to the protrusions, namely, the diameter, spacing, and height of the hemispheres, were considered during fabrication. Several specimens with different macroscopic surface structures were fabricated. Typical macroscopic structures are shown in Figure 4. These were fabricated by the slip-casting method, because large sized tools used in a casting line was the target application. However, the size of the protrusions was restricted by the fabrication process, in that it was difficult to form protrusions less than 0.45 mm in diameter.

Figure 4. Top view of SiN plate geometries with spherical protrusions.

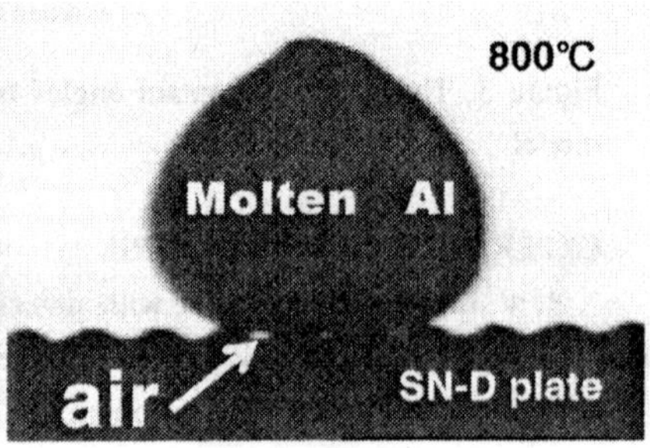

Figure 5. Typical image of a molten aluminum drop on a rough surface.

All samples were made of silicon nitride ceramics, and it was confirmed that all the surfaces were as-sintered, so that rod-like silicon nitride grains were formed. The thus obtained plates were subjected to wettability tests. A typical image observed in the sessile drop tests is shown in Figure 5. On a surface with protrusions, molten metal enters the concave portions, resulting in the formation of a curved interface; an air gap still exists between the droplet and the surface. On the micro-level, no aluminum enters into the "micro-valley", which is composed of rod-like grains that form an air gap at the valley.

Figure 6 shows the microstructure of the top of a protrusion after testing. Measurements were completed within a few minutes after ten minutes holding. As shown in Figure 6, rod-like silicon nitride grains were found. This microstructure is quite common in silicon nitrides. It should be noted that no presence of aluminum was observed between the grains. The surface of the protrusions was quite rough at the microscopic level, with grains randomly oriented on the surface. Further grain orientation control is still a problem, and the present microstructure complicates the understanding of the contact angle.

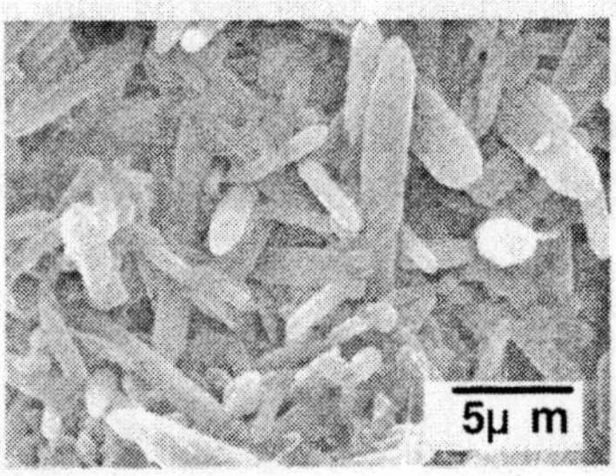

Figure 6. Microscopic structure of the top surface of a protrusion after testing.

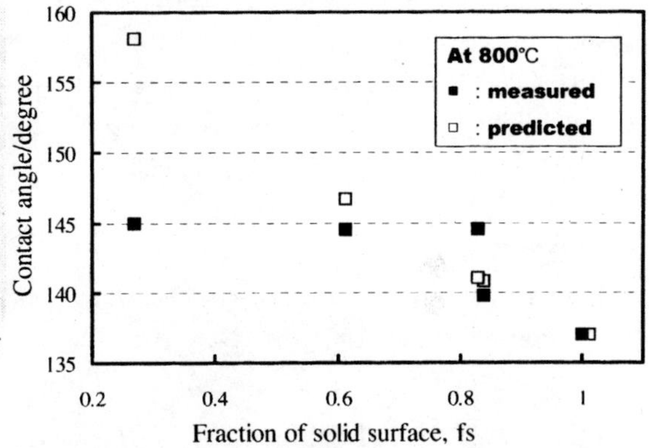

Figure 7. Contact angle for various specimens as a function of the fraction of solid surface.

Larger apparent CAs could be expected under these conditions. In the case with protrusions, air gaps were observed under all conditions, regardless of the protrusion diameter and temperature. Figure 7 shows the relationship of CAs versus f_s at a test temperature of 800 °C. In Figure 7, θ_{rc} is plotted as a function of f_s. The data plotted with open squares are predicted by Cassie's theory, and the points shown by closed squares are measured values. Measured CAs values for the surface with protrusions show higher values than those for the flat surfaces, suggesting that protrusions on the surface are effective in increasing the CA. It was shown that decreasing the contact area can reduce the adhesive force between the molten metal and the solid surface. In Figure 7, it can be seen that the experimental θ_r-f_s data has the same trend as that predicted by Cassie's theory, the values of CAs will increase when f_s decreases. However, the measured values were less subject to f_s, while f_s had a substantial influence on the predicted CAs.

Zhu et al. reported the fabrication of a controlled roughness surface with square-pillar arrays by micromachining technology, and the experimental results were closer to Cassie's theoretical predictions on super-hydrophobic surfaces than that predicted by Wenzel's theory [6)]. By contrast, this study attempted to investigate the behavior of molten metal on ceramic surfaces with large hemispherical protrusions. It was difficult to produce a complete air gap among the protrusions, as molten metal enters the concave portion, causing surface tension at the interface. At the microscopic level control, the presence of protruding grains seems effective to decrease the wettability. The effect of contact angles under these surface conditions need to be investigated with further scientific discussion.

CONCLUSIONS

Silicon nitride plates with ordered hemispherical protrusions fabricated by slip-casting were subjected to wettability tests using molten aluminum. The surfaces with protrusions exhibited a higher contact angle compared to flat surfaces. It was supposed that the contact angle depended on the fraction of solid surface, so that the experimental results were discussed using Cassie's theoretical model.

REFERENCES

[1]Information on http://www.kubota.co.jp/infra/shinsozai/default_e.html, "Ceramic Products Line Up," by New Consolidated Materials Department, KUBOTA Corporation.

[2]L. Mouradoff, A. Lachau-Durand, J. Desmaison, J. L. Labbe, O. Grisot and R. Rezakhanlou "Study of the Interaction between Liquid Aluminum and Silicon Nitride," *J. Euro. Ceram. Soc.*, **13**, 323-328, (1994).

[3]M. Yan and Z. Fan, "Review : Durability of Materials in Molten Aluminum Alloys," *J. Mater. Sci.*, **36**, 285-295, (2001).

[4]Y. Inoue, "Recent Technological Trends with Ultra-Hydrophobic Surfaces," *Tribologist*, **47** (5) 353-358, (2001), (in Japanese).

[5]K. Tsujii, "Wetting of Solid Surface and Its Fractal Structure," *Tribologist*, **44** (12) 941-946, (1999), (in Japanese).

[6]L.Zhu,y.y.Feng,X.Y.Ye and Z.Y.Zhon, Tuning Wettability and Getting Superhydrophobic Surface by Controlling Surface Roughness with Well-Designed Microstructures, *The 13th International Conference on Solid-State, Actuators and Microsystems*, Korea, 3E4.46, 1481-1484, (2005).

[7]H. John and H. Hausner, "Influence of Oxygen Partial Pressure on Wetting Behavior in the System Al/Al_2O_3," *J. Mater. Sci. Letters*, **5**, 549-551, (1986).

[8]K. Nogi, K. Ikeda, S. Shimada and K. Ogino, "Wettability of α-SiC by Liquid Pure Metals" *J. Japan Inst. Metals* , **52**, 7, 663-669, (1988).

CERAMIC SURFACE ROUGHNESS MODIFICATION USING A POLYMETHYLSILSESQUIOXANE AND SILICON OXYCARBIDE FILM COATING

Manabu FUKUSHIMA, Seiji NAKANO and Hideki KITA
National Institute of Advanced Industrial Science and Technology (AIST), 2266-98 Shimo-Shidami, Moriyama-ku, Nagoya, Aichi, 463-8560 Japan

ABSTRACT

The Roughness of a ceramic nozzle was modified using a sol-gel derived and pyrolyzed polymethylsilsesquioxane ceramic film coating. The addition of titanium acetylacetonate to the precursor, a high sol concentration, and repeating the coating process were all effective for improving roughness. These combined processes produced a smooth and flat coating film with a roughness of Ra=50nm.

1 INTRODUCTION

Small size ceramic products such as nozzles and ferrules have recently gained considerable attention due to their use in optical communication and for processing semiconductor materials. In particular, for the connection between optical fibers, the spray nozzle for the cleaning of semiconductor wafers and for the injection of gas and chemicals in a waste processing plant are main applications. These products have a micrometer or submillimeter sized through-hole, for example, to accommodate an optical fiber or fluid flow. For industrial products, high dimensional accuracy and a surface roughness below 100 nm are required for the through-hole. With a rough surface, fluid flow will be disturbed when used as spray, or the optical fiber will be damaged or fractured when inserting and connecting the fiber. Thus, a smooth through-hole is very important to prevent aforementioned problems.

For these products, zirconia is generally used, because zirconia ceramic has the good properties of low thermal conduction, machinability, wear resistance, and chemical resistance. A typical ceramic fabrication process is shown in Fig.1. In this process, cutting and polishing are very important operations that directly influence the performance of the product (e.g., to prevent the fracture of an optical fiber, or disturbed fluid flow). Additionally, polishing the through-hole is time consuming and expensive, because polishing must be

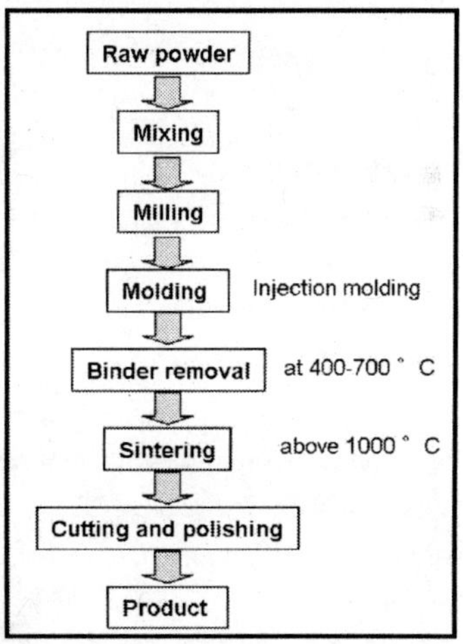

Figure 1. A typical ceramic fabrication process.

completed manually by a skilled craftsman. The polishing can account for 60-70% of the cost of these small-sized products, which makes it an important issue.

In this report, we propose an alternative method to make the through-hole surface smooth and flat using a sol-gel derived film coating [1]. The development of a smooth film coating is expected to lead to quicker delivery and significant cost reduction. Previous reports about sol-gel film coating by Kozuka et al. reported the relationship between film thickness, cracking, and processing factors such as the ratio of the alkoxide in the mixture, and additive and firing temperature [2-4]. In their reports, shape-stable thick films were obtained with the addition of organic chelating or interactive agents, and a low ratio of water/alkoxide; however, surface roughness was not measured. This previous work indicates that the formation of a shape-stable smooth film can be affected by processing factors. Among the sol-gel derived coating materials, polymethylsilsesquioxane (PMSQ) composed of unit structures in $CH_3Si(OSi)_nOR_{3-n}$ (n=0-3 and R=H or OC_2H_5) has flexible functional organic groups that are relatively stable to deformation during drying, such that, crack free self-supported films can be produced even after pyrolysis [5-6]. On the other hand, as a catalyst, titanium acetylacetonate promotes the condensation of a unit structure in PMSQ, and is likely to form well-condensed ladder-like siloxane network, which may be a suitable coating precursor to produce a smooth surface [6-7]. However, the effect of the processing factors on the surface roughness of the coating film has not been systematically investigated, and in particular, the role of catalyst or solvent and the number of coating have not been determined.

In this study, a titanium acetylacetonate chelating complex containing PMSQ and its pyrolyzed silicon oxycarbide (SiCO) ceramic coating films on zirconia nozzle were prepared, and the relationship between surface roughness and processing factors was investigated.

2 EXPERIMENTAL

Methyltriethoxysilane (MTES) provided by Shin-Etsu Chemical Co., Ltd. was used as the precursor silicon alkoxide. As the titanium source, titanium isopropoxide (TIP) $Ti(OCH(CH_3)_2)_4$ from Kojundo Chemical Laboratory Co., Ltd. was used, and was modified into a chelate complex by reacting with acetylacetone (AcAc) beforehand. A mixture of TIP and AcAc with a 1:2 molar ratio was stirred for 1 hr to obtain titanium acetylacetonate. Hereafter, this titanium source is referred to as TIP-AcAc. MTES was added to the resultant modified alkoxide solution. The mixture was stirred for co-hydrolyzing in an ethanol solution with MTES/ TIP-AcAc =1/0 or 1/0.05, MTES/ethanol = 1/3 – 1/30 and MTES/H_2O = 1/2. The sample notation is summarized in Table 1. After mixing for 0.5 hr, a zirconia nozzle or a zirconia plate with same surface roughness (Ra) of around 0.13-0.20 μm was dipped into sol solution. Then, the PMSQ coated zirconia was dried at 80°C for 1 hr, and some species were pyrolyzed at 600-800°C to convert silicon oxycarbide under flowing Ar. Multiple coatings were produced by repeatedly drying at 80°C. After mixing for 0.5 hr, a zirconia nozzle or a zirconia plate with same surface roughness (Ra) of around 0.13-0.20 μm was dipped into sol solution. Then, the PMSQ coated zirconia was dried at 80°C for 1 hr, and some species were pyrolyzed at 600-800°C to convert silicon oxycarbide under flowing Ar. Multiple coatings were produced by repeatedly drying at 80°C. To measure surface roughness, the arithmetic average roughness (Ra; μm, JIS B 0601) of the film coating on the zirconia plate was characterized using a Mitsutoyo SURFPAC. Scanning electron microscopy (SEM) was used to characterize cross sectioned through-holes of zirconia nozzles with film coatings.

Table 1 The sample notation and the ratio of mixture

Sample	Number of coating	Maximum heating temperature ( )	MTES	TIP-AcAc	HCl	Ethanol	H_2O
1	1	80	1	0	0.005	3	2
2	2	80	1	0	0.005	3	2
3	1	80	1	0.05	0.005	3	2
4	2	80	1	0.05	0.005	3	2
5	3	80	1	0.05	0.005	3	2
6	1	80	1	0.05	0.005	15	2
7	3	80	1	0.05	0.005	15	2
8	1	600	1	0.05	0.005	3	2
9	1	700	1	0.05	0.005	3	2
10	1	800	1	0.05	0.005	3	2
11	4	600	1	0.05	0.005	3	2
12	4	700	1	0.05	0.005	3	2
13	4	800	1	0.05	0.005	3	2
14	1	700	1	0.05	0.005	15	2
15	1	700	1	0.05	0.005	30	2

3 RESULTS AND DISCUSSION

3-1 The effect of condition of sol solution on surface roughness

Figure 2 shows the surface roughness of PMSQ films with or without TIP-AcAc after drying at 80°C. For reference, the roughness of a sintered surface of zirconia is also illustrated. This result shows three important points about the processing factors of (1) the addition of TIP-AcAc, (2) the number of coatings, and (3) sol concentration on surface roughness. First, the addition of TIP-AcAc has a significant effect on surface roughness and its repeatability (see sample1-2 for PMSQ and 3-7 for PMSQ with TIP-AcAc). Secondly, with an increase in the number of coating, the roughness of PMSQ

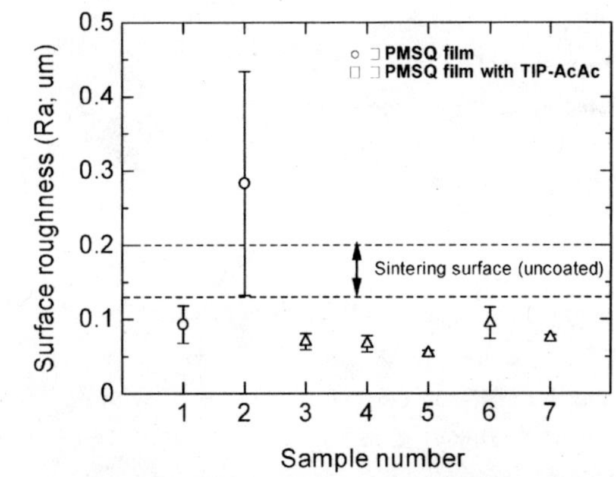

Figure 2. Surface roughness of film coating on zirconia after drying at 80°C.

film increased, while that of the PMSQ with TIP-AcAc slightly decreased. Finally, sol concentration was also found to affect surface roughness, with a lower concentration resulting in a rougher surface (compare with sample 3-5 and sample 6-7). Sample 5, with all of the above processing factors shows a decrease in surface roughness to the lowest value around 0.05 μm. These results are considered to be due to the degree of condensation of siloxane in PMSQ. The viscosity of 2.9 mm^2/s in the PMSQ with TIP-AcAc was larger than that of 2.3 mm^2/s in the PMSQ. Viscosity and molecular weight (in other words, condensation degree) are each proportional relation [8]. In fact, during the second-step coating of PMSQ on a single-coated film, the re-dissolution of initially coated film was seen. This phenomenon that the gelled PMSQ film showed re-dissolution is because of less condensation of siloxane in PMSQ. The siloxane network with residual terminal groups of silanol and ethoxy groups may interact with the ethanol solvent in next step coating solution to produce the resultant rough surface.

In contrast, for the PMSQ films with TIP-AcAc, a well-condensed siloxane network was found to be formed. The interaction between the well-condensed solution and substrate, namely the wetability, may be larger than that with less condensation, suggesting that solution with better wetability can provide the smooth surface [6, 8]. Then, it should be notable that repeatability was improved by the addition of TIP-AcAc. This is also due to the effect of well-condensation, namely, siloxane network with less reactive residual group is rigid and not significantly affected by ambient atmosphere as humidity and moisture. In addition, several coating cycles by TIP-AcAc could make a smoother surface since the repeatable coatings diminish the defect such as unexpected pore.

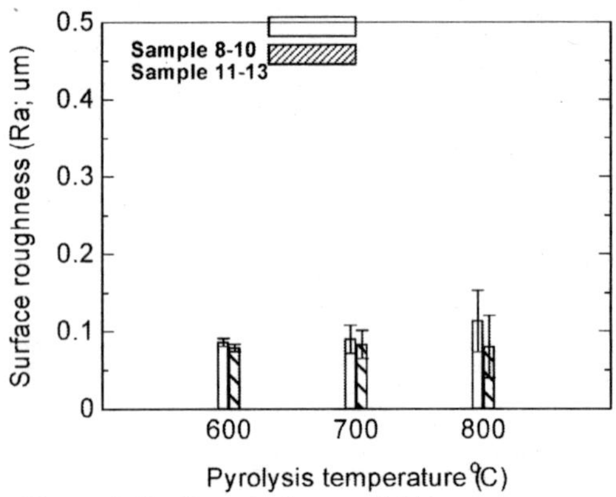

Figure 3. Surface roughness of film coatings on zirconia.

3-2 The effect of pyrolysis on surface roughness

Figure 3 shows the surface roughness of the PMSQ films with TIP-AcAc as a function of pyrolysis temperature from 600-800°C. Compared with the roughness of the unfired film with an Ra around 0.05-0.07 (in Fig.2), the roughness of the pyrolyzed film slightly increases to around 0.08-0.1. This is considered to be due to the formed micro/meso porosity due to the evolution of gas during pyrolysis [5-6]. Repeating the coating process does not show significant changes regardless of pyrolysis temperature (see sample 11-13), and the roughness is lower than with a single film coating.

3-3 Microstructural observations

Figure 4 shows SEM micrographs of (A-B) an uncoated zirconia through-hole in a nozzle, (C-D) samples 9 -10 and (E-F) samples 14-15.

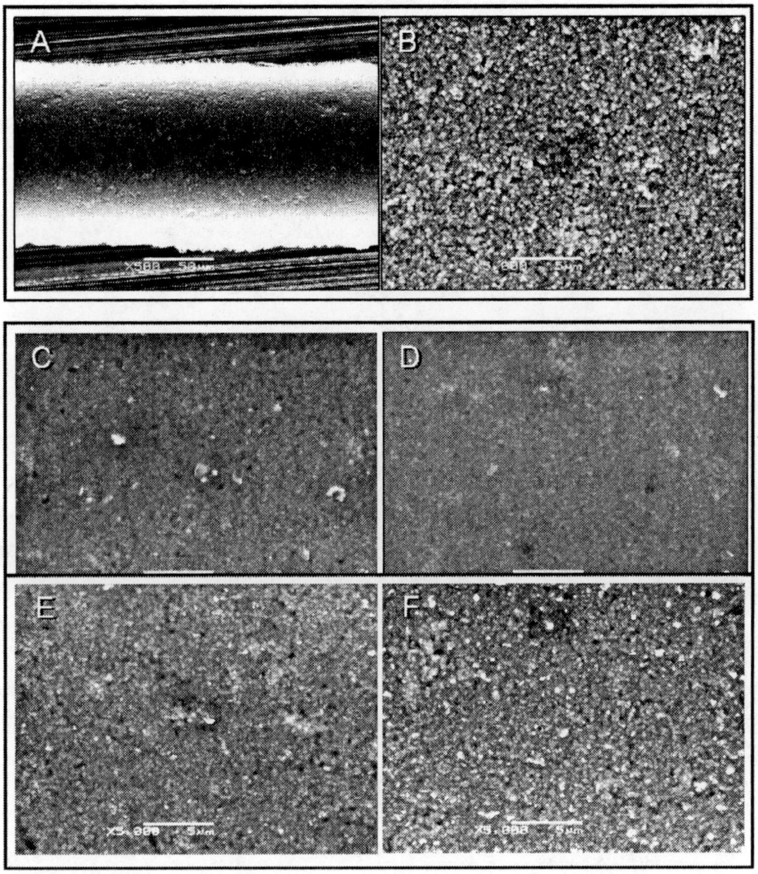

Figure 4 SEM observations of (A-B) through-hole of uncoated zirconia nozzle, (C-D) samples 9-10 and (E-F) samples 14-15.

The rough surface due to the zirconia particles was observed in the micrograph of the uncoated through-hole (A-B). In contrast, roughness of coated through-hole (C-D) was found to be modified, and no cracks or deformations are observed in the film coating, though the particles under the film seemed to slightly show through. Whether a coating film is formed or not is not distinguishable from the images for samples (E-F). A film thickness which can sufficiently fill up or cover the asperities is required for improving surface roughness. To further improve surface roughness, a thick film should be fabricated.

4 CONCLUSIONS

The roughness of a through-hole in a zirconia ceramic nozzle was modified using a pyrolyzed, sol-gel derived polymethylsilsesquioxane a ceramic coating film. The following processing factors were effective to decrease surface roughness; (1) the addition of titanium acetylacetonate, (2) a high ratio of alkoxide/solvent, and (3) multiple coatings. The addition of titanium acetylacetonate, and a high concentration of alkoxide resulted in a smoother coating film. In addition, multiple coatings also decreased surface roughness, which suggests the reduction of defects. The aforementioned processing factors were effective to decrease surface roughness and form a smooth and flat coating with an Ra=50nm.

ACKNOWLEGEMENT

The authors thank to Dr. Y.Yoshizawa and Dr. H.Hyuga at AIST for their useful discussion.

REFERENCES

[1]H.Kita, M.Fukushima and H.Hyuga, Application for Japanese patent, (2005).

[2]H.Kozuka, S.Takenaka, H.Tokita, T.Hirano, Y.Higashi and T.Hamatani, "Crack-free, thick ceramic coating films via non-repetitive dip-coating using polyvinylpyrrolidone as stress-relaxing agent", J.Sol-Gel Sci. Technol., 19, 205-209 (2000).

[3]H.Kozuka and S.Takenaka, "Single-step deposition of gel-derived PZT films: critical thickness and gel-to-ceramic film conversion", J.Am.Ceram.Soc., 85, 11, 2696-2702 (2002).

[4]H.Kozuka and A.Higuchi, "Stabilization of PVP-containing alkoxide solutions for thick sol-gel BaTiO3 films", J.Am.Ceram.Soc., 86, 1, 33-38 (2003).

[5]N.Takamura, K.Taguchi, T.Gunj and Y.Abe, " Preparation of Silicon Oxycarbide ceramic films by pyrolysis of polymethyl- and polyvinylsilsesquioxanes, J.Sol-Gel Sci.Tech., 16, 227–234 (1999).

[6]M.Fukushima, E.Yasuda, Y.Nakamura, H.Kita, H.Kawabata and Y.Tanabe, "Preparation of a transition metal containing polymethylsilsesquioxane hybrids and silicon oxycarbide ceramics -The fabrication of coating and self supported films-, J.Ceram.Soc.Jpn, 113, 210-215 (2005).

[7]S.Katayama, I.Yoshinaga, Y.Kubo, and N.Yamada, "Catalytic effect of chemically modified metal alkoxides on hydrolysis and condensation reactions of methyl- and phenyl-ethoxysilanes", J.Ceram.Soc.Jpn., 111, 743-748 (2003).

[8]Y.Abe, K.Kagayama, N.Takamura, T.Gunji, T.Yoshihara and N.Takahashi, "Preparation and properties of polysilsesquioxanes. Function and characterization of coating agents and films", J.Non-Cryst.Solids, 261, 39-51 (2000).

AN AFM STUDY OF THE INTERACTION BETWEEN AN α-ALUMINA PARTICLE AND A FLAT SAPPHIRE SURFACE IN HIGH IONIC STRENGTH ELECTROLYTE SOLUTIONS

Huseyin YILMAZ[1,2], Kimiyasu SATO[1] and Koji WATARI[1]
[1]National Institute of Advanced Industrial Science and Technology (AIST)
Anagahora 2266-98, Shimoshidami, Moriyama-ku, Nagoya-shi, Aichi 463-8560 JAPAN
[2]Gebze Institute of Technology, Materials Science and Eng. Department, Gebze –Kocaeli, TURKEY

ABSTRACT

A colloid probe method has been used to characterize the normal and lateral forces between a hydrophilic α-alumina particle and a c-sapphire surface (0001) at pH 11 in various KCl electrolyte concentrations. The normal forces are repulsive at all ionic strengths measured. The electrical double layer thickness is compressed at high electrolyte concentrations, which is evident from the decrease in the interaction distances observed in the force-distance curves. At high electrolyte concentrations, the friction forces decrease between the alumina surfaces due to the lubricating effect of the hydrated cations.

INTRODUCTION

When oxide surfaces are brought into contact with an aqueous solution, they acquire surface charges depending on the pH of the environment. Oxide colloidal particles in an aqueous solution can become charged by protonation and deprotonation reactions [1]. For α-Al_2O_3 the surface charging reaction can be written as [2]:

$$AlOH \leftrightarrow AlO^- + H^+ \qquad \text{(basic pH)}$$
$$AlOH + H^+ \leftrightarrow AlOH_2^+ \qquad \text{(acidic pH)}$$

The surface charge is balanced by dissolved counter-ions which re-distribute themselves and spread away from the surface, forming an electrical double layer (EDL). The EDL thickness, usually named as the Debye length, is strongly dependent on the electrolyte concentration. At low ionic strength the Debye length extends far enough to cause repulsion and dipersion of powders, whereas at high ionic strength due to the suppression of the double layer, the attractive van der Waals forces dominate the slurry behavior and thus flocculation occurs. For monovalent salt solution (like alkali halides, KCl) the Debye length is calculated as [1]:

$$\kappa^{-1} = \frac{0.304}{[KCl]^{\frac{1}{2}}} \qquad (1)$$

where [KCl] is the molarity (mol/l) of alkali halide in the bulk solution.

Beside the two forces mentioned above which the DLVO theory accounts for, non-DLVO forces, like hydration forces, hydrophobic forces etc, have also been reported in the literature [1-3]. The hydration force is the repulsive force observed in removing the hydration layer between interfaces and hydration layer is the structuring of water molecules at the interface [1,3]. Since the structuring is confined to a few nanometers or less [3], it is effective at very short separation distances. In addition, the presence of water molecules tightly bound to ions lead to a strong repulsion when two surfaces approach each other [3]. At high electrolyte salt concentrations it can dominate the attractive van der Waals and repulsive double-layer forces, especially at high salt concentrations (>0.1 M) [3].

Alumina colloids showed unusual stability behavior which was attributed to the non-DLVO forces operating at distances <5 nm [4]. There is a doubt about the origin of this repulsive force. It might be due to the hydration layer or due to the gel layer formation at the aqueous interface [4]. Depending on the aging conditions, a gel layer thickness ~15 nm have been reported at pH~8 [4].

Ducker and Colic et al. were able to prepare weakly flocculated alumina suspensions with tuned rheological properties [5,6]. The rheology of the flocculated alumina suspension was different from that of coagulated one due to the modified hydration layer by the presence too much hydrated cations at the interface [7]. The more hydrated cations of indifferent electrolyte at the hydrophilic interfaces, the more lubrication effect was observed [5,6]. Therefore, lower viscosity, shear modulus and yield stress was observed for alumina particle network [5,6]. The rheology of alumina slurry and powder compact must be directly related to the pairwise interaction of particles, that is the normal and frictional forces among them [8].

In this study, ionic strength dependence of normal and lateral forces between α-alumina particle and c-sapphire surface (0001) were investigated at basic pH conditions using the colloid probe method. Scanning Probe Microscope (SPM) cantilever modified with an α-alumina particle was used in measuring the normal and lateral interaction forces between a sphere and a flat surface. The aim was to show that the rheology of alumina slurry and powder compact is directly related to the normal and friction forces between particles [8]. The rheological study is in progress.

MATERIALS AND METHODS

Colloid Probe Preparation

The as received α-alumina (Admatechs Co., Japan; d_{mean}=9.9 μm) particles were first washed repeatedly in a 1 wt% sodium hexametaphosphate ($Na_6P_6O_{18}$) solution in an ultrasonic bath in PE bottles to separate out unwanted small particles and then re-suspended and stored at pH 3.5 for further use [9]. Colloid probes were prepared by picking a spherical α-alumina particle of size ~12.8 μm by a micromanipulator (Model M501-1202-M; Suruga Seiki Co. Ltd., Japan) which is then glued (Araldite AR-R30-Nichiban Co. Ltd., Japan) to the tipless AFM cantilever (TL-FM-50; Nanosensors, Switzerland) with a stiffness of 1.9 N/m. Just prior to force measurements, the alumina colloid probe, c-sapphire substrate (As One Corporation, Japan) and the SPM liquid cell were UV-plasma (PL16-110D-Sen Light Corp., Japan) treated to remove surface organics and make the surface hydrophilic. KCl was used to adjust the strength of the electrolyte solution. KOH was used to adjust the pH of the electrolyte.

A Scanning Probe Microscope (SPA 400, Seiko, Japan) with a liquid cell was used for normal and lateral force measurements. The α-alumina colloid probe and c-sapphire surface were equilibrated in the electrolyte for 30 mins before taking any measurement. Lateral friction forces over a trace distance of 500 nm were measured and were analyzed as described in literature [10].

RESULTS AND DISCUSSION

Force Curves on Approach

The measured normal forces between α-alumina probe and c-sapphire surface at pH 11 in 10^{-3} to 10^{0} M KCl electrolyte solution were shown in Figure 1. It is more convenient to express the interaction as normalized force curves ($F_N/2\pi R$) because the area over which the interaction takes place affects the force, where F_N is the normal force and R is the probe radius [11]. Only when the separation distance was less than ~20 nm a strong dispersion force emerged.

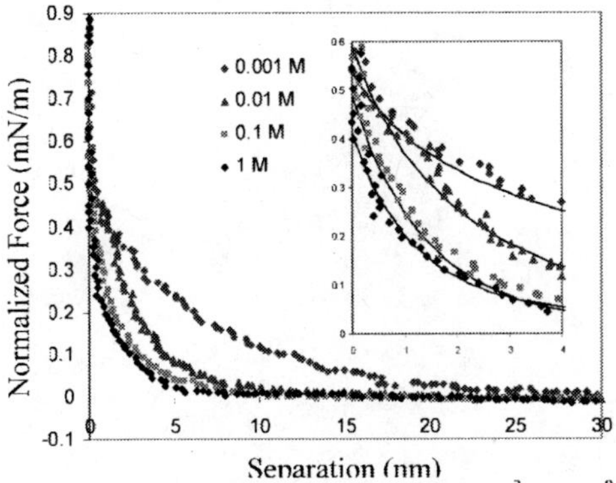

Figure 1. Normalized Force vs. distance curves at pH 11 in 10^{-3} to 10^{0} M KCl electrolyte solution. Insert shows the fit to the double exponential function.

The interaction distance of colloidal particles on approach strongly depends on the electrolyte concentration as shown in Figure 1. As the electrolyte concentration increased from 10^{-3} to 10^{0} M the repulsive interaction distance decreased from ~20 nm to ~8 nm. On normal approach repulsive force dominated the attractive van der Waals forces either due to the hydration layer or gel layer as reported previously [2,5].

The electrostatic interaction force between the probe tip and sapphire surface was modeled by using a linearized form of Poisson-Boltzmann equation for a symmetric 1:1 electrolyte, in which diffuse double layer potential and Debye lengths were used as a fitting parameter. Models used to fit experimental force-distance curves for constant surface charge condition was:

$$\frac{F_N}{2\pi R} = \varepsilon\varepsilon_0\kappa\left[\frac{2\Psi_1\Psi_2 e^{-\kappa H}}{1-e^{-\kappa H}} + (\Psi_1 - \Psi_2)^2 \frac{e^{-2\kappa H}}{1-e^{-2\kappa H}}\right] \tag{2}$$

where Ψ_1 and Ψ_2 are surface potentials of the alumina probe and c-sapphire surface respectively, and H is the separation distance [12]. The non-retarded van der Waals force (F_{vdW}) [1] was calculated using Equation 3:

$$\frac{F_{vdW}}{2\pi R} = -\frac{A}{6H^2} \tag{3}$$

where A is the non-retarded Hamaker constant which was taken as 5.1×10^{-20} J [11,13].

The results shown in Figure 2 suggest that the DLVO theory described the interaction forces between surfaces for separation distances higher than ~4 nm, as long as the electrolyte concentration was <0.1 M KCl. The constant charge model fitting results were given in Table I along with the fitted Debye lengths. The observed Debye lengths were found to be in a reasonable agreement with that calculated for the electrolyte level. The Debye lengths were found to decrease from 9 to 2 nm with the increase of electrolyte concentration.

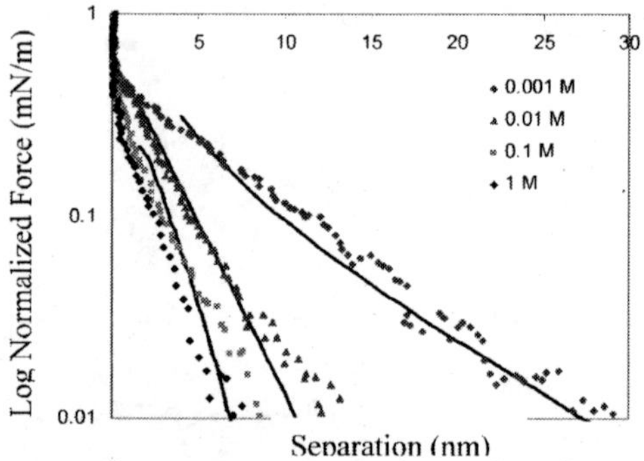

Figure 2. Normalized force vs. distance curves for the interaction of an SPM cantilever probe with a flat c-sapphire substrate immersed in KCl electrolyte solution. DLVO fit to the same data using the constant-charge boundary condition.

Table I. Constant charge boundary conditions fitting parameters and calculated Debye lengths from electrolyte concentrations.

Electrolyte Conc. (M)	Ψ (mV)	κ^{-1} (nm, from fit)	κ^{-1} (nm, calculated)
10^{-3}	-35	9	9.6
10^{-2}	-32	3	3.04
10^{-1}	-32	2	0.96

Short Range Forces

Although the classical DLVO theory fits the measured force distance curves very well down to separations of ~4 nm, the discrepancy between the theory and the experimental data become more and more significant at very close separations. Since the classical DLVO theory considers only two component, i.e., van der Waals force and electrostatic force in the system, this discrepancy may come from another force which exists only in short range. The experimental data were fitted to the extended DLVO theory, by including hydration force as the third force component into the classical DLVO theory.

Table II gives the exponential decay lengths (λ_1), and pre-exponential constants obtained (C_1) from the force distance curves fitted to the empirical double exponential function [1]:

$$F_H = C_1 \exp\left(-\frac{H}{\lambda_1}\right) + C_2 \exp\left(-\frac{H}{\lambda_2}\right) \tag{4}$$

Since only short distance structural force is assumed, the second terms were neglected.

The exponential decay lengths (λ_1) did not change with ionic strengths and stayed constant as expected. Whereas, the pre-exponential factor (C_1) calculated as 0.15 mN/m in 10^{-3} M KCl electrolyte solution increased and leveled off at 0.33 mN/m in 10^{0} M one. The pre-exponential constants given in Table II for the 10^{-1} and 10^{0} M electrolyte case did not differ much. The reason was that the force distance curves for the two cases were very close to each other.

Table II. Exponential decay lengths (λ_1), and pre-exponential constants obtained (C_1) from the force distance curves fitted to the empirical double exponential function

Electrolyte Conc. (M)	C_1 (mN/m)	λ_1 (nm, from fit)
10^{-3}	0.15	1.1
10^{-2}	0.24	1.1
10^{-1}	0.31	1.1
10^{0}	0.33	1.1

Lateral Forces

Shear interaction between α-alumina particle and c-sapphire surface (0001) were investigated from 10^{-3} to 10^{0} M KCl electrolyte concentrations at pH=11. The observed asymmetry in the friction loops indicate that there was contribution from surface roughness to the friction force [10]. All friction measurements were averaged in both scan directions.

The lateral friction (F_L) vs. applied normal (F_N) force curves at four different electrolyte concentrations were given in Figure 3. The friction behaviors were analyzed assuming that the friction force between two surfaces were proportional to the normal force obeying Amantons Law.

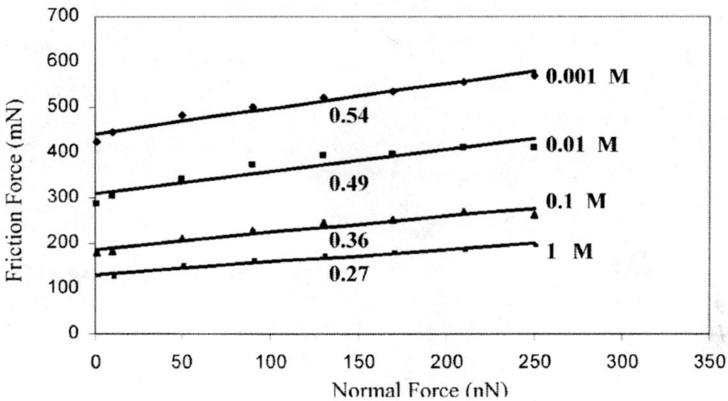

Figure 3. The lateral friction (F_L) vs. applied normal (F_N) force curves at four different electrolyte concentrations.

The micro-scale friction coefficients between alumina surfaces were calculated by fitting the Amantons law to the friction force vs. normal force data, where effective friction coefficient μ_{eff} was defined as $\mu_{eff}=(F_L/F_N)$. The effective friction coefficients were calculated as 0.54, 0.49, 0.36, and 0.27 at 10^{-3}, 10^{-2}, 10^{-1}, 10^{-0} M KCl electrolyte concentrations, respectively The friction curves at higher electrolyte concentration showed better fits to the Amantons Law. A significant lubrication effect was clear for solutions of high electrolyte concentrations which is consistent with the data given in literature [6].

CONCLUSIONS

Normal and lateral friction forces between an α-alumina colloid particle and a flat c-sapphire surface in a KCl electrolyte solution at pH=11 were measured using a colloid probe technique. At all separation distances greater than 4 nm, repulsive forces were measured that are described well by DLVO theory. At smaller separation distances, an additional repulsive force was measured that is attributed to a hydration and/or gel like layer on the surface of alumina. Analysis of friction behavior was done assuming Amanton's law was obeyed. Friction coefficients are lower in concentrated KCl electrolyte solutions compared to dilute ones.

REFERENCES

[1]Intermolecular and Surface Forces, J. N. Israelachvili, Academic Press: London, Second ed. (1991).

[2]B. K. Hordern, "Chemistry of Alumina, Reactions in Aqueous Solution and its Application in Water Treatment," Advances in Colloid and Interface Science, **110**, 19-48 (2004).

[3]J. N. Israelachvili, G. E. Adams, "Measurement of Forces Between Two Mica Surfaces in Aqueous Electrolyte Solutions in the Range 0-100 nm," J. Chem. Soc., Faraday Trans. I, **74**, 975-1001 (1978).

[4]M. E. Karaman, R. M. Pashley, T. D. Waite, S. J. Hatch, and H. Bustamante, "A Comparison of the Interaction Forces Between Model Alumina Surfaces and their Colloidal Properties", Colloids and Surface A: Physicochemical and Engineering Aspect, **129-130**, 239-55 (1997).

[5]W. A. Ducker, Z. Xu, D. R. Clarke, and J. N. Israelachvili, "Surface Forces between Alumina Surfaces in Salt Solutions: Non-DLVO Forces and the Implications for Colloidal Processing," Journal of the American Ceramic Society, **77**, 437-43 (1994).

[6]M. Colic, and M. L. Fisher, "Effect of the Size and Affinity for Water of Excluded Ions on Interparticle Forces," Chemical Physics Letters, **291**, 24-30 (1998).

[7]B. V. Velamakanni, J. C. Chang, F. F. Lange, and D. S. Pearson "New Method for Efficient Colloidal Particle Packing via Modulation of Repulsive Lubricating Hydration Forces," Langmuir, **6**, 1323-25 (1990).

[8]G. V. Franks, and F. F. Lange, "Plastic Flow of Saturated Alumina Powder Compacts: Pair potential and Strain Rate," Particle technology and Fluidization, **45**, 1830-35 (1999).

[9]Polat, M, Sato, K., Nagaoka, T., and Watari, K., "Measurement of Interaction Forces between Oxide Surfaces by the Atomic Force Microscopy-Colloid Probe Method and Comparison with the Theory," Journal of Ore Dressing, to be published.

[10]R. G. Cain, S. Biggs, and N. W. Page, "Force Calibration in Lateral Force Microscopy," Journal of Colloid and Interface Science, **227**, 55-65 (2000).

[11]Derjaguin, B.V., "Untersuchungen über die Reibung und Adhäsion," Kolloid Zeitschrift, **69**, 155-164 (1934).

[12]J. Drelich, J. Long, Z. Xu, J. Masliyah, J. Nalaskowski, R. Beauchamp, and Y. Liu, "AFM Colloid Force Measured Between Microscopic Probes and Flat Substrates in Nanoparticle Suspensions", Journal of Colloid and Interface Science, in press (2006).

[13]L. Meagher, G. W. Franks, M/ L. Gee, and P. J. Scales, "Interaction Forces Between α-Alumina Fibers in Aqueous Electrolyte Measured with an atomic Force Microscope," Colloids and Surfaces A: Physicochemical and Engineering Aspect, **146**, 123-37 (1999).

ENERGY AND ENVIRONMENT

COHESIVE DUST CAKE FORMATION ON A CERAMIC TUBE FILTER AS THE MAIN CAUSE OF A PRESSURE DROP INCREASE IN A HIGH TEMPERATURE GAS CLEANING PROCESS

Nobuhiro MISAWA
Electric Power Development Co., Ltd.(J-POWER), 6-15-1, Ginza, Chuo-ku, Tokyo 104-8165, JAPAN

Hiroshi SASATSU*1, Shinichi SAKUNO*1 and Hidehiro KAMIYA*2
*1 Electric Power Development Co., Ltd.(J-POWER), 6-15-1, Ginza, Chuo-ku, Tokyo 104-8165, JAPAN
*2 Institute of Symbiotic Science and Technology, Tokyo University of Agriculture & Technology, Koganei, Tokyo 184-8588, JAPAN

ABSTRACT

In order to determine the cause of the increase in the pressure drop across a ceramic tube filters (CTF) when changing the coal feed in a pressurized fluidized bed combustor (PFBC), the microstructure and chemical compound distribution of the remnant dust cake on the filter, sampled just after the plant shut-down phase of the operation, was observed using scanning electron microscopy (SEM) and electron probe microanalysis (EPMA). The particle size distribution of the dust cake was compared with that of the ash particles from the combustion of different types of coal during stable operation. Using EPMA, it was found that the main components of the eutectic phase with low melting points, such as Na and K, were not concentrated in the dust cake layer. We focused on the volume based ratio of fine particles (between 1 and 10 μm) to coarse particles (larger than 20 μm), and this ratio was calculated for each ash particle. In the case of Blair Athol (BA) coal, in which the fine particle ratio was relatively low, the pressure drop was low, and the PFBC was operated stably. When Drayton (DT) coal was combusted, in which the fine particle volume ratio was twice as high as that of BA coal, the pressure drop increased, and a cohesive dust cake was formed on the filter surface.

INTRODUCTION

Ceramic tube filters (CTF) were installed in a pressurized fluidized bed combined cycle (PFBC) demonstration plant located in the former Wakamatsu Coal Utilization Research Center of J-POWER in order to thoroughly remove fine dust particles from the high temperature and high pressure flue gas. The demonstration test was carried out for five years, which were divided into two phases—Phase-1 (1994–1997) and Phase-2 (1998–1999)—depending on the configuration of the dust removal system[1]. In Phase-1, the CTF was located downstream of the primary cyclones. Initially, a high and unstable pressure drop (30 kPa) across the CTF, which exceeded the design value, was observed. To solve this issue, a "cyclone by-pass system," which sends the coarser ash to the CTF, was installed. As a result, the pressure drop was successfully reduced and was stabilized over an extended duration[2]. In Phase-2, the primary cyclones were replaced by ash re-circulation cyclones which returned the collected fly ash to the bottom of the fluidized bed to improve the boiler performance. Four kinds of bituminous coal, a type of sub-bituminous coal, and petroleum coke were tested in Phase-2. When Drayton (DT) coal and Australian bituminous coal were burned, a continuous increase in the pressure drop across the CTF was observed during the partial load operation. The temperature of the flue gas at the inlet of the CTF was approximately 630°C. Although the melting point of fly ash—mainly composed of SiO_2, Al_2O_3, Fe_2O_3 and $CaSO_4$—is normally more than 1,000°C, many studies were conducted on the unusual pressure drop phenomena below the above mentioned temperature (630°C) caused by the dust collection in the high temperature dust removal system. These studies pointed out that the mechanism of pressure drop increase depended on the physical and chemical properties of the dust even in relatively low temperature conditions in the range of approximately 600–700°C[3,4,5].

In this study, we focused on the increase in pressure drop caused by changing the coal feed in the CTF demonstration plant. The increasing pattern of the pressure drop is discussed using the microscopic observations of the accumulated dust cake sampled just after the plant shut-down phase of the operation. We attempted to identify the features of the coal that contributes to form a cohesive dust cake on the CTF tube surface by analyzing the sampled dust cake and considering the formation behavior.

EXPERIMENT

1. PFBC System Flow

Figure 1 shows the system flow of the Phase-2 configuration. A fluidized bed boiler was installed inside a pressure vessel. Crushed coal with a size of below 6 mm (average: 500–600 μm) was mixed with limestone and water to make a fuel paste and was injected into the boiler. The boiler contained 60 t of bed material (BM) at a rated load. The BM mainly consisted of injected limestone with a size of less than 5 mm as the desulfurization sorbent. The combustion temperature of the operation was approximately 860°C independent of the load. The combustion pressure was approximately 1 MPa at the rated load and 0.6 MPa at 50% load.

Fly ash particles with sizes larger than 100 μm were collected by ash re-circulating cyclones, and were returned to the bottom of the fluidized bed. The particles that were smaller than 100 μm passed through the cyclones and were collected by the CTF.

2. Outline of the CTF System

Figure 2 shows the CTF system. This system basically consisted of two pressure vessels, ceramic tube filters in the six compartments, and reverse cleaning and other auxiliary equipment. The flue gas sent into the top chamber was distributed among 81 tubes. It flowed down along the filtrating tubes and then was sent to a gas turbine through the regenerators and ejectors and then through a hot gas pipe. The filter tube supplied by the Asahi Glass Company consisted of a β-cordierite filter material. The reverse cleaning equipment was installed in order to remove the dust accumulated on the internal surface of the filters. Compressed air was supplied as a cleaning media to the reservoir, which was automatically controlled to be approximately 1.5 MPa higher than the process pressure. The interval time for the six compartments was usually 9 min.

3. Dust Cake and Fly Ash Sampling and Analysis

The dust cakes adhered to the surface of the CTF tubes were sampled just after the plant shut-down phase of the operation, during which the DT coal was tested and the increase in pressure drop was observed. The dust cake was sampled from the tube that was located horizontally near the center of the upper compartment of the vessel and vertically near the top-end of the tube (Figure 2).

For reference purposes, we analyzed the fly ash of five types of coal feed—BA coal, DT coal, mixture of DT and BA coals, mixture of BA and Adaro (AD) coals, and mixture of BA coal and petroleum coke (PC) —collected by the CTF during the same operation. The collected fly ash on the CTF was detached from the filter surface by a reverse pulse flow and was transported by the flue gas and cooled to approximately 80°C in ash coolers. The pressure was then reduced in a pressure reduction facility and cooled again in a cooler to atmospheric conditions. The fly ash was sampled at the outlet of the cooler. The fly ash flow was measured by a load cell installed to a fly ash hopper located downstream of the cooler. The particle-size distribution was analyzed by the laser diffracting method. The dust cakes were crushed roughly by using mortar before the measurement of particle-size distribution.

RESULTS AND DISCUSSION

1. Pressure Drop Increase during DT Coal Operation

Figure 3 shows an example of the pressure drop and gas temperature profiles at the CTF inlet, when a typical pressure drop increase was observed by changing of the coal feed. When the fuel coal feed was changed from BA alone to a blend of BA + DT, the pressure drop increased by approximately 3 kPa. It is estimated that this increase of pressure drop occurred because of the increase in the fly ash flow rate. When the coal feed was changed from BA + DT to DT alone, the pressure drop increased by approximately 5 kPa. Because the measured fly ash flow rate did not increase by changing the coal feed from BA+DT to DT alone, it seems that the pressure drop increase was not caused by the increase in the fly ash load. After the DT coal feed, BA coal was tested again. However, the pressure drop was still 5 kPa higher when compared with the BA coal test performed earlier (on Sep. 27, 1999). This 5 kPa increase caused by changing the coal feed is almost identical to the pressure drop increase in the coal test when the feed was changed from a mixture of BA + DT to DT alone. Since this pressure drop increase of 5 kPa when compared with the normal operation had continued till the plant shut-down, it seems that the dust layer of DT coal ash continued to remain on the filter and caused the increase in the pressure drop of approximately 5 kPa.

2. Analysis of the Dust Cake Property

Table 1 shows the result of a componential analysis with the data of fly ash for five types of coal obtained during the same operation. Although the dust cake contained less $CaCO_3$ and CaO and more $CaSO_4$ in percentage base, the percentage of alkali compounds (Na_2O and K_2O) or Fe_2O_3, which is generally considered to cause the decrease in the ash melting point, was almost the same as that of the other fly ash samples. From the result of X-ray diffraction analysis, SiO_2 (quartz) and $CaSO_4$ (anhydride) were detected, but both compounds normally exist in fly ash.

Figure 4 shows the comparison between the particle size distributions of the dust cake on the filter, after over a long period operation, and that of the original fly ash. The original fly ash particles of each coal (BA, BA + AD, DT, BA + DT, BA + PC) were obtained by the sampling of ash after being cooled and depressurized. The mode of particle diameter of the fly ash samples of each coal type was approximately 60 μm, and the size distribution showed a relative wide peak ranging from 30 to 100 μm. On the other hand, the size distribution of the dust cake has showed a bimodal pattern in which the smaller peak was

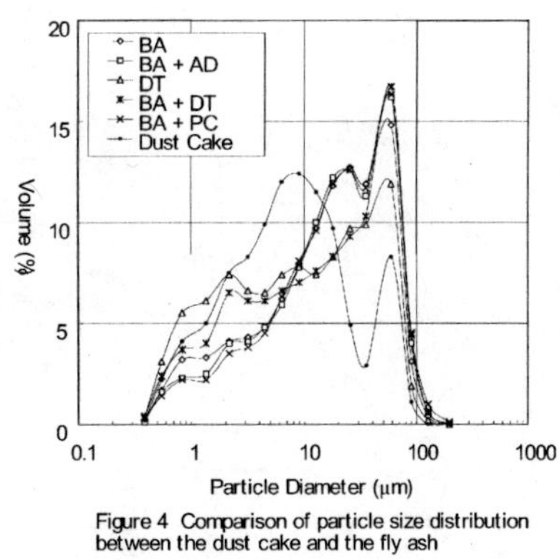

Figure 4 Comparison of particle size distribution between the dust cake and the fly ash

less than 10 μm and the larger peak was approximately 60 μm. The volume fraction of the fine particles with diameters of less than 10 μm was higher than that of the coarse particles with diameters ranging from 20 to 60 μm. Fine particles with diameters of less than 10 μm were separated and condensed in the accumulated dust layer on the filter surface after dust detachment. Figure 4 also shows that the particle size distribution depends on the types of coal. The fly ash of DT coal combustion included a higher ratio of fine particles with diameters of less than 10 μm as compared to BA. Since the fine particle ratio of DT is higher than that of other types of coal, it seems that the pressure drop with the change of coal feed from BA to DT and the fine particle ratio in the accumulated dust layer were increased by the fine particles in the DT ash.

3. Morphology of the dust cake

Both sides of the surface of the dust cake were observed by using a SEM to investigate the morphology of the particles as shown in Figure 5. On the free surface, the particle shape and morphology was maintained, and the size distribution of the observed particles was almost the same as that of the fly ash in Figure 4. On the other hand, at the interface between the dust cake and the filter surface, the shape of the particles was so unclear that it was difficult to identify each particle. Almost all the particles had sizes of less than several micrometers and each particle had a packed dense structure with few inter-particle pores.

Figure 6 shows the SEM analysis on the cross section of the dust cake. In the full view of the image 6(A), the thickness of the dust cake was observed to be approximately 170 μm, and the particles were attached at a thickness of 140 μm from the free surface. It is observed that each particle is clearly the same as that on the free surface shown in Figure 5(A). However, in an area of 10 μm from the interface of the filter, fine particles were observed identical to those shown in Figure 5(B). In the expanded view 6(B), near the filter side, a dense layer composed of agglomerated fine particles with diameters of less than 2 μm was observed.

An EPMA analysis of the dust cake was conducted on the elements such as Si, Al, Ca, S, Fe, Na and K at the cross-section and the surface of the dust layer at the interface of the filter. The main elements of the eutectic phase with relatively low melting points, such as Na or K, were not concentrated in the dust layer near the interface of the filter.

Although the gas temperature at the CTF inlet was in the range 620–630°C during the DT coal test, sintering of the fly ash particles from the PFBC boiler at such temperatures was

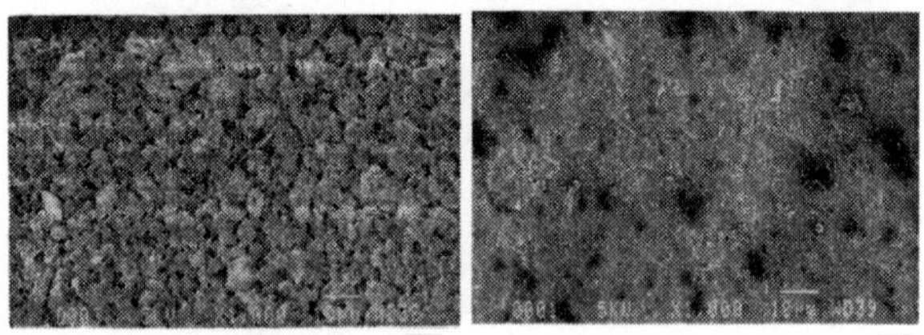

(A) Free surface 10μ m (B) Filter surface 10μ m

Figure 5 SEM images of dust cake

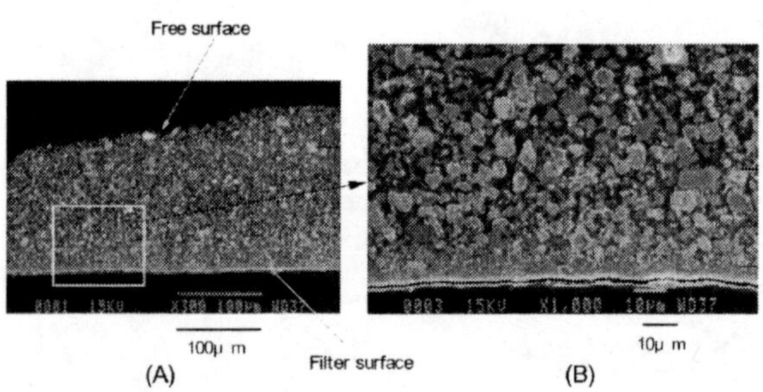

Figure 6 SEM images of dust cake (cross section)

reported[6]. In addition, as shown in Table 1, the dust cake contained a higher ratio of $CaSO_4$ as compared to other fly ash samples. The formation of a liquid phase on the tube surface in the presence of sulfate was also considered to be among the factors that promote sintering [3,4].

3. Dust cake formation

The effect of coarse particles on the pressure drop across the CTF has already been observed in the demonstration plant and studied. By the addition of coarse particles with sizes larger than several tens of micrometers to the fine fly ash, the effect of reverse cleaning on dust detachment was improved[2]. To prevent a strong dust layer formation and to improve the stability of the CTF operation, it is important to increase the ratio of coarse particles in ash.

The volume fraction ratios of fine/coarse particles in the fly ash were calculated and shown in Table 2 (for the fly ash of the BA and DT coal test). Since this ratio of BA was relatively low—0.75 for particles with sizes below 10 μm—the pressure drop was low as shown in Figure 3. On the contrary, for DT coal, this ratio was approximately 1.51. According to this result, as shown in Figure 4 and Figure 6, the ratio of fine particles to coarse particles was considered to show the degree of potential for a dust cake formation.

Table 2 Ratio of fine particles in fly ash

	BA	DT
$\frac{<1\mu m}{>20\mu m}$	0.11	0.26
$\frac{<5\mu m}{>20\mu m}$	0.48	1.06
$\frac{<10\mu m}{>20\mu m}$	0.75	1.51

Load: 50%
Volume Basis

CONCLUSION

To analyze the pressure drop profiles by changing the coal feed from BA to DT, the microstructure of dust cake sampled from a CTF tube surface was observed. The major difference between the dust cake and fly ash in normal operation was with respect to the particle size distribution. The dust cake was mainly composed of particles with sizes in the range 2–20 μm, having the mode at 10 μm. The primary particle size in the dust layer at the interface of the filter surface was smaller than 2 μm. Fine particles were considered to be agglomerated on the tube surface because the ratio of fine particles to coarse particles in the fly ash of the DT coal test was larger than those of the other fly ash samples. Consequently, these fine particles were considered to be sintered mainly in the presence of sulfate.

REFERENCES

[1]N. Misawa, S. Sakuno, H. Goto, H. Sasatsu, H.Ueda, and K. Tazawa, Development of an Ash Recirculation PFBC technology, Proceedings of the 16th International Conference on Fluidized Bed Combustion, ASME, No.0071 (2001)

[2]H. Sasatsu, R. Abe, T. Harada, N. Misawa, and I. Mochida, Countermeasures to Decrease and Stabilize Pressure Drop across Ceramic Tube Filter, JIE, **78**, 988 (1999)

[3]P. Hancock, Cracking, Spalling and Failure of Ash Deposits in Filtration Systems, Proceedings of 4th International Symposium on High Temperature Gas Cleaning, **2**, 3 (1999)

[4]J. P. Hurley, B. A. Dockter, S. E. Allan and V. Kühnel, Predicting the Relative Stickiness of Ash in PFBC Hot-Gas Filter System, Proceedings of 4th International Symposium on High Temperature Gas Cleaning, **2**, 25 (1999)

[5]H.Kamiya, A. Kimura, T. Yokoyama, M. Naito and G Jimbo, Development of a split-type tensile-strength tester and analysis of mechanism of increase of adhesion behavior of inorganic fine powder bed at high-temperature conditions, Powder Technology **127**, 239-245 (2002)
[6]B-M. Steenary, O. Lindqvist and V. Langer, Ash sintering and deposit formation in PFBC, Fuel, **77**, No.5, 407 (1998)

SIMULATION OF THE FILTRATION PROCESS FOR A FLAT FABRIC FILTER

*[1]Yuping Yao, *[2]Ning Mao, *[3]Masashi Wada, *[4]Hidehiro Kamiya and *[3]Chikao Kanaoka
*[1]Zhejiang Feida Environmental Engineering Co., Ltd., Zhuji, Zhejiang, 311800 China
*[2] Northeastern University, Shenyang, 110004 China
*[3]Ishikawa National College of Technology, Ishikawa, 929-0392 Japan
*[4]Graduate School of Bio-Applications and Systems Engineering, Tokyo University of Agriculture and Technology, Tokyo, 184-8588 Japan

ABSTRACT

A model is proposed to simulate the filtration process for a flat fabric filter. In contrast with previous models, the proposed model provides a new expression for the residual dust distribution on the filter, and the cleaning performance of a fabric filter in which dust accumulation on the surfaced and inside the filter is considered. The pressure drop through the fabric filter during the filtration process can be simulated by the model, and the filter performance can also be described and predicted. Simulation results are compared with experimental data, and it is shown that the simulations agree well with the measured changes in pressure drop, and the filtration cycle time for a filtration process.

INTRODUCTION

Fabric filtration is widely used as one of the efficient measures for controlling particulate emissions. When particles are separated from a gas stream by a fabric filter, the captured particles accumulate both on the filter surface and inside the filter, resulting in an increase in the pressure drop across the filter. Therefore, it is necessary to clean the filter periodically in order to maintain continuous operation. Pulse jet type cleaning is one of the most popular methods for this purpose.

Several models have been proposed to simulate the filtration process, and to predict the performance of filters. Due[1] presented a model to describe the pressure drop across rigid gas filters in non-steady-state. Ju[2~3] developed a simplified model to calculate the flow rate through a fabric filter in steady state. Both of the above mentioned models assume that the residual dust just accumulates on the surface of a filter, and that the dust cake on part of the filter surface is not removed during cleaning, which is called patchy cleaning.

For fabric filters, filter cleaning usually results from the movement of filter itself, such as by the deceleration of filter subjected to a pulse jet. It has been observed that the dust cake on the surface of fabric filters is cleaned more or less in laboratory experiments, and that some dust can penetrate into the inner part of a fabric filter. Based on the above observed phenomena, an improved model has been proposed to simulate the filtration process for fabric filters in this paper. Results from the proposed model are compared with the filtration experimental data. The simulation results are consistent with the experimental results.

SIMULATION OF THE FILTRATION PROCESS FOR A FLAT FABRIC FILTER

Description of a filtration process

In a previous paper[4], the authors reported that filtration behavior in one filtration cycle can be divided into 3 stages:, the inner, transient, and surface filtration stages. The inner filtration stage occurs just after dust cleaning. Particles mostly penetrate into the fabric, and are captured and accumulate in it. As a result, this narrows the flow space for gas through the fabric, and pressure drop of the filter increases non-linearly with filtration time. In the surface filtration stage, particles collect and accumulate on the filter surface, and the pressure drop through the filter increases linearly with time since captured particles form a uniform dust layer. The transient stage occurs between the inner and surface filtration stages, and particles are captured both inside and on the surface of the fabric simultaneously. Hence, the pressure drop gradually shifts from a non-linear to a linear increase with time. Periodic filling gaps resulting from patchy cleaning of particles from the filter surface are also included in this stage.

Assumptions for simulating the filtration process

To simulate the filtration process, the following assumptions are made:

1) the filter medium is uniform,

2) gas flow through the filter is constant,

3) the pressure drop across the filter is homogeneous, and .

4) after the filter is cleaned, some areas of the filter are completely cleaned and other areas are only partly cleaned (i.e., some residual dust remains on/in the filter).

The ratio of reduced mass of dust cake to the mass before cleaning is defined as α. The ratio of completely cleaned area to total area is defined as f (as shown in Fig.1). L represents the thickness of dust inside the filter.

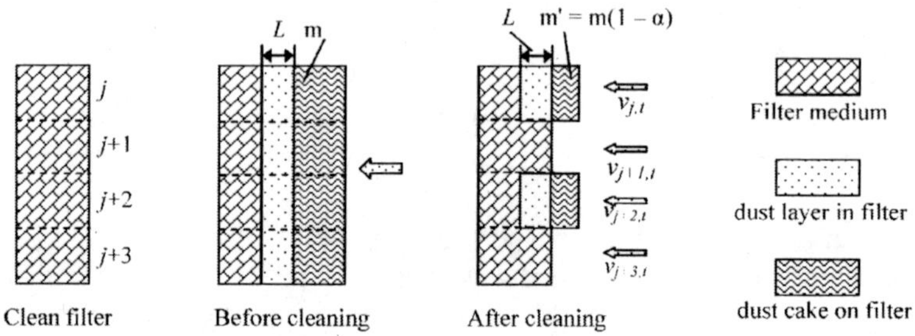

Fig.1. Schematic diagram of filtration process

The pressure drop during a filtration cycle

The pressure drop through a filter Δp can be expressed as Eq.1

$$\Delta p = \Delta p_0 + \Delta p_d + \Delta p_c \tag{1}$$

In it, Δp_0 represents the pressure drop through a clean filter medium, which is expressed as:

$$\Delta p_0 = k_m v \mu \tag{2}$$

Δp_d represents the pressure drop through the dust layer inside the filter, which is expressed by the Equation of Kozeny-Carman (Eq.3).

$$\Delta p_d = 5 S_v^{\,2} \mu v L \frac{(1-\varepsilon)^2}{\varepsilon^3} \tag{3}$$

Considering the packing ratio of the fiber, ε can then be calculated by the following equation:

$$\varepsilon = 1 - \frac{cvt}{(1-\alpha_f)\rho_p L} \tag{4}$$

Δp_c represents the pressure drop through dust cake on the filter surface, which is expressed by Eq.5.

$$\Delta p_c = k_c \mu mv = k_c \mu cv^2 t \tag{5}$$

Combining Eqs.1~5, the pressure drop across the filter at time t, Δp_t can be expressed by Eq.6.

$$\Delta p_t = (k_m + k_d(\varepsilon) + k_c \cdot cvt) v\mu \tag{6}$$

where

$$k_d(\varepsilon) = 5S_v^{\ 2} L \frac{(1-\varepsilon)^2}{\varepsilon^3} \tag{7}$$

Parameters in Eq.6 can be experimentally determined.

k_m can be calculated from Eq.2 using the pressure drop across the filter medium. k_c is the slope of the pressure drop curve in the third filtration stage. T_{in} is the time of the first filtration stage[4]. L can be calculated by combining Eqs.3 and 4.

α is determined by a trial calculation. Different values of α are put into the pressure drop calculation program, and the optimum value of α is determined when the standard deviation between the calculation results and the experimentally measured pressure drop reaches its minimum value.

Based on the hypothesis that dust will be completely removed from part of the filter surface while totally remaining on other parts of the filter surface, f can be expressed by Eq. 8[1].

$$f = \frac{\Delta p_0 (\Delta p_s - \Delta p_r)}{\Delta p_r (\Delta p_s - \Delta p_0)} \tag{8}$$

Since it is assumed that some areas are completely cleaned, but that dust remains on other areas, f decreases proportionally with α, and can then be calculated from Eq.9.

$$f = \frac{\Delta p_0 \left[\Delta p_s - \alpha(\Delta p_s - \Delta p_0) - \Delta p_r\right]}{\Delta p_r (1-\alpha)(\Delta p_s - \Delta p_0)} \tag{9}$$

It can be seen that Eqs.8 and 9 are the same when α=0. f usually varies with the filtration cycles.

Numerical solution of the model

Because the residual dust is not uniformly distributed on the filter surface, the resistance to gas flow through the filter becomes uneven. The areas with a higher residual dust load have a higher resistance to flow and a lower flow velocity. To calculate the pressure drop of the filter with an uneven residual dust distribution, we equally divide the filter surface into n elements. It is assumed that the dust load on one element is homogeneous, although dust loads on different elements may have different values.

Subscript j is used to express the number of the element, t is used to express time (Fig.1). $_{j,t}$ represents the porosity of dust layer in the element j at time t, $m_{j,t}$ is the dust mass on the surface of element j at time t, $v_{j,t}$ is the filtration velocity of element j at time t, and Δp_t is the pressure drop across the filter at time t. Δp_t can be expressed as Eq.10.

$$\Delta p_t = (k_m + k_d(\varepsilon_{j,t}) + k_c \cdot m_{j,t})v_{j,t}\mu \tag{10}$$

Where

$$\varepsilon_{j,t} = 1 - \frac{cv_{j,t}t}{(1-\alpha_f)\rho_p L} \tag{11}$$

$$k_d(\varepsilon_{j,t}) = 5S_v^{\,2} L \frac{(1-\varepsilon_{j,t})^2}{\varepsilon_{j,t}^{\,3}} \tag{12}$$

The average filtration velocity is:

$$v = \frac{1}{n}\sum_{j=1}^{n} v_{j,t} \tag{13}$$

According to Eqs.10~13, the pressure drop of filter at time t can be expressed as:

$$\Delta p_t = n \cdot v \cdot \mu \Big/ \sum_{j=1}^{n} \frac{1}{k_m + k_d(\varepsilon_{j,t}) + k_c \cdot m_{j,t}} \tag{14}$$

The filtration velocity of each element is:

$$v_{j,t} = \Delta p_t / [\mu(k_m + k_d(\varepsilon_{j,t}) + k_c m_{j,t})] \tag{15}$$

After time dt, the change of the porosity of dust layer in the filter, $d\varepsilon$, and the change of the mass of dust cake on the filter, dm, can be calculated by Eqs.16 and 17 respectively.

$$d\varepsilon = -\frac{cv}{(1-\alpha)\rho_p L}dt \tag{16}$$

$$dm_{j,t} = cv_{j,t}dt \tag{17}$$

Then

$$\varepsilon_{j,t+\Delta t} = \varepsilon_{j,t} + d\varepsilon \tag{18}$$

$$m_{j,t+\Delta t} = m_{j,t} + d\varepsilon \tag{19}$$

Using $\varepsilon_{j,t+\Delta t}$ and $m_{j,t+\Delta t}$, $\Delta p_{t+\Delta t}$ can be calculated from Eq.10. When Δp_t reaches the preset value, the dust mass of each element is reset to simulate the cleaning process. All elements are divided into two types, completely cleaned areas or partly cleaned areas. The number of completely cleaned elements is decided by f. To decide that an element is completely cleaned or not, a random number table is utilized. The residual dust mass on not completely cleaned elements is calculated by the following equation:

$$m' = (1-\alpha)m \tag{20}$$

COMPARISON OF SIMULATION RESULTS AND EXPERIMENT

The experiments were carried out according to VDI3926. The test conditions are shown in Table 1.

Table 1. Experimental conditions

Parameter	Value	Parameter	Value
Filtration velocity	3 m/min	Filter material	PPS
Dust type	Fly ash, JIS №10	Type	Needle felt
mean diameter	3.4 μm	Fiber diameter	15 μm
Concentration	5 g/m^3	Weight	530 g/m^2
Tank pressure	150, 300, 500 kPa	Porosity	0.78
Pressure drop before cleaning	1000Pa	Permeability	18 (cm^3/s)/cm^2
		Surface treatment	Singed
Pulse duration	0.25s	Filtration number	100

The parameters used to simulate the filtration process are shown in Table 2.

The experimental results indicate that T_{in} decrease with cycle number in the initial stage, but after about 20 filtration cycles, T_{in} becomes stable. To simplify the calculation, the stable values are used.

The value of f is shown in Fig.2. The curves of f indicate that the residual dust increases with increasing filtration cycles and the decreasing tank pressure. When tank pressure is set at 150kPa, f becomes nearly zero at the end of the experiment. This means that a tank pressure of 150kPa is too low to clean the filter in this case.

Table 2. Experimentally determined constants

P_t(kPa)	k_c (m kg^{-1})	k_m (m^{-1})	α(-)	L(μm)	t_{in} (s)
150			0.48	0	0
300	6.7×10^9	3.9×10^7	0.60	20	15
500			0.68	30	23

Simulation results are shown in Figs.3~5 and compared with the experimental results.

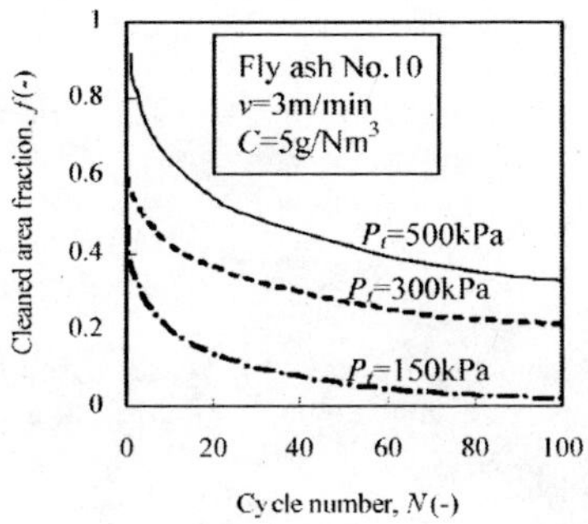

Fig.2. Change oin f with cycle number

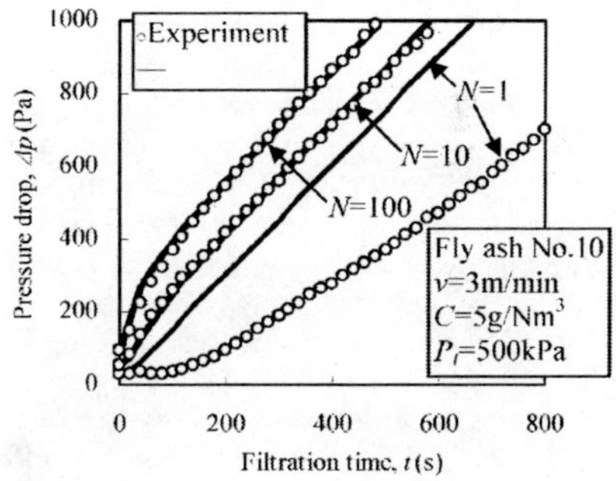

Fig. 3. Comparison of pressure drop curves

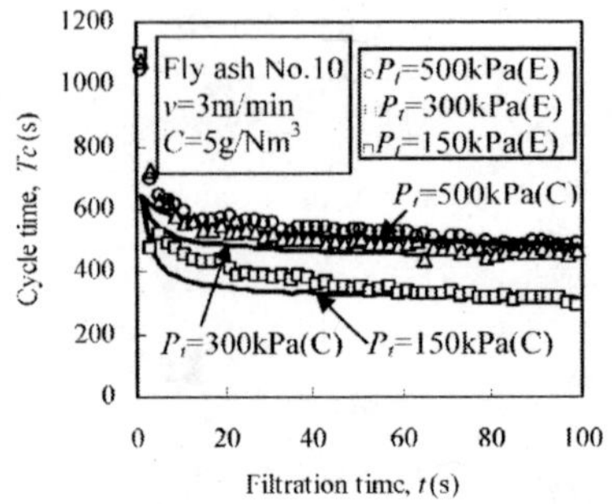

Fig.4. Comparison of filtration cycle time at tank pressures of 150, 300, and 500kPa (E: Experiment, C: Calculation)

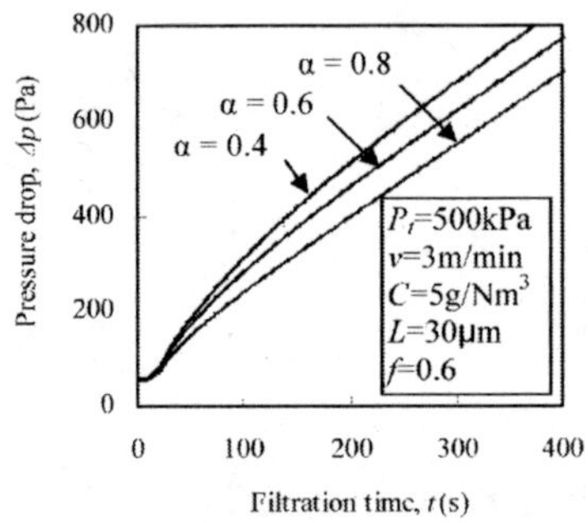

Fig.5. Influence of α on pressure drop

Fig.3 shows the pressure drop curves of the 1st, 10th, 100th filtration cycle for a tank pressure of 500kPa for both simulation and experiment. Except for the 1st cycle, the simulated and experimental curves are nearly identical.

The filtration cycle time for both simulation and experiment are shown in Fig.4. Filtration cycle time decreases with the increasing cycle number and decreasing tank pressure. During the initial stage of the experiment, especially for the first cycle, the difference between the simulation and the experiment is not so small; however, with increasing cycle number, the difference gradually decreases, and there is nearly no difference between the simulation results and the experimental results after 10 cycles.

The influence of αon the pressure drop is shown in Fig.5. When αdecreases, if all other conditions are the same, the residual pressure drop does not change much, but the filtration cycle time decreases. According to the previous models, for this case, the filtration cycle time will not change. But in fact, it is often appears that different filter media having approximately similar residual pressure drops have much different filtration cycle times. The change of α can explain the above phenomena.

CONCLUSION

The filtration process for a flat fabric filter has been discussed in detail. A new model has been proposed. Based on the model, the simulation results agree well with the measured pressure drop and filtration cycle time in a filtration process. This makes it possible to use short time experimental results to predict long-term-filter behavior.

NOMENCLATURE

C=inlet dust concentration [g·m^{-3}]
f=cleaned ratio of filter area [-]
k_c=resistance constant of dust cake [m·kg^{-1}]
k_m=resistance constant of filter medium [m^{-1}]
L=thickness of dust layer in filter [m]
m=dust mass of filter [g·m^{-2}]
m'=dust mass after cleaning [g·m^{-2}]
$m_{j,t}$=dust load of element j at time t [g·m^{-2}]
n=total number of element [-]
N=number of filtration cycle [-]
P_t=tank pressure [kPa]
S_v=specific surface area of dust [m^{-1}]
t=filtration time [s]
t_{in}=duration time of dust accumulating in the inner of filter medium [s]
T_c=filtration cycle time [s]
v=filtration velocity [m·s^{-1}]
$v_{j,t}$=gas velocity of element j at time t [m·s^{-1}]
p_0 =pressure drop through filter medium [Pa]
p_c=pressure drop through dust cake on filter [Pa]
p_d=pressure drop through dust layer in filter [Pa]
p_r=residual pressure drop [Pa]
p_s=pressure drop before cleaning [Pa]
p_t=pressure drop through filter at time t [Pa]
α=ratio of the mass of cleaned dust cake to the mass before cleaning [-]
α_f=packing ratio of filter medium [-]
ε=porosity of dust layer in filter [-]
$\varepsilon_{j,t}$=porosity of dust layer of element j at time t [-]
μ=gas viscosity [Pa·s]
ρ_p=density of particulate [kg·m^{-3}]

REFERENCES

[1]Duo W., N. F. Kirkby, J. P. K. Seville and R.Clift, "Patchy cleaning of rigid gas filters -----II. Experiments and model validation," *Chemical Engineering Science*, **52**, 153-164(1997)

[2]Ju J., M.S. Chiu and C. Tien, "A model for pulse jet fabric filters, " *J.Air Waste Manage. Assoc.* **50**, 600-612 (2000)

[3]Ju J., M.S. Chiu and C. Tien, "Further work on pulse-jet fabric filtration modeling," *Powder technology*, **118**, 79-89 (2001)

[4]Kanaoka, C. and Y. Yao; "The time dependency of pressure drop in a flat type pulse jet fabric filter, " *Kagaku Kogaku Ronbunshu*, **29**, 267-271(2003)

EVALUATION OF THE LONG TERM STABILITY OF LSM/ScSZ COMPOSITE POWDER MATERIALS FOR SOFC CATHODES

Akifusa Hagiwara, Natsuro Hobara, Hiroyuki Shimada, Koichi Takizawa,
R&D Center, Tokyo Electric Power Company
4-1, Egasaki-cho, Tsurumi-ku,
Yokohama 230-8510, Japan

Kazuyoshi Sato, Hiroya Abe and Makio Naito
Joining and Welding Research Institute, Osaka University
11-1,Mihogaoka, Ibaraki,
Osaka 567-0047, Japan

ABSTRACT

A composite electrode/electrolyte powder material with application to cathodes for solid oxide fuel cells (SOFC) was prepared using a mechanochemical process, and the long term stability of the powder was evaluated. The powder is composed of strontium doped lanthanum manganite oxide (LSM) and scandium stabilized zirconia (ScSZ) , and exhibits superior performance when used to fabricate a composite cathode microstructure. In the present study, half-cell structured cathode samples were fabricated with the powder as well as conventional powder materials for comparison purpose, and were heat treated at 1000°C in air for 500-1000 h. The interfacial area specific resistance measurements revealed that performance degradation was significantly reduced when compared with conventional LSM single phase cathodes. The present results showed the effectiveness of the mechanically prepared composite powder upon improving the long term stability of SOFC cathodes.

INTRODUCTION

Solid oxide fuel cells are expected to achieve high efficiency power generation due to effective utilization of high temperature exhaust gas through a combined gas turbine cycle.[1] However, one of the crucial issues for such systems is that the long term stability of cell performance must be maintained while assuring high power density operation. In particular, high power density operation is vital for the economy of SOFC power generation system. For this reason, various functional composite powder materials are being developed in order to increase electrode performance for both anode and cathode.[2] The electrode/electrolyte composite powder material examined here was prepared by a mechanical method.[3,4] We used this composite powder to fabricate a composite cathode in order to effectively increase the triple phase boundary (TPB) region where oxygen reduction reaction takes place.

In general, composite cathodes are fabricated simply by mixing electrode and electrolyte materials, and the mass ratio of electrolyte and electrode materials, particle diameter distribution and sintering temperature are key parameters to control electrode microstructure. The prefabricated composite powder investigated here provides an additional tool for the microstructure control, and is prepared in such a way that individual particles consist of two material components; i.e. electrode catalyst and electrolyte; thus providing prefabricated elements of a composite electrode structure. Its superior performance has been demonstrated elsewhere.[4] The present study concerns its long term stability.

EXPERIMENTAL

Composite powder preparation

The electrolyte-electrode composite powder was prepared by an attrition type particle composing machine.[4] For starting materials, we used stoichiometric $La_{0.8}Sr_{0.2}MnO_3$ ($LSM_{1.00}$) powder with a median particle diameter of 1.0μm as the electrode catalyst material, and 10Sc1CeSZ (or ScSZ) powder with a median particle diameter of 0.5μm as the electrolyte material. Note that the primary particle diameter of ScSZ is approximately 0.1μm according to electron-microscopic observation. The resultant composite powder has structure with relatively large LSM(~1μm) particles coated with fine ScSZ particles.

Preparation of half cell samples

Various cathode samples were prepared for comparison purpose. Cathode material in slurry was screen printed on a ScSZ electrolyte pellet of 15mm in diameter and 2.2mm in thickness, and the pellet was then sintered at 1200°C for three hours to form a cathode layer. Pt paste was used for the counter electrode, and a reference electrode was set around the radial surface of the electrolyte pellet.

The samples are denoted as LSM1.00 and LSM1.07 for simple LSM cathodes, CC64 and CC82 for conventional electrode/electrolyte composite cathodes, and MF for a composite cathode with prefabricated LSM/ScSZ powder material. Their detailed specifications are given in Table 1. For all the samples, we used a starting LSM powder with a median particle diameter of 1.0μm in order to maintain comparable electrode microstructure. Since the CC type cathodes were fabricated simply by mixing LSM and ScSZ powders, their parallel conduction networks within electrode are less controlled and their TPB region is accordingly smaller when compared with MF type composite cathodes with an identical LSM/ScSZ mass ratio.

Experimental method

The cathode samples were placed under an operating condition at OCV and 1000°C in air for a typical duration time of 500-1000hr, and AC impedance measurement was carried out with appropriate time intervals. We determined interfacial area specific resistance or ASR from the

Table 1. Sample specifications and the result of interfacial resistance analyses

ID	Cathode type	Specification	Time h	ASR Ω cm^2	Equivalent circuit model analysis R_1 Ω cm^2	R_2 Ω cm^2
LSM 1.00	Simple	$La_{0.8}Sr_{0.2}MnO3_{1.00}$($D_{50}$=1.0μm) Stoichiometric MnO3	0 480	0.55 0.77	0.06 0.09	0.48 0.68
LSM 1.07	Simple	$La_{0.8}Sr_{0.2}MnO3_{1.07}$($D_{50}$=1.0μm) Excessive MnO3	0 480	0.27 0.44	0.03 0.07	0.23 0.37
CC 64	Composite	LSM(D_{50}=1.0μm)/ScSZ(D_{50}=0.5μm) 60/40 mass %	0 482	0.16 0.21	0.10 0.13	0.06 0.08
CC 82	Composite	LSM(D_{50}=1.0μm)/ScSZ(D_{50}=0.5μm) 80/20 mass%	0 482	0.21 0.26	0.07 0.06	0.14 0.23
MF	Prefabricated composite	LSM(D_{50}=1.0μm)/ScSZ(D_{50}=0.5μm) 80/20 mass %	0 480	0.09 0.20	0.06 0.06	0.06 0.15

MF: mechanically fabricated,
ASR: Area specific resistance for the cathode-electrolyte interface

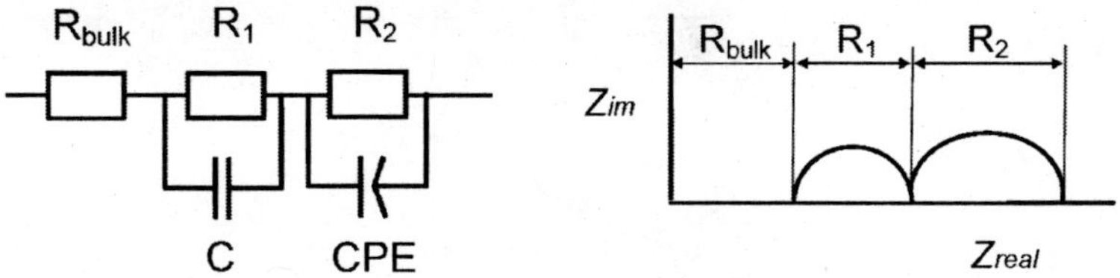

Figure 1. The equivalent circuit model for impedance spectrum analyses

impedance measurements, and the impedance spectrums were analyzed in terms of the equivalent circuit model described in Figure 1.

RESULTS

Simple LSM cathodes

Figure 2 shows the variation of ASR with elapsed time for the cathode samples. As seen here, the initial ASR for LSM1.00 is the highest among the samples. The use of lanthanum deficient LSM(LSM1.07) results in a significant reduction in ASR when compared with that of LSM1.00. This improved performance can be generally explained in terms of the deterred formation of highly resistive $La_2Zr_2O_7$ which is formed at the interface between LSM and ScSZ. Previous literatures[5,6,7] have reported that La deficient LSM is less reactive with ZrO. However, the ASRs for LSM1.00 and LSM1.07 increase with time. In particular, the rate of ASR increase for LSM1.07 after 500hr is faster than that for LSM1.00, suggesting that the formation of resistive secondary phase continues at a significant rate.

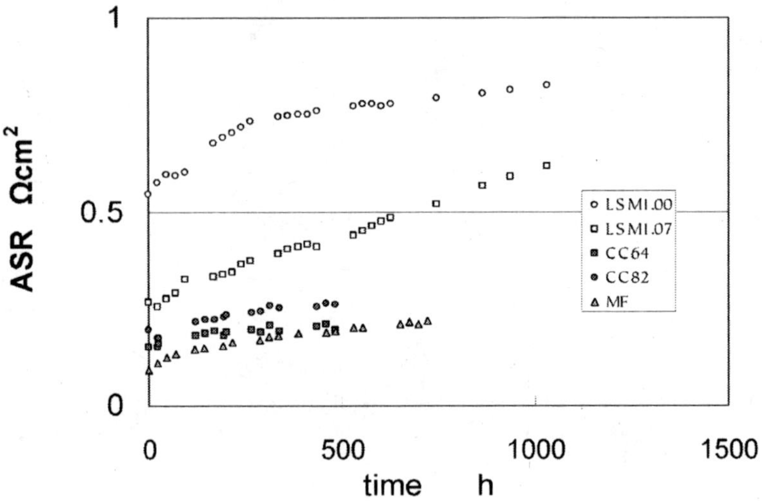

Figure 2. Variation of the ASRs measured at OCV and 1000°C in air as a function of time.

Figure 3 shows the impedance spectrums for LSM1.00 measured before and after 480h operation. Note that the bulk resistance for the spectrums is subtracted in the figure. It can be seen that the impedance spectrums consist of two semi-circles. According to the equivalent circuit model in figure 1, the semi-circle at higher frequency range, R_1, is related to the interfacial resistance such as contact resistance at the interface between electrolyte and electrode layer and is independent of oxygen partial pressure. On the other hand, the semi-circle at lower frequency range, R_2, is related to the oxygen reduction process and is strongly affected by oxygen partial pressure[8]. In particular, R_2 is directly connected to the broadness of triple phase boundary region where charge transfer reaction takes place.

The equivalent circuit model analysis reveals that R_1 and R_2 for LSM1.07 are smaller than those for LSM1.00, respectively. This indicates that the formation of resistive $La_2Zr_2O_7$ layer at the electrode/electrolyte interface. In addition, it can be noted that R_1 for LSM1.07 after 480h, is increased, with the value approaching to the revel of R_1 for LSM1.00. This confirms that the effect of La deficiency in LSM on R_1 is only to extend the lead time for $La_2Zr_2O_7$ formation. The observed increase in R_2 for both LSM1.00 and LSM1.07 can be interpreted as a decrease in the triple phase boundary region due to the growth of elemental LSM grains and the formation of $La_2Zr_2O_7$ at the TPB region in the electrode microstructure.

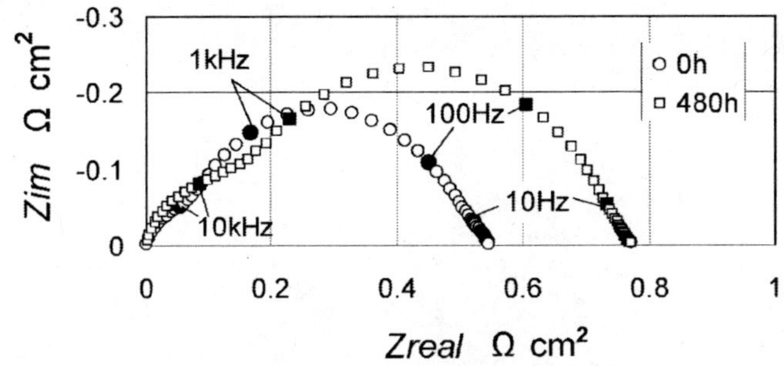

Figure 3. Variation of the impedance spectrum for LSM1.00 sample after 480h operation.

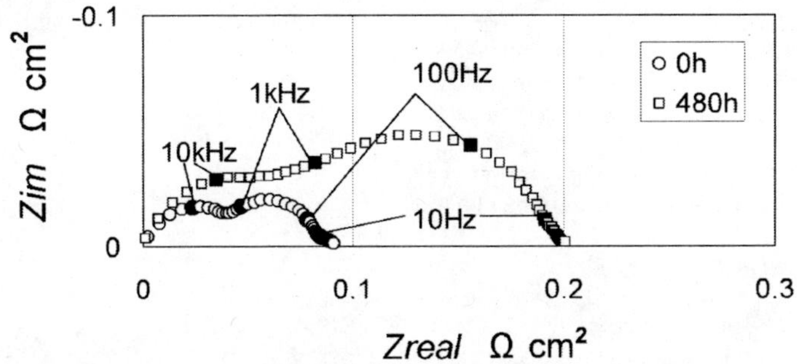

Figure 4. Variation of the impedance spectrum for MF sample after 480h operation

Composite cathodes with prefabricated powder

The present data confirms that composite cathodes, in general, exhibit better performance in ASR than simple LSM cathodes as shown in Figure 2. For the conventional composite cathodes, CC64 and CC82, their initial ASRs were measured as 0.16 Ωcm^2 and 0.21Ωcm^2, respectively. In particular, the initial ASR for MF sample is lower than any other samples examined here, with the value measured as 0.09Ωcm^2. Therefore, the use of prefabricated LSM/ScSZ composite powder is apparently effective to improve cathode performance.

Since both MF and CC82 samples are prepared under the identical starting powder condition except that the starting powder is mechanically processed for MF, we can compare these two samples directly in order to analyze the effect of prefabricated powder. According to the equivalent circuit model analysis in Table 1, R_1 and R_2 are smaller for MF than for CC82, with the improvement in R_2 being comparatively significant. This indicates that the triple phase boundary region is more effectively created for MF than for CC82.

The long term performance stability data shown in Figure 2 and Table 1 reveals that ASR increases with time. It can be noted that for both MF and CC82, R_1 does not change so much with time, whilst R_2 increases substantially. The performance difference between MF and CC82 is largerly attributed to the initial R_2 which is smaller for MF than for CC82. The increase in R_2

for CC82 can be interpreted as the loss of parallel conduction network that leads to the reduction of TPB region.

The impedance spectrum for MF sample after 480h operation is compared with the initial one in Figure 4, where a spectrum plot having two semi-circles similar to the equivalent circuit model assigned in the present study can be clearly identified. The equivalent circuit model analysis reveals that R_1 after 480h stayed at the same level as the initial one while R_2 increased from $0.06\Omega cm^2$ to $0.15\Omega cm^2$. This possibly indicates that the $La_2Zr_2O_7$ formation at the electrode/electrolyte interface is deterred when compared with the simple LSM cathodes, and that some reduction in TPB region takes place due to the particle grain growth.

DISCUSSION

According to the previous investigations on LSM/YSZ composite cathodes [8], it is generally known that R_1 is controlled by a sintering temperature at which the cathode is fabricated. Choi et al.[8] found that the contact resistance improves as the sintering temperature becomes higher and that the effect of $La_2Zr_2O_7$ formation plays negligible contribution to the magnitude of R_1. In addition, R_2 is controlled by a degree of grain growth as they found when the sintering temperature is raised. TPB region is accordingly reduced as the grain growth develops.

In the present study, we observed performance which is consistent with these previous findings. The advantage of using the prefabricated composite powder is to provide much wider TPB region when compared with conventional composite electrodes. This allows an additional margin for performance degradation in a long term operation. In particular, when the amount of electrolyte material in a composite cathode is small, i.e. LSM/ScSZ mass ratio of 80:20, this advantages can be fully utilized.

In the other aspect, TPB region can be expanded by optimizing the LSM/ScSZ mass ratio. For instance, the performance of CC64 is better than that of CC82. This is generally observed in the previous literatures and can be interpreted as the increase of chance to have parallel conduction paths based on the percolation theory. In this aspect, the prefabricated powder material can be further optimized by changing either the LSM/ScSZ weight ratio or the particle diameter ratio.

CONCLUSION

The present study revealed that the long term stability in cathode performance can be improved by using the prefabricated LSM/ScSZ powder material by an attrition type particle composing machine. The improved cathode performance obtained here encourages the application of the composite powder technology to cathode fabrication. Also, the mechanism for conventional composite cathodes can be ascertained more systematically, with possibility to control the morphology of electrode microstructure.

REFERENCES

[1]S.C.Sihnghal and K.Kendall, "High temperature solid oxide fuel cells; fundamentals, design and applications," Elsevier, 363-392 (2003).

[2]T.Fukui, K.Murata, S.Ohara, H.Abe, M.Naito and K.Nogi, "Morphology control of Ni-YSZ cermet anode for lower temperature operation of SOFCs," J.Power Soruces **125**, 17-21 (2004).

[3]M.Naito, A.Kondo and T.Yokoyama, "Application of comminution techniques for the surface modification of powder materials," ISIJ International **33**, 915-924 (1993).

[4]N. Hobara, K.Takizawa, A.Hagiwara, K.Sato, H.Abe, M.Naito, "Preparation and evaluation of mechanically composed particles for SOFC cathode," The 14th Symposium on Solid Oxide Fuel Cells in Japan, 214-217 (2005).

[5]T.Kenjo and M.Nishiya, "$LaMnO_3$ air electrodes containing ZrO_2 electrolyte for high temperature solid oxide fuel cells," Solid State Ionics **57**, 259-302 (1992).

[6]A.Mitterdorfer and L.J. Gauckler, "$La_2Zr_2O_7$ formation and oxygen reduction kinetics of the $La_{0.85}Sr_{0.15}MnyO_3$(g)/YSZ system," Solid State Ionics **111**, 185-218 (1998).

[7]S.B.Alder, "Factors governing oxygen reduction in solid oxide fuel cell cathodes," Chemical Review **104**, 4791-4843 (2004).

[8]J.H.Choi, J.H.Jang and S.M.Oh, "Microstructure and cathodic performance of $La_{0.9}Sr_{0.1}MnO_3$/yttria-stabilized zirconia," Electrochimica Acta **46**, 867-874 (2001).

A NOVEL CONCEPT AND APPROACH TO FABRICATE PROTECTIVE KB/PVDF COMPOSITE FILMS ON METALLIC BIPOLAR PLATES FOR PEM FUEL CELLS

Jingtian Yin, Takehisa Fukui, Kenji Murata
Hosokawa Powder Technology Research Institute (HPTRI), 1-9, Shoudai, Tajika, Hirakata, Osaka 573-1132, Japan

T. Hirabayashi, S. Yamamuro
Kurimoto, LTD., 2-8-45, Shibatani, Suminoe-Ku, Osaka 559-0021, Japan

M. Matsuda, M. Miyake
Graduate School of Environmental Science, Okayama University, 3-1-1, Tsushima-Naka, Okayama 700-8530, Japan

ABSTRACT

Composite coatings consisting of highly conductive fine carbon particles (Ketjen Black, KB) and corrosion resistant polyvinylidene fluoride (PVdF) were successfully fabricated on stainless steel substrates by electrophoretic deposition (EPD). The morphologies and microstructures of the coatings were characterized by SEM and TEM. The characteristics of electrical resistance and corrosion resistance of the coatings were also evaluated. The study indicated that the synthesis of composite particles and their homogeneous distribution in the suspension was a prerequisite to obtain uniform green coatings; while post heat treatment could further improve the quality of the coatings. The stainless steel substrates coated with KB/PVdF composites exhibited both high electrical conductivity and good corrosion resistance, suggesting the possibility that they serve as the bipolar plates in PEM fuel cells. In addition, it is expected that the simple but effective method employed in this work could be extended to other coating systems.

INTRODUCTION

Polymer electrolyte membrane (PEM) fuel cells are most prospective for mobile devices and transportation applications owing to their low operating temperature, low CO_2 emission as well as a combination of high power density and high-energy conversion efficiency. Two major obstacles that limit PEM fuel cells to be developed into commercial products are their high cost and relatively large volume. It is well known that these problems are significantly related to the bipolar plates which are conventionally made from graphite. The brittleness of graphite not only leads to an expensive machining cost but also a difficulty in producing lightweight compact stacks.

Stainless steels have been investigating extensively as a possible alternative to the presently used graphite, because they can provide such advantages over graphite bipolar plates as low cost, high strength and manufacture capability as well as gas impermeability.[1-7] However, corrosion of stainless steel bipolar plates under fuel cell operating conditions is a severe problem to be solved for practical applications. The stainless steels with passive layer (dense oxide film) show good corrosion resistance in acid or PEMFC environment, but their large interfacial contact resistance (ICR) with carbon paper would lead to an unacceptable loss of output voltage. Although the stainless steels with freshly abraded surface could satisfy the requirement of initial I-V characteristic, the gradual formation of metal oxides would result in the increase of ICR and the dissolution of metal cations could cause contamination and deterioration of the membrane electrode assembly (MEA). This would affect the performance and lifetime of fuel cells. Various efforts have been recently made to protect stainless steel bipolar plates in PEM fuel cell environment, among which are involved metal surface modification and coating with noble metals, nitrides, diamond-like carbon, conductive polymer, carbon/polymer composites and so on. Apart from traditional coating methods, a lot of novel techniques, such as physical and chemical vacuum evaporation, electron-beam deposition and sputtering, have also been developed to complete the above attempts. Unfortunately, these laborious efforts have achieved limited success because of unsatisfactory properties, complex fabricating processes or increased material and manufacturing costs.

Conductive carbon black (CB)-filled polymer films appear to be quite attractive as protective coatings for metal bipolar plates. In such system, CB serves as a conductor to increase conductivity of the coatings; whereas polymer binder provides chemical resistance. However, the fabricating technique for the composite coatings is a critical challenge. The complex shape of flow channel and the introduction of CB make it difficult to obtain high-quality coating layers using conventional techniques (hot molding, spray, screen-printing, electrostatic deposition, etc.). Thus, it is of particular interest to develop a new composite coating technology.

In this work, we present a simple but effective approach to deposit carbon/polymer composites on stainless steel substrates. Electrophoretic deposition (EPD) was adopted as fabricating technique due to its advantages of short deposition time, simple deposition apparatus, suitability for mass production, easiness of thickness control and capability to complex shapes and patterns. With a novel suspension system and synthesizing method of composite particles, uniform composite coatings having both high electrical conductivity and good corrosion resistance were successfully fabricated. It is considered that the versatility of this approach could be extended to other functional coatings.

EXPERIMENTAL

The suspension for EPD was made of fine carbon particles (ketjen black, KB, conductive filler), polyvinylidene fluoride (PVdF, binder), N-methyl-2-pyrrolidinone (NMP, solvent for PVdF) and high purity ethanol (dispersion medium). The following procedure was employed to prepare stable suspension. KB powder was firstly added to NMP solution of PVdF. After ultrasonic treatment and subsequent mechanical stirring, the solution suspended with KB particles was poured at an appropriate rate into a large amount of ethanol under stirring. In this step, PVdF was precipitated from the non-solvent ethanol and formed into composite particles with surrounding carbon particles. The mixed solution was subsequently emulsified and distributed at a rotation speed of 6000-7000 rpm for 10 min using a homogenizer (Physcotron, Microtec Co., Ltd). Thus, an ethanol colloidal suspension containing submicro-scaled KB/PVdF composite particles was prepared. The procedure of the preparation of suspensions is schematically illustrated in Fig. 1. The zeta-potential of the composite particles was measured using Zetasizer 2000, Malvern Instruments. The analysis indicated that the KB/PVdF composite particles formed in the suspension were negatively charged; its zeta potential is around -14 mV when the suspension was diluted with 99 times ethanol.

The EPD was carried out in a two-electrode beaker cell using a dc power source (0-1000V, 2A). Stainless steel (SS316L) disks (Φ20mm×t2mm) or SS316L bipolar plates with serpentine flow channels were used as both working and counter electrodes. The passive layers on the surface of the substrates were polished away with abrasive paper prior to use. The distance between two electrodes was set at 10 mm. The EPD was performed under constant voltage conditions (in a range of 50-100V) using variable deposition times (from 10-60s). The composite films deposited on the SS316L substrates were dried at room temperature for 2h and subsequently heat-treated at 180-250 °C under either air or Ar atmosphere for 1 h. Figure 2(a) illustrates the EPD process, where the composite particles migrate to the positive electrode (anode) and undergo electrophoretic deposition. Figure 2(b) shows the appearance of the KB/PVdF film deposited on the SS316L bipolar plates with serpentine flow channels.

The thickness, morphology and microstructure of the deposited layers were characterized by laser microscope, scanning electron microscope (SEM) and transmission electron microscope (TEM).

The ICR of the EPD substrates (Φ20mm×t2mm) with carbon paper was measured using four-terminal dc method under a constant contact pressure of 10 kgf/cm^2 with a current scanning range from -1 A/cm^2 to 1 A/cm^2.

Acid immersion tests for the composite coatings on Φ20mm×t2mm SS316L substrates were performed at 80 °C in 600 ml sulfuric acid of pH 1-2. After certain periods of immersion time, around 10 ml H_2SO_4 was taken out for the measurement of metal ions concentration using inductively coupled plasma (ICP) mass spectrometer.

Fuel cell tests were performed on JARI (Japan Automobile Research Institute) standard cells in which bare or KB/PVdF coated SS316L bipolar plates were used, respectively.[8]

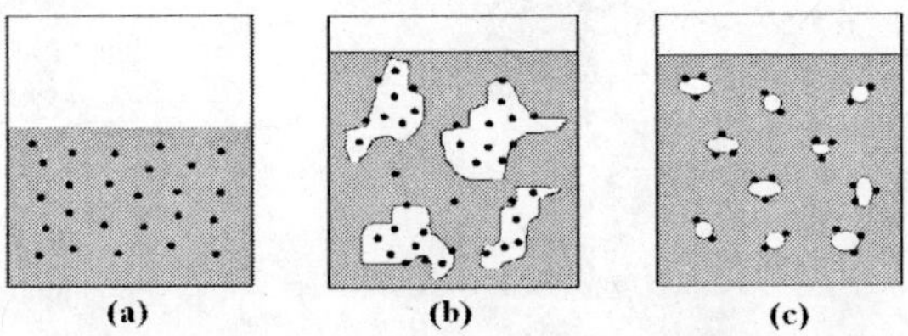

Figure 1. Schematic diagram describing the preparation of suspension: (a) PVdF/NMP solution suspended with KB particles; (b) precipitation of KB/PVdF composite particles in ethanol; and (c) emulsification and distribution of KB/PVdF composite particles.

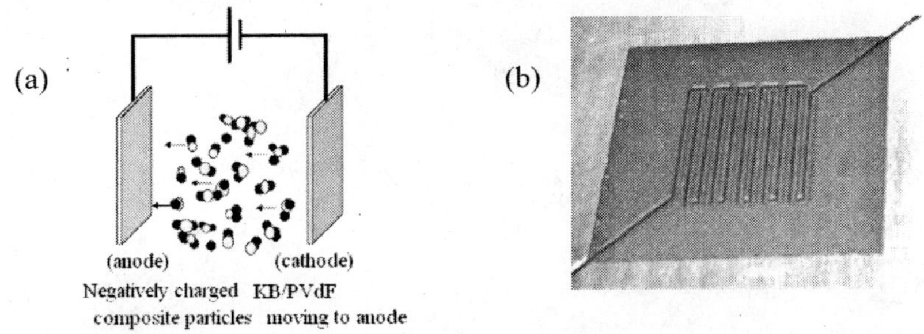

Figure 2. Schematic illustration of electrophoretic deposition (EPD) process (a) and the appearance of the KB/PVdF composite film deposited on the SS316L bipolar plate (80×80×10mm) with serpentine flow channels (area: $5cm^2$; pitch: 1.0 mm; depth: 0.5mm) (b).

RESULTS AND DISCUSSION

Figures 3(a)-(b) show the surface morphologies of KB/PVdF (=22/78 v/v) coatings after heat-treatment at different temperatures. It can be seen that the sub-micrometer particles are homogeneously deposited on the stainless steel substrate by electrophoretic force. No obvious macroscopic pinhole defects can be observed. Comparing Fig. 3(a) with Fig. 3(b) shows that the coating heat-treated at 245 °C is evidently denser than that heat-treated at 185 °C, meaning that the density of the composite coatings can be improved by increasing heat-treatment temperature. In addition, the laser microscopy analysis indicates that the thickness of the coatings is around 10-20 μm after heat-treatment.

Figures 3(c)-(d) show the microstructure in interior of the composite film heat-treated at 185 °C for one hour. It is clear that two different phases coexist in the visual area: bright PVdF matrix and hollow KB particles with a size of several tens of nanometers or their agglomerates. In such microstructure, PVdF forms a dense matrix layer that would play a protective role in PEM fuel cell environment. While KB particles embedded in PVdF matrix connect intimately with each other to form a conductive net-work. The geometric shape of these KB particles remains unchanged during EPD and subsequent annealing processes, in agreement with their original shape.

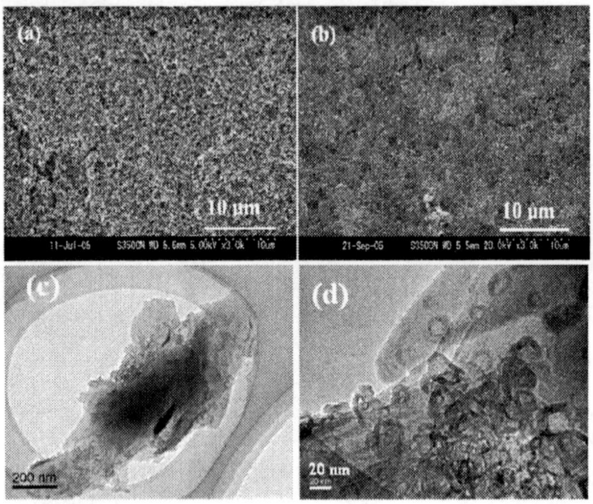

Figure 3. Surface SEM images of deposition layers (KB/PVdF=22/78 (v/v)) after heat-treating at 185°C (a) and 245 °C (b) in air for 1h; overall TEM morphology of a segment of the composite film (c) and enlarged morphology along the edge of the segment (d).

The ICR values of the composite coatings (Φ20mm×t2mm substrates) under different heat-treating conditions are given in table 1 and compared with those of KB-free PVdF film and SS316L substrates with and without passive layers. One can see that the coatings with 22 vol% KB addition show a remarkable improvement in contact resistance compared to KB-free PVdF film. This is normally attributable to the formation of conductive net-work by dispersed KB particles in PVdF matrix. The ICR of the coatings decreases with elevating the heat-treating temperature. The reason for this is associated with the improvement in density and integration of the composite coatings as shown in Fig. 3. It is also noted that the composite coatings heat-treated in Ar atmosphere exhibit lower ICR than those heat-treated in air do. This is due to the fact that the inert atmosphere reduces the formation of metal oxides between the coatings and stainless steel substrates during heat-treating. The result suggests that the KB/PVdF composite coatings prepared by EPD and subsequent heat-treatment under suitable conditions can offer a satisfactory ICR, which is lower than the required value of 30 $m\Omega \cdot cm^2$ in PEM fuel cells.

Table 1. Comparison of ICR of the composite coatings (22vol% KB) with carbon paper (pressure: 10 kgf/cm^2; current density: 1 A/cm^2).

Items	Heat treatment Conditions	ICR ($m\Omega \cdot cm^2$)	Adhesion
KB/PVdF composite films (22 vol% KB)	200°C, air	57.1	ordinary
	245°C, air	28.3	strong
	245°C, Ar	15.5	strong
	250°C, Ar	13.2	strong
	280°C, Ar	19.3	strong
PVdF film (KB-free)	200°C, air	455	strong
SS316L substrate	(Freshly abraded surface)	29	-
SS316L substrate	(Passivated surface)	131	-

In order to examine the corrosion protection provided by the composite coatings, immersion tests were carried out in pH 1-2 sulfuric acid at 80 °C. SS316L substrate (Φ20mm×t2mm) coated with KB/PVdF and heat-treated at 245 °C in air was employed as a sample. The area exposed to sulfuric acid is 78.5 mm^2. Figure 4 shows the concentrations of dissolved metal ions in sulfuric acid as a function of immersion time. The composite coating exhibits quite good corrosion resistance. After immersion for 1036h, the concentrations of dissolved Fe, Cr and Ni for the composite coating are 0.13, 0.03 and 0.02 ppm, respectively. It is considered that such little ion concentrations could not cause contamination of the membrane electrode assembly (MEA). It should be pointed out that the sample heat-treated at 250 °C in Ar, which had the most appropriate ICR value, delivered similar corrosion resistance. The relevant data will be reported in the later paper.

To verify our assumption, fuel cell tests were performed on JARI (Japan Automobile Research Institute) standard cells assembled using bare and KB/PVdF coated SS316L bipolar plates, respectively. As shown in Fig. 5, both fuel cells demonstrate an acceptable I-V behavior at the initial operation stage. After operation for 100h, the fuel cell with bare SS316L bipolar plates shows an obvious voltage loss; while the cell with KB/PVdF coated SS316L bipolar plates remains unchanged in its I-V performance. This implies that the KB/PVdF coated SS316L bipolar plates possess better corrosion resistance; neither passivation nor corrosion occurred on the interfaces between SS316L substrates and KB/PVdF coatings during a long-term operation.

CONCLUSIONS

Electrophoretic deposition was adopted to prepare KB/PVdF composite films on stainless steel substrates. The synthesis of composite particles and their homogeneous distribution in the suspension was a prerequisite to obtain uniform green coatings. Post heat-treatment was capable of improving the quality of the prepared coatings. The stainless substrates coated with KB/PVdF composites exhibited an interfacial contact resistance lower than 30 $m\Omega \cdot cm^2$ and good corrosion resistance, suggesting the possibility that they are serving as the bipolar plates in PEM fuel cells.

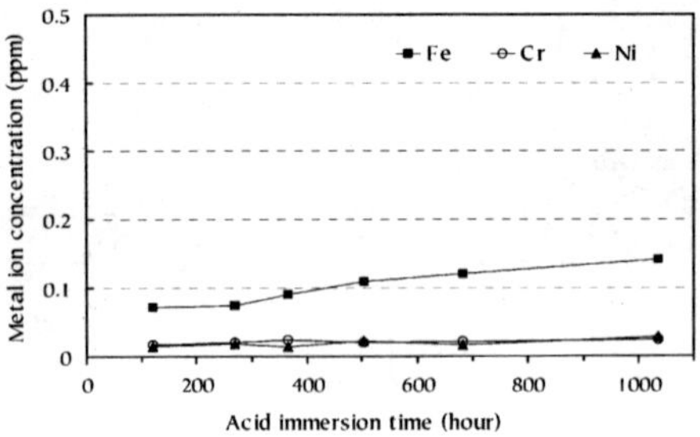

Figure 4. The concentrations of dissolved metal ions in sulfuric acid as a function of immersion time. SS316L substrate (Φ20mm×t2mm) coated with KB/PVdF composite (22 vol% KB) was used as a sample. The area exposed to sulfuric acid is about 78.5 mm^2.

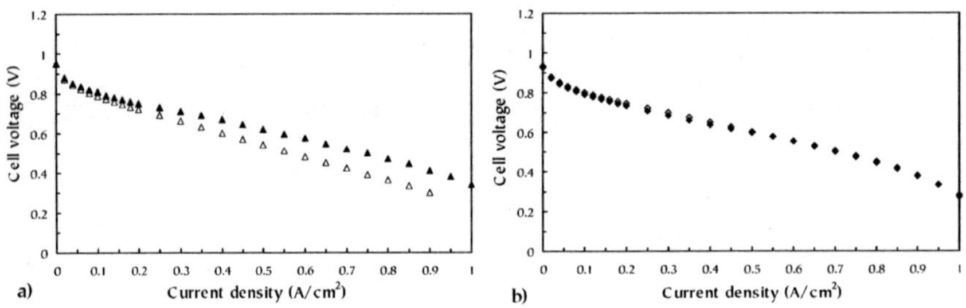

Figure 5. Comparison of I-V characteristics of fuel cells assembled using bare and KB/PVdF (22 vol% KB) coated SS316L bipolar plates, respectively. Solid and open symbols represent the initial performance and the performance after operation for 100 h, respectively.

REFERENCES

[1]J. Wind, R. Späh, W. Kaiser, G. Böhm, Metallic bipolar plates for PEM fuel cells, *J. Power Sources,* **105(2)**, 256-260 (2002).
[2]N. Cunningham, D. Guay, J. P. Dodelet, Y. Meng, A. R. Hlil, A. S. Hay, New Materials and Procedures to Protect Metallic PEM Fuel Cell Bipolar Plates, *J. Electrochem. Soc.,* **149**, A905-A911 (2002).
[3]S. Joseph, J.C. McClure, R. Chianelli, P. Pich, P.J. Sebastian, Conducting polymer-coated stainless steel bipolar plates for proton exchange membrane fuel cells (PEMFC), *International Journal of Hydrogen Energy,* **30**(12), 1339-1344 (2005).
[4]H. Wang, G. Teeter, J. Turner, Investigation of a Duplex Stainless Steel as Polymer Electrolyte Membrane Fuel Cell Bipolar Plate Material, *J. Electrochem. Soc.*, **152**, B99-B104 (2005).
[5]S.J. Lee, C.H. Huang, J.J. Lai, Y.P. Chen, Corrosion-resistant component for PEM fuel cells, *J. Power Sources,* **131**(1-2), 162-168 (2004).
[6]B.R. Padhy, R.G. Reddy, Performance of DMFC with SS316 bipolar/end plates, *J. Power Sources,* **153**(1), 125-129 (2006).
[7]M.A. Lucio García, Mascha A. Smit, Study of electrodeposited polypyrrole coatings for the corrosion protection of stainless steel bipolar plates for the PEM fuel cell, *J. Power Sources,* **158**(1), 397-402 (2006).
[8]H. Tomioka, Y. Hashimasa, N. Yoshimura, M. Akai, S. Watanabe, JARI Standard Single Cell Testing Protocol, *JARI Research Journal*, **28**(7), 247-252 (2006).

STUDY OF AN ALUM-CERAMIC HEAT STORAGE MATERIAL

Xuetan Ren[1], Anze Shui[1], Lingke Zeng[1], Yanchun Liu[1] and Junsheng Wu[2]
[1]College of Materials Science and Engineering, South China University of Technology, 510640, PR China
[2]Guangzhou senior light industry technical school, 510220, PR China

ABSTRACT

Alum ($KAl(SO_4)_2 \cdot 12H_2O$) has a high latent heat (232.4J/g) and a low melting point (91°C), which makes it very suitable as a low-temperature heat storage material. In this study, a >96%, porous body made of mullite ceramic fiber was filled with alum using gravity and capillary force. Additionally a small quantity of carbon was added to improve thermal conductivity. The resulting body, which includes 80~86% of the alum salt heat storage material, can keep its shape without any leakage during heating. The heat storage property was examined by putting the heat storage body on an electric heater. The electric energy consumption is decreased from 24.61wh compared to 60.43wh without the PCM.

INTRODUCTION

Phase Change Materials (PCMs) have been a hot topic in research in recent years as important heat energy storage media. The solid-liquid phase transformation is of great importance because most classes of materials undergo this type of transformation, releasing or absorbing a relatively large quantity of energy, without a large volume change while. Ideal PCMs have the following characteristics[1,2,3]:

1. High fusion heat per unit of mass and volume,
2. Proper melting point and reversible solid-liquid phase transition,
3. High thermal conductivity, density, and specific heat,
4. Low vapor pressure,
5. Be non-corrosive and nontoxic,
6. Low thermal expansion coefficient,
7. High flash point,
8. Good wetting characteristics, and
9. Be stable and pure.

With decreasing energy resources and an increasing demand for energy, thermal storage systems have become an important part of many industrial processes. Due to the large specific heat energy storage capacity in a phase change, the volume of the latent-heat-fusion energy storage materials is substantially smaller than that of alternative heat storage solutions. Furthermore, the phase change process is isothermal, which is advantageous in the design and efficient operation of many thermal systems.

Shape-stabilized PCMs are made up of a supporting material and a phase change material. It can keep its shape even when the temperature exceeds the melting point of the phase change material. Thus, the shape-stabilized PCM can be brought into contact with the heat transfer media directly, and there is no heat resistance due to an encapsulant between them[4,5]. Inorganic hydrate salt is not expensive, and has a good latent heat, which makes it suitable as a phase change material in many fields. The apparent porosity of the supporting material structure determines the content of the phase change salt in PCM and determines the heat storage capability.

In this study, a shape-stabilized PCM is prepared with alum salt and mullite ceramic fiber, and its heat storage property is examined by setting on an electric heater.

EXPERIMENT

Preparation of acid phosphate binder

An acid phosphate binder was prepared according to Jeng-Maw et al, and Luo[6,7] by mixing aluminum hydroxide ($Al(OH)_3$, 99%, A.R. Tianjin Hongyan Chemical Reagent Factory) and phosphoric acid (H_3PO_4, 85% A.R. Guangzhou donghong chemical Co.), so that the solution has a calculated atom ratio of P/Al of 23. The aluminum hydroxide ($Al(OH)_3$) powder was added gradually to the proper amount of phosphoric acid (H_3PO_4) while the mixture was heated to ~100°C while being stirred. It was held at 100°C until all the solids were fully dissolved, after which the solution became transparent. The associated reaction is explained by the following equation:

$$Al(OH)_3 + 3H_3PO_4 = Al(H_2PO_4)_3 + 3H_2O \qquad (1)$$

Because of excessive phosphoric acid, the resultant solution consisted of mono-aluminium phosphate $Al(H_2PO_4)_3$ and phosphoric acid H_3PO_4. It was diluted with distilled water to a concentration of 10 mass%.

Preparation of porous fiber ceramic

Mullite ceramic fiber (Hangzhou Hongda Crystal Fiber Co., Ltd.) was added to the prepared binder solution and stirred, and the resultant mixture was poured into a container. The unwanted solution was expelled from the bottom of the container ,where it passed through a 60-mesh sieve at the correct pressure. The experimental process is shown in Figure 1. The body was heated to 650°C at a rate of 10°C /min, and held for 40 minutes.

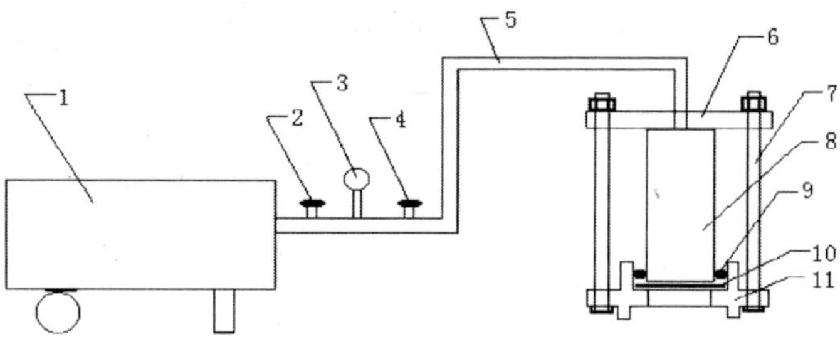

1 air compressor 2 pressure regulation valve 3 pressure gauge 4 gate 5 compressed-air pipe

6 cover 7 screw 8 shape of a bucket body 9 seal ring 10 filtrate mesh screen 11 base

Figure 1. Schematic of the apparatus used to prepare the fibrous porous material

Preparation of alum-ceramic heat storage material

Experiment 1: 200g Alum ($KAl(SO_4)_2 \cdot 12H_2O$, A.R., 99.5%,Guangzhou Chemical Factory) was heated to 100°C until the salt was completely melted inside the container. Then the fiber ceramic body was immersed in the Alum solution for half an hour while the temperature remained at ~100°C. The salt solution infiltrated the porous ceramic body under the force of capillary action. Afterwards the body was cooled to room temperature, and unnecessary salt was removed from the surface of the fiber ceramic body. The sample was named PCM1#

Experiment 2: First, 1g of superfine black carbon was mixed with 40ml of acid phosphate binder (10 mass%), while 0.1 mass% (relative to the black carbon) titanate resin NDZ-311W (Nanjing Shuguang Chemical Factory) was added, so that the carbon could be dispersed in the solution. Then the fiber ceramic was immersed in the mixture, and the excess slurry was removed using a vacuum filter. The body was dried in the oven using the same process described for experiment1. The sample was named PCM2#

The resultant Phase Change Materials (1#, 2#) were set on an electric heater (shown in Figure 2) to test their heat storage properties.

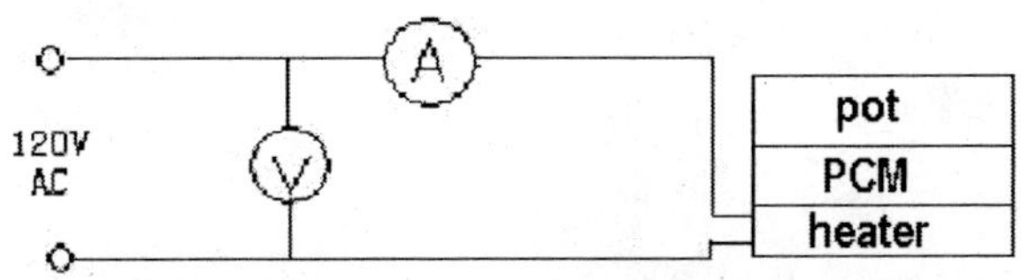

Figure 2. Schematic of test apparatus to characterize heat storage properties.

RESULT AND DISCUSSION

The melting point of alum is about 91°C, and its latent heat of fusion is 232.4 J/g, which is larger than $Mg(NO_3)_2 \cdot 6H_2O$ and $MgCl_2 \cdot 6H_2O$[8]. All these results match the properties of an ideal PCM for heat storage. The differential scanning calorimetry (DSC) pattern is shown in Figure 3. From the pattern, we can see that the endothermic peak occurs at 90.4°C and 94.4°C which shows a solid-liquid phase change has occurred, and the melting point is a peritectic point. In the experiment we found that the alum can be dissolved in water without a measured mass loss throughout the process.

The porosity of mullite fiber porous ceramic exceeded 96%, which is larger than traditional porous ceramic (porosity around 40~60%). The average pore size is about 19μm (by liquid displacement porometry) and 31μm (by mercury intrusion porosimetry)[9]. In this experiment the pore size was appropriate to enable liquid infiltration under the force of capillary action. The SEM photograph (Figure 4.) indicates that the porous ceramic has a three-dimensional, homogeneous, netty structure.

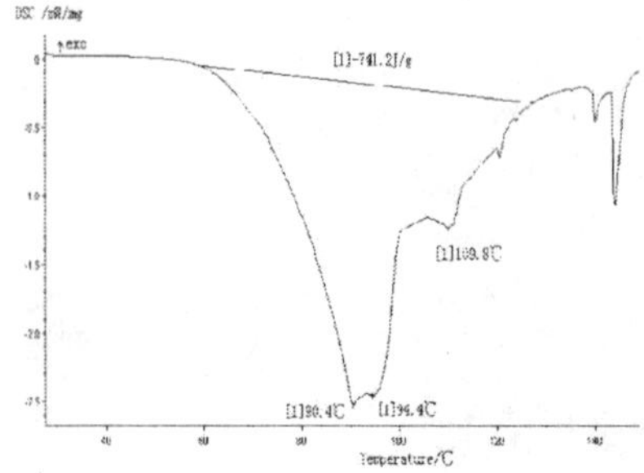

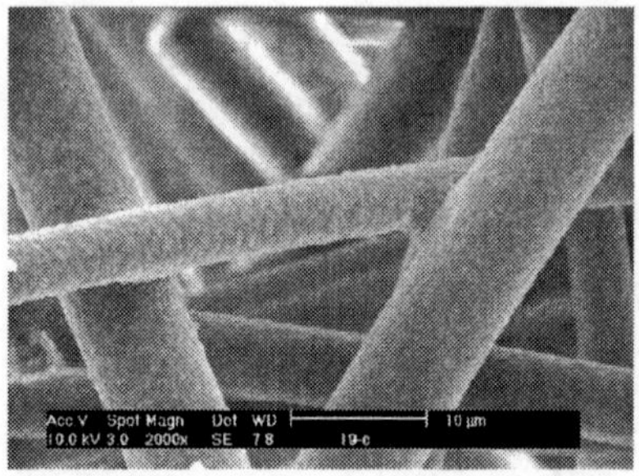

Figure 3. DSC pattern for alum. Figure 4. SEM photograph of the porous ceramic.[8]

The results of the immersion tests are shown in Table 1.

Table 1. Results of immersion test

Sample	mass before immersion	mass after immersion
PCM1#	20.6 (g)	147 (g)
PCM2#	23.1 (g)	117 (g)

Sample "PCM1#" displays a good absorption power, as the weight changed from 20.6 g to 147 g after infiltration. The shape-stabilized material is composed of the 86 mass% alum salt and 14 mass%, fiber ceramic, which is high and rarely seen in a traditional shape-stabilized ceramic/salt PCM (traditional ~50%)[10]. However the weight of the sample "PCM2#"only changed from 23.1 g to 117 g, and with only 80.3mass% of the alum salt, which might be due to carbon plugging the pores.

The PCM was characterized on an electric heater to verify the heat storage properties. A voltage gauge and an electric current meter were connected with the electric heater so that changes in voltage and current could be observed while the heater was turned on. The heating time on the electric heater was 1 hour at the temperature shown in Table 2.

Table 2. Influence of heat storage materials set on an electric heater

Sample	power on/off times/h	Voltage(V)	Current(A)	heating time(s)
0#	22	118	7.3	11.48
PCM1#	9	118	7.3	11.43
PCM2#	12	118	7.3	11.44

In this case, 0# means no PCM was placed inside the porous ceramic on the electric heater. The data in Table 2 indicates that the electric heater work time is reduced greatly when the PCM is used. 0#'s result was 22 times, compared to 9 times for the 1# and 12 times for 2# for the same heating time. The energy consumption for the different conditions was 60.43wh (0#), 24.61wh (PCM1#), 32.84wh (PCM2#). Obviously the electric energy was reduced, and the longevity of electric heater should be prolonged to some degree. Additionally, the surface temperature of the metal board over the electric heater is distinctly different, with 0# the highest, and PCM1# the lowest. As we know, the coefficient of thermal conductivity of an inorganic material is low, and carbon doping of the heat storage material can increase the heat transmission; however, our results did not produce a satisfactory increase in thermal conductivity for commercialization.

CONCLUSIONS

(1) Alum has a high latent heat (232.4J/g) and a low melting point (91°C), which allows it to be a good heating storage material.

(2) The shape-stabilized material has a good heat storage capacity, and it contain 86 mass% alum salt. The results of experiments with an electric heater indicate that electric energy consumption can be reduced to 24.61wh compared with 60.43wh without the PCM.

(3) The fibrous porous ceramic material has high porosity (> 96%), and can be an efficient carrier for a phase-change salt.

REFERENCES

[1]Ali S. Siahpush. Performance enhancement of solid/liquid phase-change thermal energy storage systems through the use of a high conductivity porous metal matrix, Degree of Doctor of Philosophy, University of Idaho, 2001.

[2]Wang, Zh., Cao, M., Gong, A., Su,Q., Variety Application and Future of the Phased-Change Materials to Store the Heat, Anhui Chem, No.2, pp.8-11(2005).

[3]Xu, Y., Liu, D., The development of research on the thermal storage technology with solid-liquid PCMs, Energy Conservation, No.12, pp.3~7(2000).

[4]Ye,H., Ge,X., The structural and chemical analyses of a kind of shape-stabilized paraffin, Acta. Energ .Solar. Sin. Vol.21, No.4, pp.417-421 (2000).

[5]H Inaba, P Tu. Evaluation of thermophysical characteristics on shape-stabilized paraffin as a solid-liquid phase change material, Heat and Mass Transfer, Vol.32, pp. 307-312 (1997).

[6]Jeng-Maw, C.□D.D.L. Chung., Improvement of the temperature resistance of aluminium matrix composites using an acid phosphate binder Part I Binders, J.Mater.Sci, Vol.28, pp.1435-1446 (1993).

[7]Jeng-Maw, C.□D.D.L. Chung., Improvement of the temperature resistance of aluminium matrix composites using an acid phosphate binder Part II Preforms , J.Mater.Sci, Vol.28, pp.1447-1470 (1993).

[8]Lu, G., Hu, L., The Theoretical Study of Thermal Storage Heat Pump System , Refrigeration. Vol.51, No.2, pp.50-54(1995).

[9]Luo,M., Study on Processing, Structure Characterization and Properties of the Mullite Fiber Porous Ceramic, Degree of Doctor of Philosophy, South China University of Technology, 2003.

[10]Huang,J., Study on Composite Phase Change Energy Storage Materials by Molten Salt Spontaneous Infiltrating Porous Ceramic Performs, Degree of Doctor of Philosophy, Guangdong University of Technology,2005

DEVELOPMENT OF FILTRATION TECHNOLOGY FOR PM2.5 IN DIESEL EXHAUST

Masahiro Yoshikawa, Tomihiko Uemura, Daiki Kan, Takehisa Fukui
Hosokawa Powder Technology Research Institute
No.9,1-Chome,Shoudai,Tajika,Hirakata-shi,Osaka 573-1132,Japan

Tawatchai Charinpanitkul, Wiwut Tanthapanichakoon
Department of Chemical Engineering, Faculty of Engineering, Chulalongkorn University
Payathai Rd.,Patumwan,Bangkok 10330,Thailand

Makio Naito
Joining and Welding Research Institute Osaka University
11-1,Mihogaoka,Ibaraki,Osaka 567-0047,Japan

ABSTRACT

Particulate matter less than 2.5 μm (PM2.5) in diameter has one of the worst impacts on human health. Almost all of the particulates in diesel exhaust are less than 2.5 μm. In this paper, in order to remove PM2.5 particulates completely from exhaust gases at elevated temperature (250 to 400°C), we aim to create light ceramic filters with a micro-pore controlled structure, and to assess their high-temperature filtration performance.

A ceramic fiber candle filter having porosity gradient structure was fabricated from ceramic fiber and colloidal silica sol solution by suction casting. The porosity varies from approximately 79% at the outer surface to 93% on the inside of the candle filter. This porosity gradient structure and the high porosity result in a light weight ceramic filter. The pressure drop across the ceramic filter in a twin-filter filtration test with carbon black as the PM2.5 dust indicates a constant value is achieved within a few hours of operation. The dust collection efficiency of the ceramic filter dried at 100°C was estimated to be almost 100%. The controlled surface pore structure of the ceramic filter might be responsible for the superior filtration performance for PM2.5.

INTRODUCTION

Particulate matter (PM) exhausted from power generators such as a diesel engine and a boiler is one of the primary air pollutants. In particular, particulate matter less than 2.5μm (PM2.5) could contain certain toxic components hazardous to human health. Dust in engine and boiler power generation are mainly composed of particulate matter, and several high temperature dust collection technologies have been proposed and investigated for this.[1-3] For instance, at the power systems development facility (PSDF), a variety of different ceramic filter and metal filter have been tested in a hot-gas filter collecting gasification ash at high temperature (higher than 500°C).[4] However, the exhaust gas in a diesel power generator or engine contains mainly fine dust (PM2.5), and this requires the development of filtration technology to collect PM2.5.

The present study is focused on the development of filtration technology to remove PM2.5 from diesel exhaust gas at intermediate temperatures from 250 and 400°C. Bag filter elements made of organic felt cannot be used higher than 250°C. To establish a ceramic filter element and a new filtration technology, we need not only high efficiency dust collection of fine dust, but also high permeability of exhaust gas. Moreover, the filter element should be light and have high durability and strength. In this study, we aim to create a ceramic filter with controlled micro-pore structure for PM2.5 filtration, and to experimentally confirm the possibility to perfectly collect PM2.5 dust. The ceramic fiber candle filter for the filtration technology was fabricated by the suction casting using silica inorganic binder. The morphology of the ceramic fiber filter is characterized and the filter's filtration performance is evaluated.

EXPERIMENT

Preparation of ceramic fiber filter

Figure 1 shows a fabrication process flow sheet for ceramic fiber filter. Ceramic fiber (Shinnikka Thermal Ceramics Corp., Alumina-silica fiber SC bulk1260), silica colloidal sol solution (Nissan Chemical Industries, Ltd. Colloidal silica Snowtex ST-40), and organic binder (Sanwa Cornstarch Corp., Ltd Cornstarch Cornα) were used as the starting materials for the ceramic candle filter. They were mixed with water for 6hrs using an agitation type mixer to make starting fiber slurry for casting. The weight ratio of the ceramic fiber and silica sol in the slurry was 1.0 and 16.0 wt%, respectively. The slurry was sucked into a stainless steel cast to form a candle filter which was 1m in length (see Fig. 1). The caste candle filter was dried at 40, 70 and 100°C for 44 to 20 hrs. The silica sol bonded the ceramic fibers during the de-hydride reaction during the drying process to form the porous ceramic filter. The bulk density of dried filter was measured using the Archimedes method, and the surface morphology of the filter was observed using a scanning electron microscope (SEM, Model S-3500N, Hitachi Ltd., Japan)

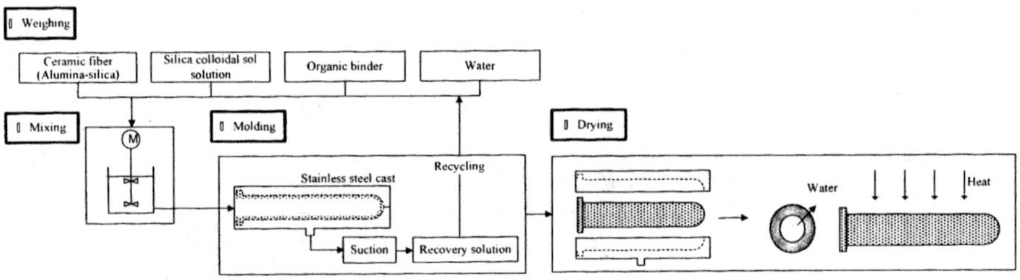

Figure 1. fabrication process for ceramic fiber candle filter

Evaluation of filtration performance of the ceramic fiber candle filter

Figure 2 illustrates the twin-candle filter unit used to test the filtration performance of the ceramic fiber candle filter. The filtration unit consists of two ceramic candles in the vessel, dust

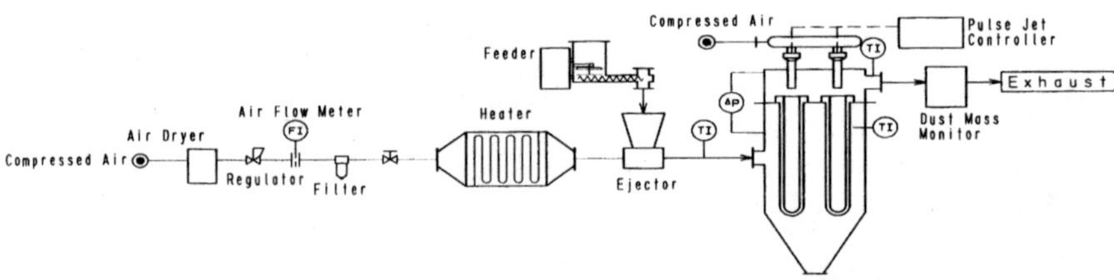

Figure 2. Twin-candle filter units for evaluation of pressure drop and dust concentration.

Table 1. Test conditions for evaluation of filtration performance

Test dust	Carbon black(Class 12/JAPANESE INDUSTRIAL STANDARD JIS Z 8901)	
	Particle size range 0.03 to 0.20 μm	
Dust concentration in feed gas	5 g/m^3 (Actual condition)	
Filtration speed and gas temp.	For 15hours	0.5m/min at 300°C
	For 15hours at a stretch	1.0m/min at 400°C
Pulse jet air for cleaning of filters	Pressure	0.6MPa
	Operation time	80msec
	Interval time	60sec
		(Pulse jet operated alternately for each filter)

feeder, pulse jet system, and hot-gas supply unit. Table 1 shows the filtration test conditions. Carbon-black powder (JIS-12) was used as fine dust for PM2.5 in the tests. The pressure drop and dust concentration after filter cleaning were measured to evaluate the filtration performance.

RESULTS AND DISCUSSION

Morphology of the ceramic fiber filter

The ceramic fiber candle filter was fabricated using a silica sol de-hydration reaction. Figure 3 shows a 1m of long and 120mmϕ outer diameter ceramic fiber candle. The ceramic fiber filter was cut to measure the bulk density of the layers between the surface and the inside of filter (see Fig.4). Figure 5 shows the bulk density distribution between the surface and the inside of the ceramic filter dried at (a)40, (b)70 and (c)100°C. It was determined, that in all of the ceramic filters, the surface has a higher density in comparison with the inside of the filter, and the bulk density decreases from the surface inwards. The porosity at the surface and inside of the ceramic filter was estimated to be ~79 and 93%, respectively, based on the bulk density. Therefore, we successfully created a gradient porosity structure in the ceramic filter. This gradient porosity structure and the high porosity result in a very light weigh ceramic filter. The weight of the ceramic fiber filter is about two thirds that of a ceramic candle filter on the market. The gradient porosity depends on the drying temperature, and the decrease in density of the filter dried at lower temperature is also bigger than that dried at higher temperature. Drying at lower temperature leads to a bigger porosity gradient in the ceramic fiber filter. Figure 6 shows a SEM photograph of the surface of the ceramic fiber filter dried at 70°C. In Fig. 6, it can be seen that the ceramic fibers are bound by silica ceramic. The silica was formed from the silica sol by the de-hydride reaction during the drying process. It is predicted that a slow dry (long time) treatment at lower temperature causes the silica sol to concentrate at the surface of the filter. Therefore, the silica sol concentration might result in the higher bulk density at the filter surface.

Figure 3. Ceramic fiber candle filter

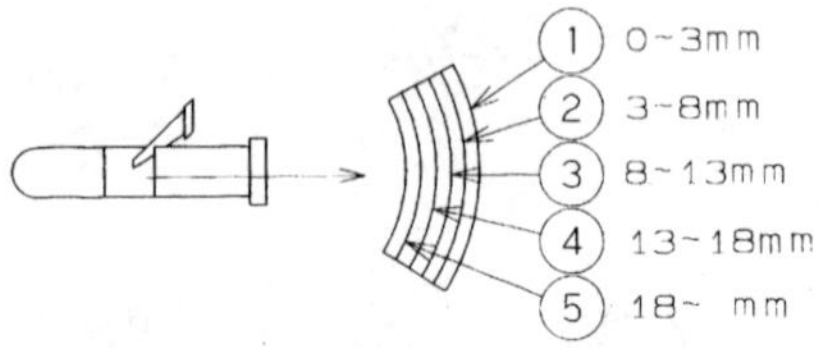

Figure 4. Sampling of filter to measure the bulk density

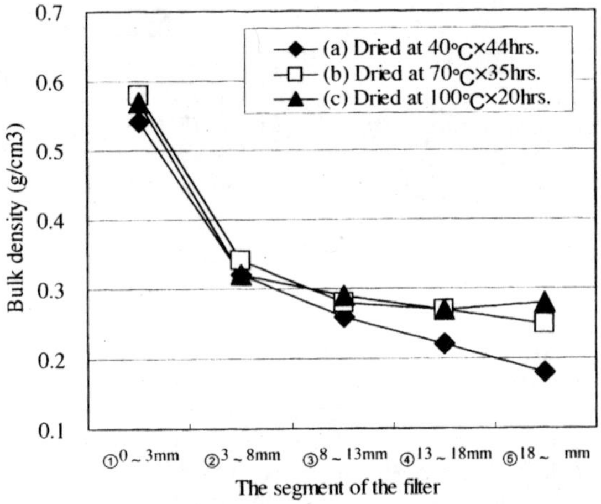

Figure 5. Bulk density from the surface to the inside of the filter

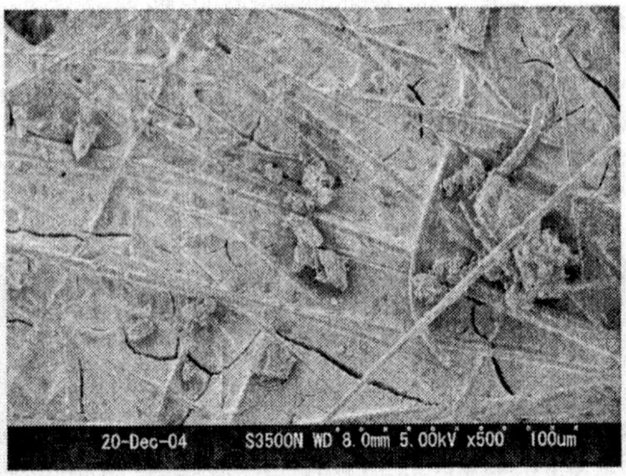

Figure 6. Surface morphology of the ceramic filter dried at 70°C

Filtration performance

Figure 7 shows the time dependence of the pressure drop across the ceramic fiber filter dried at 100°C. At times less than 1 hr, the pressure drop at a filtration speed of 0.5 m/min gradually increases during the filtration test. After 1 hr, it slightly increases, and then becomes almost constant. The pressure drop increases at a filtration speed of 1.0 m/min, and then also becomes almost constant. The other filters showed the same behavior. Table 2 shows results of the pressure drop of all of the filters at a filtration speed of 0.5 and 1.0 m/min. The pressure drop across the ceramic filter dried at 100°C is lower than that of the other filters, and the filter dried at 100°C shows higher filtration performance. Figure 5 shows that the surface bulk density of all of the ceramic filters is nearly the same. However, this surface bulk density reflects the average density of surface of the filter. We suspect that the real surface porosity of the ceramic filter dried at 100°C is higher compared with the other filters, possibly because rapid drying at 100°C results in a small silica concentration gradient in the direction vertical to the filter surface.

Table 3 shows the average dust concentration in the ceramic filter dried at 100°C after cleaning, as measured with a particle mass monitor (Sibata Scientific Technology Ltd. Particle Mass Monitor Model GT-331). The dust collection efficiency of the ceramic filter is estimated to be almost 100% from this table. The controlled surface porosity structure of the ceramic filter may be responsible for the superior filtration performance of fine dust like PM2.5.

Sample	(c) Dried 100°C×20hrs		
Operation time	hr	0-15	15-30
Filtration speed	m/min	0.6	1.0
Average of gas temp.	°C	347	387
Dust concentration	g/m3	4.5	5.0

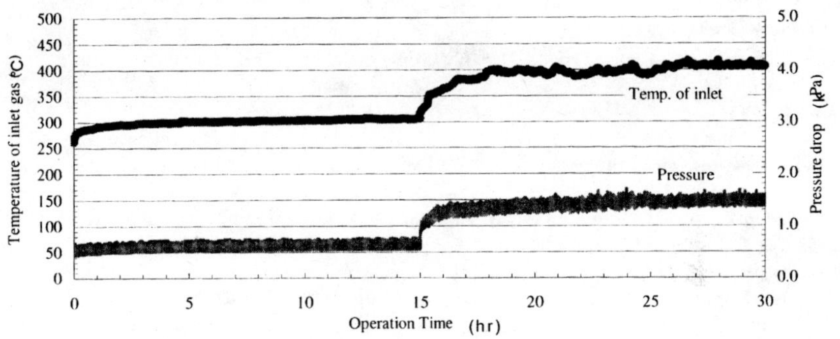

Figure 7. The time dependence of the pressure drop across the ceramic filter dried at 100°C

Table 2. Results of the pressure drop for all of the filters

	Operation time	Sample(a)	Sample(b)	Sample(c)
Pressure drops(F=0.5m/min) [kPa]	initial	0.21	0.21	0.24
Pressure drops(F=0.5m/min) [kPa]	after 15hrs	0.63	0.51	0.61
Pressure drops(F=1.0m/min) [kPa]	after 15hrs	1.81	1.67	1.44

Table 3. The average dust concentration after cleaning, as measured by the particle mass monitor

	Opetation time (hrs)	The average of dust concentration (μ g/m³)		
		Range 0.5 ~ 2.5μ	Range 0.5 ~ 7μ	Range 0.5 ~ 10μ
Filtration speed=0.5m/min	0-1	15.2	108.9	225.8
	1-15	3.7	24.0	43.2
Filtration speed=1.0m/min	15-30	2.5	17.0	28.4

CONCLUSIONS

A 1m long and 120mmϕ outer diameter ceramic fiber candle filter was fabricated from ceramic fiber and a colloidal silica sol solution by suction casting. The silica was formed from the silica sol by the de-hydride reaction during the drying process, and the ceramic fiber were bound by the silica ceramic. The structure of the ceramic filter can be controlled by drying temperature. The ceramic filter has a porosity gradient structure, and its porosity at the surface and the inside are estimated to be ~79 and 93%, respectively. The porosity gradient structure and the high porosity result in a light weight ceramic filter.

The measured pressure drop across the ceramic filter in the filtration test with carbon black as the PM2.5 dust shows a constant value is achieved after a few hours of operation. The dust collection efficiency of the ceramic filter dried at 100°C was estimated to be almost 100%. The controlled surface pore structure of the ceramic filter may be responsible for the superior filtration performance for PM2.5.

ACKNOWLEDGMENT

The present work is supported by NEDO (New Energy Development Organization).

REFERENCES

[1]Kamiya,Hidehiro, Aozasa,Takashi, Tsukada,Mayumi, Matsuda,Hiromitsu, Makino,Hisao, "Direct measurement of the adhesion force for single ash particle at high temperature", *Ceramic Transactions*, **146**, 375-381,(2005)

[2]Hurley,John P., Weber,Greg F., Tests of an oxide dispersion-strengthened high-temperature heat exchanger for indirectly fired combined cycle, "*Proceedings-Annual International Pittsburgh Coal Conference*",21st,**49**.1/1-49.1/15,(2004)

[3]Ito S., Nakayama T, Tanaka T, Kawamura S, "Long Term Characteristics of Porous Ceramic Filter Applied to Hot Coal Gas Cleaning", *Proceedings of World Congress Particle Technology (2nd)*, Kyoto, Japan Part4 144-151,(1990)

[4]Leonard,R.,T.Pinkston,L.Rogers,R.Rush,and J.Wheeldon, "The PSDF-Commercial Readiness for Coal Power Revisited", *Proceedings of the 19th Annual Pittsburgh Coal Conference*, Pittsburgh, Pennsylvania September 23-27,(2002)

SMART PROCESSING TECHNOLOGY

NANOSCALE PARTICLE PROCESSING THROUGH AEROSOL ROUTES

Olivera B. Milosevic, Lidija T. Mancic
Institute of Technical Sciences of Serbian Academy of Sciences and Arts, K.Mihajlova 35/IV, 11000 Belgrade, Serbia

Maria Eugenia Rabanal, Luz Gomez, Jose Manuel Torralba
Universidad Carlos III de Madrid, Avda de la Universidad 30, 28911 Leganés, Madrid, Spain

ABSTRACT

There is a wide range of processing methods being applied in nanosynthesis many of them relating to sophisticated chemical aerosol routes for making clusters, nanoparticles or nanostructured materials. The aerosol process involves formation of discrete droplets of precursor solution and control over their thermally induced decomposition and phase transformation. Generally, it represents a "bottom-up" approach and provides control over a variety of important parameters for nanoscaled particle processing. Moreover, the extreme synthesis conditions, i.e. high heating and cooling rates and short residence time, associated with the aerosol routes, favors to the formation of metastable crystalline and amorphous phases with nanometric grain size. This will likely have a huge impact in the search for advanced functional materials having novel and unique structures and properties. In this article, the aerosol synthesis of uniform submicron spherical particles with a nano-clustered inner structure of two different multicomponent functional oxide systems, Pt-doped ZnO and Gd_2O_3:Eu, will be presented. The local structure of the non-equilibrium phases will be evaluated through a combination of different analysis techniques, including analytical and high resolution transmission electron microscopy (HR-TEM) and x-ray powder diffraction (XRPD).

INTRODUCTION

The field of nanoscience and nanotechnology has had exciting progress in recent years, particularly regarding the control of solid-state functional materials and device synthesis[1]. The potential of aerosol routes for making solid-state structures at the nano level are virtually unlimited, providing possibilities for new and unique applications[2]. The diversity of aerosol routes can be classified based on the thermal energy source for aerosol decomposition and method of aerosol droplet /particle size distribution formation, as presented in Table I.

The degree of particle morphology control relates to the control over the interfaces and solute nucleus distribution, their bonding and assembling into clusters and coalescence to particles in the dispersed state. Since the droplets evaporation/decomposition mechanisms occur rapidly during spray pyrolysis (few miliseconds) and is spatially limited to a few microns for a droplet, long range diffusion is typically constrained. Consequently, the equilibrium phases can be kinetically suppressed, leading to the formation of metastable structures[19]. This excess energy stored in the system results in the phase field to be extended allowing for significant non-stoichiometry which will likely lead to unique functionality. Several attempts have been reported for the synthesis of metastable structures *via* spray pyrolysis, however most of them relating to the synthesis in flames[20-24]. This work discloses the capability of the aerosol hot wall reactor route for metastable phase synthesis. Non-equilibrium phase processing will be demonstrated for the case of nanostructured Pt-doped ZnO and Gd_2O_3:Eu multicomponent oxides.

Table I. Aerosol routes for nanoscaled/nanostructured materials synthesis

A. The manner of aerosol decomposition				
Hot Wall Reactors[3-8]	Flame[9-11]	Self-sustaining flame[12]	Self-combustion aerosol synthesis[13,14]	Aerosol assisted thin film preparation[15]
B. Droplet/particle size distribution				
Monodispersed droplets/particles		Polydispersed droplets/particles		
Evaporation-Condensation[16]	Electrospray aerosol route[17] Differential mobility classifier[18]	Twin-fluid atomization	Ultrasonic atomization	

EXPERIMENTAL

Pt-doped ZnO particles were prepared by combining $0.08 \cdot 10^3$ mol m^{-3} (water solution) of $Zn(NO_3)_2 \cdot 6H_2O$ with 100mg of synthesized pdda-Pt(IV) complex in order to obtain a low (<1wt%) dopant concentration. Details of the complex synthesis are presented elsewhere[25]. Precursor solutions were atomized ultrasonically (atomizing frequency, $1.7 \cdot 10^6 s^{-1}$) into a tubular flow reactor at 600°C with the residence time of 5s for decomposition. The $Gd_2O_3{:}Eu^{3+}$ were prepared as described previously[8] from $Gd(NO_3)_3.x6H_2O$ and $Eu(NO_3)_3.x5H_2O$ common solutions in order to obtain 0.080:0.020 Gd/Eu molar ratio. The solutions were atomized ultrasonically (atomizing frequency, $2.1 \cdot 10^6 s^{-1}$) into the high temperature tubular flow reactor with air as carrier gas (flow rate 1.5 l/min) and decomposed at 700°C. As-prepared powders were post-annealed at temperatures ranging from 800 to 1200°C for 12 hours.

The morphology evolution and phase identification were carried out by x-ray powder diffraction (XRPD, Philips PW 1050) in combination with analytical and high resolution transmission electron microscopy (TEM, JEOL-100CX and TECNAI 20 Philips, operated at 200 kV, the latter equipped with EDX TEM Quant and a point resolution of 2.0Å). XRD patterns were created with CuKα (1.5418Å) radiation in a X`Pert Philips automatic diffractometer. Rietveld based structural refinements were carried out using the programs Fullprof and Koalariet-Xfit[26]. The 2Θ range analysed was 10° to 100° with a step size of 0.02° and a residence time of 11s for each step. Both programs provided the phase composition and cell parameters.

RESULTS AND DISCUSSION

Fig.1 represents the typical particle morphology obtained by the hot-wall aerosol synthesis (spray pyrolysis), particularly emphasizing the structural hierarchy. As-prepared particles are in submicronic range with a spherical morphology as revealed by the TEM image (Fig.1a). The particles are a composite structure and represent an aggregate of clusters or so called "primary particles" arranged in a spherical assembly-"secondary particle". The size of primary and secondary particles is time-temperature history dependent[27] and precursor and concentration (flux density) dependent as well. Primary particles, mostly below 20nm, arise during the thermally induced precipitation and decomposition within the evaporating droplet[3]. The primary particles can then coalesce, densify and sinter[27]. Depending on the particle's temperature profile the product can be either polycrystalline, monocrystalline or an assemblage of amorphous particles[4].

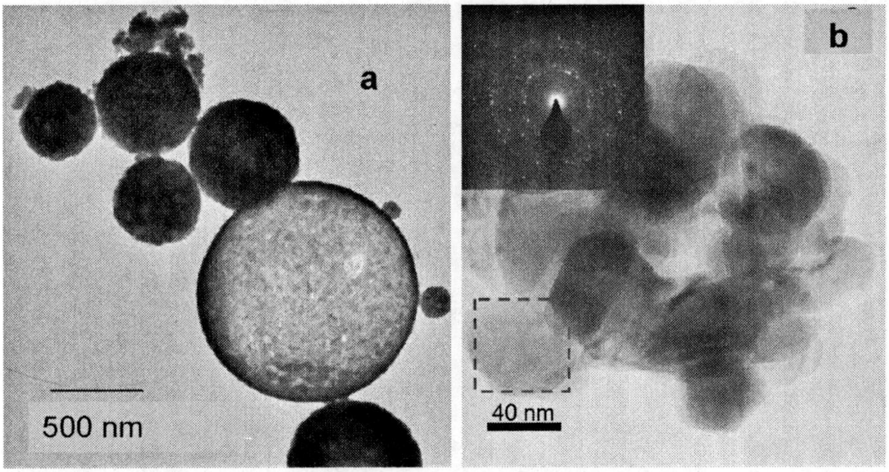

Figure 1. Low(a) and high magnification TEM image with the corresponding SAED (b) for the Gd_2O_3: Eu^{3+} and Pt-doped ZnO particles, respectively.

High magnification TEM analysis (Fig.1b) reveals the secondary particles in the Pt-doped ZnO system are composed of nanoscaled (<40 nm) "primary particles" with interparticle spacings that can be regarded as nanoporosity. Primary particle bonding and neck formation is also evident. The shape of the "primary particles" may be regarded either as irregular spheres or hexagons. Selected area electron diffraction (SAED) analysis of the marked boundary region implies a polycrystalline structure. There also appears to be an amorphous halo. Based on XRPD, a well-crystallized ZnO structure with S.G. $P6_{3mc}$ (JCPDS card 36-1451), having the lattice parameters a=3.2510(1) Å, c=5.2077(1) Å is detected in the powder sample. The peak broadening is evident due to the nanosized crystallites. The crystallite size estimated 24.5 nm from Debye-Scherrer analysis. However, the interplanar spacing estimated based on SAED in the boundary region suggest the presence of one more phase corresponding to the cubic Zn_2PtO_4 phase, *Fd3m* space group (JCPDS file card 23-1499) (Table II). This phase belongs to the inverse spinel family and according to the literature[28], the structure with a=8.549 Å was obtained previously at 900°C, under a high oxygen pressure of 200 atm. The low temperature appearance of the metastable phases, like Zn_2PtO_4, will be discussed later, and is probably associated with the extreme conditions during spray pyrolysis: high heating and cooling rates and short residence time.

XRD analysis in combination with TEM/SAED patterns proved the presence of the cubic *Ia3* phase in the Gd_2O_3: Eu^{3+} powder samples calcined/crystallized in the range 800-1200°C, as previously reported[8]. However, detailed TEM observations revealed the existence of an additional monoclinic well ordered phase for the samples treated at 1100°C/ 12 hours (which was not observed in the XRPD patterns). The bright field HR-TEM image of Gd_2O_3 : Eu^{3+} calcined to 1100°C / 12 hours that reveals the minority monoclinic phase is shown in Fig.2a.

Table II. The values of interplanar spacings (d) for the corresponding metastable phases

Zn_2PtO_4,	**d(Å)$_{(JCPDS\ 23\text{-}1499)}$**	3.023	2.578	1.961	1.745	1.645	1.445	1.428	1.113
S.G. *Fd3m*	**d(Å)$_{EXP^*}$**	2.996	2.596	2.026	1.705	1.695	1.487	1.418	1.147
Gd_2O_3: Eu^{3+},	**d(Å)$_{JPDS\ 43\text{-}1015}$**	8.6	4.31	3.15	2.86	2.739	2.43	2.1	1.91
S.G. *I2 C2/m*,	**d(Å)$_{EXP}$**	8.62	4.312	3.158	2.875	2.755	2.431	2.131	1.913

*Calculated for the beam intensities (Digital Micrograph TEM JEOL, Peabody, MA) λ 0.0025 nm, chamber length 730 mm

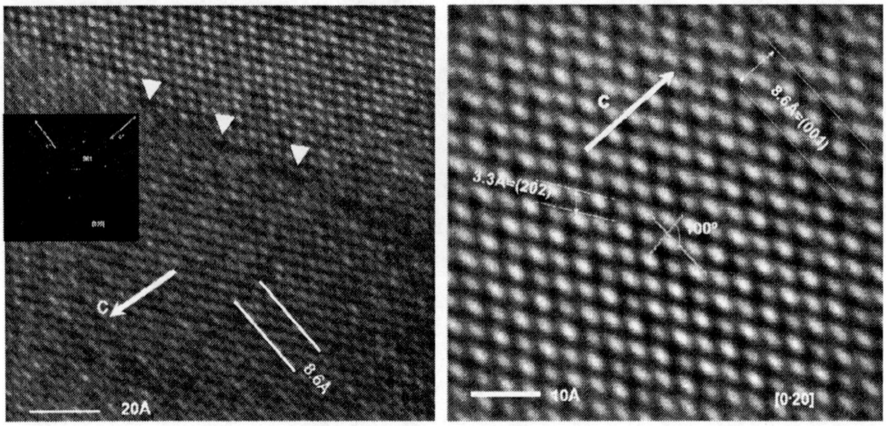

Figure 2. HR-TEM image in bright field of the Gd_2O_3:Eu^{3+} treated at 1100ºC/12 hours (a); FFT calculated for the lower area indexed based on *C 2/m* phase along the zone axis [002] (inset); Fourier Filtered HR-TEM (b)

The Fast Fourier Transform (FFT), calculated for the lower area and indexed based on a *C2/m* symmetry along the zone axis [002], is presented in the inset. The approximate unit cell parameters of this phase based on SAED pattern are: a=13.6 Å, b=3.56 Å, c=8.6 Å (a:b:c= 3.82:1:2.41), ß= 100.6°, and correspond to a monoclinic structure (JCPDS= 43-1015: a=14.06 Å, b=3.55 Å, c= 8.56 Å; α= γ=90° ß=100.1) (Table II). The image clearly resolved a grain boundary (marked with arrowheads) with small defects, such as vacancies and atom displacements.

A HR-TEM Fourier filtered enlargement of the image is shown in the Fig.2b. The (001) and (202) planes and the c axis direction are shown that correspond to a monoclinic phase with *C 2/m* symmetry along the [0¯20] axis zone. The angle between the a* and c* axis corresponds to 100°. The small atomic shifts that are resolved in Fig 2a, confirm the presence of defects near to the grain boundary.

For the case of the rare earth sesquioxides, a monoclinic phase is a metastable at ambient conditions and could be obtained by quenching from high temperatures and under high pressure[22]. However, the short residence time and high cooling rates associated with the aerosol route may favored to the formation of the metastable monoclinic structure at temperatures ≤1100°C. Similar observations are reported in the case of Y_2O_3: Eu^{3+} phosphor system obtained through the gas-phase, where the monoclinic phase was resolved in the particles less than 20nm[22, 23]. The stabilization of the higher density metastable monoclinic structure is probably attributed to the Gibbs-Thomson effect in which the increased surface tension, as a characteristic of nano particles, converts the particle to the denser metastable phase[22].This phase appeares locally in the examined Gd_2O_3: Eu^{3+} system, having short distance order and being unresolvable by XRD, however visible on TEM observations.

CONCLUSION

Aerosol route is applied for the synthesis of nanostructured submicronic-sized spherical particles in two multicomponent systems: Pt-doped ZnO and $Gd_2O_3:Eu^{3+}$. The particle morphology and phase content are evaluated with analytical and HR-TEM in combination with XRPD. Based on XRPD, only a well-crystallized ZnO structure with S.G. $P6_{3mc}$ (JCPDS card 36-1451), having the lattice parameters a=3.2510(1) Å, c=5.2077(1) Å is detected in the Pt-doped ZnO particles. However, detailed TEM/SAED analysis revealed a metastable cubic Zn_2PtO_4 inverse spinel phase (S.G.*Fd3m,* JCPDS file card 23-1499), located in the boundary region. In the 1100°C annealed europia-doped-gadolinia, HR-TEM investigations revealed the presence of a metastable monoclinic *C2/m* structure, although XRPD patterns implied the presence of only the cubic *Ia3* phase.

ACKNOWLEDGEMENT

The research is supported though The Ministry of Science and Environment Protection, Republic of Serbia (# 142010) and COST 539. The authors thank to G.del Rosario for the part of TEM analysis. The first author, O.M. gratefully acknowledges the support of the Ministry for Education and Science, Spain for the Sabatic Grant SAB 2004-0035.

REFERENCES

[1]Tuller, H., "Solid State Electrochemical Systems-Opportunities for Nanofabricated or Nanostructured Materials", *J. Electroceram.*, 1:3, 211-218, (1997)

[2]L.Madler, "Liquid-fed Aerosol Reactors for One-step Synthesis of Nano-structured Particles", *KONA*, 22, 107-119 (2004)

[3]G. L .Messing, S. - C. Zhang and G. V. Jayanthi, "Ceramic powder synthesis by spray pyrolysis", *J.Am.Ceram.Soc.***76** (11): 2707 (1993).

[4]T.T.Kodas and M.J.Hampden-Smith, Aerosol Processing of Materials, Wiley-VCH, New York, first edition, (1999)

[5]K.Okuyama and I.W.Lengorro, "Preparation of Nanoparticles *via* Spray Route", *Chem.Eng. Sci.,* **58**, 537-547 (2003)

[6]Gurav, A., Kodas, T., Kauppinen, E., Joutsensaari, J. and R. Zilliacus, "Phase evolution and gas-phase particle size distributions during spray pyrolysis of (Bi,Pb)-Sr-Ca-Cu-O and Ag-(Bi,Pb)-Sr-Ca-Cu-O powders", *Nanostr. Mater.,* **4 (5),** 583–589 (1994)

[7]T.Fukui, T.Oobuchi, Y.Ikuhara, S.Ohara and K.Kodera, "Synthesis of (La,Sr)MnO_3 - YSZ composite particles by spray pyrolysis", *J.Am.Ceram.Soc.* **80**(1): 261 (1997)

[8]O. Milosevic, L. Mancic, M.E. Rabanal, B. Yang, and P.D.Townsend, "Structural and Luminescence Properties of $Gd_2O_3:Eu^{3+}$ and $Y_3Al_5O_{12}:Ce^{3+}$ Phosphor Particles Synthesized *via* Aerosol", *J.Electrochem. Soc.*, **152,** 707-713, (2005)

[9]S. E.Pratsinis and S.Vemury, *Powder Techn.* **88**, 267-273 (1996)

[10]M.R.Zachariah and S. Huzarewicz, "Aerosol processing of YBaCuO superconductors in a flame reactor", *J.Mater.Res.*, 6(2), 264-269 (1991)

[11]J.J.Helble, "Combustion aerosol synthesis of nanoscale ceramic powders", *J.Aerosol Sci.*, 29 (5-6), 721-736 (1998)

[12]T.Tani, N.Watanabe and K.Takatori, "Emulsion combustion and flame spray synthesis of zinc oxide/silica particles", *J.Nanopart.Res.*, 5(1-2), 39-46 (2003)

[13]L.Mancic, O.Milosevic, B.Marinkovic, M.F.de Silva Lopes and F.Rizzo, "The influence of urea on the formation process of Bi-Pb-Sr-Ca-Cu-O superconducting ceramics synthesized by spray pyrolysis method" *Mat. Sci. Eng.: B*, Vol **76(2)** 127-132, (2000)

[14]S.Lee, T. Son, J.Yun, H.Kwon, G.L.Messing and B.Jun, "Preparation of $BaTiO_3$ nanoparticles by combustion spray pyrolysis", *Mater.Lett.*, 58, 2932-2936 (2004)

[15]P.S.Patil, "Versatility of chemical spray pyrolysis technique, *Mater. Chem. Phys.*, **59**, 185-198 (1999)

[16] M. Ocaña and E. Matijevic, "Preparation of uniform colloidal dispersions by chemical reactions in aerosols—V. Tin(IV) oxide", *J.Aerosol Sci.*, **21**(6), 811-20 (1990)

[17]I.W.Lenggoro, K.Okuyama, J.Fernandez de la Mora and N.Tohge, "Preparation of ZnS nanoparticles by electrospray pyrolysis", *J.Aerosol Sci.*, **31** (1), 121-136 (2000)

[18]F.Iskandar, H.Chang and K.Okuyama, "Preparation of microencapsulated powders by an aerosol spray method and their optical properties", *Adv.Powder Technol.*, **14**, 349-367 (2003)

[19]C.G.Levi, "Metastability and microstructure evolution in the synthesis of inorganics from precursors", *Acta.Mater.*, **46**(3), 787-800 (1998)

[20]A.S.Gandhi, V.Jayaram and A.H.Chokshi, "Low temperature densification behaviour of metastable phases in Zr_2O_3–Al_2O_3 powders produced by spray pyrolysis", *Mat.Sci.Eng.*, A304-306, 785-789 (2001)

[21]K.K.Lee, Y.C.Kang, K.Y.Jung and J.H. Kim, "Preparation of nano-sized BaTiO particle by citric acid-assisted spray pyrolysis", *J.AlloysComp.*, 395, 280-285 (2005)

[22]D.Dosev, B. Guo and I.M.Kennedy, "Photoluminescence of Eu^{3+}:Y_2O_3 as an indication of crystal structure and particle size in nanoparticles synthesized by flame spray pyrolysis", *Aerosol Sci.*, **37,** 402-412, (2006)

[23]A.Camenzind, R. Strobel and E. Pratsinis, "Cubic or monoclinic Y_2O_3 : Eu^{3+} nanoparticles by one step flame spray pyrolysis", *Chem. Phys. Lett.*, **415**, 193-97, (2005)

[24]R M. Laine, T. Hinklin, G. Williams and S.C. Rand, *J.Metastable Nanocryst. Mater.*, 8, 500, (2000)

[25]V. M. Djinovic, L. T. Mancic, G. A. Bogdanovic, P. J. Vulic, G. del Rosario, T. J. Sabo and O. B. Milosevic, "Aerosol synthesis of pure and Pt-doped ZnO particles using nitrate and pdda-Pt(IV) complex solutions", *J. Mater. Res.*, **20** (1) 102-113, (2005)

[26]R.W.Cheary, and A.A.Coelho, Programs XFIT and FOURYA, CCP14 Powder diffraction Library-Daresbury Laboratory, Warrington, England (1996).

[27]W. Koch and S. K. Friedlander, "Particle growth by coalescence and agglomeration", *J.Aerosol Sci.*, **21**(1),S73-S76, (1990)

[28]O. Muller and R. Roy, "Synthesis and crystal structure of Mg_2PtO_4 and Zn_2PtO_4", *Mat.Res.Bull.* ,**4** (1), 39-43 (1969)

A TREATMENT OF CARBONACEOUS WASTE CONTAINING METALS BY STEAM PLASMA

Hiroshi NISIKAWA[1], Sumihiro HIGUCHI[2], Manabu TANAKA[1] and Tadashi TAKEMOTO[1]

[1] Joining and Welding Research Institute, Osaka University
11-1 Mihogaoka, Ibaraki, Osaka 567-0047, JAPAN

[2] Graduate School of Engineering, Osaka University
2-1 Yamadaoka, Suita, Osaka 565-0871, JAPAN

ABSTRACT

A feasibility study has been conducted to determine whether steam plasma can be used for the treatment of carbonized wastes after pyrolysis. The gasification of carbonized wastes using steam plasma has been studied to reduce the weight and volume of them and to produce the combustible gas like hydrogen from them. The gasification applied to this study is often called "water gas reaction ($C + H_2O \rightarrow CO + H_2$)".

In this study the thermal plasma generated by DC plasma discharge was adopted as a heat source, where steam was added to react with carbon. Graphite containing metal wire was used as a test piece instead of carbonaceous waste. The effect of metals in graphite on the weight reduction of graphite and the gas generation was investigated. The result indicates that the weight reduction of graphite and the production of hydrogen by steam plasma were confirmed and clearly increased by containing metals in graphite.

INTRODUCTION

In Japan each inhabitant is estimated to produce about 1.1kg/day of domiciliary solid waste and the total quantity is about 5.2×10^7 tons/yr. Moreover, total quantity of industrial waste is about 4.0×10^8 tons/yr. Landfill and combustion are mainly used for disposal of these wastes. Recently, combustion has become an important treating method for combustible solid wastes, especially in small counties like Japan due to the difficulty to get suitable sites for landfill. However, due to the growing awareness regarding environmental issues, various issues of combustion and incineration are pointed out. For example, some toxic waste materials such as some organic compounds of chlorite do not break up and new toxic substrates such as dioxins, furans and others can be formed when they are burned at low temperature. On the other hand, incinerated ashes contain a considerable amount of hazardous materials such as dioxins and heavy metals and it is an important issue to dispose of them[1,2)]. If these hazardous materials are improperly treated, they shall cause detrimental secondary contamination. Therefore, the hazardous wastes need appropriate treatment to become reasonably safe for environment.

In these years, plasma treatment has been widely studied and partially put into practical use. Plasma, which is known as the fourth state of matter, is a quasi-neutral gas consisting of a large amount of charged and neutral active species. Arc plasma has been widely used in a number of fields such as ceramics and metallurgy industry because of its two peculiar characteristics, i.e. high temperature and high densities of ions and electrons. Therefore, plasma treatment is a high-energy technology to be able to treat a wide variety of very toxic wastes and can either destroy toxicity or produce a new product from the treatment of waste materials[3-7)].

In some cases, steam plasma which is the plasma with steam is used for the treatment of carbonaceous materials[8,9)]. Due to economical and chemical reasons, steam plasma is best suitable method for the decomposition process and the gasification process. In the environmental area, a reactive plasma containing radials like H, OH and O is desired to suppose chemical process. Gasification of solid carbonaceous materials means converting carbon and hydrogen contained in the organic materials into fuel gases, carbon monoxide and hydrogen gas. Some researches into gasification have been conducted to effectively produce fuel gases from waste organic materials.

A feasibility study has been carried out to determine whether thermal plasma with steam, i.e. steam plasma, can be applied for the treatment of carbonized wastes. In this study thermal plasma generated by DC plasma discharge was adopted as a heat source, where steam was added to react with carbon. Graphite containing metal wire was used as a test piece instead of carbonized waste. Graphite was suitable to investigate the reaction between carbon and steam plasma because graphite was made from carbon and it was relatively stable at high temperature. The effect of metals in graphite on the weight reduction of graphite and the gas generation was mainly investigated.

EXPERIMENT

Figure 1 shows the schematic diagram of experimental apparatus used in this study. The apparatus consists of a DC plasma torch, a plasma gas control system, a steam generator, an exhaust system and a reaction chamber. By using these systems, plasma jet and plasma arc can be arbitrarily generated. All experiments in this study were performed in the reaction chamber at Ar atmosphere. In this study, Ar plasma and steam plasma were applied. Argon gas was used as a plasma gas in every experiment. Steam plasma was generated by spraying Ar plasma with steam. Graphite was used as a test piece instead of carbonized waste. In this study, two kinds of graphite test piece were prepared, i.e. normal graphite and graphite with metal. Regarding the graphite with metal, some Fe wires, which was 10 mm long and 1 mm in diameter, was stung in the graphite as shown in Fig. 2 to investigate the effect of metals in carbonized waste on the weight reduction and the gas generation. Heat treatment of graphite was conducted by two kinds of thermal plasma, that is Ar plasma and steam plasma. Experimental conditions are shown in Table 1. Before and after the experiment, the weight of the test piece was measured to decide the weight reduction during the experiment. The gas produced during the experiment was analyzed to confirm the concentration of the combustible gas in the output gas and heating value of product gas was calculated.

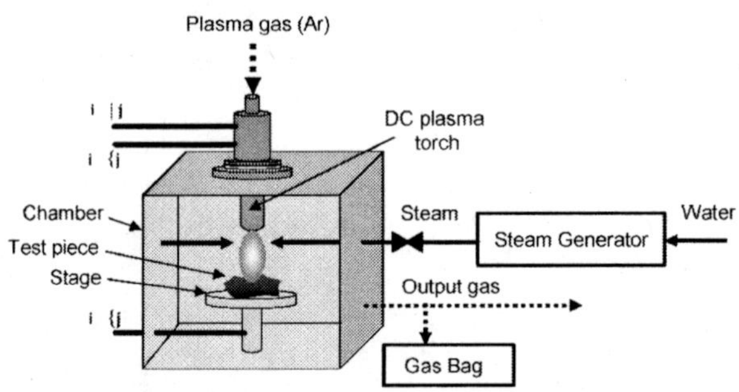

Figure 1. Schematic diagram of experimental apparatus for Ar plasma and steam plasma.

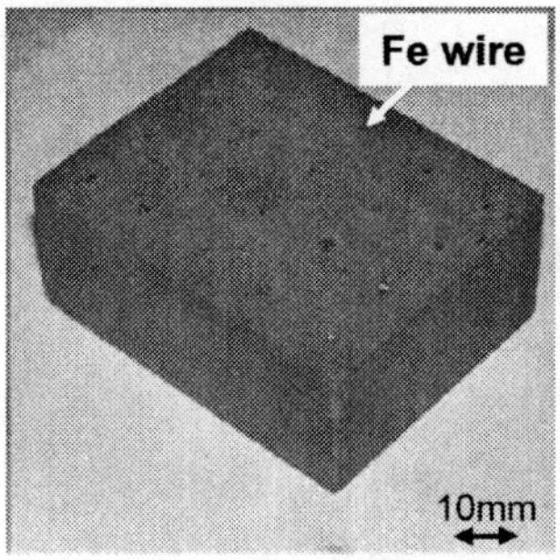

Figure 2 . Appearance of graphite with Fe wire as a test piece.

Table 1. Experimental conditions for gasification experiment.

	Ar plasma	Steam plasma
Test piece	Graphite (40mm×50mm×20mm)	
Plasma type	Plasma arc	
Process time [min]	20	
Plasma current [A]	60	
Arc length [mm]	20	
Ar flow rate [l/min]	Total 30 (Plasma gas:15 + Purge gas:15)	
Atmosphere in chamber	Ar atmosphere	
Steam flow rate [l/min]		85
Steam temperature [°C]		170

RESULTS AND DISCUSSION

Figure 3 shows steam plasma arc in operation on graphite. Usually in the case of Ar plasma arc, cylindrical plasma discharge was stably formed between plasma torch and anode material. In the case of steam plasma arc, as shown in Fig. 3, plasma discharge was slightly unstable and plasma column was diffused on graphite due to the steam sprayed from both sides.

Figure 4 shows the typical change of arc current and arc voltage with time. Fig. 4(a) shows the change of arc current and arc voltage with time for Ar plasma and Fig. 4(b) shows the change of arc current and arc voltage with time for steam plasma. In the case of Ar plasma, both arc current and arc voltage were almost stable. In the case of steam plasma, arc current was almost stable and, on the other hand, arc voltage was higher than that in the case of Ar plasma and fluctuated quickly. This is why the discharge for steam plasma is unstable itself as shown in Fig.2.

Table 2 shows the weight reduction and the reduction rate of graphite during the experiment. In the case of Ar plasma, the weight reduction of graphite was very small regardless

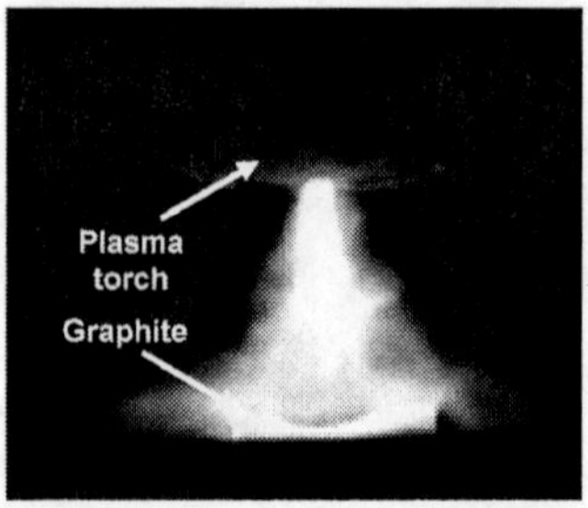

Figure 3. Steam plasma arc in operation on graphite. (Arc current: 100A, arc length: 45mm, steam flow rate: 85l/min, steam temperature: 170□)

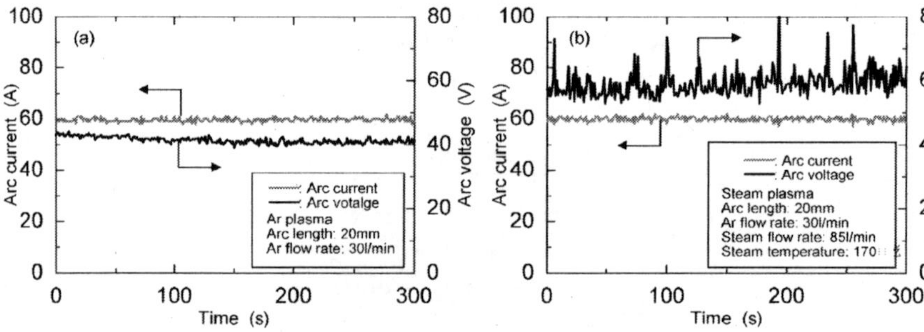

Figure 4. Typical change of arc current and arc voltage with time. (a) Ar plasma, (b) Steam plasma

Table 2. Effect of plasma type and Fe wire in graphite on weight reduction of graphite.

Plasma type	Smaple	Weight reduction (g)	Reduction rate (%)
Ar plasma	Graphite	0.68	0.96
	Graphite with Fe wire	1.22	1.56
Steam plasma	Graphite	2.40	3.26
	Graphite with Fe wire	21.56	28.06

of the existence of Fe wire. On the other hand, in the case of steam plasma, the graphite with Fe wire had large weight reduction compared to the graphite without Fe wire. From the aspect of reduction rate, reduction rate for graphite with Fe wire was about ten times greater than that for graphite without Fe wire. It was clear that the Fe wire in graphite had beneficial effect on the weight reduction for steam plasma.

Concerning the effect of steam on the weight reduction for graphite, the weight reduction of graphite for steam plasma was much larger than that for Ar plasma, especially when Fe wires were stung in the graphite. When organic compounds are heated without oxygen, it is commonly said that pyrolysis takes place without combustion. So in Ar plasma, the weight reduction took place as a result of pyrolysis. On the other hand, in steam plasma, the weight reduction occurred as a result of pyrolysis and gasification. The gasification of carbon with steam plasma is mainly indicated as follows.

$$C + H_2O \rightarrow CO + H_2 \qquad (1)$$

To investigate the cause of the difference in the weight reduction for steam plasma, the surface temperatures of graphite for steam plasma were measured using the infrared thermometer. The surface temperature of graphite with Fe wire ranged from 1000□ to 1200□and the temperature of graphite without Fr wire was almost 1000□. The surface temperature of graphite with Fe wire was certainly higher than that of graphite without Fe wire. It assumed that whole temperature of graphite was difference between the graphite with Fe wire and the graphite without Fe wire. It was clear that the graphite with Fe wire had large weight reduction compared to the graphite without Fe wire because of the difference in the whole temperature of graphite and that the reaction temperature was very important for the gasification.

The samples of the output gas were taken during experiments and were analyzed by gas chromatograph. Table 3 represents H_2, CO, CO_2 and HC concentrations in the output gas. There was little difference in HC concentration between Ar plasma treatment and steam plasma treatment. Usually HC is produced in the pyrolysis process. On the other hand, in the case of H_2 and CO concentrations, when the graphite with Fe wire was treated with steam plasma, fairly high concentrations of H_2 and CO were measured compared to the case of Ar plasma treatment. This result indicates that the gasification of graphite wire took place effectively between steam plasma and graphite.

Table 4 shows the heating value of product gas and energy recovery rate for each process.

Table 3. H_2, CO, CO2 and HC concentrations in the output gas.

Plasma type	Sample	H_2	CO	CO_2	HC
		vol.%	vol.%	vol.%	ppm
Ar plasma	Graphite	0.019	0.015	0.17	6.3
	Graphite with Fe wire	0.0019	0.040	0.6	8.3
Steam plasma	Graphite	0.53	0.037	0.4	5.2
	Graphite with Fe wire	3.3	1.9	2.1	9.8

Table 4. Heating values of product gas and energy recovery rates for each process.

Plasma type	Smaple	Heating value (kJ)	Energy recovery rate (%)
Ar plasma	Graphite	2.87	0.09
	Graphite with Fe wire	1.70	0.05
Steam plasma	Graphite	63.59	1.59
	Graphite with Fe wire	601.27	12.35

The heating values are evaluated under the assumption that combustion products are in the gas phase. The energy recovery rate is defined as the ration of the heating value of product gas to power consumption calculated from the arc current and the arc voltage measured. When the graphite with Fe wire was treated with steam plasma, the heating value of the product gas was much the highest of all processes. And the energy recovery rate for the steam plasma treatment with graphite with Fe wire was more than one hundred greater than that for the Ar plasma treatment. It is considered that the metal in graphite is very effective for the gasification of graphite by steam plasma.

CONCLUSION

Graphite containing metal wire was used as a test piece instead of carbonized waste. The effect of metals in graphite on the weight reduction of graphite and the gas generation was mainly investigated. The main results are summarized as follows:

(1) Plasma discharge was slightly unstable and plasma column was diffused on graphite due to the steam sprayed from both sides.
(2) In the case of steam plasma, the graphite with Fe wire had large weight reduction compared to the graphite without Fe wire. It was clear that the Fe wire in graphite had beneficial effect on the weight reduction for steam plasma and the reaction temperature was very important for the gasification.
(3) When the graphite with Fe wire was treated with steam plasma, fairly high concentrations of H_2 and CO were measured compared to the case of Ar plasma treatment and the heating value of the product gas was much the highest of all processes.
(4) It is considered that the metal in graphite is very effective for the gasification of graphite by steam plasma.

REFERENCES

[1]Y.C.Ling, P.C.C.Hou, A Taiwanese study of 2,3,7,8-substituted PCDD/DFs and coplanar PCBs in fly ashes from incinerators, *J. Hazardous Mater.* , 58, 83-91 (1998).

[2]A.Yazawa, S.Nakazawa N.Menad, Thermodynamic evaluations on the formation of dioxins and furans in combustion gas, *Chemosphere*, 39, 2419-2432 (1999).

[3]T.W.Cheng, J.P.Chu, C.C.Tzeng, Y.S.Chen, Treatment and recycling of incinerated ash using thermal plasma technology, *Waste Management*, 22, 485-490 (2002).

[4]B.Glocker, G.Nentwig, E.Messerschmid, 1-40kW steam respectively multi gas thermal plasma torch system, *Vacuum*, 59, 35-46 (2000).

[5]J.P.Chu, I.J.Hwang, C.C.Tzeng, Y.Y.Kuo, Y.J.Yu, Characterization of vitrified slag from mixed medical waste surrogates treated by a thermal plasma system, *J. Hazardous Mater.*, 58, 179-194 (1998).

[6]J.Qiu, X.He, T.Sun, Z.Zhao, Y.Zhou, S.Guo, J.Zhang, T.Ma, Coal gasification in steam and air medium under plasma conditions: a preliminary study, *Fuel Process. Technol.*, 85, 969-982 (2004).

[7]I.Mohai, J.Szepvolgyi, Treatment of particulate metallurgical wastes in thermal plasmas, *Chem. Eng. Process.*, 44, 225-229 (2005).

[8]P.Grosdidier de Matons W.H.Gauvin, Gasification of peat in a steam plasma reactor, *The Canadian J. Chem. Eng.*, 63, 93-98 (1985).

[9]S.W.Kim, H.S.Park H.J.Kim, 100kW steam plasma process for treatment of PCBs (polychlorinated biphenyls) waste, *Vacuum*, 70, 59-66 (2003).

EFFECT OF ULTRASONICATION ON DISPERSION AND AGGREGATE SIZE OF TiO_2 NANOPARTICLES IN CONCENTRATED AQUEOUS SUSPENSION

Kimitoshi Sato, Ji-Guang Li and Takamasa Ishigaki
National Institute for Materials Science
1-1, Namiki, Tsukuba, Ibaraki, Japan, 305-0044

Hidehiro Kamiya
Tokyo University of Agriculture and Technology, BASE
Koganei, Tokyo, Japan, 184-8588

ABSTRACT

The dispersion and aggregate size of TiO_2 nanoparticles in a concentrated aqueous suspension was controlled by adding polyacrylic acids with various molecular weights ranging from 1,200 to 30,000, and by ultrasonicating the suspension for 0.5 hours. The ultrasonicated slurry, which was combined with polymer dispersants with a molecular weight of 8,000, showed the lowest viscosity below 100 mPa·s for 15 - 30-vol% solid fractions for all of the dispersants used in this study. Also, the aggregate size of the TiO_2 nanoparticles in the concentrated suspension was close to the particle size determined from specific surface area using this dispersant. In particular, the aggregate size of plasma-synthesized spherical TiO_2 nanoparticles was almost equal to the particle size determined by the surface area data.

INTRODUCTION

A steric repulsion interaction caused by adsorption of polymer dispersants enables control of dispersion in a concentrated nanoparticle suspension. The molecular weight of the polymer dispersants required to achieve the minimum viscosity in a nanoparticle suspension decreases as the solid fraction increases because of a short surface distance[1]. The slurries in this work were prepared by ball milling with mm-size balls. Agglomerated nanoparticles do not fragment easily, and should be difficult to disperse perfectly using mm-size-ball milling. Although bead milling with ~30 micron balls has been developed, to our knowledge, no studies have been reported on nanoparticle dispersion. For submicron-sized particles, ultrasonication of suspensions provides an effective tool for eliminating agglomeration. A relationship between the ultrasonication treatment and measured particle size has been reported for submicron-sized particles[2]. Ultrasonication of a slurry led to shock waves by collapsing cavitations, and the shock wave led to collision among particles. The agglomerated particles were eroded and split by the collision[3].

The present work focused on the effect of ultrasonication of a slurry with dispersants on the dispersion and aggregate size. To our knowledge, nanoparticle dispersion in liquid by ultrasonication has not been reported systematically. The effects of the molecular weight of polymer dispersants, and the solid volume fractions were examined on the dispersion and aggregate size in ultrasonically treated suspensions with dispersants.

EXPERIMENTS

Three kinds of TiO_2 as powders were used in this work, nanoparticulate P25 (46.5 m^2/g surface area, Degussa, Germany), plasma-synthesized nanoparticles[4] (42.0 m^2/g surface area), and submicron-sized particles (HT0514, 5.8 m^2/g surface area, Tohotitankogyo, Japan). The concentrated aqueous suspension was prepared by adding various commercial polyacrylic acids (Aldrich, Germany) with various molecular weights ranging from 1,200 to 30,000 and by ultrasonicating the suspension for 0.5 hours. Also, the suspension was prepared by φ5-mm Al_2O_3 ball milling for 24 hours. The content of polymer dispersants and slurry pH were fixed at 0.5 mg/m^2 and pH=8.5 - 9.0, respectively. The solid fraction was varied from 15 to 35 vol%. The flow curve (shear stress against shear rate) was obtained by a concentric cylinder viscometer (550, HAAKE, Germany) at a shear rate ranging from 1 s^{-1} to 300 s^{-1}. The viscosity was determined at the shear rate of 300 s^{-1}. The aggregate size distribution in the slurry was determined by two kinds of methods. An x-ray particle size analyzer (BI-XDC, Nikkisoh, Japan) was used for the low solid fraction of 1 vol%. An ultrasonic attenuation spectrometer (DT1200, Nihonlufuto, Japan) was used for high solid fractions ranging from 15 to 35 vol%.

RESULTS AND DISCUSSION

Figure 1 shows the effect of mixing methods and solid fraction on the flow curve of the slurry. For submicron-sized powders (HT0514, Fig. 1 (b)), viscosity of the suspension made by ultrasonication was almost the same as that made by ball milling. On the contrary, for nanometer-sized powder (P25, in Fig. 1 (a)) the viscosity of suspension after ultrasonication was significantly lower than that of the suspension prepared by ball milling. In the case of ultrasonication, the hysteresis of flow curve was only observed at high solid fraction (35 vol%). The apparent viscosity of the ultrasonicated slurry of a 30-vol% solid fraction was 89 mPa·s at 300 s^{-1} in shear rate.

On the contrary, in the case of ball-milled slurry, the hysteresis of flow curve was observed at a lower solid fraction of 20 vol%, and the apparent viscosity reached 1,000 mPa·s at 300 s^{-1} in shear rate.

The effects of molecular weight and mixing method on apparent viscosity of

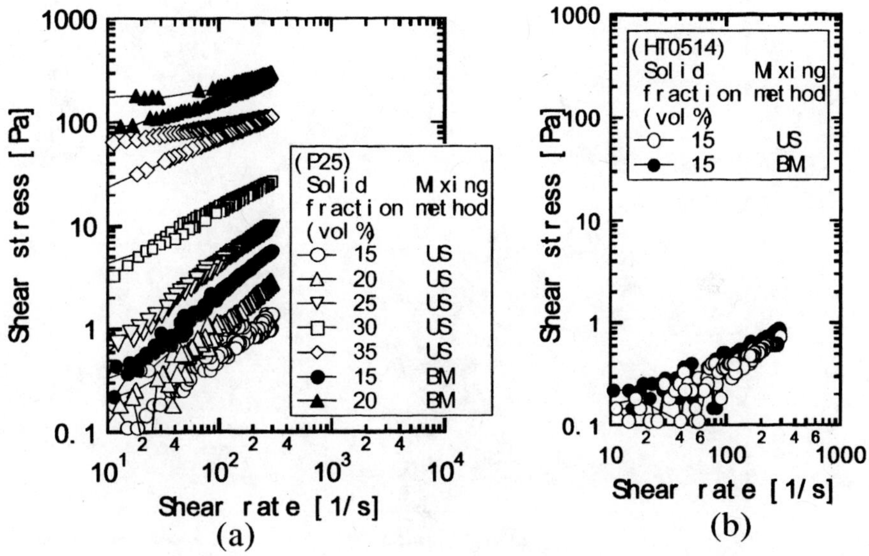

Fig. 1 Flow curves of nano-particle (P25), and sub-micron particle (HT0514) slurries, in which polymer dispersants with 8,000 molecular weights were added. US and BM denote ultrasonication andφ5-mm ball milling, respectively.

nanometer sized powder, P25, were shown in Fig. 2. For the ball-milled slurry with 20 vol% solid fraction, the viscosity with dispersants of any molecular weight was almost the same, and was much higher than that of the ultrasonicated slurry.

On the contrary, for the ultrasonicated slurry, the viscosity depended on the molecular weight of polymer dispersant. By using 8,000 molecular weights dispersant, the concentrated slurry with 35 vol% was still fluid and its viscosity was as small as 300 mPa·s. However, by using other dispersants of 2,100, 15,000, and 30,000 molecular weight, the suspension became a viscous paste at high solid fraction ≧30 vol%. The optimal molecular weight to obtain the minimum viscosity was 8,000.

Figure 3 shows the aggregate-size distribution in suspension with different solid fraction and mixing method. The primary particle size, dp_s, was estimated from the specific surface area, which has values of 46.5, 5.8 and 42.0 m^2/g for P25, HT0514

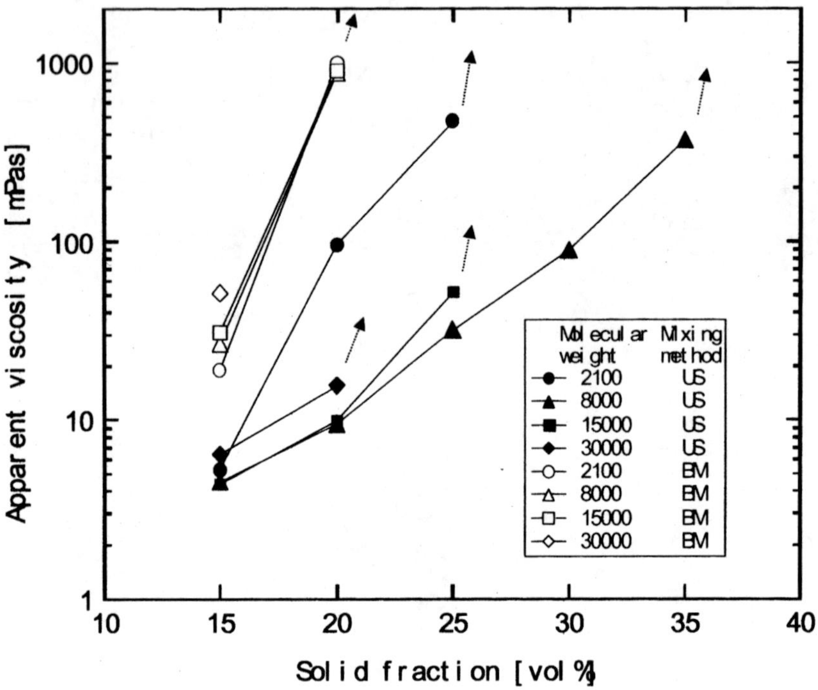

Fig. 2 Viscosity of P25 slurries, in which polymer dispersants with various molecular weights were added. US and BM denote ultrasonication and φ 5-mm ball milling, respectively.

and plasma-synthesized nanoparticle, respectively, by using the following equation,

$$dp_s = \frac{6}{SSA \times \rho p}, \tag{1}$$

where, SSA is specific surface area (m^2/g) and ρ_p is theoretical density (g/cm^3).

For submicron-sized particles, HT0514, the aggregate-size distribution in suspensions prepared by both mixing methods was almost the same, and almost agrees with the estimated primary particle diameter, dp_s. Based on this result, submicron-sized particles (HT0514) by either ball milling or ultrasonication was almost completely dispersed to primary particles in suspension.

The aggregate size of P25 and the plasma-synthesized nanoparticles in the ultrasonicated slurry was appreciably smaller than that in the ball-milled slurry. Aggregates of both nanoparticles could not be completely dispersed by the ball milling

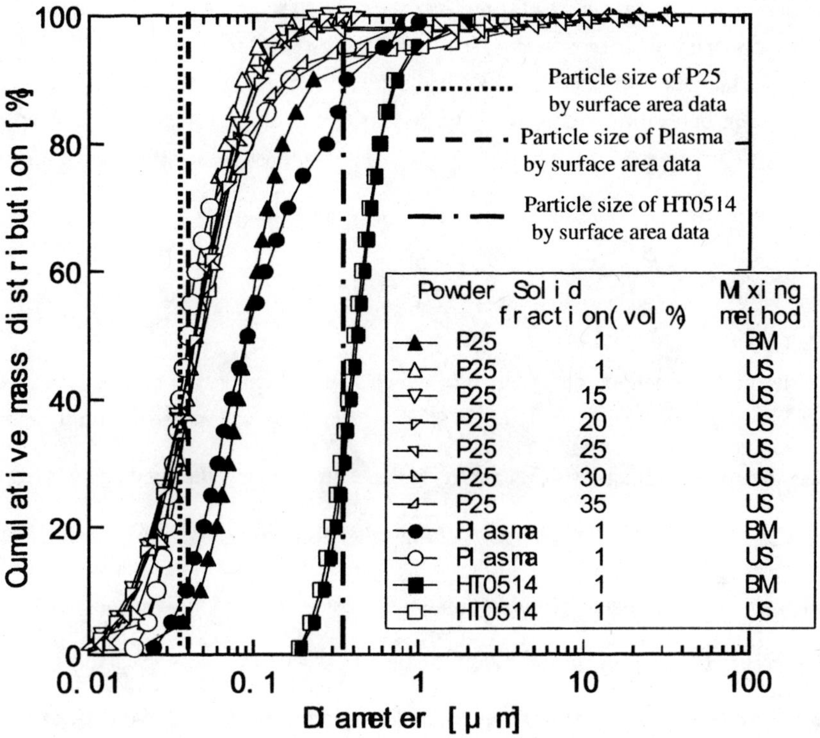

Fig. 3 Particle distribution in slurry. The molecular weights of polymer dispersants that were used were as follows: 8000 for ultrasonicated nanopowders (P25, plasma), 2100 for ball-milled P25, 8000 for both ultrasonicated and ball- milled submicron powder (HT0514). US and BM denote ultrasonication and φ5-mm ball-milling, respectively.

method. Both nanoparticles in suspension prepared by ultrasonication were almost dispersed to primary particles without aggregation. In particular, the aggregate size of the plasma-synthesized spherical nanoparticles was almost equal to the particle size evaluated from the surface are data.

The aggregate size of P25 suspension was close to the primary particle size estimated by specific surface area for slurries with solid fractions ranging from 15 to 30 vol%. It seems that P25 nanoparticle was dispersed to primary particle even if solid fractions increased up to 30 vol%. The size distribution of P25 was independent on solid fraction up to 30 vol% and the mean size is close to the estimated primary particle size, dp_S. However, in the case of 35 vol% solid fraction of P25 suspension, the portion of large aggregate with size over 100 nm increased. Such large aggregates promoted the increase of suspension viscosity at higher than 35 vol% solid fraction.

Summarizing the results obtained in the present work, ultrasonication was more effective than ball milling to decrease the viscosity and aggregate size in nanoparticle slurries with dispersants. Ultrasonication results in homogeneous and frequent collisions among nanoparticles, while ball milling with φ5-mm balls gives only inhomogeneous and insufficient shear stress on nanoparticles.

CONCLUSIONS

Ultrasonicated nanoparticle slurry, in which appropriate polymer dispersants were added, showed very low viscosity and a small aggregate size, compared with the slurry treated by ball milling with φ5-mm balls. Surprisingly, the aggregate size of plasma-synthesized TiO_2 nanoparticles in aqueous suspension was almost equal to the particle size determined from the surface area data. The optimal molecular weight of the polymer dispersants to achieve the minimum viscosity in the ultrasonicated slurry with 15 - 35-vol% solid fractions was 8,000. The slurries with 15 - 30-vol% solid fractions showed low viscosities below 100 mPa·s, and the aggregate size of nanoparticles in the concentrated suspension was close to the particle size determined from the surface area data.

REFERENCES

[1] T. Kakui, Ph.D thesis, “Analysis of action mechanism of polymer dispersant and dispersion control of fine particles in liquid phase for dense suspension”, Tokyo University of Agriculture and Technology, p.p.55-103 (2004)

[2] O. Hayakawa, K. Nakahira, M. Naito and J. Tsubaki, “Experimental analysis of sample preparation conditions for particle size measurement”, Powder Technology, 100, 61 (1998)

[3] S.J. Doktycz and K.S. Suslick, "Interparticle Collisions Driven by Ultrasound", Science, 247, 1066, (1990)

[4] J.-G. Li, H. Kamiyama, X. Wang, Y. Moriyoshi and T. Ishigaki, "TiO_2 nanopowders via radio-frequency thermal plasma oxidation of organic liquid precursors: Synthesis and characterization", J. Euro. Ceram. Soc., 26, 423 (2006)

[5] L. V. Woodcock, "Developments in the non-newtonian rheology of glass forming systems", Proc. of a Workshop on Glass Forming Liquid, 277, 113 (1985)

INVESTIGATION OF BONDING MECHANISM BETWEEN PLASMA SPRAYED Al_2O_3 TOP COATING AND Ti-Al UNDERCOATING ON STEEL SUBSTRATE

Shinichiro Adachi[1] and Kazuhiro Nakata[2]

[1]Technology Research Institute of Osaka Prefecture,
Osaka, Japan, 594-1157

[2] Joining and Welding Research Institute Osaka University
Osaka, Japan, 567-0047

ABSTRACT

Bonding mechanism of plasma sprayed Al_2O_3 top coating onto Ti-Al undercoating made with Ti and Al mixed powders on mild steel substrate was investigated by measuring adhesive strength and observing fractured surfaces of the dual coating system in comparison with single phase undercoatings of Ti or Al. The Ti-Al coating consisted of Al phase and Ti compound phases, but its surface layer was composed mainly of Al splats covered with thin oxide layer of Al_2O. The Al_2O_3 top coating would bond to the Ti-Al undercoating by mechanical anchoring effect.

1. INTRODUCTION

Industrial world of today is required to decrease the usage of hazardous substances for environmental problems. In the surface finishing industry, chrome plating, widely used in many cases, is obtained from the hazardous hexavalent chromium solution. While a plasma spraying is so-called a dry coating process without exhausting any waste solution. And the Plasma sprayed Al_2O_3 coating possesses high hardness, high wear resistance and high corrosion resistance, and also the coating does not contain any hazardous substance.

For the reduction of the environmental burden, we consider applying the Al_2O_3 coating as the alternatives to the chrome plating. However, the Al_2O_3 coating is inferior in quality of adhesive strength, and an undercoating is often deposited on substrate prior to spraying of the Al_2O_3 coating to improve the adhesive strength. Ti and Al are considered to be safe to environment in comparison with heavy metals, so we have developed the Ti-Al coating as the undercoating for the Al_2O_3 coating to achieve highly adhesive and corrosion protection performance. We showed already that the sprayed Ti-Al coating with use of blended Ti-50mass% Al powder consisted of $TiN_{0.3}$, TiNO (TiN-TiO solid solution), Al and small amounts of Ti_3Al by X-ray diffraction, and Ti compound phases of the Ti-Al coating were synthesized during particle flight in plasma jet, and further the adhesive strength of the sprayed Ti-Al coating onto mild steel substrate was 60MPa[1].

As the adhesion mechanism of the sprayed Al_2O_3 top coating on the Ti-Al undercoating, the mechanical and metallurgical bonding would be expected. In this study, as for the mechanical bonding, the effect of surface roughness (*Ra*) of the Ti-Al coating on the bonding strength was examined, and also for the metallurgical bonding the coating surface conditions were characterized by using SEM and ESCA.

2. EXPERIMENTAL PROCEDURE

2.1 PLASMA SPRAYING OF Ti-Al COATING

Ti powder with powder size distributions of 60 - 80μm from Sumitomo Titanium Corporation, and Al powder with powder size distributions of 40 - 60μm from Hikari Sozai Corporation were used as starting materials. The Ti powder was blended uniformly with the 50mass% Al powder in advance, and the blended powder applied for the spraying of Ti-Al coating. The plasma spraying was performed with Argon gas and compressed air as plasma gas in the air atmosphere using an Aeroplasma Limited Company APS7050 system. End face of mild steel bar (JIS SS400, corresponding to ISO E275A) of 25mm diameter was conditioned by grit blasting before the spraying, and then the Ti-Al coatings with thickness of 100-150μm were fabricated. As spray parameters, the input electrical power to the plasma torch of 27kW and the spray distance of 0.1m were selected.

In addition, sprayed Ti coating and sprayed Al coating were fabricated with the same spray parameters. For comparison, the mirror-polished Ti-Al coating surface was prepared by metallographic finish of 0.3μm alumina polishing.

2.2 PLASMA SPRAYING OF Al_2O_3 COATING

Al_2O_3 powder with powder size distributions of 20-40μm from Fujimi Incorporated was used for spraying. The Al_2O_3 coatings with thickness of about 300μm were deposited onto the Ti-Al coating, Ti coating and Al coating, which were formed on the mild steel bar as undercoating. The spraying was performed under the condition of the input electrical power to the plasma torch of 100kW and spray distance of 0.1m using the same spray system as the one used for the Ti-Al coating. Argon gas and carbon dioxide gas were used as plasma gas.

2.3 ADHESIVE TEST

Tensile adhesive strength measurements of the coatings were carried out with an Instron type testing machine at cross head speed of 1mm/min. Coated face of the test specimen was bonded to the end face of the same diameter mild steel bar by an epoxy resin adhesive. Self-aligning device was used for applying the tensile load to the assembly of the coating and fixtures[1].

Number of test samples was 3 to 5.

2.4 CHARACTERIZATION OF COATING

Surface roughness of the sprayed Ti-Al coating, Ti coating and Al coating was evaluated with use of Form Talysurf S5, (Taylor Hobson Limited Company).

Electron Spectroscopy for Chemical Analysis (ULVAC-PHI, INC, ESCA5700) was used to detect the elements and their bonding states on the surface of the sprayed Ti-Al coating with the analysis area of 2mm in diameter.

The Ti-Al coating surface and the cross-section of the Al_2O_3 coating on the Ti-Al undercoating were observed by SEM.

3. RESULTS AND DISCUSSION

3.1 SURFACE STRUCTURE OF Ti-Al COATING

Fig. 1 shows surface images of the sprayed Ti-Al coatings by SEM. Point A with bright contrast was the Ti compound phase, and point B with dark contrast was the Al phase. That is,

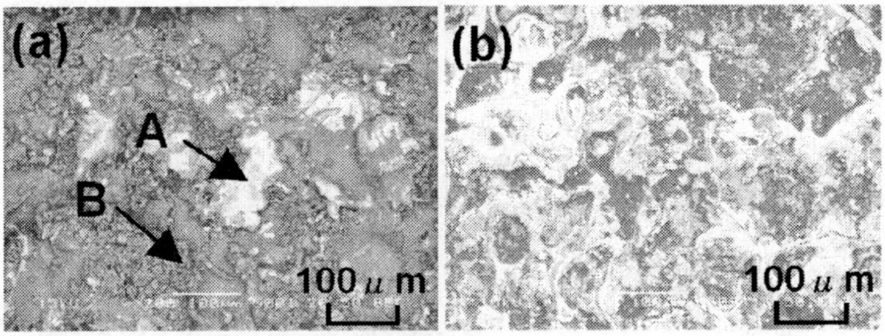

Fig. 1. SEM images of sprayed Ti-Al coating surfaces;
(a) As-sprayed Ti-Al; (b) Mirror-polished Ti-Al.

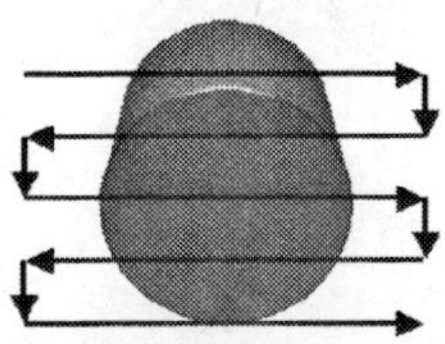

Fig. 2. Traverse of plasma torch on end face of sample bar.

the coating surface (a) composed of mainly Al splats. This phenomenon would be caused by considered that the separation of blended powders of Ti particle and Al particle by difference in particle mass during spraying. The plasma torch traversed on the end face of the sample bar from side to side in horizontal direction, and moved down from the upper to the lower area of the face as shown in Fig. 2. Since the Al particle was smaller mass than the Ti particle, the particles deposited lastly on the coating surface should be mainly the Al particles.

Fig. 3 shows the depth profiles of Ti, Al, O and N by ESCA. From the compositional concentrations at etching time near 0, the top surface of the sprayed Ti-Al coating would be seen to consist mainly of Al_2O.

3.2 STRUCTURE OF INTERFACE BETWEEN Al_2O_3 COATING AND Ti-Al COATING

Fig. 4 shows cross-sections of interface between the Ti-Al coating and the Al_2O_3 coating by SEM. Dual phase structure of Ti-Al coatings is clearly seen in (a). At the interface between the both areas of the Al phase (b) and the Ti compound phase (c), glaring cracks or pores were not observed.

By Suganuma et al, at the interface between pure Al and α- Al_2O_3, reaction layer was not formed[2] even after the heating at 923K for 1h. In this work, thin Al oxide layer covered the Al splat surface, and Ti splat consisted mainly of Ti oxide and Ti nitride. Therefore, it would be reasonable to consider that the reaction layer would not be synthesized at the interface between Al_2O_3 top coating and Ti-Al undercoating.

3.3 BONDING STRENGTH OF SPRAYED Al_2O_3 TOP COATING ONTO Ti-Al UNDERCOATING

Table 1 shows average roughness (*Ra*) of the undercoating surfaces and adhesive strength of sprayed Al_2O_3 coatings. Fig. 5 shows fractured surfaces of the undercoating by the adhesive test, which consists of 3 areas; point A with bright contrast is the Ti compound phase, and point B with half dark contrast is the Al phase, and point C is Al_2O_3 flake.

In case the Al_2O_3 coating delaminated at the interface between the Al_2O_3 coating and the undercoating by adhesive test, the measured adhesive strength should correspond to the bonding strength between them. Whereas the Al_2O_3 coating delaminated inside the undercoating, the measured adhesive strength should correspond to the tensile strength of the undercoating itself. The adhesive strength of the sprayed Al_2O_3 coating onto the Ti-Al undercoating was 50MPa. The Al_2O_3 coating delaminated mainly at the interface between the Al_2O_3 coating and the Ti-Al coating. And some flakes of Al_2O_3 coating adhered to the Ti-Al coating as shown in (a). Therefore the measured adhesive strength 50MPa should correspond to the bonding strength between the Al_2O_3 coating and the Ti-Al coating.

Bonding mechanism of the sprayed Al_2O_3 coating onto metal substrate is generally

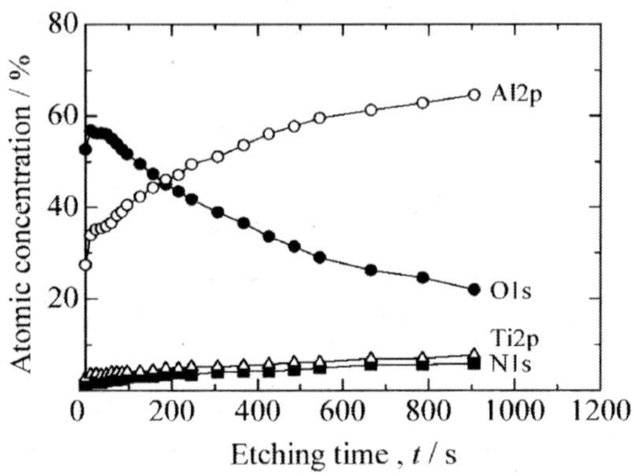

Fig. 3. Depth profiles of compositional elements of sprayed Ti-Al coating surface by ESCA.

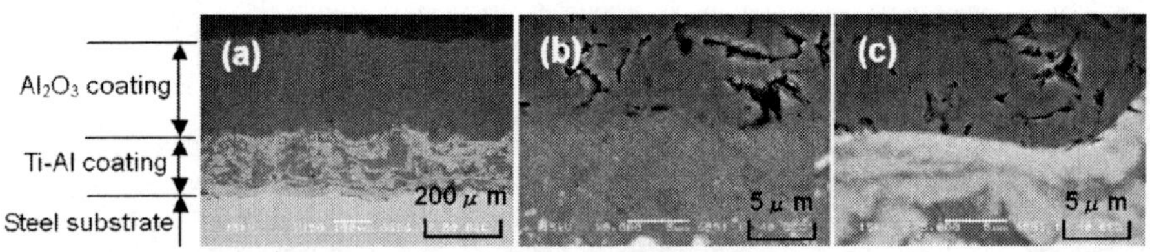

Fig. 4. Cross sections of the interface between Al_2O_3 coating and Ti-Al coating; (a) Overall view; (b) Al_2O_3 / Al phase; (c) Al_2O_3 / Ti compound phase.

Table 1 *Ra* and adhesive strength of samples.

Undercoating sample	Surface state	*Ra* (μm)	Adhesive strength (MPa)	Fracture location
Ti-50mass%Al	As-sprayed	12	50	Al_2O_3/Ti-Al
Ti-50mass%Al	Mirror-polished	0.5	14	Al_2O_3/Ti-Al
Ti	As-sprayed	13	12	Ti coating
Al	As-sprayed	10	43	Al_2O_3/Al
Al	Mirror-polished	0.2	28	Al_2O_3/Al

considered to the mechanical anchoring effect[3]. As for the mirror-polished Ti-Al undercoating, the adhesive strength of the sprayed Al_2O_3 coating onto it was very low to be 14MPa, and the coating delaminated at the interface between the Al_2O_3 coating and the Ti-Al coating as shown in (b). And also, the adhesive strength of the Al_2O_3 coating on the mirror-polished Al undercoating was lower than that of the Al_2O_3 coating on the as-sprayed Al undercoating. These results suggest apparently that the mirror-polished undercoating would not contribute to bond the Al_2O_3 coating by mechanical anchoring due to small *Ra*.

However in this work, it's necessary to consider the effect of the phase of the undercoating surface on the bonding strength. The Ti-Al coating was covered with Al splat layer, in contrast the mirror-polished Ti-Al coating surface consisted of the Ti compound phases and the Al phase

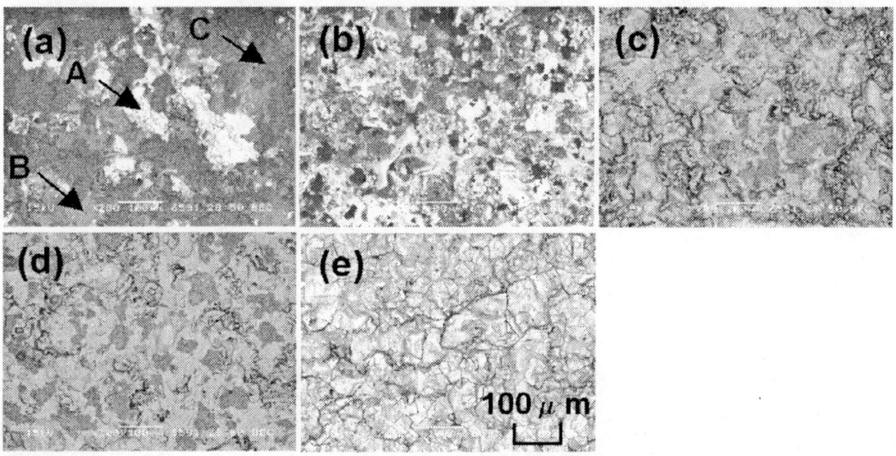

Fig. 5. SEM images of fracture surfaces of sprayed Al_2O_3 coatings onto various undercoating; (a)As-sprayed Ti-Al; (b) Mirror-polished Ti-Al; (c) As-sprayed Al; (d) Mirror-polished Al; (e) As-sprayed Ti.

as shown in Fig. 1 (b).

Since the sprayed Al_2O_3 coating onto the Ti undercoating fractured inside the Ti coating as shown in Fig.5 (e), the measured adhesive strength of 12MPa should correspond to the tensile strength of the Ti coating. Therefore it is difficult to discuss the effect of the surface phase of the Ti-Al coating on the bonding strength by comparing with the adhesive strength of the Al_2O_3 coating on as-sprayed undercoating of Al or Ti.

In Fig. 5 (b) and (d), the sprayed Al_2O_3 coatings onto the mirror-polished undercoating of Ti-Al and Al were delaminated at the interface between the Al_2O_3 coating and the undercoating by the adhesive test. The adhesive strength of the Al_2O_3 coating on the mirror-polished Al undercoating was higher than that of the Al_2O_3 coating on the mirror-polished Ti-Al undercoating as shown in Table 1. These results suggest that the Al phase would bond to the Al_2O_3 coating with higher bonding strength than the Ti compound phases bonding.

4. CONCLUSION

The sprayed Ti-Al coating surface consisted mainly of Al splats covered with thin oxide layer of Al_2O by SEM and ESCA. And the bonding strength of the Al_2O_3 coating on the mirror-polished Al undercoating was higher than that of the Al_2O_3 coating on the mirror-polished Ti-Al undercoating. This indicates the Al phase would have higher affinity with the Al_2O_3 coating than the Ti compound phases.

In addition, bonding strength of Al_2O_3 coating on the mirror-polished undercoating was much lower than that of the Al_2O_3 coating on the as-sprayed undercoating with large roughness for each undercoating. This result suggests that the mirror-polished coating surface would not contribute to bond the Al_2O_3 coating by mechanical anchoring effect due to small *Ra*.

5. REFERENCES

[1] S.Adachi and K.Nakata, “Improvement of adhesive strength of Ti-Al plasma sprayed coating” *Surf. Coat. Technol*, (in press).

[2]K.Suganuma, E.Saiz and A.P.Tomisa, “Microstructure and strength of interface between pure aluminum and α-alumina” *J.Japan Inst.Metals*, **62**, 92-97 (1998).

[3]M.Mellali, P.Fauchais and A.Girmaund, “Influence of the surface roughness and temperature on alumina coating adhesion” *VDI Ber (Ver DtschIng)*,**1166**, 387-393 (1995).

MECHANOCHEMICAL SYNTHESIS OF BARIUM TITANATE FROM NANOCRYSTALLINE $BaCO_3$ and TiO_2

Akira Kondo, Kazuyoshi Sato, Hiroya Abe, and Makio Naito
Joining and Welding Research Institute, Osaka University
11-1, Mihogaoka, Ibaraki, Osaka, 567-0047, Japan

Hirofumi Shimoda
Institute of Nanotechnology and Material Science, Kurimoto,Ltd.
8-45, Shibatani 2-chome Suminoe-Ku, Osaka, 559-0021, Japan

ABSTRACT

Mechanochemical synthesis of $BaTiO_3$ from nanocrystalline powder mixture of $BaCO_3$(50nm) and TiO_2(7nm) was investigated. The XRD peak intensities of $BaCO_3$ and TiO_2 drastically decreased with an increase in milling time at early stage of the milling (<7min). Concomitant with the decrease, the peaks corresponding to $BaTiO_3$ appeared after 10min. After 15 min, the peaks corresponding to $BaCO_3$ and TiO_2 disappeared, and the peaks for $BaTiO_3$ were only identified. The mean particle size of the synthesized $BaTiO_3$ was calculated as 200nm from specific surface area. The results indicated that fine $BaTiO_3$ powder can be mechanochemically synthesized from the nanocrystalline powder mixture in very short times, without external heating.

INTRODUCTION

Barium titanate ($BaTiO_3$) finds extensive application in the electronic industry due to high relative dielectric constant and low losses. Recently, since continued advances in microelectronics and communications are leading to the miniaturization of ferroelectric components, the interest in high quality $BaTiO_3$ powders with small size is increasing. In conventional solid state reactions, difficulty exists due to the growth of grain during calcinations in high temperatures. To prepare fine powders, chemistry-based processing methods[1,2] have been investigated. However, most of these methods are more expensive than the solid state reaction and some of them have significant drawbacks. In particular, powders produced by hydrothermal synthesis or solution precipitation methods often contain lattice hydroxyl groups that lead to the formation of undesired porosity during the sintering.[3] A satisfactory alternative may be a mechanochemical method.[4]

The mechanochemical method is characterized by the repeated welding, deformation and fracture of the constituent powder materials.[5] Chemical reactions occur at the interfaces of the particles that are continuously re-generated during milling. As a consequence, the solid-state reactions are promoted in the milling apparatus without any need for external heating. Stojanovic et al. reported that mechanochemical reaction led to the gradual formation of $BaTiO_3$ phase from the powder mixture of BaO and TiO_2.[6] Xue et al. synthesized single phase $BaTiO_3$ from the powder mixture of BaO and TiO_2 under nitrogen atmosphere.[7] When the powder mixture was milled in air, the formation of $BaCO_3$ was triggered and $BaTiO_3$ phase was not achieved. Gomez-Yanez et al. analyzed the effect of long-term milling times up to 78h on the structure and chemical behavior of the initial constituent $BaCO_3$ and TiO_2.[8] It was shown that mechanical activation accelerated the formation of $BaTiO_3$. Otherwise, long-term milling provoked the formation of Ba_2TiO_4 in addition to the $BaTiO_3$ phase. To date, fine $BaTiO_3$ powder has not been obtained from the powder mixture of $BaCO_3$ and TiO_2 by mechanical activation without additional heat treatment after milling.

In this study, mechanochemical synthesis of $BaTiO_3$ from nanocrystalline powder mixture of $BaCO_3$(50nm) and TiO_2(7nm) was investigated. Although high quality nanocrystalline powders are currently available on the market in large quantities and exhibit a potential advantage for solid state diffusion,[9] they have not been used into mechanochemical synthesis. It was found that fine $BaTiO_3$ powder was synthesized after milling only for 15min from the nanocrystalline powder mixture.

EXPERIMENTAL

Raw materials were nanocrystalline $BaCO_3$ (BW-KS20, Sakai Kagaku, Japan) and nanocrystalline TiO_2 (ST-01, Ishihara Sangyo, Japan). The mean particle sizes calculated from the specific surface area (SSA) were 50nm and 7nm for the $BaCO_3$ and TiO_2, respectively. These two powders were mixed with equimolar ratio, and the powder mixture of 60g placed into the chamber of an attrition type apparatus, as illustrated in Figure 1. This apparatus was previously employed for the mechanochemical synthesis of $LaMnO_{3+\delta}$.[10] Its main components are a fixed chamber and a rotor set with a certain clearance against the inside wall of the chamber. Both the chamber and the rotor were made from stainless steel. When the rotor rotates, the powder mixture is compressed into the clearance (1 mm in gap) and receives various kinds of mechanical forces including compression and shearing. No media balls were used in this milling, and the ambient atmosphere in the chamber was not controlled (i.e., the milling was conducted in air). The rotating speed was 4000 rpm in the present set of experiments. The milled powders were characterized by X-ray diffraction (XRD; JDX-3530M, JEOL,Japan) using Ni filtered Cu-Kα radiation. SSA of the samples was measured by a nitrogen gas adsorption instrument (micromeritics Flowsorb 2300, Shimadzu, Japan) based on the BET method. Particle morphology was observed with a SEM (ERA-8800FE, Elionics, Japan).

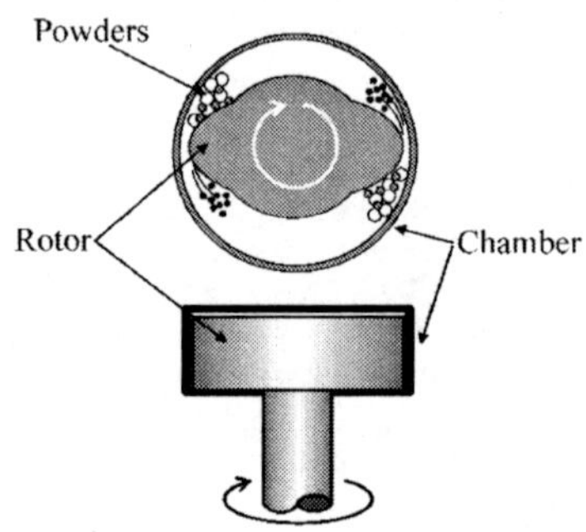

Figure 1. Schematic illustration of apparatus for mechanochemical synthesis of $BaTiO_3$.

RESULTS AND DISCUSSION

Figure 2 shows the XRD patterns of the powder mixture that have been milled for various time periods: 0, 3, 7, 10 and 15min. Only peaks corresponding to $BaCO_3$ and TiO_2 were observed for the powder mixture before milling. The peak intensities drastically decreased with an increase in milling time at early stages of the milling (<7min). Concomitant with the decrease, the peaks corresponding to $BaTiO_3$ appeared after 10min. After 15 min, the peaks corresponding to $BaCO_3$ and TiO_2 almost disappeared, and only the peak intensity for $BaTiO_3$ increased. The crystal structure of the synthesized powder was identified as the $BaTiO_3$ with a pseudo-cubic structure.

The hydrothermally synthesized $BaTiO_3$ powders show the similar crystalline phase, due to the strain on the lattice by the presence of hydroxyl defects in the lattice, and the hydroxyl defect concentration of the commercial hydrothermal $BaTiO_3$ powders has been reported to be 0.75mass% and more.[11] In the present $BaTiO_3$ powder, it was measured as only 0.18mass% in thermogravimetric analysis, suggesting that the present pseudo-cubic phase may be attributed to the defects obtained during the mechanochemical process and/or the fine crystalline domain sizes. This will be addressed in our next paper.

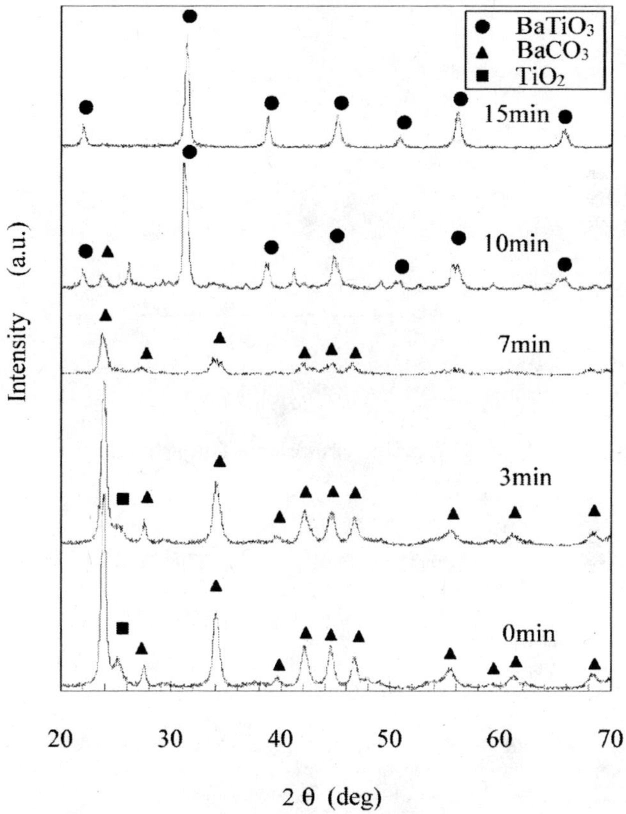

Figure 2. XRD patterns of the powder mixture as a function of milling time.

Figure 3 shows the change of SSA of the powder mixture with increase in milling time. The starting SSA of the powder mixture was $110m^2/g$. The SSA significantly decreased until 7min ($110 \rightarrow 5.5m^2/g$). The mechanical treatment at the early stage of milling resulted particle bonding or aggregation of the powder mixture. In further milling, the SSA remained almost constant, leading to the $BaTiO_3$ phase. The SSA of the powder mixture milled in 15min was about $5m^2/g$, with an equivalent diameter calculated to be 200nm. Figure 4 shows the FE-SEM image of the synthesized $BaTiO_3$ powder. An aggregated powder structure is observed, which consisted of finer, nanosized, particles.

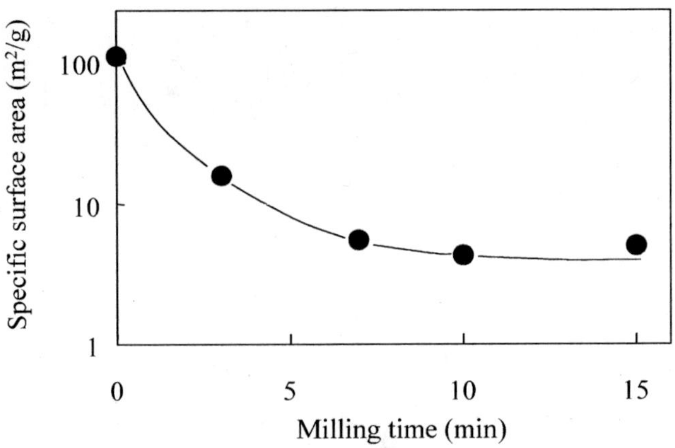

Figure 3. Specific surface area of the powder mixture as a function of milling time.

In this study, it was shown that the single phase $BaTiO_3$ powder was mechanochemically synthesized from the constituent nanocrystalline powder mixture in 15min. The present mechanical milling employed no media balls. It is concluded from Figures 2 and 3 that the following steps proceeded: (1) the aggregation with amorphization of the raw material in the early stage of the milling (<7min), and (2) solid state reaction, led to the formation of crystalline $BaTiO_3$ phase. Any structural defects and disorder together with a degree of amorphization favor diffusion and atomic rearrangement at a reduced temperature.[12]

Localized heating and pressure at regions of contact between the reactant grains may be a contributing factor for the phase formation in the activated matrix.[13] Our results indicated considerable activation took place in the nanocrystalline powder mixture. The phase formation among nanosized grains occurred in the present mechanical milling without any media balls. We also found that the $BaTiO_3$ phase was not realized in 15min when the larger TiO_2 (70nm) was mixed with the same $BaCO_3$ powder, emphasizing the importance of TiO_2 size in the present mechanochemical reaction. The phase evolution shown in Figure 2 is, however, a complicated process, and the detail for the two step reaction will be explained in future reports.

Figure 4. SEM image of the synthesized $BaTiO_3$.

CONCLUSIONS

Fine but agglomerated, $BaTiO_3$ powder was mechanochemically synthesized from the constituent nanocrystalline powder mixture in short times. The mechanical treatment at the early stage of milling (<7min) resulted nano particle bonding and aggregation of the powder mixture, with amorphization of the raw materials. In the further milling, the SSA remained constant, leading to the $BaTiO_3$ phase, with the single phase $BaTiO_3$ identified after milling for only 15min.

ACKOWLEGEMENT

This work was partly supported by Grand-in-Aid for Cooperative Research Project of Nationwide Joint-Use Research Institute on Development Base of Joining Technology for New Metallic Glasses and Inorganic Materials from The Ministry of Education, Culture, Sport, Science and Technology, Japan.

REFERENCES

[1]D.Hennings, "Review of Chemical Preparation Routes for Barium Titanates", *Br.Ceram.Proc.*, **42**, 1-10 (1989).

[2]P.Nanni, M.Viviani, and V.Buscagla, "Synthesis of Dielectric Ceramic Materials", pp.429-55 in *Handbook of Low and High Dielectric Constant Materials and Their Applications*, Edited by H.S.Nalwa, Academic Press, San Diego, CA, 1999.

[3]D.F.K.Hennings, C.Metzmacher, and B.S.Schreinemaher, "Defect Chemistry and Microstructure of Hydrothermal Barium Titanate", *J. Am. Ceram. Soc.*, **84**, 179-82 (2001).

[4]B.D.Stojanovic, "Mechanochemical Synthesis of Ceramic Powders with Perovskite Structure", *J. Mater. Processing Technol.*, **143-144**, 78-81 (2003).

[5]T.Tsuzuki and P.G.McCormick, "Mechanochemical Synthesis of Nanoparticles", *J. Mater. Sci.*, **39**, 5143-46 (2004).

[6]B.D.Stojanovic, A.Z.Simoes, C.O.Paiva-Santos, C.Jovalekic, V.V.Mitic, and J.A.Varela, "Mechnochemical synthesis of Barium Titanate" *J. Eur. Ceram. Soc.*, **25**, 1985-89 (2005).

[7]J.Xue, J. Wang, and D. Wan,"Nanosized Barium Titanate Powder by Mechanical Activation", *J.Am.Ceram.Soc.* , **83**, 232-34 (2000).

[8]C.Gomez-Yanez, C.Benitez, and H.Balmori-Ramirez, "Mechanical Activation of the Synthesis Reaction of $BaTiO_3$ from a Mixture of $BaCO_3$ and TiO_2 powders", *Ceram. Inter.*, **26**, 271-77 (2000).

[9]M.T.Buscaglia, M.Bassoli, and V.Buscaglia, "Solid-State Synthesis of Ultrafine $BaTiO_3$ Powders from Nanocrystalline $BaCO_3$ and TiO_2", *J. Am. Ceram. Soc.*, **88**,2374-79 (2005).

[10]K.Sato, J.Chaichanawong, H.Abe, and M.Naito, "Mechanochemical Synthesis of $LaMnO_{3+\delta}$ Fine Powder Assisted with Water Vapor", *Mater. Lett.* , **60**, 1399-402 (2006).

[11]T.J.Yosenick, D.V.Miller, R.Kumar, J.A.Nelson, C.A.Randall, and J.H.Adair, "Synthesis of Nanotabular Barium Titanate via a Hydrothermal Route", *J.Mater.Res.*,**20**, 837-43 (2005).

[12]N.N.Thadhan, "Shock-induced and Shock-assisted Solid-state Chemical Reaction in Powder Mixtures", *J. Appl. Phys.*, **76**, 2129-38 (1994).

[13]M.L.Trudeau, R.Schulz, R.Dussault, and A.V.Neste, "Structural Changes During High-Energy Milling of Ion-Based Amorphous Alloys: Is High-Energy Ball Milling Equivalent to a Thermal Process? ", *Phys. Rev. Lett.*, **65**, 99-102 (1990).

THE EFFECT OF DIRECT DIODE LASER BEAM SIZE IN HEAT CONDUCTION LAP WELDING OF A THIN FILM ON A THICK SUBSTRATE

Nobuyuki Abe[1], Naoyuki Nakamura[1], Yoshinori Funada[2], and Masahiro Tsukamoto[1]
[1]Joining and Welding Research Institute, Osaka University
11-1 Mihogaoka Ibaraki, Osaka, 567-0047, Japan
[2]Industrial Research Institute of Ishikawa
2-1 Kuratsuki Kanazawa, Ishikawa, 920-8203, Japan

ABSTRACT

A finely focused elliptical beam of a direct diode laser was applied to lap welding of thin foils on thick substrates. This beam shape enabled welding with a narrow bead at a high speed and low power. The welding phenomena were very calm because of its low power density. An elliptical beam of 230μm x 1800μm successfully welded thin stainless steel foil of 5μm in thickness on 1mm thick stainless steel substrate. These characteristics are caused by the heat conduction type welding with the low power density elliptical beam. In this report, characteristics of heat conduction lap welding of thin foil on thick substrate with a finely focused elliptical beam of direct diode laser will be discussed with welding parameters such as power, power density and welding speed.

INTRODUCTION

Recently there has been a growing trend toward miniaturization of electronics and mechanical parts to reduce energy consumption and achieve greater compactness. In order to respond to such demands, the development of an advanced welding method is required for welding thin foils of 50μm or less in thickness. In our previous reports[1, 2], we examined the application of a finely focused elliptical beam of 360μm x 1800μm for edge-, butt-, corner-, and lap-welding of thin foils. This elliptical beam shape enabled welding with a narrow bead at a high speed and low power with very calm welding phenomena. Stainless steel foil of 20μm in thickness could be successfully lap-welded onto 2mm thick substrate with preheating process.

These characteristics are caused by the heat conduction type welding with a low power density elliptical beam. Therefore, it was expected that much narrower beam could be applied to lap-welding of thinner foil. In this report, characteristics of heat conduction lap welding of thin foils on thick substrate with a finely focused elliptical beam of 230μm x 1800μm will be examined for thinner foils with various welding parameters such as power, power density and welding speed. The effect of beam size of a finely focused elliptical beam of direct diode laser on heat conduction lap welding of thin foil on thick substrate will be discussed, together with finite element analysis.

EXPERIMENTAL APPARATUS

Direct Diode Laser Micro-Welding Systems

Two micro-welding systems with different beam sizes, 360μm x 1800μm and 230μm x 1800μm, were used for lap welding examinations of thin foils on thick substrate. They consist of a direct diode laser (GTS500, Spectra-Pysics) with a maximum output power of 500W and a wavelength of 808nm and a feed table with a fixing jig. The size of the laser head is 65mm in diameter and 300mm in length and 2kg in weight. A micro-positioning stage with fixing jig was used for precise alignment and tight fixing of foil and substrate. Specimens were irradiated with a laser beam at a normal angle. Figure 1 shows the outline of the system.

Figure 2 shows the beam profiles at the focal point measured using a beam profiler (UFF100, PROMETEC). The shape of the beam was elliptical and one beam size was 360μm x 1800μm in length and the other beam size was 230μm x 1800μm in length. The mean power density at the maximum power of 500W was 77kW/cm^2 and 121kW/cm^2, respectively. The working distance from the laser head to the focal point was 42.5mm.

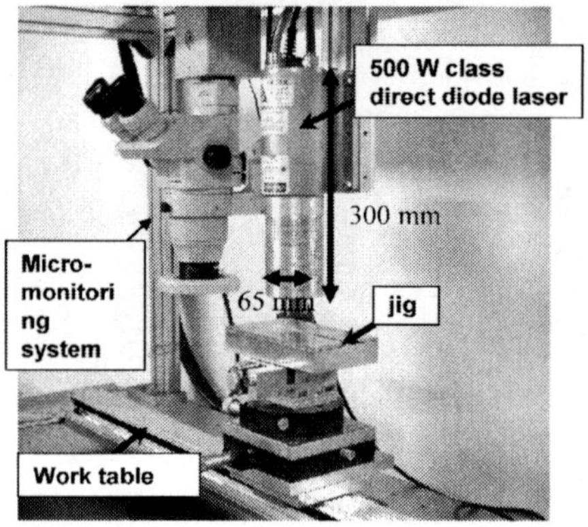

Fig.1 Direct diode laser micro-welding system

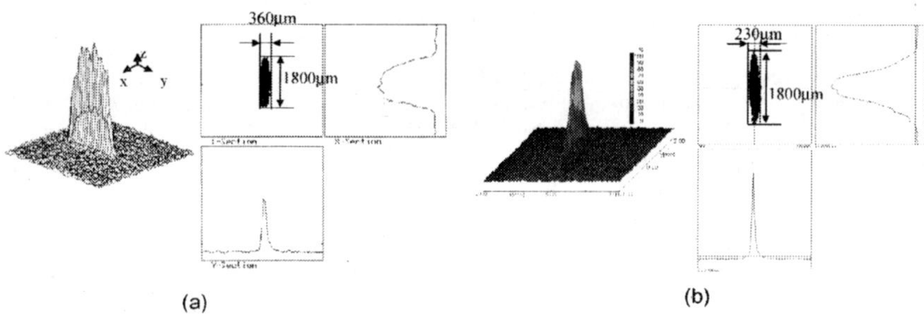

Fig.2 Beam profiles at focal position. (a) 360μm x 1800μm, (b) 230μm x 1800μm

Specimens and Welding Conditions

The specimens used were stainless steel foils of 20μm (SUS316), 10μm (SUS316) and 5μm (SUS304) in thickness. A stainless steel SUS316 plate of 1mm in thickness was used as a substrate. The thin foils were lapped on the thick substrate and fixed with a transparent plate, so that they could make enough contact with the substrate for heat conduction. The thin foils were irradiated with laser beam through the transparent plate as shown in Fig.3. The plate used for fixing the thin foils was fused silica and was able to transmit laser beam of 808nm with a transmittance more than 99%. The specimens were fixed on an Aluminum base plate mounted on the feed table as shown in Fig.4. The feed table with the specimen traveled parallel to the long axis of the elliptical beam. The surfaces of the specimens were irradiated with laser beam at the focal point under the welding conditions as shown in Table 1.

Cross sections of the weld beads were polished, electrochemically etched and observed using an optical microscope.

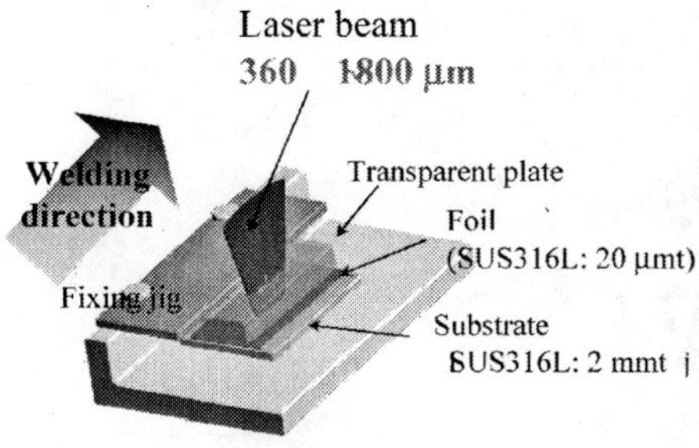

Fig.3 Experimental set-up

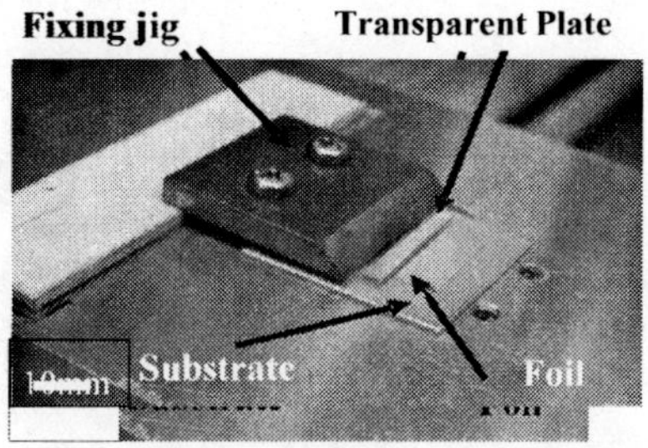

Fig.4 Photograph of specimen set-up

Table 1 Welding conditions

Foil thickness (µm)	Output power (W)	Welding speed (mm/sec)
20	75 - 400	5 - 600

Foil thickness (µm)	Output power (W)	Welding speed (mm/sec)
5, 10	75 - 200	5 - 200

RESULTS AND DISCUSSION

Surface Appearance and Cross Section of Weld Bead with 360µm Width Beam

Figure 5 shows optical microscopic photographs of surface appearances and cross sections of beads welded with 360µm x 1800µm beam at a welding speed of 175mm/sec and various laser powers under pre-heating temperature of 473K. The thickness of stainless foil and substrate were 20µm and 2mm, respectively. The weld beads had a sound appearance at output power lower than 280W. A sound bead was not formed at an output power above 280W. It is though that this was due to the excessive amount of molten pool generated by excessive heat input.

Figure 6 shows optical microscopic photographs of cross sections of the weld beads. The cross section of the weld bead showed a semicircular shape which is generally seen in the heat conduction welding.

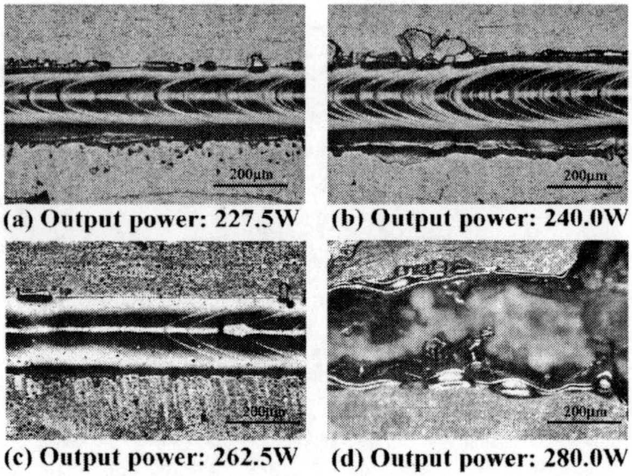

Fig.5 Surface appearance of bead welded by a 360μm width beam

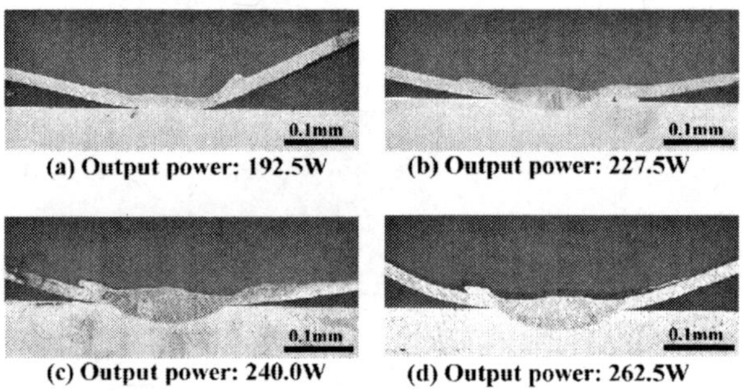

Fig.6 Cross sections of bead welded by a 360μm width beam

Surface Appearance and Cross Section of Weld Bead with 230μm Width Beam

Figure 7 shows the typical optical microscopic photographs of surface appearances of weld beads of 20μm, 10μm and 5μm foils welded by using 230μm width beam at an output power of 150W and several welding speeds. When the welding speed was low as 20mm/sec, welding beads of all specimens showed the discontinuous bead with very wide width. At the welding speed of 30mm/sec, even a thin foil of 5μm in thickness could be welded with smooth surface and narrow bead. However, at a higher welding speed as 50mm/sec, the foils could not be penetrated. All welding was performed without pre-heating.

Figure 8 shows typical optical microscopic photographs of cross sections of the weld beads of 20μm, 10μm and 5μm foils. Although, the bead shape at 20mm/sec showed very large bead width, the bead width at 30mm/sec showed very narrow bead width compared with that of 20mm/sec. The cross sections of the weld bead showed a semicircular shape that is generally seen in heat conduction welding. Although the lap welding with a direct diode laser is heat conduction type, the weld bead can penetrate the thin foil and reach to the substrate with narrow bead width at proper welding speed, because the bead width was strongly dependent on beam width and welding speed, and a direct diode laser can be precisely controlled input heat against the thickness of specimen. This is because an elliptical beam of direct diode lasers can input heat to the specimens effectively, keeping the width of the molten pool narrow when it is moving along its long axis. It is concluded that a direct diode laser is applicable to lap welding of a thin foil onto thick substrate when foil's thickness is extremely thin and fixed onto a substrate without a gap.

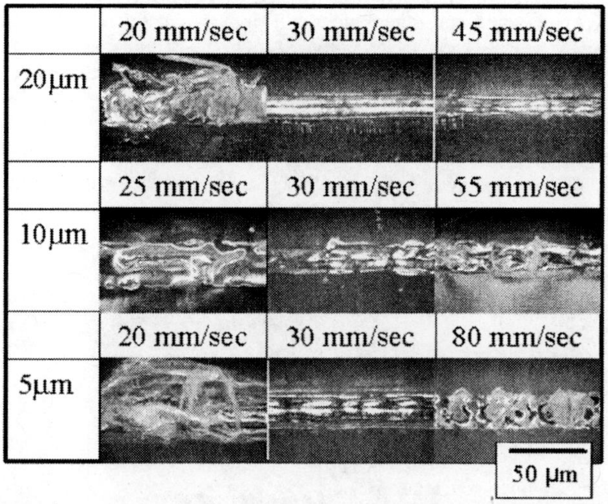

Fig.7 Surface appearance of bead welded by a 230μm width beam

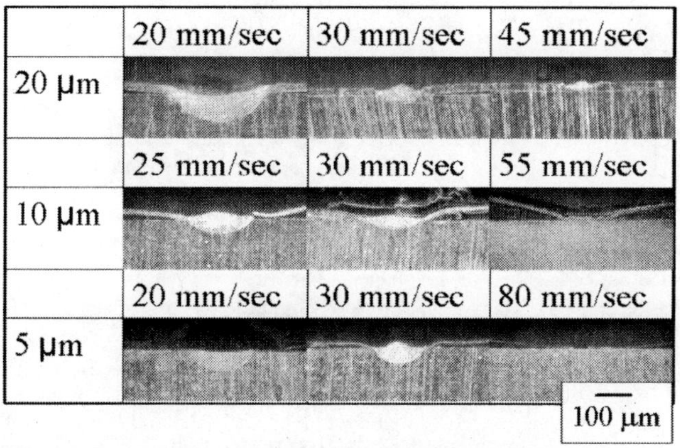

Fig.8 Cross sections of bead welded by a 230μm width beam

Weldable Condition Area for Lap Welding

Figure 9(a) shows weldable condition range of lap-welding for thin foils of 20μm in thickness onto a thick substrate of 2mm in thickness by using 360mm width beam with pre-heating at temperature of 473K. Circle marks represent a sound bead. Triangle marks represent a discontinuous bead. Cross marks represent failure of lap welding. With a welding speed of 175mm/sec, a sound weld bead was obtained in a range of laser power from 228W to 263W. With a welding speed of 500mm/sec, a sound weld bead was obtained in range of laser power from 350W to 489W. An increase in welding speed required the increase of laser power, and made the weldable condition range wider. This is because that the width of molten pool could be kept narrow as a welding speed increased. On the contrary, precise control of the laser power is required when lap-welding has to be carried out at a slow welding speed.

Figure 9(b) shows the weldable condition area of 230μm beam for lap welding of thin foils of 20μm in thickness onto a thick substrate of 1mm in thickness. The weldable area for a 230μm beam shifts to the left, that is, a lower welding speed region. This indicates that a narrower beam can use at slower speed. This is the reason why the 360μm beam welding required a preheating process.

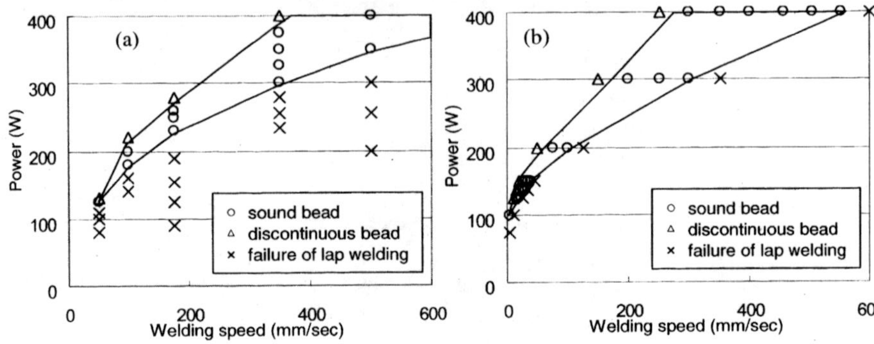

Fig.9 Weldable condition area. (a) 360mm width beam (b) 230mm width beam

Effect of Beam Width on Lap Welding of Thin Foil onto Thick Substrate

In order to estimate the effect of beam width on the lap welding of a thin foil onto a thick substrate, heat conduction lap welding with various beam widths was simulated using the finite element method. A finite element method heat analysis program (QuickTherm, RCCM) was used. Figure 10 shows the simulation model. The model size is 1.5mm in width, 2mm in thickness and 3mm in length. The laser irradiated the centerline of the specimen. A measured profile was used as the assumed beam shape. The mesh interval in the X, Y, and Z directions are 50, 2 and 500μm. The laser power was 200W, and the welding speed was 500mm/sec.

Figure 11 shows the results of simulation and actual bead cross sections. On the left are simulated bead cross sections with 360μm beam showing both with and without pre-heating. On the right is simulated bead cross section with 230μm beam. Using a wide beam of 360μm without pre-heating, the bead depth is shallow compared to the foil thickness of 20μm, and lap welding fails. Pre-heating to 473K makes a deeper bead. Using a narrower beam of 230μm, a similar bead depth is produced without pre-heating. Simulation results show good agreement with measured bead shape. It is concluded that narrower beam width makes narrower bead width when the specimen thickness is extremely thin.

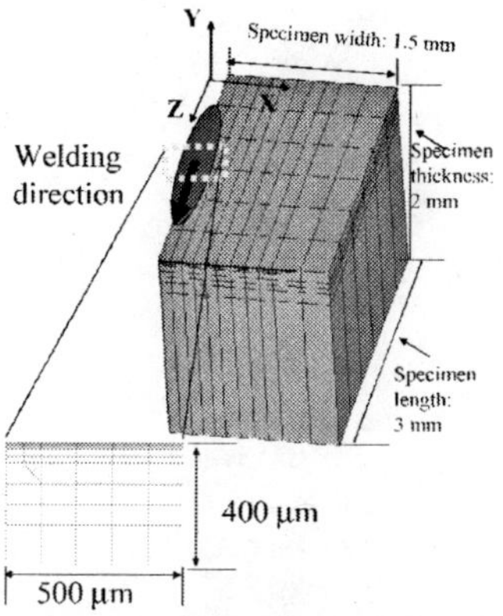

Fig.10 Simulation model

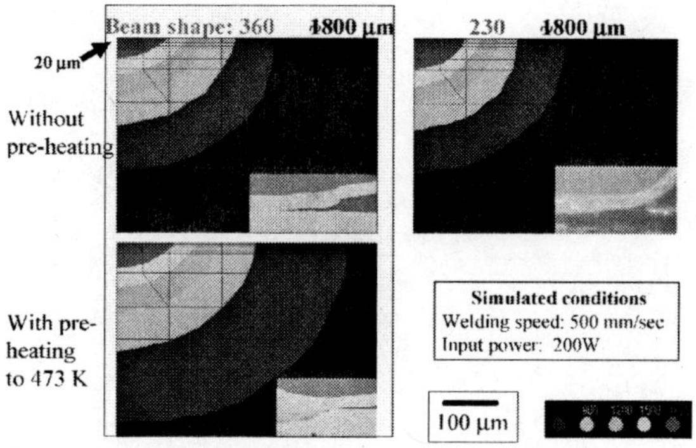

Fig.11 Results of simulation

CONCLUSION

A finely focused elliptical beam of a direct diode laser was applied to the lap welding of thin foils on thick substrates.

The following results were obtained.

(1) A narrower elliptical beam shape enabled lap-welding with a narrow bead at a slower speed and lower power region without preheating.

(2) An elliptical beam of 230μm x 1800μm successfully welded thin stainless steel foil of 5μm in thickness on 1mm thick stainless steel substrate at 5mm/sec and 100W.

(3) The narrower beam width could weld thinner foil onto thick substrate without pre-heating.

REFERENCES

[1]Funada, Y. and Abe, N., "Micro Welding of Thin Stainless Steel Foils with a Direct Diode Laser", Proceedings of the 22nd International Congress on Applications of Lasers & Electro-Optics 2003: Laser Institute of America, Jacksonville, USA, M204.

[2]Funada, Y. and Abe, N., "Lap Welding with Direct Diode Laser of Thin Foils on Thick Substrate", Proceedings of the 23rd International Congress on Applications of Lasers & Electro-Optics 2004: Laser Institute of America, San Francisco, USA, M702.

MICTOSTRUCTURAL CHARACTERIZATION AND MECHANICAL PROPERTIES OF PLASMA SPRAYED HYDROXYAPATITE COATINGS

M. F. Morks, Akira Kobayashi

Joining and Welding Research Institute, 11-1, Mihogaoka Ibaraki 567-0047 Osaka Japan

ABSTRACT

Hydroxyapatite (HA) is a bioactive material with a chemical structure that is close to that of natural bone. Gas tunnel plasma spraying (GTPS) was employed in the present study to deposit HA coatings on a 304 stainless steel substrate. Controlling the spraying parameters is the key for spraying highly crystalline HA coatings on metallic implants. The objective of this study is to investigate the influence of gun current on the microstructure and mechanical properties of HA coatings. The surface morphology and microstructure of as-sprayed coatings were examined by scanning electron microscopy. The different phases and structure of HA coatings were investigated using X-ray diffraction analysis. Hardness values of HA coatings sprayed at different arc gun currents were also characterized. HA coatings sprayed at 0, 30, and 50 A have many pores, and semi-melted particles. However, HA coatings sprayed at high gun current (70 and 100 A) are dense, and have high hardness.

KEYWORDS: Hydroxyapatite, Plasma spraying, Gun current, Microstructure, Hardness

INTRODUCTION

Hydroxyapatite (HA), $Ca_{10}(PO_4)_6(OH)_2$, has been extensively used as biomedical material due to its bioactive properties and its chemical structure which is close to that of natural bone. Clinical tests have been showed that HA material is compatible with the tissue of the human body which makes it attractive material to treat and replace the broken parts of human bones. However, HA material has poor mechanical properties such as low bending strength and fracture toughness, so it could not be used in bulk form for load bearing applications such as orthopedics. Coatings of hydroxyapatite on metallic implants such as Ti6Al4V and other biocompatible alloys are good candidate as they exploit the biocompatible and bone bonding properties, while utilizing the superior mechanical properties of the substrate. Pure HA materials can assist to build up new tissue and fix the implants inside the body.

Although there are many methods [1-5] have been used to deposit pure HA coatings, the most of current researches are interested to deposit HA coatings by using thermal spraying technologies because these methods have the ability to spray thick and highly crystalline HA coatings with good adhesion property in short time. Theses methods include HVOF, D-GUN and conventional plasma spraying [6-8]. The problem associated with thermal spraying methods is the decomposition of HA coatings during spraying to form oxyhydroxyapatite (OHA), tetracalcium phosphate (TTCP), α and/or β–tricalcium phosphate (α–TCP, β–TCP), amorphous phase and calcium oxide (CaO) materials [9-11]. The decomposition is related to the high temperature of thermal spraying flame. For that, a careful control of thermal spraying parameters is required to prevent the decomposition and get high quality HA coatings.

Gas tunnel type plasma spraying (GTPS) developed in Osaka University [12, 13] is a unique method to deposit high quality ceramics [14, 15] due to its high power and high spraying efficiency. The GTPS system composed of two sources of plasma: gun and vortex. It has two anodes and one cathode. The nozzle diameter of the vortex is 20 mm and the gun nozzle diameter is 8 mm. In this study, the vortex arc current was kept constant at 450 (A) and the gun current was changed from 0 – 100 A. The microstructure and mechanical properties of the resulting coatings were investigated.

EXPERMINTAL PROSEDURE

Materials

Commercial hydroxyapatite (HA) powder of spherical shape and particle size range of +10-45 μm was supplied by Pawlex Ltd., Tokyo, Japan, and used as spraying materials. The scanning electron micrograph and X-ray diffraction pattern of started HA powders are shown in Fig. 1. The powder is highly crystalline and composed of pure hydroxyapatite phase as illustrated from the X-ray diffraction pattern. The substrate material is SUS 304 with dimensions 50 x 50 x 2.5 mm. The substrate was grit blasted by alumina powder to clean and roughen the surface before spraying (Ra 5-7 μm).

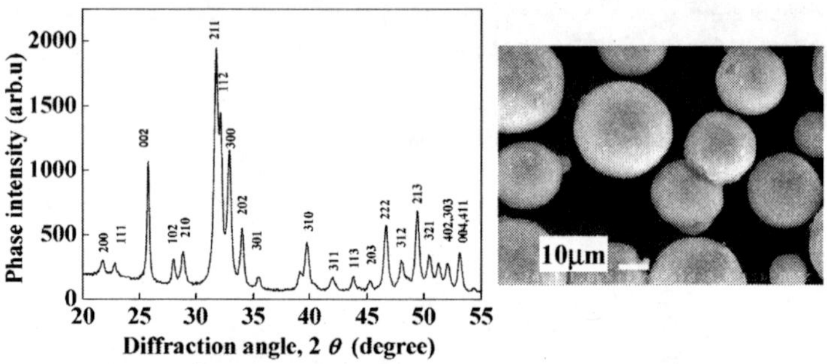

Figure 1 Morphology and X-ray diffraction pattern of starting HA powder.

Spraying process

A systematic diagram of GTPS torch is shown in Fig. 2. The external anode (vortex) is composed of a circular copper tube and the working gas (Ar) moves inside it in a rotational flow with high velocity. The gas tunnel generator is composed of a "vortex generator" and a "gas divertor." The vortex generator has many small nozzles arranged in a circle, so that a cylindrically symmetrical vortex flow can be obtained. The spraying conditions are listed in Table 1. The vortex arc current was kept constant at 450 (A) and the gun arc current was changed from 0 to 100 A.

Table 1 Spraying parameters for HA coatings

Arc gun current (A)	0-100
Arc vortex current (A)	450
Working gas flow rate (l/min)	120
Carrier gas flow rate (l/min)	5
Powder size (μm)	+10-45
Spraying distance (mm)	60
Substrate traverse speed (mm/s)	40

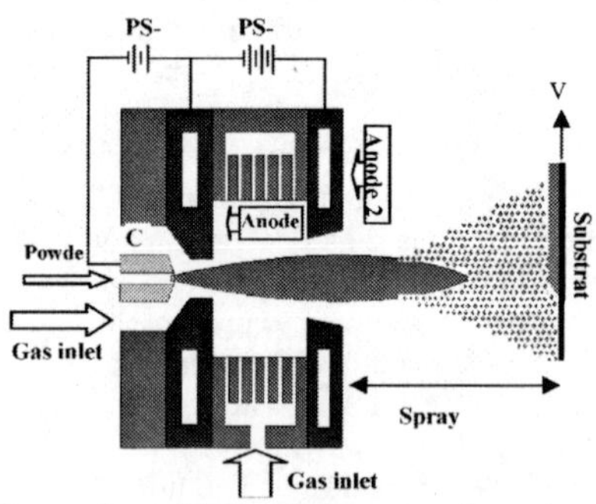

Figure 2 The gas tunnel type plasma spraying setup.

Characterization techniques

Feedstock and coatings morphologies were examined by ERA-8800FE scanning electron microscope. The examined cross-section samples were mounted in epoxy resin, polished and buffed with alumina paste to get a mirror finished surface. Phase identification of feedstock and as-sprayed coatings were examined by JEOL JDX-3530M X-ray diffractometer system with CuKα radiation source; the operating voltage was 40 kV and current 40 mA.

Hardness tests were performed on polished and buffed cross-section coated samples using Akashi AAV-500 series hardness tester. The load used was 490.3 mN and the loading time was 20 s. Each hardness value is the average of 10 readings.

Abrasive wear test was carried out using SUGA ABRASION TESTER which follow NUS-ISO-3 standard. More details about the abrasive wear test are found in a previous work [16]. The starting thickness of tested HA coatings is 200-300 μm. The abrasive wear test was performed in dry condition in air atmosphere at room temperature.

RESULTS AND DISCUSSION

Morphologies and phase structure of HA coatings

Scanning electron micrographs of HA coatings sprayed at different gun currents are shown in Fig. 3. It is clear from the figure that all coatings are porous and contains unmelted particles. However, the porosity is lower and the number of unmelted particles decreases for sprayed coatings at 70 and 100 A. HA coatings sprayed at low gun currents have many pores and unmelted particles. From the morphologies of cross-sections of sprayed HA coatings, one can recognize that the gun current affects the microstructure and coatings porosity. The presence of pores and unmelted particles is mainly due to the decrease of plasma power level.

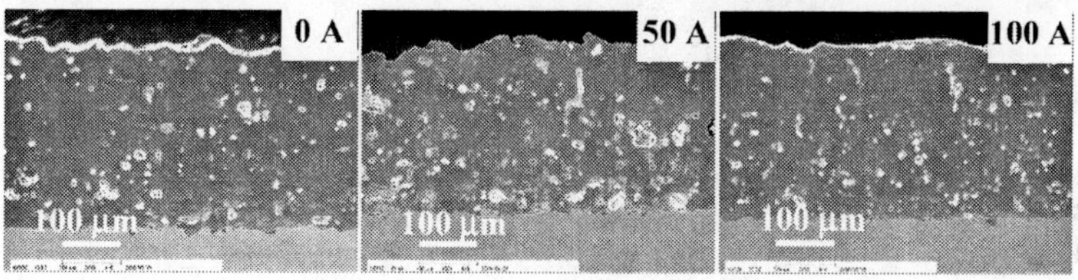

Figure 3 SEM micrographs of HA coatings sprayed at different gun currents.

For more details about the coatings morphology, the surface of HA coatings sprayed at different gun currents was observed (see Fig. 4.)

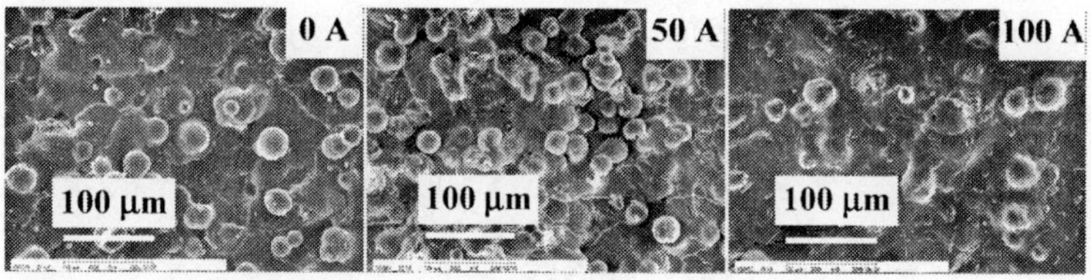

Figure 4 SEM micrographs of the surfaces of HA coatings sprayed at different arc gun currents and vortex current of 450 A.

The morphologies of HA coatings surfaces show spherical unmelted particles in all coatings. However, HA coatings sprayed at 100 A (gun current) have a few unmelted particles and the coatings are mainly formed from impact and fully melted or semi-melted particles. This indicates that, the particle temperature is higher than that sprayed at 0 A of gun current. There are no cracks or amorphous phases at the interface which indicates a good adhesive strength with the substrate.

X-ray diffraction patterns of HA coatings sprayed at different gun currents are shown in Fig. 5. All the peaks of HA coatings sprayed at 0, 30 and 50 A belong to HA phase. However, a-tri calcium phosphate was detected in case of spraying at 70 and 100A due to the increase in particle temperature. On the other hand, there are some differences in the peak intensities of the coatings sprayed at different arc gun currents. The intensities of the peaks (002 and 004) increase as the gun current increases. Also, the intensities of the peaks (211 and 112) decrease at high arc gun currents of 70 and 100 A. The differences of peak intensities are related to the change of crystallite orientation of HA coatings due to the change of particle temperature at different gun currents.

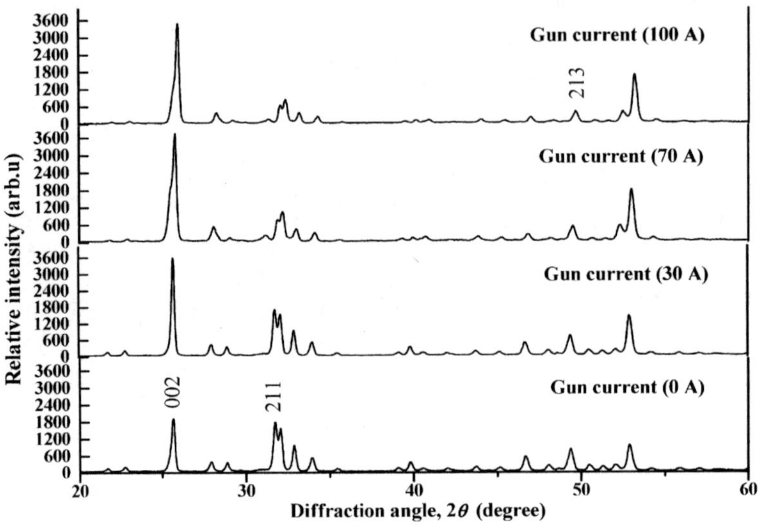

Figure 5 X-ray diffraction patterns of HA coating sprayed at different gun currents.

Vortex and gun power

The power of vortex and gun was measured at different arc gun currents and at constant vortex arc current (450 A) as shown in Fig. 6. The purpose from this measurement is to know whether the arc gun current affects the vortex power or not. It has been concluded that, the gun power increases to reach 4 kW as the arc gun current increases to 100 A because of the increase of arc gun currents. On the other hand, the vortex power level reaches a maximum value of 21.3 kW at gun current of 0 A. The voltage of vortex is 44.5 V in this case. However, at arc gun current of 100 A, the measured vortex power is 18.5 kW of voltage value 41 V. The decrease of vortex power by increasing arc gun current is related to the consumption of the gun for some working gas (Ar) which leads to a decrease of vortex voltage.

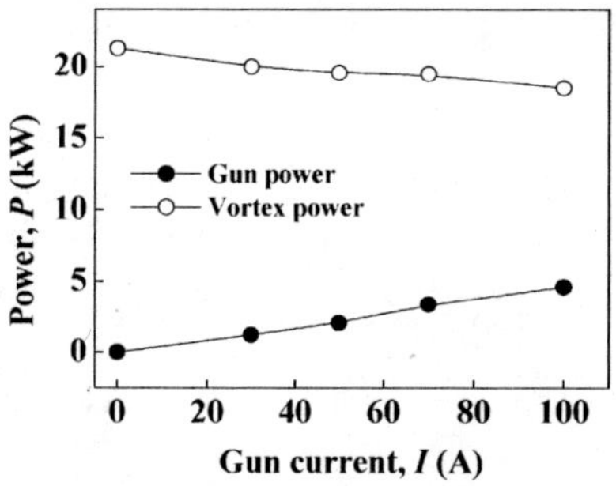

Figure 6 Vortex and gun power at different gun currents and 450 (A) vortex current.

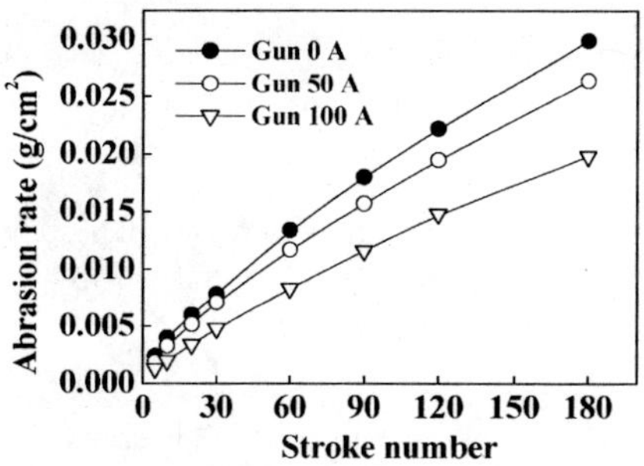

Figure 7 Abrasive wear of HA coatings.

The total amount of gas flow rate which reaches the vortex decreases by increasing the gun power. The vortex power was decreased from 21.3 to 20 kW by increasing the gun arc current from 0 to 30 A. However, the vortex power decreased by 0.5 kW for every 20-30 (A) increases of the gun current. The total power of GTPS (vortex + gun) increases from 21.3 to 23 kW as the gun current increases from 0 to 100 A.

HA powder was fed externally and was only interacted with vortex plasma jet. In this case, the properties of HA coatings depends mainly on the power and properties of vortex plasma jet. Moreover, the plasma jet produced by the gun affects the efficiency of the plasma jet produced by the vortex. As the gun power increases, the total power level of GTPS torch increases and the vortex plasma temperature is increases and as a result dense with low porosity coatings are formed.

Hardness

Hardness of HA coatings sprayed at different gun currents was measured. The results indicated that, the hardness was increased slightly as the gun current increased. The main value of hardness increases from *Hv* 420 for sprayed coating at 0 A of gun current to *Hv* 460 for sprayed coating at 100 A of gun current. The hardness property is mainly affected by the microstructure and porosity of the sprayed coatings. Spraying at low gun currents resulting in the presence of pores and unmelted particles which leads to a decrease in the hardness mainly due to the weakness of cohesion bonds of accumulated splats. The coatings sprayed at high gun current exhibit better particle melting and flatting; thus, the coating become denser and its hardness increases.

Abrasive wear

The abrasion behavior of HA coatings sprayed at different arc gun currents is shown in Fig. 7. HA coating sprayed at 100 A has the lowest abrasive wear rate. However, the abrasive wear rate increases as the gun current decreases. The abrasion resistance of HA coatings is mainly depends on the bonding strength among the splats which accumulate to form the coatings. The higher the bonding strength, the lower the abrasive wear. On the other hand, the presence of pores and unmelted particles leads to the decrease in abrasion resistance of the coatings. HA coating sprayed at high gun current (100A) exhibit low porosity and high hardness and as a result its abrasion resistance increases.

CONCLUSIONS

HA coatings were prepared using gas tunnel type plasma spraying with different arc gun currents, and a constant vortex arc current of 450 A. The results can be summarized as follows:

(1) As the arc gun current increases, the pores and unmelted particles decrease in the HA coatings, mainly due to the increase in the total plasma power.

(2) HA coating hardness increases slightly as the gun current increases due to the formation of a dense structure, and low porosity coatings.

(3) The abrasive wear resistance of HA coatings increases as the arc gun current increases due to the increase in the total plasma power, the decrease in porosity, and the bonding strength of the accumulated particles.

REFERENCES

[1]V.W. Raemdonck, P. Ducheyne, and D.P. Meester, *J. Am. Ceram. Soc.*, **67**, 381-384 (1984).

[2]K. Yamashita, T. Arashi, K. Kitagaki, S. Yamada, T. Umegaki, and K.J. Ogawa, Preparation of Apatite Thin Films Through RF-Sputtering from Calcium Phosphate Glasses, *Am. Ceram. Soc.*, **77**, 2401-2407 (1994).

[3]C.S. Kim and P. Ducheyne, Compositional Variations in the Surface and Interface of Calcium Phosphate Ceramic Coatings on Ti and Ti-6Al-4V due to Sintering and Immersion, *Biomaterials*, **12**, 461-469 (1991).

[4]X. Nie, A. Leyland and A. Matthews, Deposition of Layered Bioceramic Hydroxyapatite/TiO_2 Coatings on Titanium Alloys Using a Hybrid Technique of Micro-Arc Oxidation and Electrophoresis, *Surf. Coat. Technol.* **125**, 407-414 (2000).

[5]M. Shirkhanzadeh, Bioactive Calcium Phosphate Coatings Prepared by Electrodeposition, *J. Mater. Sci. Lett*, **10**, 1415-1417 (1991).

[6]H.C. Gledhill, I.G. Turner, and C. Doyle, "Direct Morphological Comparison of Vacuum Plasma Sprayed and Detonation Gun Sprayed Hydroxyapatite Coatings for Orthopaedic Applications," *Biomaterials*, **20**, 315-322 (1999).

[7]P.L. Silva, J.D. Santos, F.J. Monteiro, and J.C. Knowles, *Surf. Coat. Technol.*, **102**, 191-196 (1998).

[8]R.S. Lima, K.A. Khor, H. Li, P. Cheang, and B.R. Marple, "HVOF Spraying of Nanostructured Hydroxyapatite for Biomedical Applications," *Mater. Sci. Eng.*, **A 396**, 181-187 (2005).

[9]R. Mc Pherson, N. Gane, and T.J. Bastow, *J. Mater. Sci. Mater. Med.*, **6**, 327 (1995).

[10]K.A. Gross and C.C. Berndt, *J. Biomed. Mater. Res.*, **39** (1998) 580.

[11]K.A. Gross, C.C. Berndt, and H. Herman, *J. Biomed. Mater. Res.*, **39**, 407 (1998).

[12]A. Kobayashi, Y. Arata, and Y. Habara, *J. High Tem. Soc.*, **13** (3), 116-124 (1987).

[13]Y. Arata and A. Kobayashi, "Application of a gas tunnel to high-energy-density plasma beams," *J. Appl. Phys.*, **59** (9), 3038-3044 (1986).

[14]S.Sharafat, A.Kobayashi, Y. Chen, and N.M. Ghoniem, "Plasma Spraying of Micro-Composite Thermal Barrier Coatings," *Vacuum*, **65**, 415–425 (2002).

[15]A. Kobayashi, T. Kitamura, "Effect of Heat Treatment on High-Hardness Zirconia Coatings Formed by Gas Tunnel Type Plasma Spraying," *Vacuum*, **59**, 194-202 (2000).

[16]M. F. Morks, Akira Kobayashi, N.F. Fahim, "Abrasive Wear Behavior of Sprayed Hydroxyapitite Coatings by Gas Tunnel Type Plasma Spraying," *Wear*, **262**, 204-209 (2007).

MEASUREMENTS OF CATHODE SURFACE TEMPERATURE OF PLASMA TORCH

Shinichi Tashiro, Hiroshi Nishikawa and Manabu Tanaka
Joining and Welding Research Institute, Osaka University
11-1, Mihogaoka,
Ibaraki, Osaka, JAPAN 567-0047

ABSTRACT

A plasma torch can easily control high temperature arc plasma and, additionally, requires low cost for equipment investment. Therefore, it is widely utilized as a heat source device, for example, for thermal plasma spraying, material processing such as melting, cutting and welding, or decomposition, volume reduction and detoxification of toxic waste and so on. For these processes, the heat source properties strongly depend on properties of the arc plasma. Since the current in the arc plasma is originated mainly from thermionic emission on the high temperature cathode surface, it is significantly important to know the temperature of the cathode surface to realize required properties of the arc plasma. This paper describes the experimental result of the cathode temperature measurement of a helium plasma torch with two-wavelength pyrometry technique. The result is compared with that of an argon plasma torch.

INTRODUCTION

Since a plasma torch can stabilize high temperature arc plasma by employing shielding gas, it is used as a heat source. Gas Tungsten Arc (GTA) is the most widely employed type of the plasma torch. It produces the arc plasma between a tungsten cathode and an anode material. GTA has high heating efficiency and highly controllable characteristics. Additionally, GTA requires low cost for equipment investment. Therefore, it is widely utilized, for example, for material processing such as melting, cutting and welding[1], or decomposition, volume reduction and detoxification of toxic waste[2] and so on.

For these processes, the heat source properties of GTA strongly depend on physical properties of the shielding gas. For example argon, helium, oxygen and carbon dioxide and so on are generally employed as the shielding gas. Recently, a novel heat source for oxygen and carbon dioxide was developed[3]. It has a double gas shielded system which consists of carbon dioxide or oxygen and inert gas to avoid the consumption of the tungsten electrode by oxidation. Since helium with low electrical conductivity[4] or carbon dioxide with high specific heat[5] causes the constriction of the current channel in the arc plasma, these gases enable to produce the arc plasma with high energy density and are suitable especially for processes to require concentrating the heat input at a point. Furthermore, the target material can be heated without any chemical reaction by using an inert gas such as argon or helium as the shielding gas. On the other hand, oxygen or carbon dioxide is chosen if the oxidation of the target material is required, for example, for cutting. Thus, suitable shielding gas composition for each process is selected.

Since the arc current is originated mainly from thermionic emission of electrons on the high temperature cathode surface, it is significantly important to know the temperature distribution on the cathode surface to realize required properties of the arc plasma. The cathode temperature is determined by energy balance between heating and cooling processes such as ion collision, heat transfer, joule heating, radiation, thermionic emission of electrons and water cooling etc. A number of experimental investigations of the temperature distribution on the cathode surface of argon GTA have been conducted mainly with two-wavelength pyrometry technique. Recently, it was found that reflection of the plasma radiation by the cathode can be avoided by measuring the cathode emission near its edges relative to the line of the sight[6]. Measurement of dependence of the cathode temperature on arc current revealed that although the temperature near the cathode tip hardly depends on the arc current, the temperature at position away from the tip rises with increase of the arc current mainly due to joule heating[7]. The cathode temperature for various cone angles was also investigated and it was found that the temperature is maximized at the tip and is, conversely, minimized at position away from the tip at the cone angle of 60 degrees[8]. Furthermore, the temperature of pure tungsten cathode was compared with that of the tungsten cathode adding an emitter material such as thorium oxide or lanthanum oxide, and it was found that the temperature of the pure tungsten cathode becomes higher than that adding the emitter material. The heating caused by the ion collision from the arc plasma is prompted in the case of pure tungsten cathode, because ion current increases due to suppressed thermionic emission of electrons owing to large work function of the pure tungsten[9].

In addition to the above investigations, it is required also to investigate the effect of difference of shielding gas composition on the cathode temperature, because heating due to the ion collision and the heat transfer from the arc plasma strongly depend on physical properties of the arc plasma. However the cathode temperature for the shielding gas except for argon is hardly understood. This paper describes the experimental result of the cathode temperature measurement of helium GTA with two-wavelength pyrometry technique. The result is compared with that of argon GTA.

EXPERIMENTAL APPARATUS

Figure 1 shows schematic illustration of experimental apparatus consisting of GTA torch and a video camera with a spectroscope. An emitter material with low work function such as cerium oxide, thorium oxide or lanthanum oxide and so on is added to the tungsten cathode in order to improve emission characteristics of electrons from the cathode surface. In this experiment, a tungsten cathode adding 2% cerium oxide is employed. The cathode has the diameter of 3.2mm and the cone angle of 60 degrees, and the anode is water-cooled copper. The electrode gap is set to be 5mm. Helium is selected as shielding gas composition and is introduced at the flow rate of 30L/min.. An arc current is DC 200A.

The two-wavelength pyrometry technique has an advantage to reduce an error due to unknown electrode emissivity by selecting adjacent two wavelengths. In order to measure precise cathode temperature with this technique, it is necessary to eliminate the arc plasma radiation surrounding the cathode. In this study, wavelengths of 950nm and 980nm in infrared ray range are selected. This combination enables to measure the cathode temperature eliminating the arc plasma radiation which is mostly in ultra-violet and visible ray range. The video camera with the spectroscope has an ability to record two images for both wavelengths simultaneously. Figure 2 shows schematic illustration of principle of image processing. A virtual image formed in the optical system by the object lens is duplicated to two identical images with the branch mirror. These images pass through the interference filters and are recorded with the sensor as the images for wavelengths of 950nm and 980nm, respectively.

Figure 3 shows recorded images for both wavelengths at the arc current of 200A. It is found that the arc plasma radiation is eliminated completely, because of the weaker radiation of helium arc plasma compared with that of argon. The cathode temperature is calculated referring to the calibration data taken with a standard lamp.

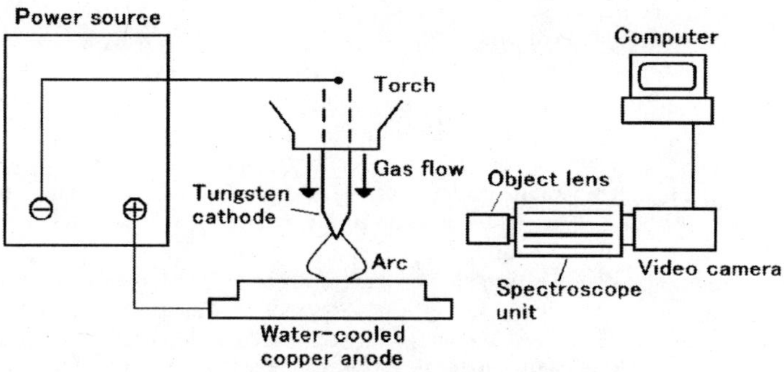

Figure 1. Schematic illustration of experimental apparatus.

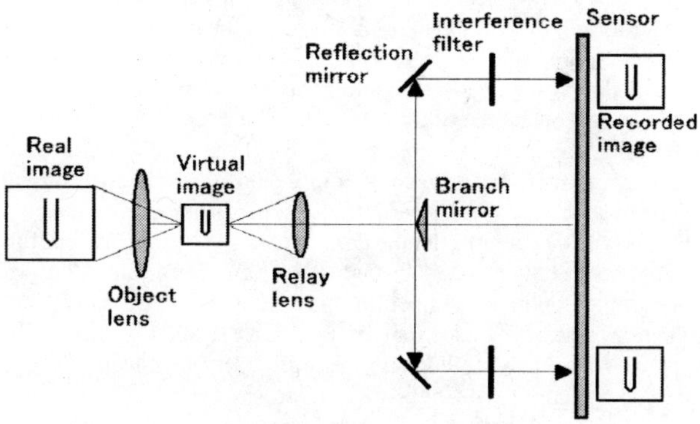

Figure 2. Schematic illustration of principle of image processing.

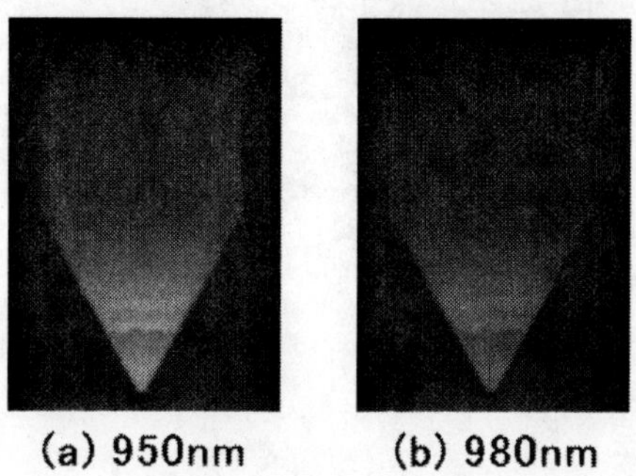

Figure 3. Recorded images for wavelengths of 950nm and 980nm at an arc current of 200A.

RESULTS AND DISCCUSION

Figure 4 shows recorded images for wavelength of 950nm at the arc current of 100A and 200A at 60sec. after starting the arc. The zone A in the figure includes an arc-cathode attachment region and becomes high temperature because of large effect of a combination of heat transfer from the arc plasma and joule heating. The emitter material with lower melting point compared with the pure tungsten melts and migrates from the low temperature region away from the cathode tip to the high temperature cathode surface near the tip along the grain boundary due to capillary action. Although a part of the melting emitter material covers the cathode surface and evaporates, it is refilled with the migrating material[10]. Because the surface of the melting emitter material covering the cathode is smooth and, therefore, has low emissivity, it is seen as relatively dark region than the zone B. The zone C is low temperature region because of heat loss due to heat transfer to low temperature shielding gas close to the cathode surface and water-cooling of the cathode. The zone B is the transition region between the zone A of high temperature and the zone C of low temperature. In this zone, the emitter material accumulates and forms a high density layer, because the temperature is not sufficient for vaporization or the vaporization rate is lower than the migration rate[10]. In the case of 200A, it is seen that the zone A-B is narrower and a little brighter compared with that of 100A and it is expected that the arc-cathode attachment region is concentrated near the cathode tip and is high temperature.

Figure 5 shows dependence of the cathode surface temperature on axial distance from the cathode tip in the same condition as figure 4. For comparison, the cathode temperature for argon shielding gas[7] are also plotted. In the case of argon, the cathode temperature at the tip reaches 3700K and decreases inversely proportional to the axial distance from the cathode tip defined as z and approaches 2100K at z=4mm. On the contrary, in the case of helium for 200A it is seen that the cathode temperature is nearly constant in the range between 3000K and 3300K at $z<1.7$mm. Although the maximum temperature lowers by approximately 400K compared with that of argon, the temperature exceeds it at $1\text{mm}<z$. On the other hand, at $1.7\text{mm}<z$ corresponding to the zone C, the cathode temperature decreases steeply compared with that of argon and becomes approximately 1000K at z=3.4mm.

As shown in Figure 6, helium has larger thermal conductivity compared with that of argon because of the smaller atomic weight[11]. It is approximately five times higher than that of argon especially below 10000K. Therefore, the heating caused by the heat transfer from the arc plasma becomes a dominant factor in the zone A close to the high temperature arc plasma and the cathode temperature increases in large extent. Consequently, it is considered that maximum temperature decreases compared with that of argon, because of the decrease of joule heating caused by lower current density due to the expansion of the region where thermionic emission of electrons occur. The cathode temperature in the zone C decreases steeply compared with that of argon, since the heat loss increases due to greater heat transfer to the cold shielding gas close to the cathode surface. Furthermore, in the case of helium for 100A the zone A-B expands and becomes low temperature compared with that of 200A. This tendency agrees with that shown in figure 4. For a reason it is supposed that the heat loss due to the heat transfer to the cold shielding gas close to the cathode surface in the zone C decreases because the cathode jet velocity drops in lower arc current. It also leads the decrease of the temperature in the zone A-B due to the lower current density.

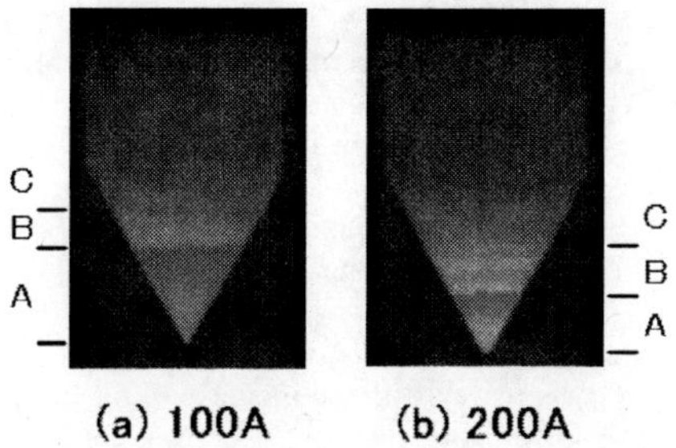

Figure 4. Recorded images for wavelengths of 950nm at an arc current of 100A and 200A.

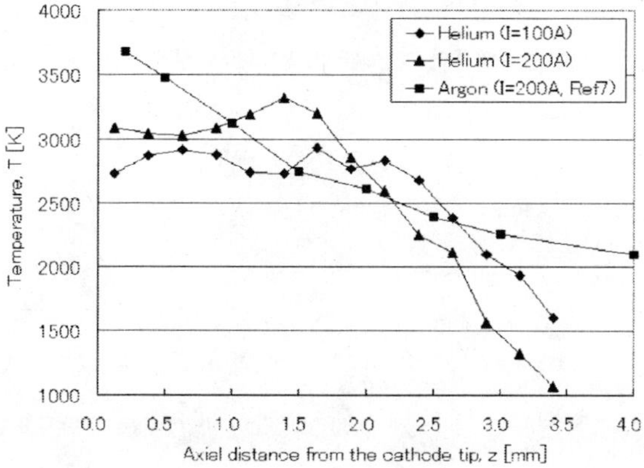

Figure 5. Dependence of temperature on axial distance from the cathode tip.

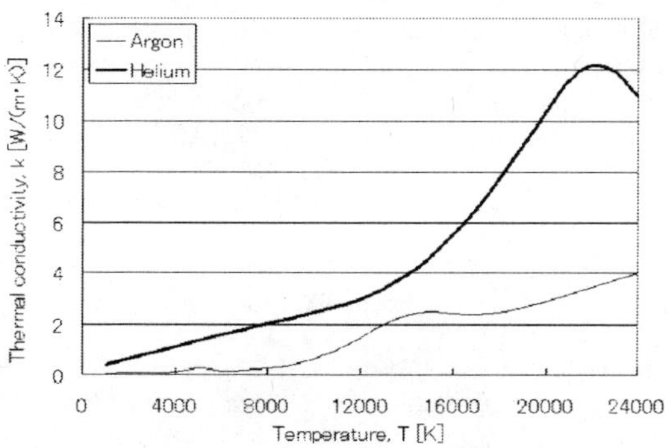

Figure 6. Dependence of thermal conductivity on temperature for argon and helium.

CONCLUSIONS

The experimental result of the cathode temperature measurement of helium GTA with two-wavelength pyrometry technique was presented and the result was compared with that of argon GTA. The main conclusions are summarized as follows:

1) The maximum cathode temperature becomes lower compared with that of argon.
2) The cathode temperature is nearly constant in large extent in the arc-cathode attachment region.
3) The cathode temperature outside the arc-cathode attachment region is lower compared with that of argon and it decreases with the increase of the arc current.
4) The area of the arc-cathode attachment region decreases and the temperature in this region rises with the increase of the arc current.

ACNOWLEDHEMENT

This work is supported by a grant-aid for Scientific Research (B) (2005) from the Japan Society for Promotion of Science (JSPS).

REFERENCES

[1]M. Ushio, M. Tanaka and J. J. Lowke, Anode Melting From Free-Burning Argon Arcs, *IEEE Trans P S*, **32,** 108-17 (2004).
[2]T. Inaba and T. Iwao, Treatment of Waste by dc Arc Discharge Plasmas, *IEEE Trans D E I*, **7**, 684-695 (2000).
[3]M. Tanaka, S. Tashiro, M. Ushio, T. Mita, A. B. Murphy and J. J. Lowke, High Intensity Heat Source using CO2 Gas Shielded Plasma Arc, *Advances in applied plasma science*, **5**, 329-33 (2005).
[4]M. Tanaka, M. Ushio and J. J. Lowke, Numerical study of gas tungsten arc plasma with anode melting, *Vacuum*, **73**, 381-89 (2004).
[5]S. Tashiro, M. Tanaka, M. Ushio, A. B. Murphy and J. J. Lowke, Prediction of Energy Source Properties of Free-Burning Arcs, *Advances in applied plasma science*, **5**, 103-8 (2005).
[6]J. Haidar and A. J. D. Farmer, A method for measurement of the cathode surface temperature for a high-current free-burning arc, *Rev. Sci. Instrum.*, **64**, 542-47 (1993).
[7]J. Haidar and A. J. D. Farmer, Surface temperature measurements for tungsten-based cathodes of high-current free-burning arcs, *J. Phys. D: Appl. Phys*., **28**, 2089-94 (1995).
[8]J. Haidar and A. J. D. Farmer, Large effect of cathode shape on plasma temperature in high-current free-burning arcs, *J. Phys. D: Appl. Phys.*, **27**, 555-60 (1994).
[9]M. Tanaka, M. Ushio, M. Ikeuchi and Y. Kagebayashi, In situ measurements of electrode work functions in free-burning arcs during operation at atmospheric pressure, *J. Phys. D: Appl. Phys.*, **38**, 29-35 (2005).
[10]M. Ushio, A. A. Sadek and F. Matsuda, Comparison of Temperature and Work Function Measurements of Obtained with Different GTA Electrodes, *Plasma Chem. Plasma Process*, **11**, 81-101 (1991).
[11]M. I. Boulos, P. Fauchais and E. Pfender, THERMAL PLASMAS Fundamentals and Applications Volume 1, Plenum Publishing Corporation, New York (1994).

TOPOLOGY ANALYSIS OF THE CU_3SN PHASE IN ELECTRONIC INTERCONNECTIONS

Feng Gao[1], Hiroshi Nishikawa[1], Tadashi Takemoto[1]
[1] Joining and Welding Research Institute, Osaka University
11-1 Mihogaoka, Ibaraki-shi, Osaka 567-0047 Japan

ABSTRACT

The topology of the Cu_3Sn phase during liquid soldering and isothermal aging was characterized to explore the close correlation with the Kirkendall void formation. For Sn-3.5Ag/Cu during the liquid soldering, the growth of the Cu_3Sn phase is closely dependent on the geometry of Cu_6Sn_5 phase. That is, at the grain boundary domain of Cu_6Sn_5 phase, the growth rate of Cu_3Sn phase is slower than that within the Cu_6Sn_5 grain region. The undulation behavior of the Cu_3Sn phase is perfectly consistent with that of Cu_6Sn_5 phase. This phenomenon persists even after the aging at 150°C for 504h, although both Cu_6Sn_5 and Cu_3Sn phases tended to be flat. This indicates that the Cu migration through the grain boundary from the base substrate mainly contributes to the growth of Cu_6Sn_5 and Cu_3Sn phases near the solder, while volume diffusion of Cu enhances the growth of the Cu_3Sn phase. It is interesting that the minor addition of (Co, Ni) into Sn-3.5Ag eutectic solder altered this undulation correlation, and suppressed the growth of the Cu_3Sn phase remarkably, in particular during the aging process, which definitely impacted Kirkendall void formation.

INTRODUCTION

The electroless Ni(P)/immersion Au (ENIG) has been widely utilized as bonding pads for the solder joints in electronic packaging industry. However, due to the persistence of the black pad defect, the replacement of Ni(P) plating is emerging. And a couple of Cu-based candidates have been suggested. Therefore another problem comes up, that is, the faster reaction between the solders and the Cu substrate will impact the joint reliability [1]. In particular, during solid state aging, the growth of Cu_3Sn phase between the Cu_6Sn_5 intermetallic and Cu substrate will result in the Kirkendall voids. The presence of Kirkendall voids at the interface will cause the weak bonding between the solder and Cu plating, thereby will definitely influence the joint reliability [2-3]. Recently, the drop testing has revealed that the significant decrease of drop life was observed after thermal cycling aging and isothermal aging, for both Sn-3.8Ag-0.7Cu/OSP (Organic Solderability Preservative) and Sn-3.8Ag-0.7Cu/ENIG packages, which was also attributed to the formation of Kirkendall voids at the intermetallic/substrate interface [4]. Since the Kirkendall voids are usually formed at the solder/substrate interface, thus it is essential to study the interfacial behavior. In particular, for the reaction between the solders and OSP Cu plating, the Cu_3Sn/Cu interface where the voids nucleate should be emphasized. This paper is devoted to study the topology of Cu_3Sn phase during soldering and isothermal aging to reveal the formation mechanism. In the meantime, the effects of (Co, Ni) additives doped into the solder on the Cu_3Sn intermetallic growth and the subsequent Kirkendall void formation were examined.

EXPERIMENTAL PROCEDURES

The 5 × 5 array Cu-OSP pads were assembled on the Flame Resistant 4 (FR4) board. The thickness of Cu pad is 35 μm, and the diameter of Cu pad is 800 μm. Before soldering, the Cu pads were degreased in 4 % HCl solution, and then further cleaned in ethanol ultrasonically. Two sorts of lead-free solders, say, Sn-3.5Ag eutectic solder and Sn-3.5Ag-0.2Co-0.1Ni (Nihon Genma MFG Co. LTD., Japan), were selected to react with the Cu OSP plating. The commercial Rosin Middle Activated (RMA) flux was utilized during soldering. The microstructure observation was conducted by SEM (JSM-6320F, JEOL LTD., Japan) under backscattered electron mode (BSE). The liquid soldering was performed at 250°C using SMT Scope soldering equipment (SanyoSeiko Co. LTD, Japan) under N_2 atmosphere, and the subsequent isothermal aging was conducted at 110°C, 130°C and 150°C for 168 h, 504 h and 1008 h, respectively. The samples were cross-sectioned and then polished to a 0.05 μm final finish with colloidal silica suspension. The thickness of Cu_3Sn phase layer was examined by the image-acquirable software connected directly with the optical microscopy (VH-7000, Keyence Corp., Japan).

RESULTS AND DISCUSSION

Cu_3Sn phase topology during soldering

Generally, the Cu_3Sn phase is too thin to be ignored during short-time liquid soldering, e.g., less than 5 min. For the Sn-rich lead-free solder on Cu substrate at 250°C, the primary phase is Cu_6Sn_5 which possesses the larger formation driving force than that of Cu_3Sn phase. It has been reported that the ε-Cu_3Sn phase does not form immediately upon reaction of liquid Sn with Cu thin film; rather, the ε-Cu_3Sn phase forms after approximately 10 s at 280°C [5]. For the high-Pb solder, namely, 95Pb-5Sn solder on Cu substrate at higher soldering temperature, say, 350°C, the primary phase will be turned into Cu_3Sn phase [6]. However, for the Sn-rich solder after the prolonged soldering time, the Cu_3Sn phase tends to be visible. Fig. 1 depicts the topology of the Sn-3.5Ag/Cu solder joints after soldering at 250°C for 1 min and 8 days, respectively. As shown in Fig. 1(b), both Cu_6Sn_5 and Cu_3Sn phases were generated at the solder/Cu interface. The Cu_6Sn_5 exhibits the typical scallop-shape. The Cu_3Sn phase does not show the scallop morphology, however, the correlated topology with the Cu_6Sn_5 phase is revealed. That is, the growth of Cu_3Sn phase is tightly dependent on the geometry of Cu_6Sn_5 phase. The undulated layer with a wavelength generally corresponds to the spacing of the Cu_6Sn_5 phase scallops. At the grain boundary domain between the Cu_6Sn_5 phase grains, the growth of Cu_3Sn phase is depressed. While within the Cu_6Sn_5 scallop region, the enhanced Cu_3Sn phase growth is detected, which is quite consistent with the observation by R.A. Gagliano et al. [7]. Moreover, the undulation of the Cu plating surface also matches the periodicity of the Cu_6Sn_5 phase grains and the undulations in the Cu_3Sn phase. That is, the local maxima in the Cu surface correspond with the peaks in the Cu_6Sn_5 scallops, while the minima occur at the regions between the grains (or minima in the undulations).

Therefore, it can be reasoned that the undulations are the result of the rapid diffusion of Cu atoms through the grain boundary channels between the intermetallic grains and the slow volume diffusion through the intermetallic matrix. For the growth of both Cu_6Sn_5 and Cu_3Sn phases, the Cu flux for the growth of intermetallic is thus mainly supplied from the diffusion through the grain boundaries. Apparently, following with the coarsening of Cu_6Sn_5 intermetallic after the prolonged soldering time, the undulations in the Cu_3Sn phase and Cu substrate surface increase in wavelength while decrease in amplitude.

Owing to the addition (0.2Co-0.1Ni) into Sn-3.5Ag solder, the morphology of the interfacial reactive product was modified severely. As illustrated in Fig. 2 (a) and (b), the scallop-shape of intermetallic of Cu_6Sn_5 has been converted into the more flat and rod-like morphology. Based on EMPA examination and XRD spectrum, the intermetallic is verified as Cu_6Sn_5-based, and can be labeled as $(Cu_{0.81}Co_{0.10}Ni_{0.09})_6Sn_5$ on the solder side and $(Cu_{0.98}Co_{0.01}Ni_{0.01})_6Sn_5$ on the Cu side, respectively. This implies that the Co and Ni additives acts as the solubility elements in Cu_6Sn_5 phase [8]. At short time soldering, the undulation of Cu_6Sn_5 is reduced remarkably due to the suppressed coarsening of intermetallic. After longer soldering time as shown in Fig. 2(b), the surface of Cu_6Sn_5 phase tends to be more irregular. It is quite interesting that the Cu_3Sn phase is almost invisible even after soldering at 250°C for 60 min. In the meantime, the undulation of Cu substrate surface is also not obvious after the prolonged soldering time, which may be attributed to the refined intermetallic and thus the supplying channels for Cu atoms diffusion is significantly enhanced.

Cu_3Sn phase topology after isothermal aging

Considering the practical application, the isothermal aging is performed to examine the intermetallics at the solder joint. In particular, the Kirkendall voids are often detected at the Cu_3Sn/Cu interface after aging, which leads to the initiation of micro-cracks at the interface and thus the final failure of the solder joint. Thus in order to study the formation of the Kirkendall voids, it is necessary to explore the Cu_3Sn phase evolution during the isothermal aging.

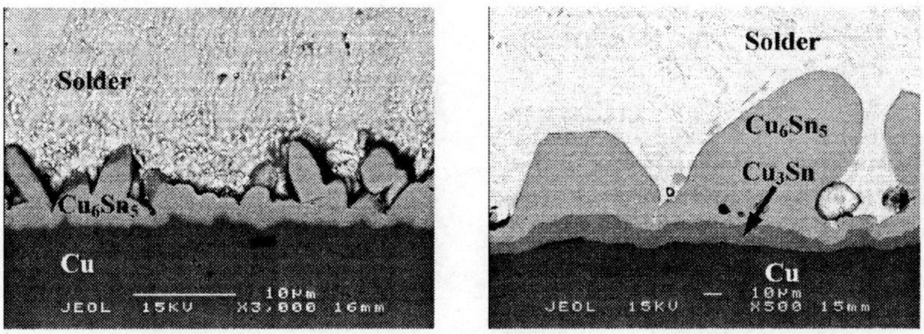

(a) (b)

Fig. 1. The topology of $Cu_6Sn_5/Cu_3Sn/Cu$ solder joint for Sn-3.5Ag/Cu after soldering at 250°C: (a) 1 min, and (b) 8 days.

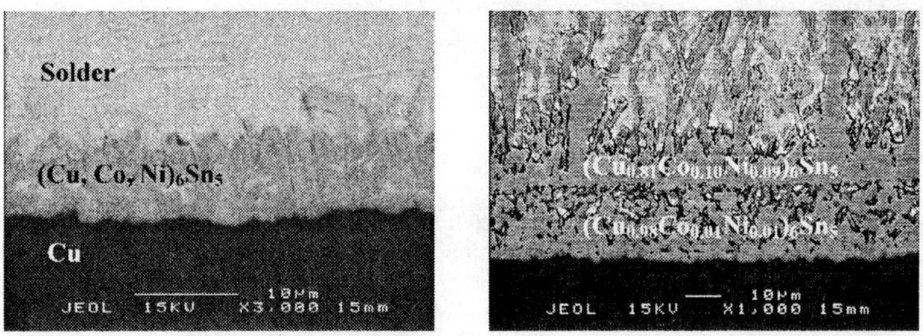

(a) (b)

Fig. 2. The topology of $Cu_6Sn_5/Cu_3Sn/Cu$ solder joint for Sn-3.5Ag-0.2Co-0.1Ni/Cu after soldering at 250°C: (a) 1min, and (b) 60 min.

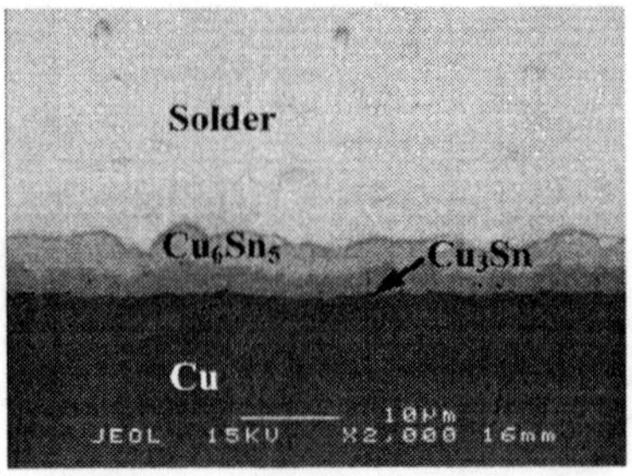

(a)

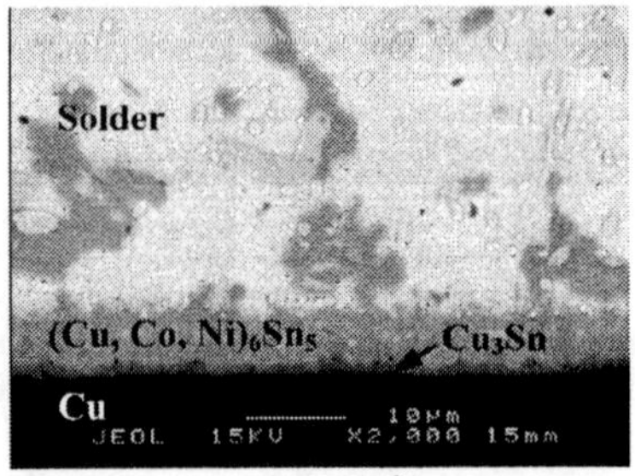

(b)

Fig. 3. The solder/Cu joints after aging at 150°C for 504 h: (a) Sn-3.5Ag/Cu, and (b) Sn-3.5Ag-0.2Co-0.1Ni/Cu.

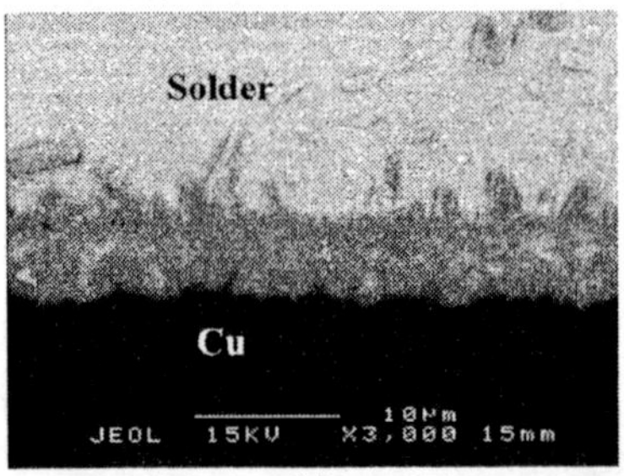

(a)

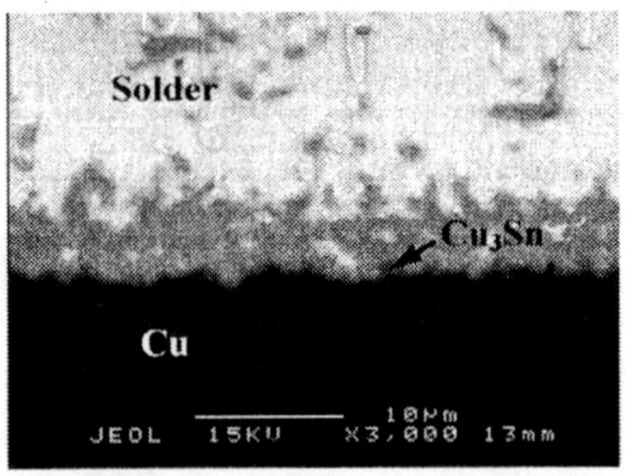

(b)

Fig. 4. The Sn-3.5Ag-0.2Co-0.1Ni/Cu joint aged for 504 h: (a) at 110°C, and (b) at 130°C.

Fig.3 illustrates the solder/Cu joints after aging at 150°C for 504 h. For Sn-3.5Ag/Cu reactive couple, the thickness of Cu_3Sn phase is comparable with the Cu_6Sn_5 layer. For both Cu_6Sn_5 and Cu_3Sn phase layers, the amplitudes of the undulation tend to decrease, which may be due to the slow diffusion of Cu atoms at isothermal aging temperature. However, the undulation of Cu_3Sn still matches the tendency of Cu_6Sn_5 intermetallic layer. In addition, the surface undulation of Cu substrate also corresponds to the tendency of Cu_6Sn_5 or Cu_3Sn phases. While for Sn-3.5Ag-0.2Co-0.1Ni/Cu reactive couple, the growth of Cu_3Sn phase was depressed dramatically. It is striking that at lower aging temperature, say, 110°C, no Cu_3Sn phase is formed at the Cu_6Sn_5/Cu interface. At 130°C, the Cu_3Sn phase is only probed at several sites, as shown in Fig. 4. The undulation of Cu_3Sn shows the discontinuity of irregular jagged shape at lower aging temperature, and then becomes a continuous layer after aging at 150°C for 504 h. The plausible reason for the suppressed Cu_3Sn growth is that the solubility of (Co, Ni) in Cu_6Sn_5 leads to the severe drop of enthalpy, and thereby stabilizes the phase. Fig. 5 displays the Cu_3Sn phase thickness as a function of aging time and aging temperature. The dramatic difference of Cu_3Sn phase formation between these two reactive couples was demonstrated.

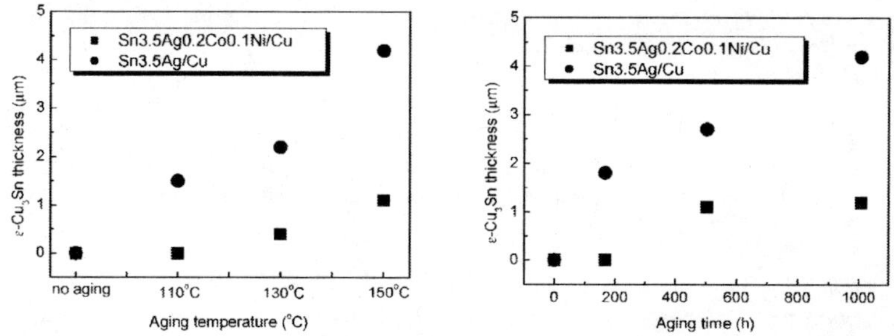

(a) (b)

Fig. 5. The Cu_3Sn phase thickness: (a) aging time of 504 h, and (b) aging temperature of 150°C.

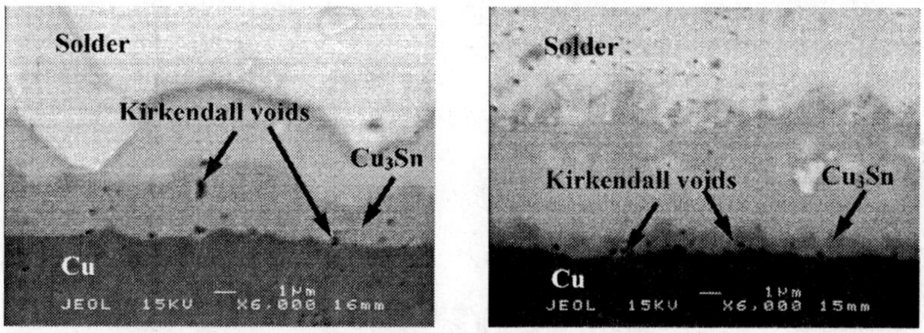

(a) (b)

Fig. 6. The Kirkendall voids at the Cu_3Sn/Cu interface after aging at 150°C for 504 h: (a) Sn-3.5Ag/Cu, and (b) Sn-3.5Ag-0.2Co-0.1Ni/Cu.

Kirkendall voids characterization

The occurrence of Kirkendall voids at the solder joint will deteriorate the reliability of the solder joint [1-4]. Apparently, the Cu-Sn interdiffusion process continues during the solid state reaction. The Cu element is the dominant diffusing species through Cu_3Sn and Cu_6Sn_5 at the isothermal aging temperature. That is, the Cu atoms reach the Cu_3Sn/Cu_6Sn_5 and solder/Cu_6Sn_5 interfaces, and thus promotes the growth of intermetallics towards the solder. However, owing to the unequal intrinsic flux of Cu and Sn through the interface, in particular, at the Cu_3Sn/Cu interface, the atomic-level vacancies due to the migrating of Cu atoms on the bare Cu surface can not be balanced by the reverse transportation of Sn atoms. Therefore, these vacancies coalesce and form Kirkendall voids. During thermal cycling, the dynamic recrystallization at the solder/Cu_6Sn_5 interface can accelerate the interdiffusion of Cu atoms [4]. However, during the isothermal aging, no dynamic recrystallization occurs and the solder grains monotonously grow up. First of all, it can be deduced that the stabilized Cu_6Sn_5 due to the solubility of (Co, Ni) additives will retard the decomposition of Cu_6Sn_5, and thus the Sn supply for the Cu_3Sn formation is suppressed. Moreover, the reduced Cu_3Sn thickness will in turn enhance the Cu flux from the substrate and lead to a significant unbalanced Cu-Sn interdiffusion through interface. Overall, after the isothermal aging, the Kirkendall voiding for Sn-3.5Ag-0.2Co-0.1Ni/Cu couple still occurs, as shown in Fig. 6. It implies that although the addition of (Co, Ni) is effective to depress the formation of Cu_3Sn phase during aging, the Kirkendall voiding is still present since the unequal diffusion flux exists. However, at lower aging temperature, say, 110°C and 130°C, the Kirkendall voids are almost vanished, which may facilitate the solder joint reliability. Two approaches are feasible to eliminate the Kirkendall voids at high aging temperature: (a) to prevent the Cu_3Sn phase formation completely; (b) try to embed or create a barrier to balance the Cu and Sn interdiffusion flux. This issue will be further investigated in future.

CONCLUSIONS

(1) During liquid soldering with Sn-3.5Ag/Cu, the undulation of the Cu_3Sn phase correlates with the morphology of Cu_6Sn_5, and the Cu surface also matches the undulation of Cu_6Sn_5 phase layer. After solid aging, the undulation of Cu_3Sn phase still exhibits this correlation the Cu_6Sn_5 phase, although the amplitude decreases. However, for Sn-3.5Ag-0.2Co-0.1Ni/Cu, this undulation correlation is not obvious.

(2) For Sn-3.5Ag-0.2Co-0.1Ni/Cu, the growth of the Cu_3Sn phase is suppressed significantly, and it completely vanishes after aging at 110°C. However, the Kirkendall voids are still present at the Cu_3Sn or Cu_3Sn/Cu interface at 150°C. The stabilized $(Cu, Co, Ni)_6Sn_5$ phase and thinner Cu_3Sn phase still trigger unbalanced interdiffusion between the Cu and Sn atoms.

ACKNOWLEDGEMENT

This work was partly supported by Grand-in-Aid for Cooperative Research Project of Nationwide Joint-Use Research Institute on Development Base of Joining Technology for New Metallic Glasses and Inorganic Materials from The Ministry of Education, Culture, Sport, Science and Technology, Japan.

REFERENCES

[1] K.J. Zeng, R. Stierman, T.-C. Chiu, D. Edvards, K. Ano, and K.N. Tu, "Kirkendall Void Formation in Eutectic SnPb Solder Joint on Bare Cu and its effect on Joint Reliability," *J. Appl. Phys.*, **97**, 024508-1 to 8 (2005).

[2] P.L. Liu, J.K. Shang, "Segregant-Induced Cavitation of Sn/Cu Reactive Interface," *Scripta. Mater.*, **53**, 631-634 (2005).

[3] Z.Q. Mei, M. Ahmad, M. Hu, and G. Ramakrishna, "Kirkendall Voids at Cu/Solder Interface and Their Effects on Solder Joint Reliability," *Proc. 55th Electron. Comp. Tech. Conf.*, Orlando, FL, May. 2005, pp. 415-420.

[4] L.H. Xu and H.L. Pang, "Effect of Intermetallic and Kirkendall Voids Growth on Board Level Drop Reliability for SnAgCu Lead-free BGA Solder Joint," *Prof. 56th Electron. Comp. Tech. Conf.*, San Diego, CA, May. 2006, pp. 275-282.

[5] F. Bartels, J.W. Morris, Jr., G. Dalke, and W. Gust, "Intermetallic Phase Formation in Thin Solid-Liquid Diffusion Couples," *J. Electron. Mater.*, **23**, 787-790 (1994).

[6] H.Q. Wang, X. Ma, F. Gao and Y.Y. Qian, "Sn Concentration on the Reactive Wetting of High-Pb Solder on Cu Substrate," *Mater. Chem. Phys.*, **99**, 202-205 (2006).

[7] R.A. Gagliano and M.E. Fine, "Growth of η Phase Scallops and Whisker in Liquid Tin-Solid Copper Reaction Couples," *JOM*, **53**, 33-38 (2001).

[8] F. Gao, T. Takemoto, H. Nishikawa, A. Komatsu, "Microstructure and Mechanical Properties Evolution of Intermetallics between Cu and Sn-3.5Ag Solder Doped by Ni-Co Additives," *J. Electron. Mater.*, **35**, 905-911 (2006).

FREEFORM FABRICATION OF PHOTONIC CRYSTALS WITH 3-DIMENSIONAL DIAMOND STRUCTURE BY MICRO-STEREOLITHOGRAPHY

Weiwu Chen, Soshu Kirihara, and Yoshinari Miyamoto
Joining and Welding Research Institute, Osaka University
Osaka 567-0047, Japan

ABSTRACT

3-dimensional (3D) photonic crystals with a diamond structure were successfully formed from the 40vol%TiO_2-acrylate slurry by means of a CAD/CAM micro-stereolithography system. The designed lattice constant of the diamond unit cell was 500 μm and the obtained forming accuracy was 10 μm. They were converted to pure TiO_2 ceramics with 92% theoretical density by sintering at 1400°C. The EM wave propagation in the Γ-X <100> direction of formed and sintered samples was measured by terahertz-time-domain spectroscopy respectively. The results showed that the photonic band gap appeared at 280-360 GHz in the formed sample and at 140-160 GHz in the sintered sample, which agreed well with the calculated band gap by the plane wave expansion method.

INTODUCTION

3D photonic crystals with a diamond structure of dielectric medium can open the complete photonic band gap and inhibit the propagation of electromagnetic (EM) waves with wavelengths corresponding to the lattice periods for all directions.[1, 2] Micro-machining and soft lithography used in semiconductor process have been adopted to fabricate photonic crystals to operate in infrared and visible optical ranges.[3-5] However, these processes are difficult in ceramic materials and limited in fabrication of z-dimension.

Stereolithography is a rapid prototyping method to form complex 3D objects layer by layer using a precise laser scanning to liquid photo-curable resin with a CAD/CAM system.[6] It was adopted for fabrication of ceramic green parts by using resins filled with ceramic particles.[7] This technique can reach around 600 micron resolution and provide much more material options. We succeeded in fabrication of ceramic-epoxy 3D photonic crystals for GHz frequencies by stereolithography.[8-9] Micro-stereolithography was developed recently with a much higher resolution by using a novel DMD (Digital Micromirror Device) technique.[10] The fabrication of a pure polymer 3D microstructure with 0.6 micron resolution has been demonstrated, but there are few reports on finer 3D ceramic-containing structures.[11]

This paper reports our work on fabrication of 3D photonic crystals with a diamond structure for THz frequencies (0.1-10 THz) using TiO_2 ceramics by micro-stereolithography. The feature resolution and curing depth were examined. The photonic band gaps of formed and sintered micro 3D photonic crystals were measured and compared with the calculated band diagram.

EXPERIMENTAL

Pure TiO_2 powders with a mean particle size of 200 nm (Anatase type, Hosokawa Ltd., Osaka, Japan) were mixed at 40vol% with photo-curable acrylate resin (KC1184, D-MEC Co. Ltd., Tokyo, Japan) by using a centrifugal defoaming machine to form a bubble-free suspension for micro-stereolithography.

3 photonic crystal samples with 2, 3 and 4 periodic layers in z-dimension and each periodic layer containing 8x8 diamond unit cells (lattice constant = 500 μm) were designed by using CAD software (Thinkdesign 9.0, Toyota Keram Co. Ltd., Aichi, Japan). The designed CAD model was converted into STL files and sliced into a set of thin sections by software (Magics 9.9, Materialise Co. Ltd., Leuven, Belgium), and then transferred to a micro-stereolithography machine (Acculas SI-C1000, D-MEC Co. Ltd., Tokyo, Japan). This machine uses a 405 nm laser for exposure with 2-micron space resolution.[12]

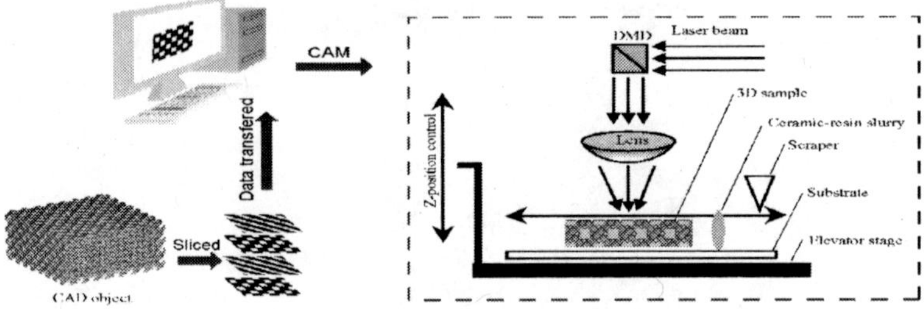

Figure 1 Schematic of a micro-stereolithography process

During the micro-stereolithography fabrication, the mixed liquid suspension was squeezed and coated layer by layer onto a working stage. The laser light illuminated on the DMD chips was shaped according to the sliced pattern and then irradiated on the coated layer surface. The illuminated area was solidified simultaneously under one exposure, while the dark regions remained liquid. After fabricating one layer, the working stage moved downward one layer thickness and then a new layer was fabricated on top of the existing structure. The time for paste squeezing and coating is about 1.3 minute to form one layer of 40mm×40mm size. In the case of 2-μm space resolution, one exposure is 1.751×1.313 mm in size and it takes about 2 s. For a 400-layer photonic crystal (4×4×2mm), the fabrication time is 11h in total. This micro-fabrication process is illustrated in Fig. 1. The formed samples were then sintered at 1400°C for 2 h to obtain pure TiO_2 photonic crystals.

The feature of the formed and sintered samples was observed by using a digital optical microscope (VH-Z100, KEYENCE Co., Osaka, Japan) and a scanning electron microscope (ESEM-2700, Nikon Co., Tokyo, Japan). The apparent density of the sintered bulk sample was measured by the Archimedes method. The photonic band gap of formed and sintered samples was measured using THz time-domain spectrometer (THz-TDS, J-Spec2001/ou, Advanced Infrared Spectroscopy Co. Ltd., Tokyo, Japan).[13-14]

RESULTS AND DISCUSSION

In micro-stereolithography of ceramic suspensions, the laser light scattering from ceramic particles influences significantly the lateral resolution and the curing depth, which are critical for forming complex 3D microstructures in high-resolution. Figure 2 shows the resolution of micro-stereolithography for the 40 vol% TiO_2 suspension with a laser intensity of 1000 mJ/cm^2 and layer thickness of 5 μm. The cured width depends on the spot size of the focused laser beam, and on the spreading of the laser radiation by side-scattering from ceramic particles. Compared to the designed samples, the solidified parts appeared around 10 μm expansion. Considering the symmetrical side scattering, one side spreading of laser radiation from TiO_2 particles scattering was around 5 μm, and the feature resolution of micro-stereolithography in this study was estimated around 10 μm.

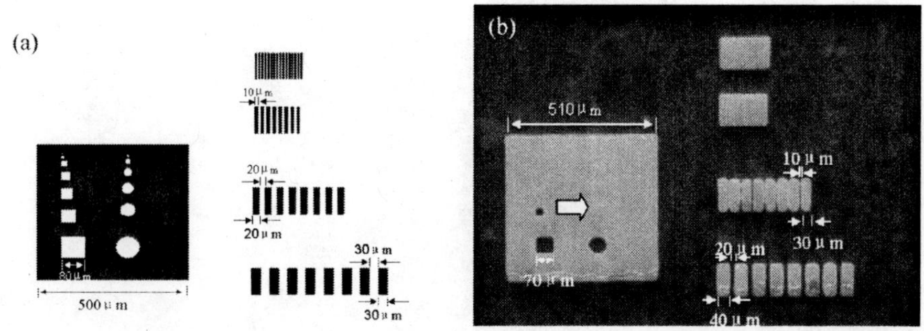

Figure 2 (a) Designed 2D samples for feature resolution check: dark regions will be irradiated by laser. (b) Corresponding samples fabricated by micro-stereolithography. Some non-irradiated areas were filled with ceramic slurry, as denoted by an arrow.

The curing depth for the ceramic suspension achieved by stereolithography was reported to be proportional to $1/\Delta n^2$, where Δn is the difference in refractive index of the photo-curable resin and ceramic particles.[7] The refractive index of TiO_2 is around 2.5, which is much higher than that of acrylate resin (around 1.4). For conventional stereolithography, it is difficult to achieve 3D ceramic object of TiO_2 due to the poor curing depth. But in micro-stereolithography, this problem may be solved by reducing the layer thickness because 2-3 μm in thickness per layer can be achieved by the current machine. As shown in Fig. 3, when adjusting the layer thickness to 5 μm, the connection between layers appeared very strong and no obvious interface was observed.

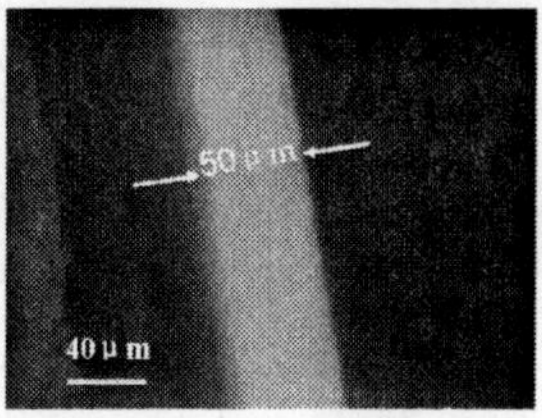

Figure 3 Side view of a fabricated TiO_2-acrylate bulk sample at 5-μm thickness/layer with 10 layers

A unit cell of diamond structure consisting of dielectric rods is shown in Fig. 4a. The lattice constant and the aspect ratio of a dielectric rod were designed as 500 μm and 1.5, respectively. Photonic crystals containing 2, 3 and 4 periodic layers and each layer containing 8×8 unit cells were designed for fabrication. As an example, a photonic crystal with 4 periodic layers as a stacking of 8×8×4 unit cells is shown in Fig.4b.

The designed photonic crystals were formed by micro-stereolithography with the laser intensity of 1000 mJ/cm^2 and the layer thickness of 5 μm. Figure 4c and d show optical micrographs of a formed photonic crystal object with 4 periodic layers. Compared to the designed structure, the fabricated sample had the same dimension and no deformation. The stacked diamond unit cell also had a lattice constant around 500 μm and clear feature resolution. After sintering with liner shrinkage of 23-25%, pure TiO_2 ceramic photonic crystals were obtained. The lattice constant became around 380 μm, as shown in Fig. 4e.

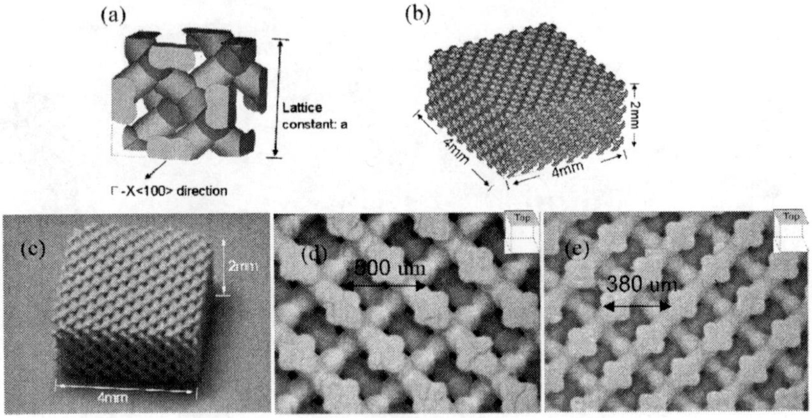

Figure 4 (a) Designed diamond unit cell with <100> direction, (b) Designed photonic crystal with diamond structure by stacking 8×8×4 unit cells, (c) Optical microscope images of fabricated photonic crystal with 4 periodic layers, (d) Top view of a formed photonic crystal, (e) Top vies of a sintered photonic crystal

When the nature surfaces of sintered sample was observed under SEM (as shown in Fig. 5), the top view image shows the grain size is around 5-8 μm and no obvious pore was observed. While the side view image indicates a porous structure with a smaller grain size. In the edge of layer at the lateral side of pre-sintered samples, the thickness is less 5 μm and there is a gap between layers. The grain growth during sintering would be suppressed in this limited volume. Therefore, the grain size in the side view is much smaller than that of top view. The apparent density of sintered bulk sample at the same conditions was 3.87 g/cm^3, around 92% theoretical density of a rutile phase.

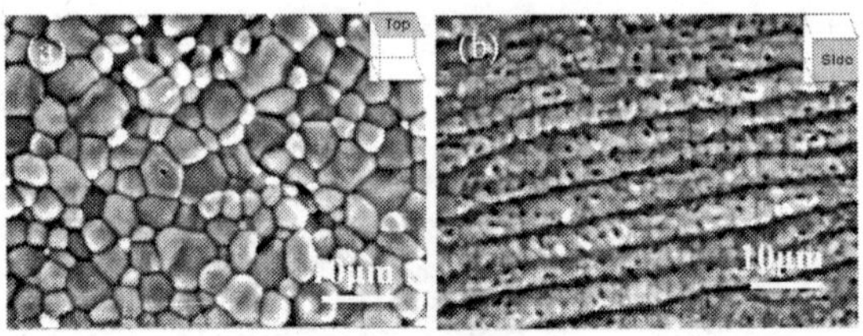

Figure 5 SEM images of a sintered photonic crystal (a) top view, (b) side view

Figure 6 shows the EM wave transmission spectra of formed and sintered photonic crystals with 2, 3 and 4 periodic layers. The transmittance-axes of them were plotted by $\log_{10}$ and liner scale respectively for easy reading. As indicated, with the increase of the number of periodic layer, the first band gap becomes deeper and wider in the formed samples, locating in 280-360

GHz. Theoretically in the band gap frequencies, the propagation of electromagnetic wave is forbidden due to the effect of standing wave formation in the periodic lattices. For the sintered samples, the transmission spectra are not so smooth, but it should be noticed there is a gap in the range of 140-160 GHz in all three samples, as denoted by arrows. Measured by THz-TDS, the dielectric constant and loss are around 16 and 0.5 for formed TiO_2-Acrylate in the range of 250-500 GHz and around 100 and 3 for sintered TiO_2 in the range of 120-250 GHz, respectively. So the rough spectra and weak transmission intensity in the sintered samples should be related to the high dielectric loss.

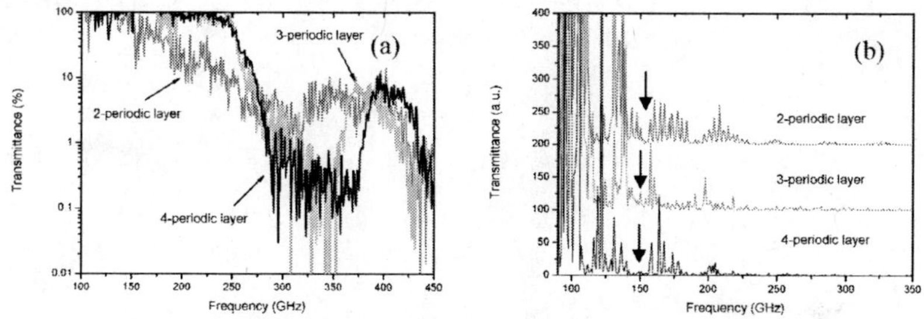

Figure 6 Measured transmission intensity as a function of frequency for (a) formed photonic crystals, (b) sintered photonic crystals

In addition, the band gap diagram of a photonic crystal with diamond structure can also be calculated based on its structure parameters and the dielectric constant of lattice materials by plane wave expansion method[15]. As shown in Fig. 7, the band gap diagrams for the formed and sintered diamond structure were calculated. In the Γ-X <100> direction, the calculated band gap located in the range of 250-360 GHz for the formed photonic crystal and in the range of 140-220 GHz for the sintered photonic crystal, which was consistent with the measured photonic band gap.

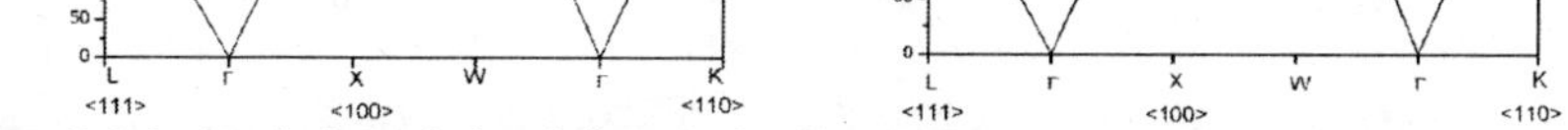

Figure 7 Calculated photonic band diagram for diamond structure (a) formed sample, (b) sintered sample. The measured photonic band gaps were shadowed.

CONCLUSIONS

Micro-stereolithography was successfully applied to fabricate 3-dimensional photonic crystals with diamond structure for THz wave control. For the TiO_2-acrylate (40 vol% TiO_2) suspension, micro-stereolithography achieved 10 μm feature resolution, and obtained strong layer joining at a layer thickness of 5 μm. The formed 3D photonic crystal has the same dimensions as the designed one. After sintering, the 3D sample has 23-25% liner shrinkage and reached 92% theoretical density of rutile phase. The photonic band gap was observed at 280-360 GHz in the formed samples and 140-160 GHz in the sintered samples, which agreed well with the calculated result.

ACKMOWLEDGEMENTS

This work is supported by the Grant for the 21st Century's COE Program "Center of Excellence for Advanced Structural and Functional Materials Design", and by the Grant-in-Aid for Scientific Research (S) No. 171067010, both under the auspice of the Ministry of Education, Culture, Sports, Science and Technology, Japan.

REFERENCES

[1]J. D. Joannopoulos, P. R. Villeneuve and S. Fan. "Photonic Crystals: Putting a New Twist on Light," *Nature*, **386,** 143–9 (1997).

[2]E Yablonovitch, "Photonic Band-gap Structures," *J. Opt. Soc. Am. B* **10,** 283–95 (1993).

[3]A. Birner, R. B. Wehrspohn, U. M. Gösele, and K. Busch, "Silicon-based Photonic Crystals," *Adv. Mater.*, **13**, 377-88 (2001).

[4] J. S. Foresi, P. R. Villeneuve, J. Ferrera, and E. R. Thoen, G. Steinmeyer, S. Fan, J. D. Joannopoulos, L. C. Kimerling, Henry I. Smith, and E. P. Ippen, " Photonic Band Gap Microcavities in Optical Waveguides," *Nature,* **390,** 143–5 (1997).

[5]J. Lee, C. Kim, Y. Kim, K. Ho, K. Constant, and C. Oh, "Three-dimensional Metallic Photonic Crystals Fabricated By Soft Lithography for Midinfrared Applications," *App. Phy. Lett.*, **88**, 181112 (2006).

[6]P. F. Jacobs, *Stereolithography and Other RP&M Technologies*, Society of Manufacturing Engineers, Dearborn, 1996.

[7]M. L. Griffith and J. W. Halloran, "Freeform Fabrication of Ceramics Via Stereolithography," *J. Am. Ceram. Soc.*, **79**, 2601-8 (1996).

[8]S. Kanehira, S. kirihara and Y. Miyamoto, "Fabrication of TiO_2-SiO_2 Photonic Crystals with Diamond Structure," *J. Am. Ceram. Soc.*, **88**, 1461-64 (2005).

[9]H. Yin, S. Kirihara and Y. Miyamoto, "Fabrication of Ceramic Photonic Crystals with Diamond Structure for Microwave Applications," *J. Am. Ceram. Soc.*, **87**, 598-601 (2004).

[10]C. Sun, N. Fang, D. M. Wu, and X. Zhang, "Projection Micro-stereolithography Using Digital Micro-mirror Dynamic Mask," *Sensors and Actuators A,* **121**, 113-20 (2005).

[11]X. Zhang, X. N. Jiang and C. Sun, "Micro-stereolithography of Polymeric and Ceramic Microstructures," *Sensor and Actuators*, **77**, 149-56 (1999).

[12]URL http://www.laser-solutions.co.jp

[13]T. Aoki, M. W. Takeda, J. W. Haus, Z. Yuan, M. Tani, K. Sakai, N. Kawai and K. Inoue, "Terahertz Time-domain Study of a Pseudo-simple-subic Photonic Lattice," *Phys. Rev.B*, **64**, 046106 (2001).

[14]H. Kitahara, N. Tsumura, H. Kono, M. W. Takeda, J. W. Haus, Z. Yuan, N. Kawai, K. Sakoda and K. Inoue, "Terahertz Wave Dispersion in Two-dimensional Photonic Crystals," *Phys. Rev.B*, **64**, 045202 (2001).

[15]K. M. Ho, C. T. Chan and C. M. Soukoulis, "Existence of a Photonic Gap in Periodic Dielectric Structures," *Phys. Rev. Lett.*, **65**, 3152-55 (1990).

PROPERTIES OF INDUCTIVETY-COUPLED RF PLASMAS SUSTAINED WITH INTERNAL ANTENNA FOR DEPOSITION OF CARBON-RELATED FILMS

Kosuke Takenaka and Yuichi Setsuhara
Joining and Welding Research Institute, Osaka University,
11-1, Mihogaoka, Ibaraki,
Osaka, Japan, 567-0047

Kazuaki Nishisaka and Akinori Ebe
EMD Corporation,
KyodaiKatura VenturePlaza, 1-36 Goryouohara-cho, Nishikyo-ku
Kyoto, Japan, 615-8245

ABSTRACT

Large-area plasma sources with high plasma density are desired for a variety of advanced plasma processes. Furthermore development of plasma sources at low plasma potential via suppression of electrostatic coupling is required for high-quality processing. To meet these requirements, we have developed inductively-coupled radio frequency plasma sources, in which multiple internal low-inductance antenna (LIA) units were installed. In this study, plasma properties were measured for an inductively coupled RF plasmas, which were sustained with the multiple LIA units in a cylindrical source chamber with 500 mm in diameter. Experiments with 8 LIA units resulted in stable source operation even at 2000 W RF powers to attain plasma densities as high as 10^{11} - 10^{12} cm^{-3} and plasma potential as low as 13V at argon pressures around 1.3 Pa at an RF power of 1000 W. Kinetic energy distribution of Ar ions was measured at argon pressures around 1.3 Pa at an RF power of 1000 W, which has peaks around 12 eV which almost correspond to the magnitude of the plasma potential. Furthermore plasma chemical vapor deposition of carbon related films for CH_4 plasma was also attempted using this cylindrical source. Deposition of polymer-like carbon film which has high hydrogen content was realized using this cylindrical source.

INTRODUCTION

Development of large-area plasma sources with high plasma density is desired for fabrication of microelectronics devises and flat panel display (FPD), which are tended toward substrate enlargement and high throughput mainly due to cost reduction of the fabrication processes.

Among various high-density plasma sources such as a capacitively-coupled plasma (CCP) [1], an inductively-coupled plasma (ICP)[2-5], electron cyclotron resonance (ECR) plasma [6-9], and surface wave plasma (SWP)[10], ICPs attract great interests as one of the promising candidates of large-area high-density plasma source.

In scaling up of ICP sources with conventional antenna configuration such as a spiral antenna placed on top of a reactor, the antenna size and hence the antenna impedance is tended to become larger for large-area plasma generation. However, the increase in the antenna size and/or the impedance raises problems associated with anomalous rise of antenna terminal voltages and formation of standing wave, which result in non-uniform RF current distribution and hence non-uniform plasma profile[11]. Furthermore the increase of the antenna size and hence the antenna impedance also raise increase of the potential in the plasma[11].

To solve these problems, we developed ICP sources with multiple low-inductance antenna (LIA) units. For plasma processing of large-area substrates such as 300 mm wafers, cylindrical plasma source of 500 mm in diameter with multiple LIA units have been developed. Previous experiments with multiple LIA units resulted in stable source operation even at 3000 W RF powers to attain plasma densities as high as 10^{11} - 10^{12} cm^{-3} and floating potentials as low as 4 - 6 V in argon [12].

In this paper, we report properties of argon plasmas sustained in the cylindrical plasma source of 500 mm in diameter with multiple LIA units. We also report the preliminary experimental results regarding plasma chemical vapor deposition (plasma CVD) of carbon-related films using this cylindrical plasma source.

EXPERIMENTAL

Schematic diagram of the chamber with a set of 8 LIA units is shown in Fig. 1. The LIA unit consists of a U-shaped antenna conductor, which was fully covered with dielectric tubing for complete isolation from the plasma. Eight LIA units with a 100 mm width and a 160 mm height were mounted on the top flange of the discharge chamber and coupled to a 2000 W RF power generator at 13.56 MHz via a matching network. Each LIA unit was connected in parallel to the matching network. This chamber had a 500 mm inner diameter and a 200 mm height, which was connected to a diffusion chamber made of stainless-steel vessel with a 500 mm inner diameter and a 400 mm height. The substrate holder was placed at a distance of 297 mm from the top flange.

Plasma parameters including plasma density and plasma potential were measured with a cylindrical Langmuir probe which was inserted radially at an axial position of 200-280 mm below the top flange. The kinetic energy distribution of ions were measured with a quadrupole mass spectrometer (Q-mass: Hiden, EQP500) mounted on sidewall as shown in Fig. 1.

The film depositions were performed with a pure methane (CH_4) plasma at the pressure of 1.3 Pa. The deposition time was 15 min. A Si substrate was placed on the substrate holder (electrode). The substrate temperature was kept at room temperature. The deposited films were analyzed by Fourier transform infrared spectroscopy (FTIR: Nicolet, Magna-IR 560) with

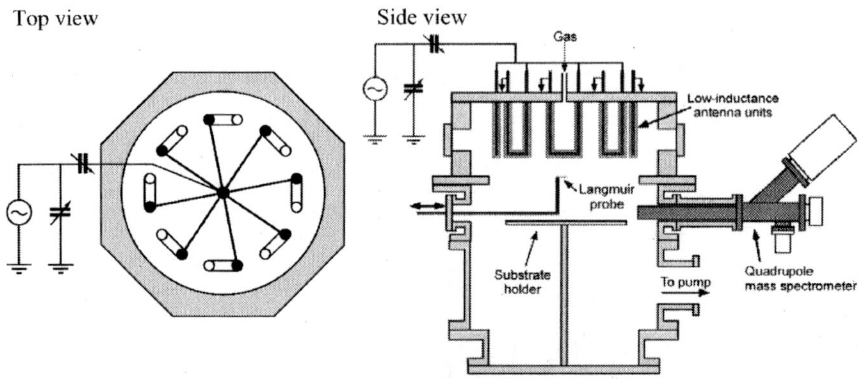

Figure 1. Schematic diagram of discharge chamber with a set of LIA units.

Attenuated Total Reflection (ATR) instrument.

RESULTS AND DISCUSSION

Plasma diagnostics

Plasma properties were measured for an inductively coupled RF plasmas, which were sustained with the 8 LIA units in a cylindrical source chamber with 500 mm in diameter. RF power dependence of plasma density measured at the radial center of the chamber is shown in Fig. 2 for argon plasmas sustained with 8 LIA units in a pressure range of 1.3 Pa. The plasmas ware stably sustained and the plasma densities increased almost linearly with increasing RF power to attain plasma density of 1.3×10^{12} cm^{-3} at an RF power of 2000 W at an argon pressure of 1.3 Pa.

Information about kinetic energy of ions impinging on the substrate surface has been obtained with Q-mass and Langmuir probe. Figure 3 shows kinetic energy distributions of Ar^+ ions at an RF power of 1000 W at an argon pressure of 1.3 Pa. The distribution has peaks around 12 eV, which almost correspond to the magnitude of plasma potential (12.9 V) measured by Langmuir probe. Therefore the plasma potential is equivalent to the representative kinetic energy of ions impinging on conductive substrates on a grounded substrate holder (electrode). Furthermore, the potential drop from the plasma potential to the floating potential, which is equivalent to the energy of ion bombardment during the film growth with plasma CVD on non-conductive substrate such as glass and polymers, was 9.4 V. Thus the result regarding the plasma potential demonstrates that ICPs driven by low-inductance antenna are suitable as plasma sources for CVD processing, which requires low-potential plasmas for formation of high-quality films with markedly reduced plasma damage. Together with the capability of the high density plasma generation, these results exhibit that simultaneous achievement of high-rate and high-quality film deposition can be feasible with the ICPs sustained with LIA units.

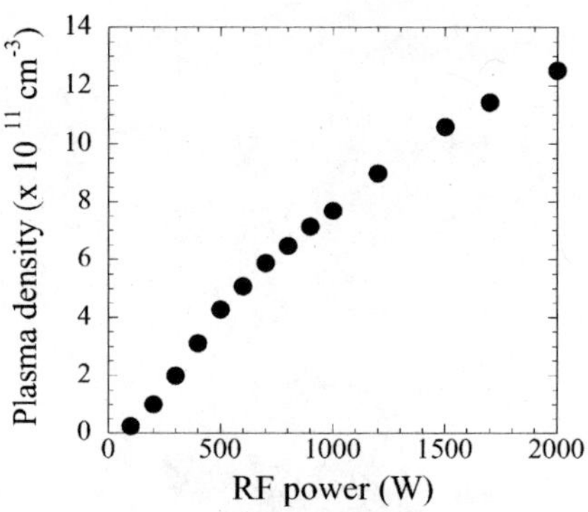

Figure 2. RF power dependence of plasma density for argon plasmas at pressure of 1.3 Pa.

Deposition of carbon-related films

Using the cylindrical source with the multiple LIA units, plasma CVD of carbon-related films was attempted for CH_4 plasma. Figure 4 shows IR absorption spectra of the film deposited at an RF power of 1000 W at a methane pressure of 1.3 Pa. The spectrum has an absorption band at 2872 cm^{-1} and 2955 cm^{-1} which corresponding to symmetrical and asymmetrical sp^3 C-H_3 stretching modes[13,14]. The spectrum also exhibits peak at 2926 cm^{-1} originated from asymmetrical sp^3 C-H_2 stretching mode[13]. This result indicates that this carbon-related film is polymer-like carbon which has high hydrogen content[15]. In general it was known that breaking of bonds between carbon and hydrogen atoms and dehydrogenating caused by ion bombardment with sufficient energy.[15] Therefore the result exhibits that energy of ion bombardment during film growth is sufficiently suppressed using the cylindrical source with the multiple LIA units.

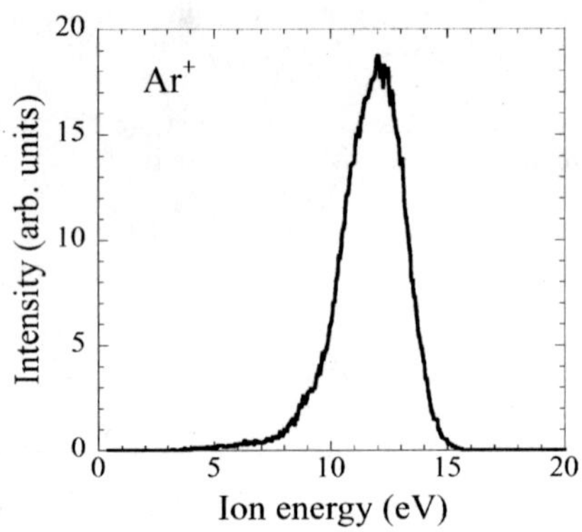

Figure 3. Kinetic energy distributions of Ar^+ ions at an RF power of 1000 W at an argon pressure of 1.3 Pa.

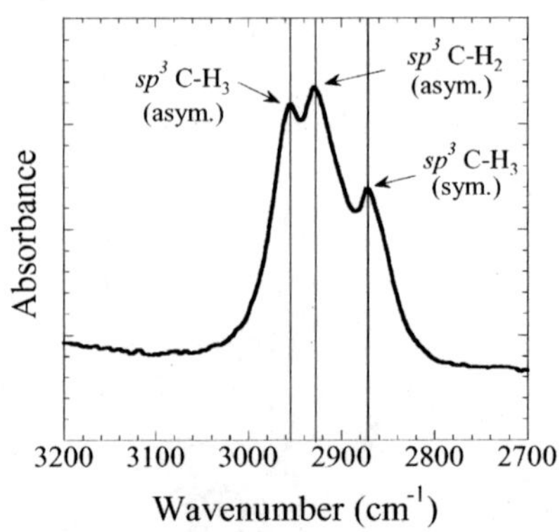

Figure 4. IR absorption spectra of the film deposited with at an RF power of 1000 W at a methane pressure of 1.3 Pa

CONCLUSIONS

Large-area plasma sources with high plasma density are desired for a variety of advanced plasma processes. Furthermore development of plasma sources at low plasma potential via suppression of electrostatic coupling is required for high-quality processing. To meet these requirements, we have developed inductively-coupled radio frequency plasma sources, in which multiple internal low-inductance antenna (LIA) units were installed. In this study, plasma properties were measured for an inductively coupled RF plasmas, which were sustained with the multiple LIA units in a cylindrical source chamber with 500 mm in diameter. Experiments with 8 LIA units resulted in stable source operation even at 2000 W RF powers to attain plasma densities as high as 10^{11} - 10^{12} cm^{-3} and plasma potential as low as 13V at argon pressures around 1.3 Pa at an RF power of 1000 W. Kinetic energy distribution of Ar ions was measured at argon pressures around 1.3 Pa at an RF power of 1000 W, which has peaks around 12 eV which almost correspond to the magnitude of plasma potential. Furthermore plasma CVD of carbon related films for CH_4 plasma was attempted using this cylindrical source. Deposition of polymer-like carbon film which has high hydrogen content was realized using this cylindrical source.

REFERENCES

[1] B. P. Wood, I. Henins, R. J. Gribble, W. A. Reass, R. J. Faehl, M. A. Nastasi and D. J. Rej, "Initial Operation of a Large-Scale Plasma Source Ion Implantation Experiment", *J. Vac. Sci. & Technol.*, B **12**, 870-74 (1994).

[2] J. Hopwood, "Review of inductively coupled plasmas for plasma processing", *Plasma Sources Sci. Technol.*, **1**, 109-116 (1992).

[3] J. Hopwood, "Ion Bombardment Energy Distributions in a Radio Frequency Induction Plasma", *Appl. Phys. Lett.* ,**62**, 940-42 (1993).

[4] W. Z. Collison, T. Q. Ni and M. S. Barnes, "Studies of The Low-Pressure Inductively-Coupled Plasma Etching for a Larger Area Wafer Using Plasma Modeling and Langmuir Probe", *J. Vac. Sci. & Technol.*, A **16**, 100-07 (1998).

[5] S. S. Kim, H. Y. Chang, C. S. Chang and N. S. Yoon, "Antenna Configuration for Uniform Large-Area Inductively Coupled Plasma Production", *Appl. Phys. Lett.*, **77**, 492-94 (2000).

[6] K. Suzuki, S. Okudaira, N. Sakudo and I. Kanomata, "Microwave Plasma Etching", *Jpn. J. Appl. Phys.*, **16**, 1979-84 (1977).

[7] S. Matsuo and Y. Adachi, "Reactive Ion Beam Etching Using a Broad Beam ECR Ion Source", *Jpn. J. Appl. Phys.*, **21**, L4-6 (1982).

[8] M. Misina, Y. Setsuhara and S. Miyake, "Electron Cyclotron Resonance Plasma Enhanced Direct Current Sputtering Discharge with Magnetic-Mirror Plasma confinement", *J. Vac. Sci. & Technol.*, A **15**, 1922-28 (1997).

[9] M. Misina, Y. Setsuhara and S. Miyake, "Ion Assisted Deposition of Crystalline TiNi Thin Films by Electron Cyclotron Resonance Plasma Enhanced Sputtering", *Jpn. J. Appl. Phys.*, **36**, 3629-34 (1997).

[10] M. Moisan, C. Baudry and P. Leprince, "Small Microwave Plasma Source for Long Column Production without Magnetic-Field", *IEEE Trans. Plasma Sci.*, **3**, 55-59 (1975).

[11] Y. Setsuhara, "Meter-Scale Large-Area Plasma Sources for Next-Generation Processes", *J. Plasma Fusion Res.*, **81,** 85-93 (2005).

[12] Y. Setsuhara, T. Shoji, A. Ebe, S, Baba, N. Yamamoto, K. Takahashi, K. Ono, and S. Miyake,

MATERIALS DESIGN

CHARACTERISTICS OF ITO FILMS DEPOSITED BY DC MAGNETRON SPUTTERING USING VARIOUS SINTERED INDIUM-TIN-OXIDE TARGETS

Joon Hong Park[1], Sang Chul Lee[2], Jin Ho Lee[3], Pung Keun Song[1*]

[1]School of Materials Science and Engineering, Pusan National University, 30 Jangjeon-dong, Geumjeong-ku, Busan 609-735, Korea
[2]Digital Information Materials Div., SAMSUNG CORNING Co., LTD., 644 Jinpyeong-dong, Gumi-si, Gyeongsangbuk-do, 730-725, Korea.
[3]Next Gen. ITO Target PJT, R&D Center, SAMSUNG CORNING Co., LTD., 472 Sin-dong, Yeongtong-ku, Suwon-si, Gyeonggi-do, 443-732, Korea.

ABSTRACT

Indium Tin Oxide (ITO) films were deposited on nonalkali glass substrate by dc magnetron sputtering using high density ITO targets which were prepared under different conditions such as conductivity and micro-pore size. In order to investigate correlation between target and film properties, depositions were carried out at total gas pressure (P_{tot}) of 0.5 Pa, substrate temperature (T_s) of RT~300 °C, oxygen flow ratio [$O_2/(O_2+Ar)$] of 0 ~ 1.5% and dc power of 100W. In the case of conductivity, high conductivity target A2 showed relatively high stability in electrical property with increasing target erosion ratio. Optimum O_2 addition ratio to obtain the lowest resistivity was decreased with increasing target erosion ratio. High conductivity ITO target could lead to decrease in micro-nodule formation on the target surface because of high cooling efficiency. In the case of micro-pore size, clear difference in resistivity of the films was not observed between two targets for 11 hours sputtering where target surface was eroded about 20%. Whereas, target B2 showed small degradation in resistivity with increasing sputtering time after 11 hours compared to large micro-pore size target. Optimum oxygen ratio to obtain the lowest resistivity was decreased with increasing substrate temperature. The lowest resistivity was 1.06×10^{-4} Ωcm for the film deposited using target B2 at $O_2/(O_2+Ar)$ ratio of 0.05% and at T_s =300 °C.

INTRODUCTION

Thin films of Sn-doped In_2O_3 (ITO), which has a high electrical conductivity, have been used as transparent electrodes for LCD, PDP, OLED and solar cells.[1,2)] The preparation processes of ITO thin films include spray[3], chemical vapor deposition (CVD)[4)], evaporation[5)], and magnetron sputtering[6)] Among the these processes, the most preferred method in production line is dc magnetron sputtering using ITO ceramic target because it has several advantages such as high controllability, high deposition rate and large area deposition[7)] Sintered high density ITO target is widely used to deposit transparent conductive thin film for various kinds of flat panel displays. Considerable efforts have been focused on depositing thin films of ITO with significantly reduced resistivity to accommodate the increasing technological demand for larger area flat panel displays with higher image quality. In this dc magnetron sputtering processes, the black crystals of a dark phase, which are called nodule, are formed on the surface of ITO target degraded film properties, i.e., resistivity, transmittance and crystallinity, etc[8,9)] This harmful phenomenon grows worse during a sputtering and finally demand stop of the sputter run and mechanical cleaning of the target surface. In order to reduce the nodule formation, ITO target has been improved in it's density[10,11]. K. Nakashima et al[12] reported that the nodule formation of the ITO target was reduced by using the fine raw powder and by applying the modified mixing process.

In this study, ITO films were deposited by dc magnetron sputtering using different types of high density ITO targets where contained different micro-pore size and conductivity. The correlation between film properties and target erosion ratio were investigated for each ITO targets.

EXPERIMENTAL PROCEDURE

ITO films with thickness of 150~160nm were deposited on nonalkali glass substrate (corning E2000) at total gas pressure of 0.5Pa by dc magnetron sputtering. Depositions were carried out various ITO ceramic disks (Samsung Corning Co., Ltd.) where ITO targets were prepared by control of conductivity and micro-pore size, respectively. In order to optimize resistivity of ITO films, oxygen gas was added to Ar gas with $[O_2/(O_2+Ar)]$ flow ratio of 0 ~ 1.5%. Water partial pressure of the residual gas was maintained to be less than $8x10^{-4}$ Pa to guarantee high reproducibility of the film properties. Depositions were also carried out at substrate temperature (T_s) of RT ~ 300 °C and dc power of 100W. Micro-arcing frequency was monitored during the sputtering time (~40 hours). Film thickness was measured using a surface profiler (DekTak^3, Veeco). Resistivity, carrier density and Hall mobility were measured by Hall effect measurements (HMS-3000, ECOPIA) at room temperature. X-ray diffraction(XRD) was carried out by 40kV-30mA CuKα radiation (X'Pert-MPD System, PHILIPS).

RESULTS AND DISCUSSION

Figure 1 shows the change of resistivity of ITO films with increasing O_2 gas flow ratio in relation to target erosion ratio. Deposition was carried out by low conductivity target (target A1) at without substrate heating. For each of target erosion ratios (5, 15, 30%), relatively low resistivities were obtained at O_2 addition ratios of 0.7, 0.5 and 0.3%, respectively. It means that optimum O_2 addition ratio to obtain the lowest resistivity was decreased with increasing target erosion ratio. This result implies that chemical state of target surface was changed with increasing sputtering time. For all the films, the lowest resistivity ($3.64x10^{-4}\Omega cm$) was obtained for the film deposited using 5% eroded target at O_2 flow ratio of 0.7%. Whereas, each film deposited at O_2 addition ratio of 0.7% showed relatively large difference in resistivity ($3.64x10^{-4}$ ~ $4.44x10^{-4}$ Ωcm) with increasing target erosion ratio. In result, relatively small difference in resistivity ($3.87x10^{-4}$ ~ $4.24x10^{-4}$ Ωcm) was obtained for the films deposited at O_2 addition ratio of 0.5%.

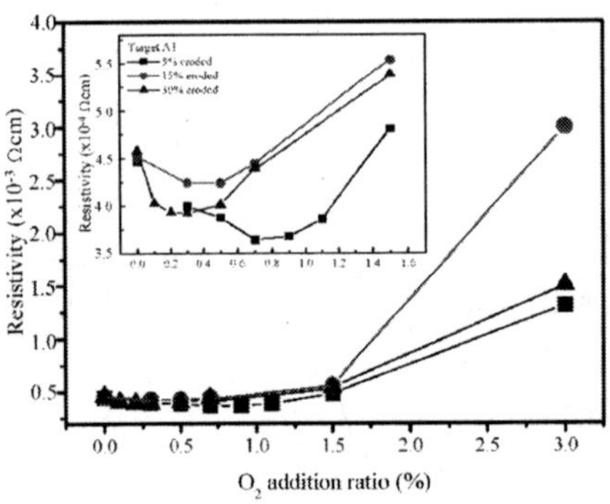

Figure 1. The change of resistivity of ITO films deposited using low conductivity ITO target in relation to O_2 addition ratio for different target erosion ratio

Figure 2 shows the change of resistivity of ITO films with increasing O_2 gas flow ratio in relation to target erosion ratio when relatively high conductivity ITO target (target A2) was used. For all the deposition conditions, the lowest resistivity (3.39×10^{-4} Ωcm) was obtained for the film deposited using 5% eroded target at O_2 addition ratio of 0.7%. With increasing sputtering time, relatively small difference in resistivity (3.61×10^{-4} ~ 3.79×10^{-4} Ωcm) was obtained for the films deposited at O_2 addition ratio of 0.5%. For each of films, optimum O_2 addition ratio to obtain the lowest resistivity was also shifted to low addition ratio with increasing sputtering time, i.e., target erosion ratio. From these results, target A2 showed relatively high stability in electrical property with increasing target erosion ratio, compared to target A1. High conductivity ITO target led to decrease in micro-nodule formation on the target surface because of high cooling efficiency.

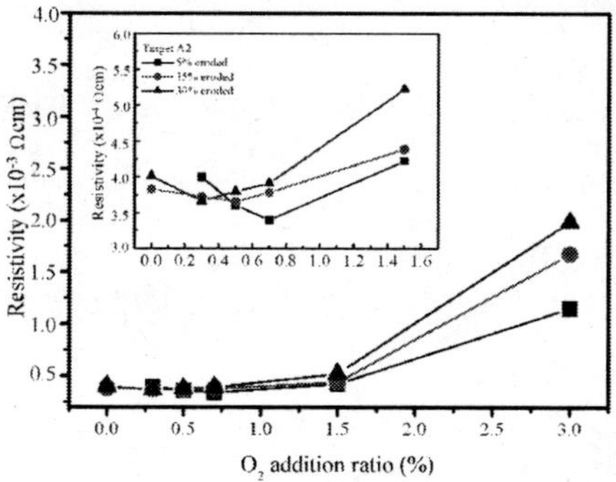

Figure 2. The change of resistivity of ITO films deposited using high conductivity ITO target in relation to O_2 addition ratio for different target erosion ratio

Figure 3 show the variation in electrical properties, (a) resistivity, (b) carrier density and (c) Hall mobility in relation to sputtering time for the ITO films deposited using different two targets without substrate heating. Used ITO targets contained different micro-pore size. Target B1 contained relatively large micro-pore size in the sintered disk, compared to target B2. Each of the ITO films was deposited after different sputtering time, i.e., different eroded state of target surface. In the case of target B1, large change in resistivity of the films was not observed for 11 hours sputtering where target surface was eroded about 20%. In this case, resistivity of the films changed from 9.72×10^{-4} to 1.03×10^{-3} Ωcm when erosion ratio of target surface increased from 15 to 20%. Whereas resistivity remarkably increased for the films deposited using the targets eroded higher than 20%.

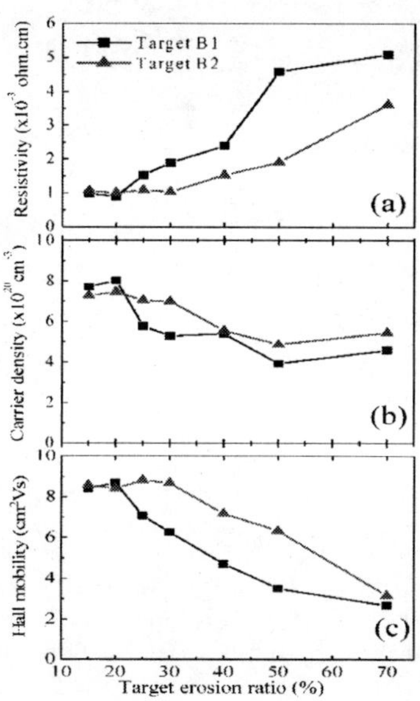

Figure 3. Resistivity(a), carrier density(b) and Hall mobility(c) of ITO films deposited using ITO targets contained different micro-pore size with increasing sputtering time

On the other hand, clear difference in resistivity of the films deposited using target

B2 was not observed for 17 hours sputtering even though target surface was eroded about 30%. Resistivity was changed from 1.06×10^{-3} Ωcm to 1.03×10^{-3} Ωcm when erosion ratio of the target surface increased from 15% to 30%, and then resistivity monotonously increased with further increase in the erosion ratio of target surface. These results could be related to the chemical state of target surface. In the case of target B2, cooling efficiency of target surface could be higher than target B1 because it's small micro-pore size. Therefore, on the target surface, target B2 has small formation ratio of sub-stoichiometric oxides (like In_2O_{3-X}) compared to target B1.

Figure 4 show the changes in resistivity, carrier density and Hall mobility with increasing erosion ratio of target surface for ITO films deposited using (a) target B1 and (b) target B2 at various substrate temperatures. Resistivity decreased with increasing T_s for the all films deposited using both target B1 and target B2. Especially, remarkable changes in resistivity were observed for the films deposited using the target eroded 70% at T_s of 150 °C. In the case of target B2, resistivity of the films showed small change with increasing erosion ratio of target surface at T_s of 150 ~ 300 °C, compared to target B1. This is might be attributed to the decrease in formation ratio of sub-stoichiometric oxides on the target surface because of high cooling efficiency of target B2, i.e., small micro-pore size.

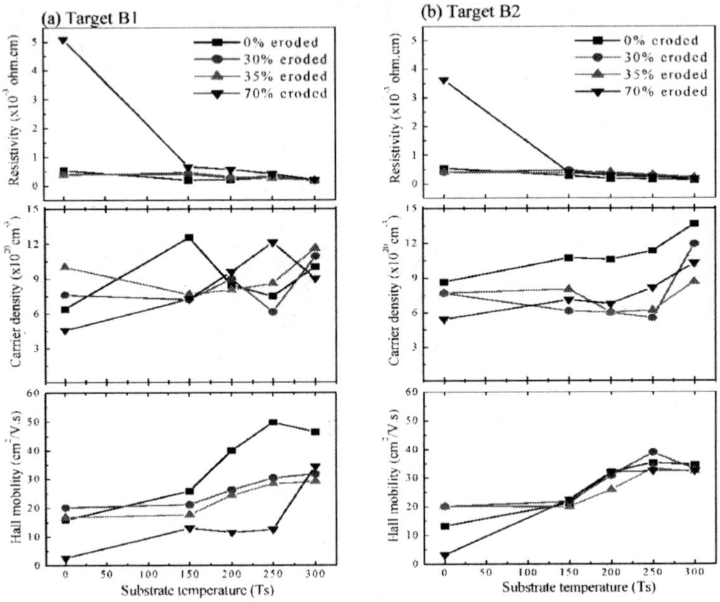

Figure 4. The changes in electrical properties with increasing erosion ratio of target surface for ITO films deposited using (a) target B1 and (b) target B2 at various substrate temperatures (T_s).

Figure 5 shows (a) electrical properties and (b) XRD patterns for the ITO films deposited using target B1 and target B2 under various oxygen flow ratios at T_s of 300 °C. Optimum oxygen ratio to obtain the lowest resistivity was decreased with increasing substrate temperature. This is attributed to enhancement of oxidation on film surface with increasing T_s. In the case of T_s=300 °C, target B2 showed relatively smaller changes in film resistivity to target B1 for $O_2/(O_2+Ar)$ flow ratio of 0 ~ 0.3%. The lowest resistivity was $1.06x10^{-4}$ Ωcm for the film deposited using target B2 at $O_2/(O_2+Ar)$ ratio of 0.05% and at T_s =300 °C. Peak intensity of (222) plan increased and peak intensity of (400) plan decreased with increasing oxygen flow ratio. This result could be due to the enhancement of oxidation on film surface.

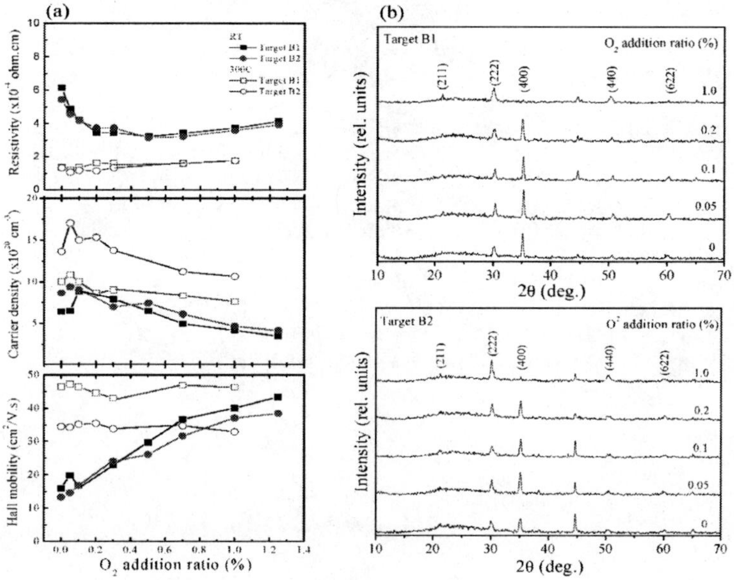

Figure 5. Electrical properties (a) and XRD patterns (b) for the ITO films deposited using target B1 and target B2 under various oxygen flow ratios at T_s of 300 °C.

CONCLUSIONS

ITO films were deposited on nonalkali glass substrate by magnetron sputtering using various ITO targets which have different conductivity and micro-pore size. In the case of conductivity, high conductivity target A2 showed relatively high stability in electrical property with increasing target erosion ratio. Optimum O_2 addition ratio to obtain the lowest resistivity was decreased with increasing target erosion ratio. High conductivity ITO target could lead to decrease in micro-nodule formation on the target surface because of it's high cooling efficiency. In the case of micro-pore size, clear difference in resistivity of the films was not observed between two targets for 11 hours sputtering where target surface was eroded about 20%. Whereas, target B2 showed small degradation in resistivity with increasing sputtering time after 11 hours compared to large micro-pore size target. Optimum oxygen ratio to obtain the lowest resistivity was decreased with increasing substrate temperature. The lowest resistivity was $1.06x10^{-4}$ Ωcm for the film deposited using target B2 at $O_2/(O_2+Ar)$ ratio of 0.05% and at T_s =300 °C.

FOOTNOTE
* Corresponding author: Tel:+82-51-510-2390, Fax: +81-51-512-0528,
E-mail: pksong@pusan.ac.kr

REFERENCES

[1] H. Koh, K. Sawada, M. Ogawara, T. Kuwata, M. Akatsuka and M. Matsuhiro, SID Dig. Tech. Pap., Vol.19 (1988), P.53.

[2] Z. C. Jin, I. Harmberg ans C. G. Granqvist, J. Appl. Phys., Vol.64 (1988), P.5517.
[3] G.Frank, H. Kostlin, Appl. Phys. A 27 (1982) 197.
[4] T. Maruyama, K. Fukui, Thin Solid Films 203 (1991) 297.
[5] I. Hamberg, C.G. Granqvist. J. Appl. Phys. 60 (1986) 11.
[6] Y. Sato, R. Tokumaru, E. Nishimura, P.K. Song, Y. Shigesato, JURNAL OF VACUUM SCIENCE & TECHNOLOGY A, Vol.23, Issue.4, pp.1167-1172.
[7] R. Latz, K. Michael, M. Scherrer, Jpn. J. Appl. Phys., Vol.30 (1991), P.149.
[8] B.L. Gehman, S. Jonsson, T. Rudolph, M. Schere, M. Weigert, R. Werner, Thin Solid Films Vol.220 (1992), P.333.
[9] H. Watanabe. Sputtering Plasma Process Vol.10 (1995), P.37.
[10] N. Nadaud, M. Nanot, P. Boch ., J. Am Ceram Soc., Vol.77 (1994), P.843.
[11] J. Son, D. Kim, J Am Ceram Soc, Vol.81 (1998), P.2489.
[12] K. Nakashima and Y. Kumahara, Vacuum, Vol.66 (2002), P.221.

SINTERING AND ENHANCED MECHANICAL PROPERTIES OF DENSIFIED MULTIWALLED CARBON NANOTUBE/TiN COMPOSITES

Lian Gao*, Linqin Jiang
State Key Laboratory of High Performance Ceramics and Superfine Microstructure, Shanghai Institute of Ceramics, Chinese Academy of Sciences, 1295 Ding Xi Road, Shanghai 200050, China.
Fax: +86-21-52413122; Tel: +86-21-52412718 E-mail: liangaoc@online.sh.cn

ABSTRACT

Densified multiwalled carbon nanotube (MWNT)/TiN composites with various MWNTs contents were successfully obtained through a spark plasma sintering (SPS) method. The influences of MWNT content on the densification of composites and mechanical properties were studied. The addition of the nanotubes can promote the sintering process of the composite. In the presence of only 0.1 wt% MWNTs, the densified MWNT/TiN composite exhibits the highest increase of about 31% in fracture toughness compared with pure TiN ceramic.

INTRODUCTION

Since the bulk production of carbon nanotubes (CNTs) has become possible and practical, there is growing interest in the fabrication and property characterization of CNT-incorporated composites [1-4]. Especially, the incorporation of CNTs into ceramic matrix has become a new focus in recent years [1,2,5,6]. CNTs should be ideal reinforcing fibers for composites due to their exceptionally high mechanical properties such as strength, stiffness and flexibility [2]. Theoretical and experimental studies [7,8] show that carbon nanotubes with very high aspect ratios (length-to-diameter ratio of 1000 or more) have exceptional mechanical characteristics. Most of the investigations on composites containing CNTs have so far focused on oxide ceramic-based and polymer-based composites with improved mechanical properties [1-6]. But few reports have been concerned on the fabrication of CNT/nitride ceramic up to now [9]. Tatami et al. developed CNT-dispersed Si_3N_4 ceramics with high density, electrical conductivity, and superior mechanical properties by using novel sintering aids [10]. Balázsi et al. prepared multi-walled CNT-silicon nitride composites by sinter-HIP method with enhanced mechanical properties as compared with other carbon filled samples [11]. In this report, titanium nitride (TiN) was chosen as a typical nitride ceramic to study the mechanical properties of CNT/TiN composite. TiN is an important nitride ceramic material because of its many superior properties including wear resistance, hardness, oxidative stability, corrosion resistance, and high conductivity [12-14].

The high strength of carbon nanotubes makes it promising for reinforcement. But the great aggregation of CNTs restricts their effective load transfer across the CNT-matrix interface. Thus a homogeneous dispersion of CNTs in the ceramic matrix and good interfacial bonding are critical for improving the mechanical properties of ceramic composites. In this communication, we used acidic CNTs to realize a good dispersion and good interfacial combination between the nanotubes and TiN in order to fabricate homogeneous CNT/TiN composites. A SPS sintering technique is employed to densify the composites. The effects of CNT contents on the sintering properties, mechanical properties including hardness and fracture toughness, and the microstructure of composites were investigated.

EXPERIMENT PROCEDURE

Multiwalled carbon nanotubes (MWNTs) with diameters ranging from 10 to 30 nm and lengths ranging from 0.5 to 500 μm were kindly provided by Shenzhen Nanotech Port Co. Ltd. MWNTs were oxidized by refluxing at 140°C in concentrated nitric acid for 6 h. After washing and centrifugation, the acidic MWNTs were obtained by further drying. Tetra-butyl titanate [$Ti(OC_4H_9)_4$], used as a starting material, was dissolved in anhydrous ethanol to form a 10 vol% solution. The mixed solution was added dropwise into the prepared suspension containg the desired MWNTs contents with vigorous stirring until the final Ti:H_2O molar ratio reached 1:150. The hydrolyzate was filtered and rinsed thoroughly with anhydrous ethanol, then dried at 100 °C for 12 h. The precipitate was subsequently calcinated at 300 °C for 2 h. The obtained MWNT/TiO_2 powder was loaded into a quartz boat and heated in flowing NH_3 (0.5 L min^{-1}) at 800 °C for 5 h in a tube furnace to obtain MWNT/TiN composite powder. The offgas was absorbed with water. According to the initial additive contents of the acidic MWNTs (in wt%: 0.5, 1, 3 and 5), the obtained composites are signed as TNC-0.1, TNC-1, TNC-3 and TNC-5.

Using as-prepared composite powders, the disks of 10 mm diameter and 2 mm thickness were obtained by spark plasma sintering in a Dr. Sinter 2080 SPS apparatus (Sumitomo Coal Mining Co., Ltd, Japan). The powder was filled into a graphite die sintered at 1250 °C, with a pressure of 50 MPa for 2 min in vacuum. The microstructure of products was observed using scanning electron microscopy (FE-SEM, JSM-6700F, JEOL, Japan) and transmission electron microscopy (TEM, JEM 2100F, JEOL, Japan). The composition of the composites was characterized by using X-ray powder diffraction (XRD, D/max 2550V, Rigaku, Japan) with Cu kα radiation (λ=0.15406 nm). Density of the sintered composites was measured using the Archimedes technique. Indentation tests were done on a Wilson-Wolpert Tukon 2100B (Instron Co., USA) hardness tester with a diamond Vickers indenter under a load of 49 N, and the fracture toughness was evaluated by measuring the crack length generated after indentation.

RESULTS AND DISCUSSION

XRD spectra of the resultant MWNT/TiN composites with various MWNT contents are shown in Figure 1. The XRD peaks of the composites correspond well with the reported values of TiN (JCPDS 87-0682). The peak at 26.06° is attributed to the characteristic peaks of the MWNTs. The results indicated relatively high purity of the synthesized composites and little loss of MWNTs in the fabrication process of the composites.

TEM and SEM images well visualize the fracture microstructure of the products with various contents of MWNTs, as shown in Figure 2. Figure 2a shows the TEM image of tiny TiN nanoparticles attached onto the MWNT, with a diameter of ca. 5-10 nm. This microstructure is critical for the fabrication of a homogeneous MWNT/TiN composite, which is in favor of strengthened interfacial combination between the nanotubes and the TiN matrix. Figure 1b-1f show the fracture surfaces of TiN ceramic and densified MWNT/TiN composites by SPS technique, which exhibits pronounced difference between MWNT-free ceramics and the composites. The size of a single hexagonal TiN grain in the pure TiN ceramic is 100-200 nm. The morphology of the cross section has clear edges and corners, which denotes the intergranular fracture mode. As MWNTs is added, the interface of grains become blurry and rugged, which indicates the mode would partly change into the intragranular fracture mode. The MWNTs are homogeneously distributed in the ceramic matrix. No hole surrounding the MWNTs is formed after fracture. These unique features suggest that tight binding between the MWNTs and ceramic matrix is achieved. Morever, it is obvious that the MWNTs protrude out of the ceramic matrix, which is very vital to take advantage of mechanical properties of MWNTs in the composite materials [6].

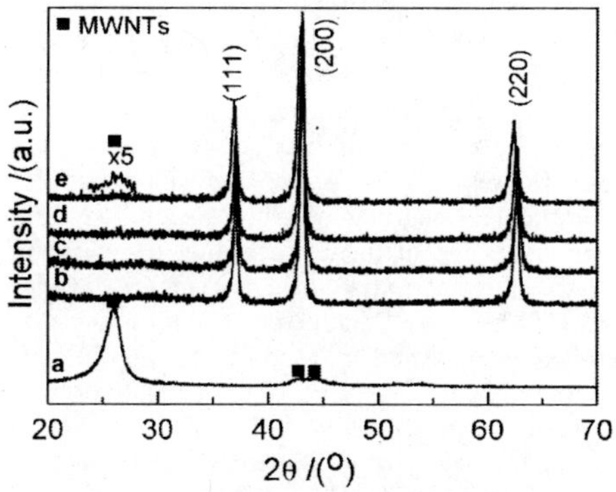

Figure 1. XRD patterns of (a) MWNTs, (b) TNC-0.1, (c) TNC-1, (d) TNC-3 and (e) TNC-5 composites nitrided at 800 °C for 5 h in an NH_3 atmosphere.

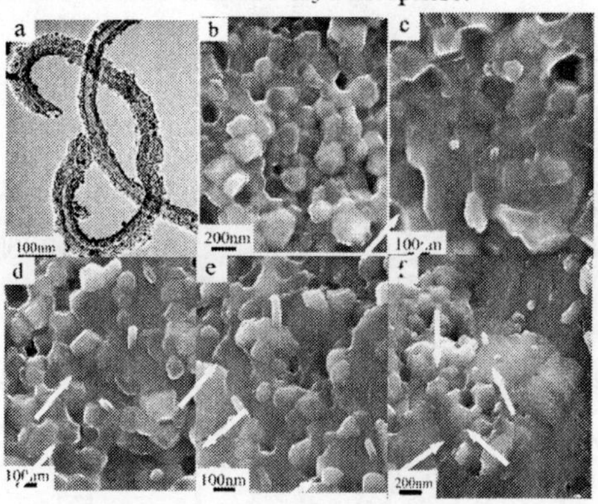

Figure 2. (a): a typical TEM image of tiny TiN nanoparticles on MWNT; and SEM images of (b) pure TiN ceramic, (c) TNC-0.1, (d) TNC-1, (e) TNC-3 and (f) TNC-5 composites.

The dependence of mechanical properties including hardness and fracture toughness on MWNT content in the MWNT/TiN composites is shown in Table 1. The pure TiN ceramic can be consolidated by SPS at 1250 °C for 2 min to theory density. Measured Vickers hardness and fracture toughness are 12.8 GPa and 2.6 MPa $m^{1/2}$, respectively. In the presence of only 0.1 wt% MWNTs, the hardness and toughness are 13.6 GPa and 3.4 MPa $m^{1/2}$, which increase by about 31% and 6% compared with that for pure TiN ceramic, respectively. However, when the content of MWNTs is larger than 0.1 wt%, the mechanical properties decreased with the increasing MWNTs. It can be suggested that excessive MWNTs would not effectively enhance the mechanical properties of the MWNT/TiN composite.

From Table I, it can be found that the mechanical property values for 0.1 wt% MWNT/TiN composite are maximal compared with other composites containing 1, 3 and 5 wt% MWNTs, which is very related with the relative density of the composites [2]. The addition of 0.1 wt% MWNTs can obtain the full-densified MWNT/TiN composite and effectively enhance the mechanical properties of TiN ceramic [6]. In general, it is necessary to study the composite fracture behavior to understand the load transfer between the matrix and MWNTs by SEM image. After adding MWNTs, some interesting microstructure features can be noted. First, the carbon nanotubes were dispersed in the TiN matrix separately. Second, some MWNTs were located at the boundary of TiN grains and some of them encapsulated nanoscale TiN grains. Third, the nanotubes protrude out of the TiN matrix. Fourth, no single long nanotubes or ropes of the MWNTs float on the surfaces of the matrix, that is to say, a tight interfacial contact is achieved. It has been reported that the phenomena described as above are possible mechanisms leading to the improvement of the mechanical properties [2,6]. With the addition of 0.1 wt% MWNTs, an effective load transfer between the nanotubes and TiN achieved and a net-like work may be formed to prevent the fracture. However, with the increasing MWNTs, the relative density of the composites (TNC-1,3,5) decreased and thus the mechanical properties decreased. Although a serious aggregation of the nanotubes cannot be found in SEM images obviously with increasing MWNTs, we cannot exclude the existence of the agglomerated MWNTs in the grain boundary or within the grain, which not only impede the densification of the composites to decrease the relative densities of the composites, but also become the defect source [15]. The strength of MWNTs/ceramic is largely attributable to sufficient load transfer from the matrix to MWNTs through the interface. A serious aggregation of MWNTs in the matrix with the increasing nanotubes made the load transfer efficiency low and cannot result in further enhanced mechnical properties. The decrease of relative density with the increasing MWNTs amounts indicated that more pores existed between the grains and thus impeded the improvement of the mechnical properties. From our results, it can be suggested that more quantity of MWNTs would be no help in improving the mechanical properties of MWNTs/TiN composites.

Table I. Resultant properties of MWNT-reinforced TiN nanocomposites consolidated by spark-plasma sintering at 1250 °C for 2 min

Materials	MWNTs content (wt%), (vol%)	Relative density (%TD)	Hardness H_v (GPa)	Fracture toughness K_{IC} (MPam$^{1/2}$)
TiN	0, 0	100	12.8	2.6
TNC-0.1	0.1, 0.2	100	13.6	3.4
TNC-1	1, 2	98.6	11.8	3.2
TNC-3	3, 7	97.8	10.8	3.0
TNC-5	5, 11	97.8	7.2	3.0

CONCLUSIONS

In conclusion, the full-densified MWNT/TiN composites with various MWNTs contents were successfully synthesized and exhibit enhanced fracture toughness compared with pure TiN ceramic. MWNTs are attractive materials for reinforcement of nitride nanoceramic, and that the effective combination of the nanotubes with the matrix plays an important role. The addition of 0.1 wt% MWNTs gets a full-densified MWNT/TiN composite and achieves maximal increase in fracture toughness and hardness compared with pure TiN ceramic and other composites containing various contents of MWNTs. Excessive MWNTs (≥ 1 wt%) would not enhance the mechanical properties of the composite effectively and decrease the relative density of the densified composite.

REFERENCES

[1]E. Flahaut, A. Peigney, Ch. Laurent, Ch. Marlière, F. Chastel, and A. Rousset, Carbon Nanotube-Metal-Oxide Nanocomposites: Microstructure, Electrical Conductivity and Mechanical Properties, *Acta. Mater.*, **48**, 3803-12 (2000).

[2]G-D. Zhan, J.D. Kuntz, J.L. Wan, and A.K. Mukherjee, Single-Wall Carbon Nanotube as Attractive Toughening Agent in Alumina-Based Nanocomposites, *Nature Mater.*, **2**, 38-42 (2003)..

[3]L. An, W. Xu, S. Rajagopalan, C. Wang, H. Wang, Y. Fan, L. Zhang, D. Jiang, J. Kapat, L .Chow, B. Guo, J. Liang and R. Vaidyanathan, Carbon-Nanotube-Reinforced Polymer-Derived Ceramic Composites, *Adv. Mater.*, **16**, 2036-40 (2004).

[4]R.W. Siegel, S.K. Chang, B.J. Ash, J. Stone, P.M. Ajayan, R.W. Doremus, and L.S. Schadler, Mechanical Behavior of Polymer and Ceramic Matrix Nanocomposites, *Scripta Mater.*, **44**, 2061-64 (2001).

[5]S.I.Cha, K.T. Kim, K.H. Lee, C.B. Mo and S.H. Hong, Strengthening and toughening of carbon nanotube reinforced alumina nanocomposite fabricated by molecular level mixing process, *Scripta Mater.*, **53**, 793-97 (2005).

[6]J. Sun, L. Gao and W. Li, Colloidal Processing of Carbon Nanotube/Alumina Composites, *Chem. Mater.* **14,** 5169-72 (2002).

[7]M.M.J. Treacy, T.W. Ebbesen and J.M. Gibson, Exceptionally High Young's Modulus Observed for Individual Carbon Nanotubes, *Nature* **381**, 678-80 (1996).

[8]R.S. Ruoff and D.C. Lorents, Mechanical and Thermal Properties of Carbon Nanotubes, *Carbon* **33**, 925-30 (1995).

[9]L.Q. Jiang and L. Gao, Carbon Nanotubes-Metal Nitride Composites: a New Class of Nanocomposites with Enhanced Electrical Properties, *J. Mater. Chem.* **15**, 260-66 (2005).

[10]J. Tatami, T. Katashima, K. Komeya, T. Meguro and T. Wakihara, Electrically Conductive CNT-Dispersed Silicon Nitride Ceramics, *J. Am. Ceram. Soc.* 88, 2889-93 (2005).

[11]Cs. Balázsi, Z. Kónya, F. Wéber, L.P. Biró and P. Arató, Preparation and Characterization of Carbon Nanotube Reinforced Silicon Nitride Composites, *Mater. Sci. Eng. C.* 23, 1133-37 (2003).

[12]A. Kato and N. Tamari, Crystal Growth of Titanium Nitride by Chemical Vapor Deposition *J. Cryst. Growth.*, **29**, 55-60 (1975).

[13]R.A. Janes, M. Aldissi and R.B. Kaner, Controlling Surface Area of Titanium Nitride Using Metathesis Reactions, *Chem. Mater.*, **15**, 4431-35 (2003).

[14]I1-S. Kim and P.N. Kunta, Hydrazide Sol-Gel Synthesis of Nanostructured Titanium Nitride: Precursor Chemistry and Phase Evolution, *J. Mater. Chem.*, **13**, 2028-35 (2003).

[15]J. Sun, L. Gao, M. Iwasa, T. Nakayama and K. Niihara, Failure Investigation of Carbon Nanotube/3Y-TZP Nanocomposites, Ceram. Inter., 31, 1131-34 (2005).

CHARACTERIZATION OF THE SURFACE MORPHOLOGY OF ORGANIC AND INORGANIC HYBRID THIN FILMS – A FIRST STEP TO FABRICATE AN ARTIFICIAL CELL MEMBRANE

Wakako MANABE, Hiroshi TAKANO, Masayuki ITOH
Department of Chemical Engineering and Materials Science Faculty of Engineering, Doshisha University

1-3 Tataramiyakodani, Kyo-Tanabe, Kyoto 610-0321 Japan

ABSTRACT

As the first step to fabricate an artificial cell membrane (ACM), we characterized the surface morphology of multi-layered hybrid thin films consisting of an organic bilayer of all-trans-retinoic acid or L-(α)-phosphatidylcholine produced using the Langmuir-Blodgett (LB) technique, and a silver metal thin film fabricated using Quantum Dot Physical Deposition (QDPD) method, which allows us to produce such silver films at room temperature on organic or biological thin films such as an LB bilayer. The membrane potential of an ACM can be emulated with the hybrid structure of O-M-O (Organic-Metal-Organic), and the permeability of ions passing through an ACM is dependent on the structure of each layer.

The major objective of this report was to investigate the latter; the relationship between the permeability and the surface morphology of and ACM. The molecular arrangement was observed and characterized with several kinds of nano-scale measurement like STM following the fabrication of permeable multi-layered hybrid thin films as a first step towards ACM fabrication.

INTRODUCTION

Because of the importance of so called *Hybrid Thin Film* in nano-technological aspects[1)], numerous studies have been reported on functional thin films with the fabrication of organic monolayer onto a metal substrate and vice versa. In these fabrication processes of such thin films, the Langmuir-Blodgett (LB) technique is one of the most fascinating techniques to prepare organic thin film of ordered structures with controlled thickness of the molecular level.[1)] For the organic compound of artificial cell membrane (ACM), Phospholipids are very suitable materials because the essential components of biological membranes[2)][2)] and retinal-based proteins (such as bacteriorhodopsin and rhodopsin) seem to be prevalent in a large variety of photobiological processes; they are ranging from light harvesting in photosensitive bacteria to color vision in human beings[3)][3)]. Therefore, the LB technique is a very convenient way to build up high-quality lamellar structures such fatty acids.

The most difficult problem in the case of production of such hybrid thin film is how to fabricate a metal thin film onto organic one without any thermal damages to the surface of the organic monolayer. To overcome such difficulty, the Quantum Dot Physical Deposition (QDPD) method, which is a nano-scale aerosol process as in the reference[4)], was applied to the production of a metal-organic multi-layer. The QDPD method is advantageous in the operating temperature; by applying the quantum size effect of nano metallic particle being a liquid droplet at normal room temperature, a metal thin film can be fabricated onto LB film by the deposition of such liquid metal particles, after the liquid spreading, mutual-fusing and self-crystallization processes without any damages on organic structures by additional substrate heating often seen in usual CVD process.

By following a fabrication of hybrid thin film consisting of O-M-O, i.e., organic bilayer of all-trans-retinoic acid or L-(α)-phosphatidylcholine accumulated with Langmuir-Blodgett (LB) technique and the silver metal thin film fabricated with Quantum Dot Physical Deposition (QDPD) method, ion permeability of the thin film regarding surface morphology and membrane potential was investigated with multi-layered hybrid thin films as the artificial cell membrane (ACM). Therefore, the major objective of this report is in the investigation of the permeability dependence of ion, which is passing through ACM according to the structural difference of each layers in ACM; the relationship between the permeability and the surface morphology, especially relating to the molecular arrangement, was observed and characterized following the fabrication of such permeable multi-layered hybrid thin films from the view point of the first trial of ACM characterization with several kinds of nano-scale measurement like as STM and so on.

EXPERIMENTALS

Materials:

L-α Phosphatidylcholine from EggYolk (PC) and all-trans Retinoic acid (RA) were dissolved in chloroform (99% pure) to prepare stock solutions with the concentration of 2mM and 1.5mM, respectively. Water was purified by means of a water purification system upto the electric resistance of 18.2 MΩcm, and used as the subphase for the LB method. Silver nano-particles (<3nm, 99.99% pure) produced by the QDPD method used to fabricate the metal film.

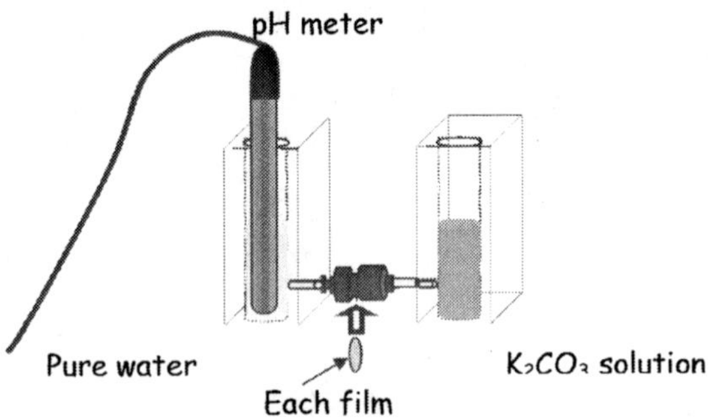

Figure.1 Experimental setup for the ion permeability test. Considering the bulk of pH meter (whose volume is 4 ml), 15 ml pure water and 19 ml $K_2CO_3^{2-}$ solution were poured into each cell.

Experimental Methods:

LB film accumulation was performed on a film balance system with a deposition apparatus (USI-SYSTEM FSD-300) constructed by Fukuoka, Japan. All the experiments were performed at 20°C. For the surface pressure-area per molecule (π-A) isotherm measurements, a stock solution containing PC or RA was spread on the subphase by a micro syringe. Each spreading volume was 50μl and 100μl, respectively. 10 minutes was allowed for the solvent evaporation. After the pre-treatments of the specimen, the monolayer was compressed symmetrically at the rate of 1mm/sec by a barrier until the surface pressure reached 50 mN/m or 13 mN/m respectively. The surface pressure was held constant throughout the remainder of the experiment by the automatic adjusting of interfacial area by the barriers. The change of the surface pressure was recorded as a function of a mean molecular area. A polycarbonate membrane (13mm in diameter: Whatman plc.) was used as substrates to sustain the LB films.

The polycarbonate membranes were dipped into the subphase with a rate of 5 mm/min to deposit the LB bilayer. After a silver thin film was fabricated on the LB bilayer by QDPD method, the LB bilayer was deposited on the hybrid film to construct O-M-O hybrid thin film. The surface structure of silver film fabricated on the micro grid was observed with a transmission electron microscope (TEM). The molecular structure of LB film deposited on a HOPG (Highly Oriented Pyrolytic Graphite) [5] substrate was checked by scanning tunneling microscope (STM).

The ion permeability of the O-M-O hybrid thin films was measured with a pH meter. The experimental setup for the ion permeability measurement is shown in Fig.1. Each mini-chamber was made of Duralumin and kept at constant voltage. To keep the chamber temperature at 20°C, they were put in the constant-temperature bath, and the constant-temperature bath was placed in the clean box encapsulated by argon gas to eradicate the influence of CO_2 [6] to prevent the contamination by surrounding aerosols and so on.

Data Reduction:

The permeability of the O-M-O hybrid thin films, organic thin films and metal thin films were calculated from the time dependent change of pH values of both chamber as followings: Diffusion of each ion is given by

$$-j_i = D_i \frac{dc_i}{dz} + z_i c_i \frac{F}{RT} \cdot \frac{d\psi}{dz} \quad . \tag{1}$$

Without an applied electric field, the equation for the two ions are easily combined and integrated to give the diffusion potential as,

$$\psi_0 - \psi_1 = \frac{RT}{|z|F} \frac{D_2 - D_1}{D_2 + D_1} \ln \frac{C_{10}}{C_{1l}} \quad , \tag{2}$$

where the subscripts 1 and 2 refer to the cation and anion, respectively, and the subscripts 0 and l are to the different side of the membrane interfaces (0, l=different side).

This, in turn, can be used to simplify the flux equation

$$-j_i = D\frac{dc_1}{dz} = \frac{2}{1/D_2 + 1/D_1} \frac{dc_1}{dz} . \tag{3}$$

This equation is easily integrated across the thin film to give

$$j_i = \frac{DH}{l}(C_{10} - C_{1l}) \tag{4}$$

where $C_{10} - C_{1l}$ is the concentration difference across the membrane, D is the diffusion coefficient averaged between the ions, H is the partition coefficient averaged between ions, and l is the membrane's thickness.[7] As D, H and l are not be able to be determined experimentally, the following relationship, $D' = \frac{DH}{l}$, was assumed for simplification. Using such simplification and assumptions,

$$\begin{aligned} D_{K^+}' &= \frac{j_{K^+}}{C_{K^+0} - C_{K^+1}} = \frac{j_{K^+}}{(1\times 2 - 0)\times 10^3} \ \frac{m}{\min} \\ &= \frac{1}{2\times 10^3[\frac{\text{mol}}{\text{m}^3}]} \times \frac{\text{Molar quantity [mol]}}{t[\min]\times A[m^2]} \\ &= \frac{15\times 10^{-3}[\text{l}]\times 0.559[\frac{mol}{l}]}{2\times 10^3 \ \frac{mol}{m^3} \ \times t[\min]\times A[m^2]} \end{aligned} \tag{5}$$

where t is the permeation time, and A is the permeation area and estimated from the relation $D''=D'{*}A$ [m^3/min] to check the permeation property of each film.

RESULTS AND DISCUSSION

Surface Pressure-area (π-A) Isotherm of PC and RA:

The π-A isotherms of monolayer formed from PC and RA are presented in Fig.2. These isotherms were recorded at 20°C. For the PC isotherm, a plateau region appeared at about 35mN/m and a collapse point found at 45 mN/m. For RA, the plateau region was about 12 mN/m and the collapse point was 15 mN/m. Then the deposition pressure of each component was determined 40mN/m and 12 mN/m, respectively.

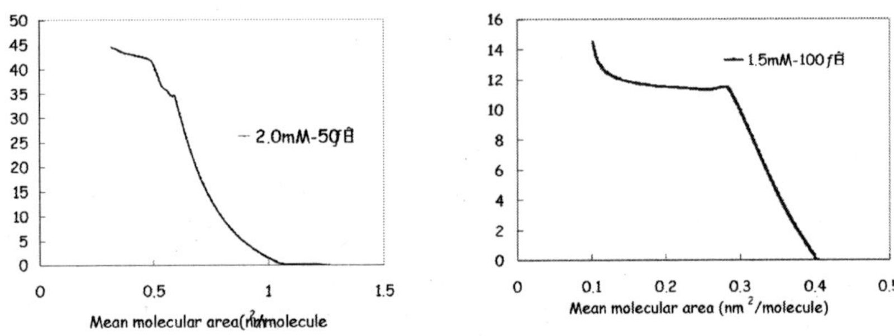

Figure.2 π-A isotherm of PC monolayer (left) and RA monolayer (right) on subphase of pure water.

Morphology of Organic and Silver Films:

Figures 3, 4 and 5 show the surface structure of PC bilayer, RA bilayer and silver thin film, respectively. The orderly arrayed organic molecules were successfully observed. The periodic sizes of 0.31nm and 0.26nm in these images corresponded to the molecular size of PC and RA drawn by CEM draw, respectively. Reticulate structure shown in the silver thin film may act as a channel of ion permeation and a good leading-in electrode for membrane potential.

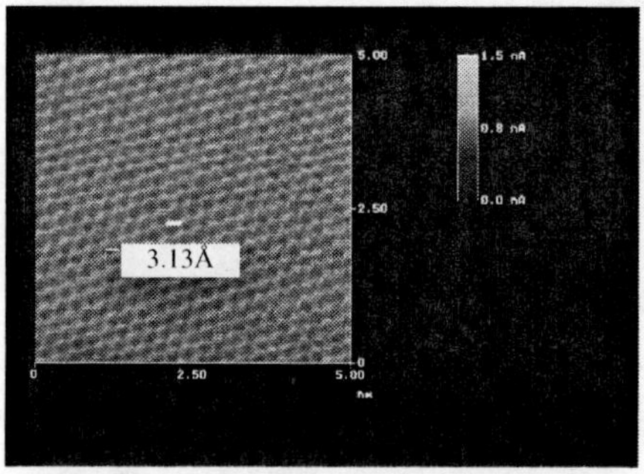

Figure.3 STM image of PC bilayer.

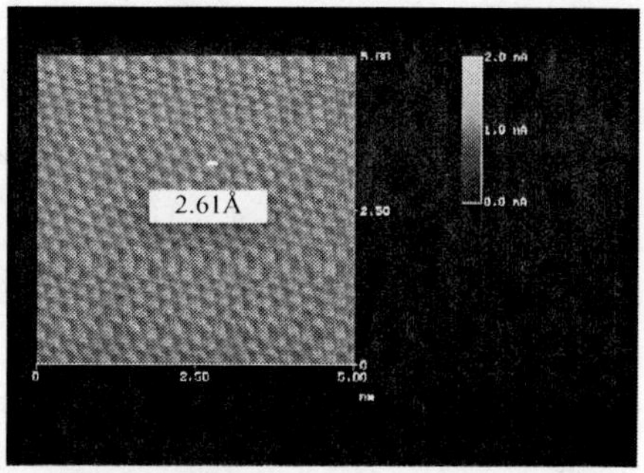

Figure.4 STM image of RA bilayer.

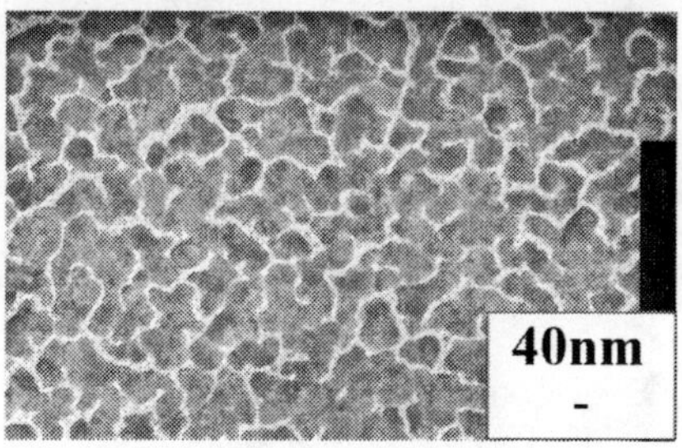

Figure.5 TEM image of silver thin film at deposition time of 60 minutes.

Ion Permeability of the Films:

As shown in Fig.1, the ultra-pure water and the K_2CO_3 solution were poured into each cell and the film was set in the membrane holder. Concentration change of the ion was calculated from the time dependent pH value in the right-side mini-chamber with the reduction method described in Eqs.(1) to (5).

Table 1 shows that the permeability of the hybrid thin films are nearly equivalent to the multiplication of those of each LB film and silver film. In case of only Polycarbonate, K^+ ions concentration reached to equilibrium about 20 s, and the holes (10μm) of polycarbonate were acting as diffuser passage. In details, RA showed lower permeability than PC even though the length of RA molecule was smaller than PC molecule as shown in Figs.3 and 4; thus the thickness of LB layer was not the major reason of the difference. Eventually under the fact that there was no difference in their array, it may be concluded that the benzene ring of RA molecule affected on the motion of ions and obstructed the passage of ions. Still to fix up these problems, much precise measurement, for example, with a high-performance liquid chromatography (HPLC) will be desired for the next stage investigation, which enables us to know what ions exist in the solution and accurate concentration of each ion. Ag film has channels as shown in Fig.5. And from Table1, it is confirmed that these channels transfer ions successfully.

Table 1 Permeability of each film for K^+ ion

	Poly carbonate	Ag film	PC LB film	PC-Ag hybrid	RA LB film	RA-Ag hybrid
Peak pH	12.55	11.92	12.13	12.21	11.36	11.92
Time [s]	20	1160	1000	1260	1920	3000
D_{K+} '' $*10^{-6}$ [m^3/min]	25.15	0.43	0.50	0.19	0.26	0.017

CONCLUSION

An O-M-O multi-layered thin film was successfully fabricated using QDPD and the LB technique. The permeability of the hybrid film was examined from the view point that the membrane potential of an ACM can be emulated with the hybrid structure of O-M-O (Organic-Metal-Organic), and the permeability of ions passing through ACM is dependent on the structure of each layer. The apparent facts obtained from the permeability experiment confirmed that the hybrid film showed a practical level of permeation in this first trial characterization of an ACM.

Good permeability of each hybrid film was observed. This means that the hybrid films are permeable, and that it should be possible to control the permeability with an applied voltage on the metallic film.

REFERENCES

[1] Michael C. Petty, *Langmuir-Blodgett films*, pp.84-86 (Cambridge University Press, 1998).
[2] J. B. Peng, M. Prakash, R. Macdonald, P. Dutta, and J. B. Ketterson, "Formation of Multilayers of Dipalmitlylphosphatidylcholine using the Langmuir-Blodgett Technique," *Langmuir.*, **3**, 1096 (1987).
[3] C.P. de Melo, M.I.M.Sanchez, " Langmuir-Blodgett Films of Retinal Derivatives," *Langmuir.*, **14**, 490(1998).
[4] S. Aono and M. Itoh , "Characterization of thin film produced by quantum dot deposition process ," *J. Aerosol Sci.*, **27**, S149 (1996).
[5] A. L. Weisenhorn, M. Egger, F. Ohnesorge, S. A. C. Gould, S.-P. Heyn, H. G. Hansma, R. L. Sinsheimer, H. E. Gaub, and P. K. Hansma, "Molecular-Resolution Images of Langmuir-Blodgett Films and DNA by Atomic Force Microscopy," *Langmuir.*, **7**, 8 (1991).
[6] K. Aota, "Synthesis and Characterization of PC-Ag-PC layered Thin Film," *Doshisha Univ.*, Master's thesis (2005).
[7] E. L. Cussler , "Diffusion mass transfer in fluid systems," pp.114-115, (Cambridge University Press, 1996).

ION PERMEABILITY AND MEMBRANE POTENTIAL OF ORGANIC AND INORGANIC HYBRID THIN FILMS- CHARACTERIZATION OF ARTIFICIAL CELL MEMBRANE

Ryuji Okamura
Department of Science of Environment and Mathematical Modeling
Faculty of Engineering, Doshisha University
1-3 Miyakodani, Tatara,
Kyo-Tanabe, Kyoto, 610-0321, Japan

Masayuki Itoh and Hiroshi Takano
Department of Chemical Engineering and Materials Science
Faculty of Engineering, Doshisha University
1-3 Miyakodani, Tatara,
Kyo-Tanabe, Kyoto, 610-0321, Japan

ABSTRACT

To emulate the concentration of ions due to a membrane potential, artificial cell membranes (ACM) were fabricated on a porous substrate of polycarbonate with a L-α-Phosphatidylcholine (PC)-Ag-PC multi-layer. A narrow mesh structure silver thin film was fabricated using the Quantum Dot Physical Deposition (QDPD) method at a room temperature, and this served as a leading-in electrode to produce the membrane potential. The PCs of both sides were produced using the Langmuir-Blodgett (LB) technique and were set as insulators. TEM analysis of the silver thin film growth process was completed as a function of the QDPD process deposition time. The total permeation of the ACM under osmotic pressure was characterized the measuring the K^+ ion concentration with K^+ ion meter. The quality of PC as an insulator was examined for several ranges of thickness; however, adequate performance was not achieved within the range used in this experiment.

INTRODUCTION

Permeable membranes have been utilized in various fields for ion separation and concentration and have been researched for many years[1]. Still, as in a heat pump for thermal energy, cell membranes concentrate ions due to their membrane potential against to the ion flow caused by the difference in chemical potential of ions separated by the membrane. The concentration process is a low energy consumption process, and is the most marked advantage of cell membrane compared with typical permeable membranes. An ion channel in a cell membrane may be caused by membrane potential or vice versa. The easiest way to emulate roughly such a membrane potential with a thin film is to use a hybrid system of organic-metal-organic-metal-organic multi-layer having permeability. The metal part will act as a lead-in electrode producing the electrical potential, and the organic parts act as insulators on both sides of the electrode. Both parts of the hybrid thin film should be permeable, and the organic part should be electrically.

The most important characteristics as artificial cell membranes (ACM) were the permeability and ability of electrical potential, i.e., insulation of the two lead-in electrodes by the organic thin films. The objective of this report is, therefore, in the experimental investigation on the above two characteristics. As L-α-Phosphatidylcholine (PC) is, in general, known as an insulation material and often used in Langmuir-Blodgett (LB) including our cases, hybrid thin film of PC-Ag-PC multi-layer was fabricated on polycarbonate with associations of LB technique[2] and QDPD method developed by our group[3-6]. After the examination on the characteristic of permeability of a leading-in electrode of silver, the total ion permeability of the hybrid film only by osmotic pressure was examined by the measurement of K^+ ion concentration change with K^+ ion meter. The quality of PC as insulator was performed with several method for wide ranges of accumulation number of the PC monolayer.

EXPERIMENTAL METHODS

Quantum Dot Physical Deposition (QDPD) method is one of the nano-scale aerosol process, and apply the phenomenon that nano-sized particle behaves like liquid droplet at room temperature due to decrement of melting point by quantum size effect. In usual, the melting point of several organic materials used in LB process is less than the boiling point of water, and typical processes fabricating metallic thin film under high temperature are generally very harmful for such organic thin films. However, the QDPD process makes it possible to produce metal thin film even for binary case[7] only by deposition and mutual-fusion of nano-particles at room temperature.

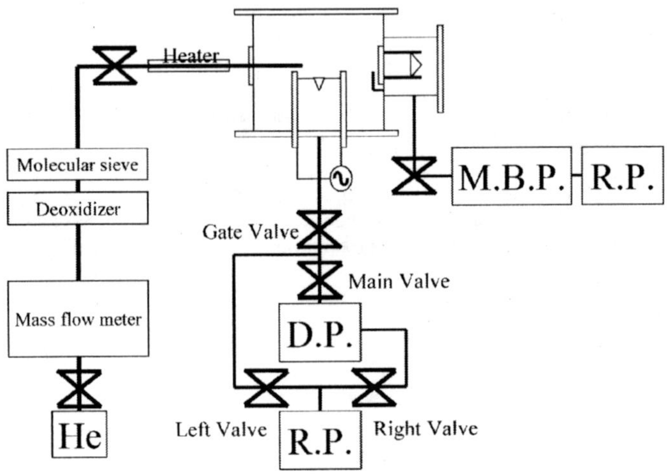

Figure 1. Schematic of QDPD equipment.

Figure 1 is the schematic diagram of QDPD equipment. It consisted of the particle manufacturing part and the film fabricating part. In the case of silver thin film fabrication, the operation procedures and parameters used in QDPD were as follows: Silver grains (99.999%) were heated up to the melting point in basket heater placed at the center of the chamber. After the homogeneous nucleation of vaporized silver atom, the silver particles (<5 nm) were transferred to substrate by helium (99.995%) carrier gas of 0.554ℓ/min to fabricate silver polycrystalline thin film via mutual-fusion. The particle size was controlled with the inner pressure (He) of chamber at 45Pa. A silver thin film having narrow mesh structure was fabricated on the substrate using the QDPD method, and it served as a lead-in electrode to produce the membrane potential.

In LB process, L-α-phosphatidylcholine (PC, from Egg Yolk, Wako Pure Chemical Industries, Ltd.) was dissolved in chloroform solution at the concentration of 2.0mM. The LB film was synthesized on polycarbonate using FSD-300 (USI-SYSTEM). The π-A isotherm of the PC solution showed the transition surface pressure at 50mN/m. L-α-phosphatidylcholine (PC) double layers accumulated on the silver thin film by LB technique, and were set as insulators. By such similar cycle, Ag-PC-Ag-PC hybrid thin film was constructed.

The substrate of ACM was polycarbonate membrane which diameter was 25mm and pore size was 10μm, still the observation by TEM for growing process of silver thin film was preceded with silver thin film on carbon micro grid. Regarding the estimation of the electric property of PC, several numbers of PC layer were accumulated on the Cu substrate. The Cu substrates were washed by "Semico Clean 23" and ultra pure water to avoid any other contaminations.

Figure 2 shows a schematic of the permeation experiment. The permeation was checked only on the osmotic pressure without any electric fields as a membrane potential. The hybrid film was placed at the center of this chamber. To ignore any other electrical influence, the chamber was made of duralumin. Relating to biological viewpoint, KCl solution of 0.066M was used as ion sample, and 30ml of solution was poured in the left cell of the chamber. In the right cell, ultra-pure water was poured. Under the no electrical field force as mentioned above, ions in the left cell moves to the right cell by the osmotic pressure via permeation of the hybrid film. The permeation of K^+ ion via ACM for osmotic pressure was determined by the measurement of K^+ ion concentration with K^+ ion meter.

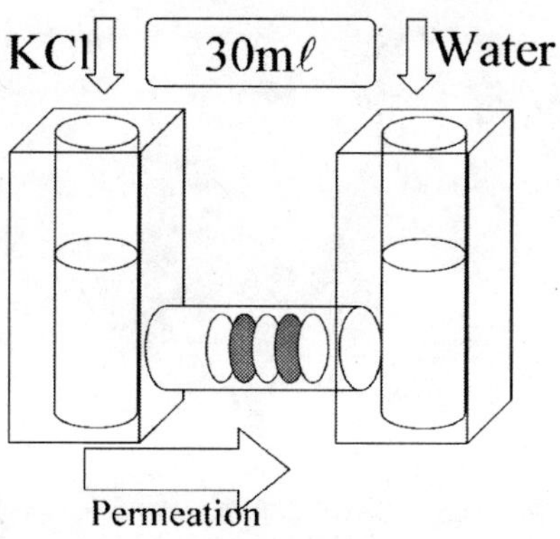

Figure 2. Schematic of permeation chamber.

RESULTS AND DISCUSSION

TEM images showing the growth process of the silver thin film having reticulated gaps are shown in Fig.3 for three different growing times. In the early stage of growth as seen Fig. 3 (a), the mutual-fusion of individual nano-particles progressed, and many small islands more than 50nm in diameter appeared according to the particle deposition. In the medium stage as seen in Fig. 3 (b), the islands grow and interconnect with each other, and there is a good reticulated slit between the fragments. Further as seen in Fig. 3 (c), produces a silver thin film with a polycrystalline structure having reticulated gaps about 50nm in the film. Case (c) looks promising; however, the degree of reticulated gaps area comparing to the case (b) is insufficient for permeation. Case (b) showed sufficient gas area and connections between the fragments, which were much better for a lead-in electrode for the ACM.

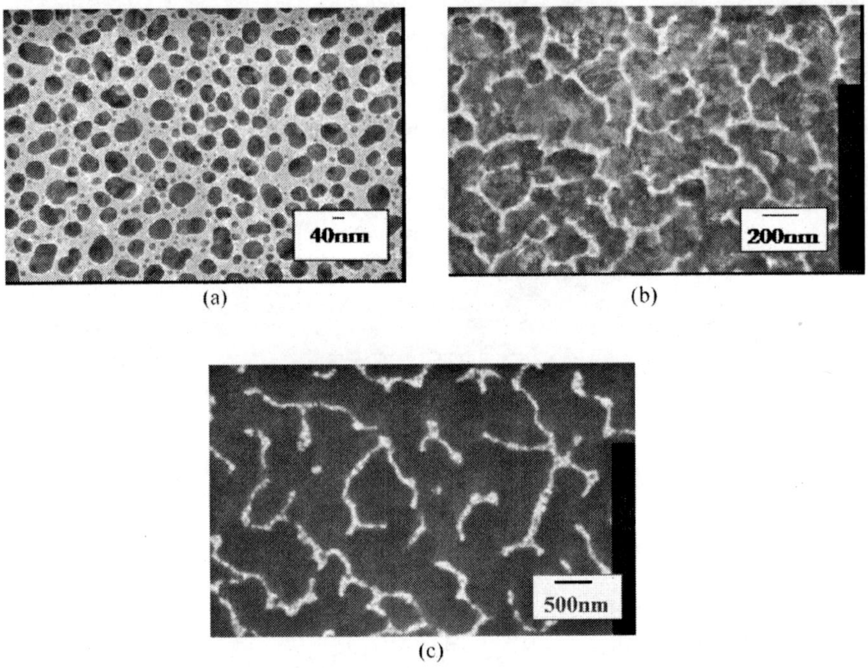

Figure 3. TEM images of Ag thin film at various deposition times.
(a) 30min, (b) 60min, and (c) 2hrs.

The total permeation of K^+ ions via our Ag-PC(bilayer)-Ag-PC(bilayer) hybrid thin film is depicted in Figure 4, where the case (b) was applied as a silver film. Permeation is good in the specific hybrid thin film due to the difference of osmotic pressure of both cells. The coincidence of mass balance between the two cells seemed not so good, which may be due to the relaxation time to uniformity and equilibrium in both cells. After sufficient time for permeation, both concentrations reach mostly same value in other experiments. The details of such permeation characteristics including surface structure and materials dependence are discussed in another report[8].

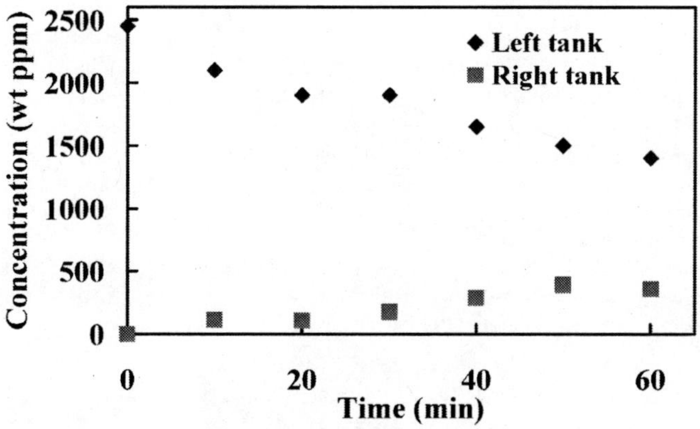

Figure 4. Time variation of K^+ concentration.

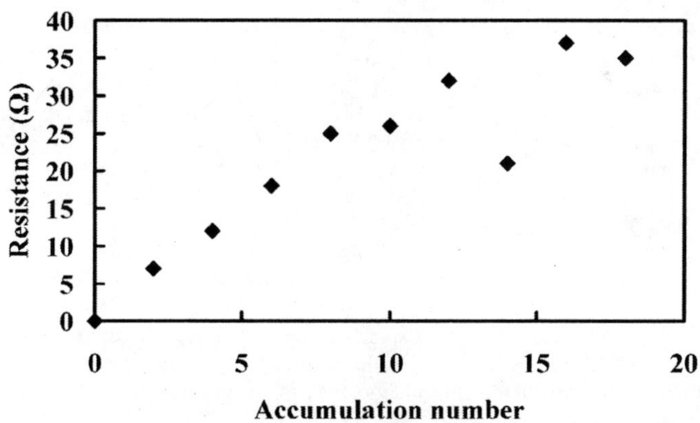

Figure 5. Relationship between PC resistance and accumulation number.

The experimental results of the total electric resistance of the PC on 10 mm diameter Cu is shown in Fig. 5 as a function of accumulation number. The electrical resistance of the PC was measured with a digital multi meter. The molecular length of egg PC is about 1.8 nm[9]. The experimental value of the thickness estimated with STM on our PC membrane was 2.5 nm for a dual layer, and 5nm for a quadric layer, and they agree well with the theoretical value. The thickness itself was proportional to the accumulation number. So the resistance was nearly proportional to the thickness of the film, except only one due maybe to any experimental error. However, such a small resistance may not be suitable for an insulator for the lead-in electrodes in an ACM. This means that some other materials, like porous polycarbonate, should be considered as an insulator of next stage.

CONCLUSION

A PC-Ag-PC-Ag hybrid thin film was fabricated on a polycarbonate using the LB technique and the QDPD method. The appropriate deposition time for Ag film fabrication using the QDPD method is 60min. The growing process of silver thin film having reticulated gaps was characterized from TEM images as a function of QDPD process deposition time.

The permeation of the ACM under osmotic pressure was examined by measuring the K^+ ion concentration with a K^+ ion meter. Permeation was good in the specific hybrid thin film due to the difference of osmotic pressure between the cells.

The quality of the PC as an insulator was examined for several ranges of thickness; however, sufficient performance was not obtained within the range used in this experiment, and some new process for the insulation will be necessary in the next stage.

REFERENCES

[1] Milton Ohring, Materials Science of Thin Films, 2nd Edition, pp-559-640, Academic Press (2002).

[2] Michael C. Petty, Langmuir-Blodgett Films, pp.234, University of Cambridge (1996).

[3] S. Aono, M. Itoh, H. Takano, “Characterization of Thin Film Produced by Qoantum Dot Deposition Process”, *J. Aerosol Sci.*, **27**, S149-S150 (1996).

[4] K. Sakiyama, M. Itoh, H. Takano, and S. Aono, “Characterization of the Thin Film Prepared at Room Temperature by QDPD”, *J. Aerosol Sci.*, **29**, S905-S906 (1998).

[5] M. Itoh, K. Sakiyama, S. Handoh, and H. Takano, Advanced Technology for Particle Processing, Vol. 1, pp.246-250, AIChE, (1998).

[6] M. Itoh, M. Izumi, S. Handoh, K. Sakiyama, and H.Takano, “Structural Analysis of Nano-Size Binary Particles Generated by the Modified Gas Evaporation Method”, *J. Aerosol Sci.*, **30**, S339-S340 (1999).

[7] K. Sakiyama, H. Takano, M. Itoh, “Binary Thin Film Synthesis at Room Temperature by the Aerosol Process Applying Quantum Size Effect of Nano-Particle”, *J.Aerosol Res. Jap.*, **16**, 133-137 (2001).

[8] W. Manabe, H. Takano and M. Itoh, “Characterization of Surface Morphorogy of Organic and Inorganic Hybrid Thin Films – A First Trial for Artificial Cell Membrane”, in this book (2007).

[9] R.Lipowsky and E.Sackmann (eds.), Handbook of Biological Physics Vol.1, p.658, Elsevier (1995).

CORROSION RESISTANCE OF TiAlN/CrN, TiAlN, AND CrN COATINGS PREPARED BY RF MAGNETRON SPUTTERING

Dong Hwan Song 1, Woo Yang Jang 2, Jong Kook Lee 1
Department of Advanced Materials Engineering, Chosun University 1
375 Seosuk-dong Dong-gu
Gwangju, Korea, 501-759

Department of Advanced Metallurgy and Materials Engineering 2
375 Seosuk-dong Dong-gu
Gwangju, Korea, 501-759

ABSTRACT

TiAlN wear resistant coatings are known to be superior to other coatings in protecting tools that may be damaged by corrosion, oxidization, and abrasion. CrN is a hard, chemically resistant substance that resists corrosion and oxidation. In this study, a TiAlN/CrN-multilayer was prepared by RF magnetron sputtering, followed by heat treatment at various temperatures. CrN and TiAlN monolayers were also prepared for comparison to the TiAlN/CrN-multilayer. Network and multiple cross link microstructures of CrN and TiAlN/CrN coatings were observed by FE-SEM. X-ray patterns of TiAlN, CrN and TiAlN/CrN show preferred orientations of TiAlN, $Ti_3Al_2N_2$, β-Cr_2N and γ-CrN.

The particle size in CrN and TiAlN/CrN coatings increases when heat treated at 700°C, have exaggerated size particles appear after heat treatment at 800°C.

INTRODUCTION

TiAlN has been developed as a coating material since the 1980s. The possible application of TiAlN for cutting tools has been reported by Leyendecker and Konig [1,2]. They demonstrated that the cutting behavior of TiAlN coatings was not only superior to that of other conventional binary coatings but also can be used at significantly higher cutting speeds. Since then, researches on preparation and applications of metal-based nitride coatings such as CrN, HfN, MoN, VN, BN, Si_3N_4, WN were intensively studied [3-6].

Of particular, CrN coatings exhibit excellent corrosion and wear properties, better oxidation resistance, and better adhesion to tools compared to TiN coatings [7,8]. Although TiAlN coatings and TiAlN/CrN multilayer have excellent adhesion and the corrosion resistance, few studies are reported on the coatings. TiAlN/CrN multilayer coatings produced by magnetron sputtering showed high oxidation and corrosion resistance and hardness [9,10]. It is known that the coating properties are significantly affected by the chemical composition, microstructures and surface morphology of the coating [11].

The purpose of this study was to prepare TiAlN/CrN multilayer by RF magnetron sputtering and evaluate the corrosion behavior of the multilayer. Furthermore, the effect of heat treatment on the corrosion resistance was investigated and compared with TiAlN and CrN coatings.

EXPERIMENTAL PROCEDURE

The targets used in this study were a chromium (¢ 2 X 0.25 $inch^2$, 99.95 %, ORYX corporation, USA) and a titanium-aluminum alloy (¢ 2 X 0.25 $inch^2$, ORYX corporation, USA) with a composition of 50:50 (at.%). Substrates (SKD11) were annealed at 1030°C, cooled down on air and then tempered at 520°C. The substrates were then finely polished to 1 μm, ultrasonically cleaned in acetone for 10 min. TiAlN/CrN multilayers were prepared by coating TiAlN and subsequent coating CrN on the SKD11 by R.F magnetron sputtering. TiAlN and CrN monolayers were also prepared for comparison. The experimental parameters of the deposition process are shown in Table 1. The coated layers obtained were heat treated at temperatures of 400-800°C under N_2 atmosphere.

Microstructures of the coatings were observed by FE-SEM (HITACHI S-4800, Japan). The phase analysis of coating was determined by X-ray diffraction (PANALYTICAL X-Pert Pro MPD, Netherland) using Cu Kα radiation, operated at 40 kV and 30 mA. The glancing incidence angle is 1° and scanning speed is 1° and 2°. For corrosion test, potentiodynamic polarization measurements were performed using EG&G 263A (USA). Samples were loaded in a Teflon sample holder and the exposed surface area to the corrosive medium was approximately 0.28 cm^2. Corrosion tests were performed in 0.9 % NaCl solution, -2.0-+3.0 V of transmission volt size under free air condition at room temperature. Samples were cleaned in distilled water before loading the sample holder. The Tafel plot was obtained after the electrochemical measurement. The E_{corr} (corrosion potential) and I_{corr} (corrosion current density) were deduced from the Tafel plot (that is, log I vs. E plot).

$$I_{corr} = [b_a b_c/\{2.303(b_a + b_c)\}](1/R_p)$$

where b_a and b_c are the Tafel slopes or the Tafel constants, expressed in V/decade and R_p is the polarization resistance expressed in Ω cm^2. The polarization resistance was calculated using the following equation.

$$R_p = (\Delta E/\Delta I)|\Delta E \rightarrow 0$$

where ΔE is the polarization potential and ΔI is the polarization current. If the polarization resistance increases the corrosion current decreases [12].

Table 1. Operating parameters of R.F magnetron sputtering.

Parameters	Values	Parameters	Values
Base pressure (torr)	1.5×10^{-7}	Gas mixture (sccm)	Ar:20(99.999%), N2:30(99.999%)
Working pressure (torr)	7.4×10^{-2}		
Substrate	SKD11	R.F power	80
Rotational speed of the substrate (rpm)	10	Distance of cathode to substrate (mm)	50

RESULTS AND DISCUSSION

The TiAlN/CrN coating consisted primarily of TiAlN and CrN phases. The phase identification by X-ray analyses was shown in Fig. 1. The X-ray pattern of CrN coating showed preferred orientations γ-CrN (111) and (200), β-CrN (112). The X-ray pattern of TiAlN coating shows preferred orientations TiAlN (111) and (200). And the X-ray pattern of TiAlN/CrN coating shown preferred orientations $Ti_3Al_2N_2$ (106) and β-CrN (112). The peaks at 24.4, 33.5, 39.5, 44.6, 65.0, 82.5° of diffraction angle 2θ dominated all the spectra, which are ascribed to substrate of SKD11.

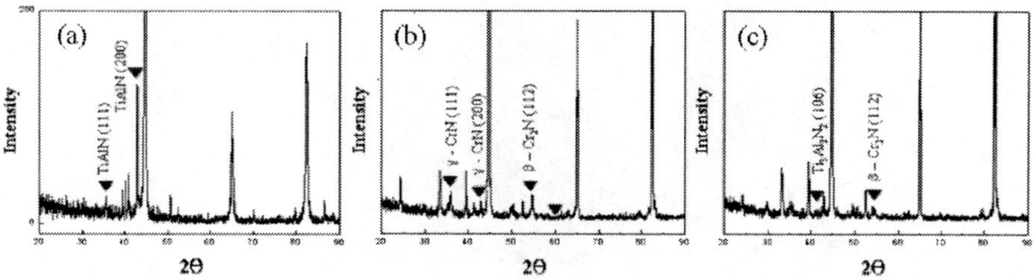

Fig. 1. X-ray diffraction patterns of (a) TiAlN, (b) CrN and (c) TiAlN/CrN coated layer after heat treatment at 800°C.

The micrographs of the coatings heat treated at 400°C revealed particles size of 20-40 nm (Fig. 2(a), Fig. 3(a) and Fig. 4(a)). As the temperature increased to 800°C, the particle size increased. TiAlN coatings showed uniform and dense microstructure. However, exaggerated particles were existed in CrN and TiAlN/CrN surfaces at 800°C. Exaggerated particles of CrN and TiAlN/CrN were 100-300 nm and considered as β-CrN.

Fig. 2. FE-SEM micrographs of TiAlN coatings heat-treated at (a) 400°C, (b) 500°C, (c) 600°C, (d) 700°C and (e) 800°C.

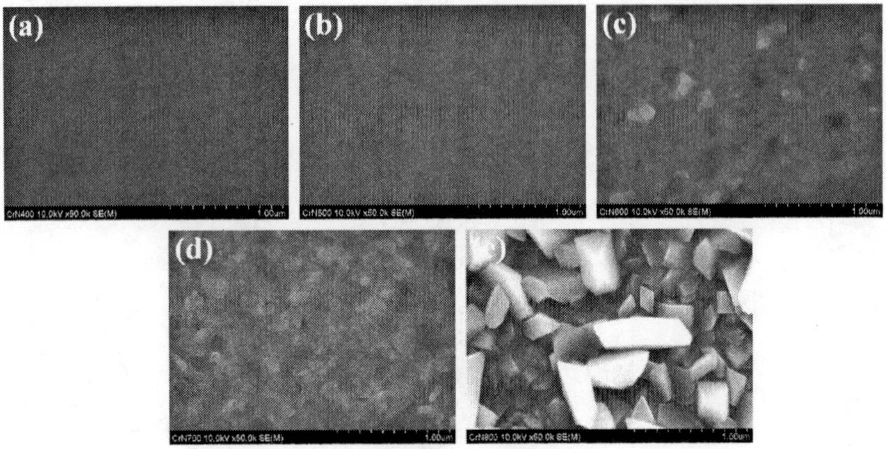

Fig. 3. FE-SEM micrographs of CrN coatings heat-treated at (a) 400°C, (b) 500°C, (c) 600°C, (d) 700°C and (e) 800°C.

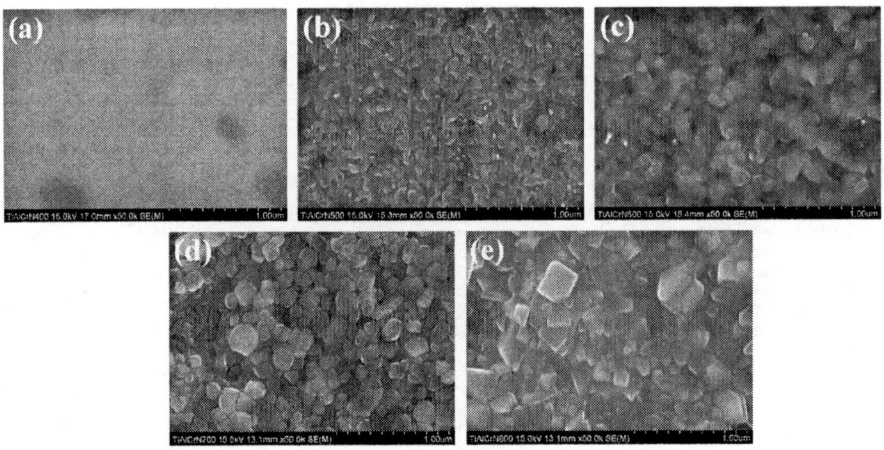

Fig. 4. FE-SEM micrographs of TiAlN/CrN coatings heat-treated at (a) 400°C, (b) 500°C, (c) 600°C, (d) 700°C and (e) 800°C.

The Tafel plots of TiAlN, CrN and TiAlN/CrN coatings in 0.9 % NaCl are shown in Fig. 5, 6 and 7, respectively. The TiAlN coatings at 450°C, 600°C, the CrN coatings at 450°C, 650°C and the TiAlN/CrN coatings at 400°C, 600°C showed lowest passivation. On the other hand, the

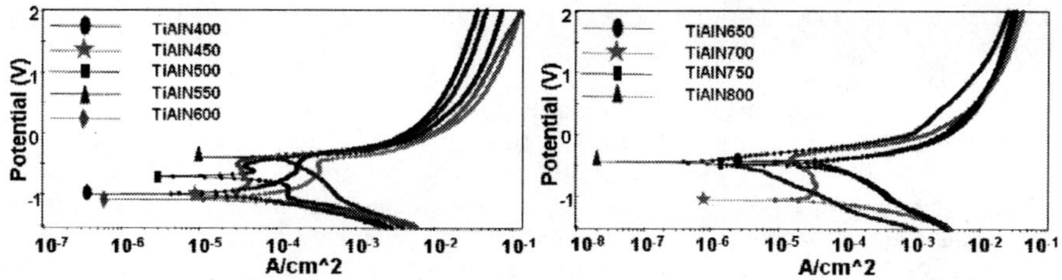

Fig. 5. Potentiodynamic polarization curves of TiAlN coating after heat treatment from 400°C to 800°C.

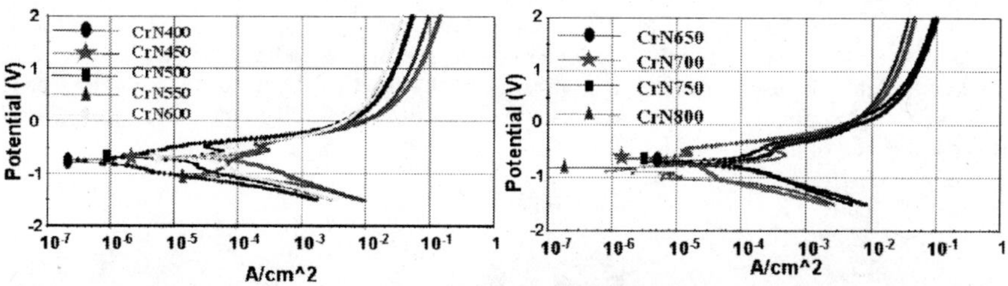

Fig. 6. Potentiodynamic polarization curves of CrN coating after heat treatment from 400°C to 800°C.

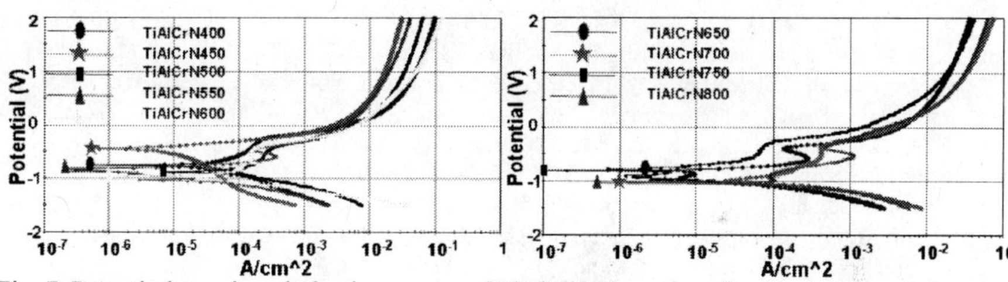

Fig. 7. Potentiodynamic polarization curves of TiAlN/CrN coating after heat treatment from 400°C to 800°C.

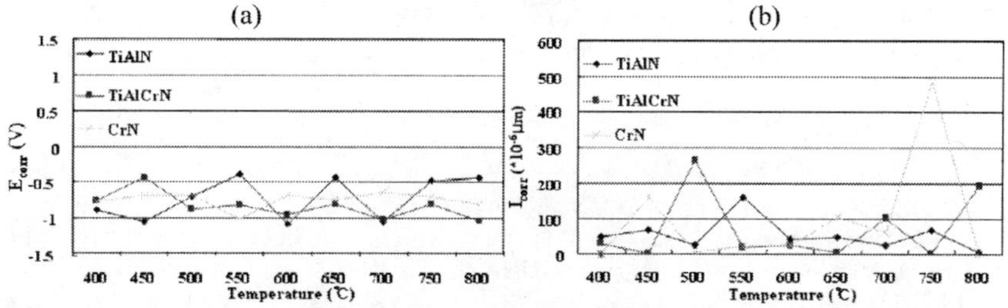

Fig. 8. (a) E_{corr} and (b) I_{corr} of the coatings with heat treatment.

TiAlN coatings at 500°C, 550°C, the CrN coatings at 500°C, 600°C and the TiAlN/CrN coatings at 450°C, 550°C showed the highest passivation value.

The corrosion potential (E_{corr}) and corrosion current (I_{corr}) for TiAlN, CrN and TiAlN/CrN coatings are given in Fig. 8. It can be seen that the corrosion resistance of 400-500°C of CrN, 550°C, 650°C of TiAlN, 450°C of TiAlN/CrN are better than other temperatures because 400-500°C of CrN, 550°C, 650°C of TiAlN, 450°C of TiAlN/CrN have the ability to form a passive layer over the surface. The CrN exhibits a dense microstructure and provides no direct diffusion path for the corrosive medium [13]. TiAlN coatings exhibit better corrosion behavior than the CrN coatings.

CONCLUSIONS

CrN coatings heat-treated at 400-700°C exhibit columns, forming a direct path for the corrosive medium to pass through the coating defects. On the other hand, TiAlN coatings have shown excellent corrosion resistance compared with the CrN. The TiAlN/CrN multilayer coatings showed superior corrosion resistance than that of the monolayer coatings of TiAlN and CrN. The surface of TiAlN/CrN coatings heat-treated at 400°C have a network a microstructure and multiple cross links. The TiAlN/CrN multilayer coatings heat-treated at 400-550°C show low corrosion current density and high polarization resistance.

REFERENCES

[1]W. Konig, R. Fritsch and D. Kammermeier, "Physically vapor deposited coatings on tools : performance and wear phenomena." Surf. Coat. Technol., 49 316 (1991)

[2]T. Leyendecker, O. Lemmer, S. Esser and J. Ebberink, "The development of the PVD coating TiAlN as a commercial coating for cutting tools." Surf. Coat. Technol., 48 175 (1991)

[3]R. Nowak and S. Maruno, "Surface deformation and electrical properties of HfN thin films deposited by reactive sputtering." Mater. Sci. Eng., A202 226 (1995)

[4]J.-N. Tu, J.-G. Duh and S.-Y. Tsai, "Morphology, mechanical properties, and oxidation behavior of reactively sputtered Cr-N films." Surf. Coat. Technol., 133/134 181 (2000)

[5]P.E. Hovsepian, D.B. Lewis, W.-D. Mvunz, A. Rouzaud and P. Juliet, "Chromium nitride/niobium nitride superlattice coatings deposited by combined cathodic-arc/unbalanced magnetron technique." Surf. Coat. Technol., 116 727 (1999)

[6]W. Xu, B. Li, T. Fujimoto and I. Kojima, "Suppressing the surface roughness and columnar growth of silicon nitride films." Surf. Coat. Technol. 135 274 (2001)

[7]B. Navinsek, P. Panjan and I. Milosev, "Industrial applications of CrN (PVD) coatings, deposited at high and low temperatures." Surf. Coat. Technol., 97 182 (1997)

[8]J. Jagielski, A.S. Khanna, J. Kucinski, D.S. Mishra, P. Racolta, P. Sioshansi, E. Tobin, J. Thereska, V. Uglov, T. Vilaithong, J. Viviente, S.-Z. Yang and A. Zalar, "Effect of chromium nitride coating on the corrosion and wear resistance of stainless steel." Appl. Surf. Sci., 156 47 (2000)

[9]H.C. Barshilia, M.S. Prakash, A, Jain and K.S. Rajam, "Structure, hardness and thermal stability of TiAlN and nanolayered TiAlN/CrN multilayer films." Vacuum., 77 169 (2005)

[10]H.C. Barshilia and K.S. Rajam, "Raman spectroscopy studies on the thermal stability of TiN, CrN, TiAlN coatings and nanolayered TiN/CrN, TiAlN/CrN multilayer coatings." J. Mater. Res., 19 3196 (2004)

[11]J. Takacoum and H. Houmd Bennani, "Influence of substrate roughness and coating thickness on adhesion, friction and wear of TiN films." Surf. Coat. Technol., 82 272 (1997)
[12]V.K. William Grips, Harish C. Barshilia, V. Ezhil Selvi, Kalavati and K.S Rajam, "Electrochemical behavior of single layer CrN, TiN, TiAlN coatings and nanolayered TiAlN/CrN multilayer coatings prepared by reactive direct current magnetron sputtering." Thin Solid Films., 3 8 (2006)
[13]C. Liu, Q. Bi and A. Matthews, "EIS comparison on corrosion performance of PVD TiN and CrN coated mild steel in 0.5 N NaCl aqueous solution." Corros. Sci., 43 1953 (2001)

PLASMA THERMAL DEPOSITION OF ALUMINUM ON Mg-Li WORK-HARDENED ALLOY

Masato Tsujikawa
Graduate School of Engineering, Osaka Prefecture University
1-1 Gakuen-cho, Naka-ku
Sakai, Osaka, 599-8531, JAPAN

Shin-ichiro Adachi
Technical Research Institute of Osaka Prefecture
2-7-1 Ayumino
Izumi-shi, Osaka, 594-1157, JAPAN

Kazuhiro Nakata
Joining and Welding Research Institute of Osaka University
11-1 Mihogaoka
Ibaraki-shi, Osaka, 567-0047, JAPAN

Masaichiro Kamita
Yamani Co. Ltd.
915-5 Kamitakeda, Ichijima-cho
Tamba-shi, Hyogo, 669-4341, JAPAN

Sachio Oki
School of Science & Engineering, Kinki University
3-4-1 Kowakae
Higashi-osaka-shi, Osaka, 577-8502, JAPAN

ABSTRACT

Magnesium-lithium alloys containing 14 mass% lithium are putatively the lightest metallic structural material that has good cold workability and high specific strength, which are attributable to the cubic crystalline structure of the lithium solid solution, which differs from the closely packed hexagonal crystalline structure of a magnesium solid solution. Unfortunately, the alloy's poor corrosion resistance hinders its application. Alloy surface modifications have been tried, but vigorous corrosion has made aqueous treatments difficult. Moreover, dry processes are restricted thermally at temperatures less than about 150°C because of lithium's low melting point and low re-crystallization temperatures of cold-worked alloy material.

Plasma-sprayed pure aluminum coatings on cold-rolled Mg-Li alloy plates were investigated. A Mg-14mass%Li-1mass%Al alloy ingot was cast in an argon atmosphere. After homogeneous heat treatment, the ingot was rolled to a 3-mm-thick plate under a reduction ratio of 95%. After grit blasting, plasma spraying of pure aluminum was performed under several conditions. The coated layer thickness was 50–300 μm. The substrates' temperature increases during spraying were small. Corrosion resistance of plasma sprayed plates, as evaluated by salt-water spraying tests, was improved through formation of a dense aluminum-coating layer.

INTRODUCTION

Mg-Li alloy

Lightweight materials are effective for reducing the mass of transportation-related devices. Magnesium alloys are the lightest of all practical metallic alloys for structures. They offer excellent mechanical properties including high specific strength and high specific rigidity. They are remarkable as materials of future-generation aircraft; studies of higher specific-strength alloys have been performed widely. Notwithstanding, high specific-strength, practically useful magnesium alloys have not attained the capabilities of high-strength aluminum alloys. It is important for expansion of magnesium usage that the specific strength of magnesium alloy be greater than the value of high-strength aluminum alloys.

Two means exist to increase the specific strength of magnesium alloys. One is the pursuit of high strength through addition of hardening elements such as yttrium and zinc, which usually increase the alloy's density. The other means is addition of lithium, which is a lighter element than magnesium. First, the addition of Li decreases the resultant alloy's density. Secondly, additional Li changes the crystal structure from CPH to BCC, thereby allowing cold-working. Using cold-working, work hardening, which is impossible for ordinary HCP magnesium alloys, is expected.

An alloy called LA141 includes 14 mass% Li and 1 mass% Al is an Mg-Li alloy. Its density is 1.326 g/cm^3. The ultimate tensile strength of the cold-worked LA141 is 194–294 MPa. depending on the reduction ratio. Its specific strength is 175–223 $MPa/(g/cm^3)$. These values are greater than the value of 125 of heat-treated aluminum alloy A6061-T6, and 150 of conventional magnesium alloy extruded AZ31 (Mg-3mass%Zn-1mass%Al alloy), and heat treated aluminum alloy A6061. The values are almost comparable to those of 235 $MPa/(g/cm^3)$ of heat-treated strong aluminum alloy A7075-T6. The alloy should soon be put to practical use [1].

Surface protection for Magnesium alloys

Corrosion resistance is an important subject for all magnesium alloys. No protective dense oxide film exists on the surface of magnesium alloys, such as that formed on an aluminum alloy surface. Furthermore, no effective sacrifice anode element exists for magnesium that would act similarly to zinc for steel surfaces. For those reasons, no hairline crack would be acceptable in a protective surface of magnesium alloys.

A diamond-like carbon (DLC) coating can contribute to excellent tribological and corrosion-protecting properties of materials. In most cases, a metallic film such as one of Cr or Si is used as an interlayer between DLC film and substrate to improve the coating's adhesion. The adhesion of DLC film on magnesium alloy is improved through the use of some kind of pretreatment instead of an interlayer [2, 3]. Similarly, low-temperature high-voltage anodic oxidation forms a layer effectively [4, 5] but it is hard and brittle on the surface. A brittle surface layer often engenders hairline cracks during impact loading. Cracks to the substrate form the local batteries in a moist environment.

In contrast, thin ductile metallic surfaces with corrosion resistance protect the substrate during impact loading because they have sufficient toughness to absorb such impact energy. Excellent protective surfaces for magnesium alloy can be formed if there were an adhesive thin aluminum layer on the magnesium alloy.

This study investigated the corrosion resistance of plasma thermal spraying of aluminum on Mg-Li alloy.

EXPERIMENTAL PROCEDURES

Preparation of Mg-Li alloy

It is very difficult to melt Mg-Li alloys in air or in an evacuated chamber because they have high reactivity and high vapor pressure. Because of their low density of 1.326 g/cm^3, flux melting is also out of the question because it causes oxide inclusion. For those reasons, it is necessary to melt and cast these alloys in a pressurized inert gas atmosphere. In this study, the alloys were melted at 953 K in a chamber with 0.15 MPa argon gas. Degasification was done for one hour using argon gas bubbling. After standing for 15 min, the melts were poured in the chamber into block moulds under a pressurized argon gas atmosphere. The resultant ingots were 250 mm high and 250 mm wide, with 50 mm thickness and weight of around 5 kg. These ingots were certified as defect free by x-ray radiography. The ingots were homogenized at 673 K for 24 h.

The ingots were cut into 45 mm × 300 mm × 250 mm slabs. Each slab was cold rolled to 2-mm thickness using a rolling mill with 380 mm roll diameter in 12 reduction passes. The rolling reduction was 95.6%. The plate's microstructure is shown in Fig. 1.

The LA141 alloy strength varies according to the degree of cold working. The maximum value of the ultimate tensile strength of this alloy is 294 MPa, as mentioned above, at the condition of 98% in reduction ratio at cold rolling. The 2-mm-thick plates prepared for this investigation have 205 MPa in ultimate tensile strength.

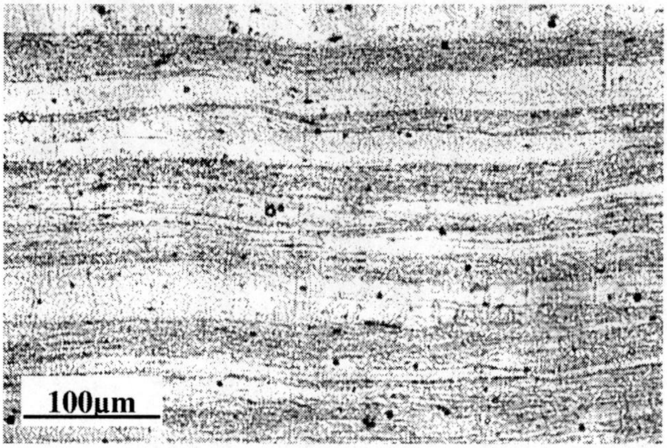

Figure 1. Microstructure of cold-rolled Mg-14mass%Li alloy.

Plasma thermal spraying of aluminum onto the Mg-Li alloy

These Mg-Li alloy plates were ground using #120 emery paper and grit-blasted using white #24 alumina particles as a pretreatment for thermal spraying. Then, the plates were plasma thermally sprayed with 99.7% pure aluminum particles of 50 μm diameter. The apparatus used in this investigation has an output of 26 kW (TA-7050; Aero Plasma Co. Ltd). The thermal spray distance was constant at 0.15 m.

Cold-rolling

A fraction of the thermally sprayed plate was slightly cold-rolled. The reduction ratio was 3%. The effect of rolling on corrosion protection was evaluated.

Salt-spray test

Corrosion resistance was evaluated using the salt spray test. A thermally sprayed sample was cut to $30 \times 50\ mm^2$, then the substrate side and end faces were polymer-coated for isolation. The test used the spray of 50 g/l NaCl solution with 7.0 pH, the spray intensity was $125\ ml/m^2/h$ and the test temperature was 35°C. A mechanically polished sample and a chemically polished sample were also tested simultaneously as references.

The duration of salt spray testing was 24 h. Surface conditions were checked after 4 h and 8 h during the test.

Figure 2. Exemplary cross section of a plasma thermally sprayed surface layer.

RESULTS AND DISCUSSION

Figure 2 shows the surface layer of thermally sprayed as-cast Mg-8mass%Li alloy. The sprayed layers were dense and the pores were not mutually linked. The sprayed layer thickness was about 500 μm and the deposit-layer hardness was 44HV0.1. The hardness of this as-cast Mg-8Li alloy is 56HV0.1; cold-rolled 3-mm-thick plates were 55HV0.1 [1]. The deposit layer is softer than the substrate.

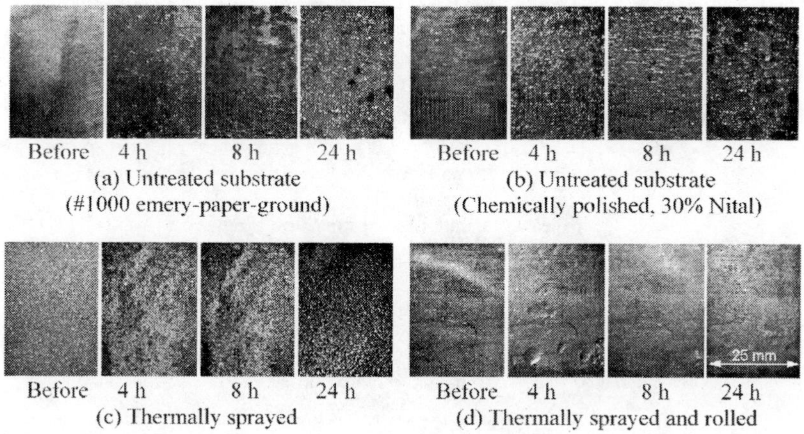

Figure 3. Salt-spray test results Mg-14mass%Li alloy with various surface conditions.

The surfaces of specimen plates during testing are shown in Fig. 3. Untreated specimens (a) and (b) started blistering before 4 h. No marked difference exists between the emery ground surface and the chemically polished surface. The thermally sprayed specimen started to stain markedly by 4 h. In contrast, the thermally sprayed. then cold-rolled specimen surface was clear until the end of testing.

Figure 4 shows specimen surfaces after 24-h testing and rinsing using deionized water in comparison to the dried state. The unsprayed surfaces depicted in Figs. 4(a) and 4(b) show vigorous corrosion. The thermally sprayed specimen portrayed in Fig. 4(c) show uneven staining. The thermally sprayed and cold-rolled surface showed no evidence of blistering or considerable staining. Results indicate that cold-rolling of the thermally sprayed surface is effective for corrosion protection of Mg-Li alloy.

The thermally sprayed specimen is easily stained because of microchannels between the specimen surface and substrate. Figure 5 shows surfaces of the as-sprayed specimen (a) and the sprayed, then rolled specimen (b). Figure 5(a) shows that the thermally sprayed surface includes small particles. A microchannel must exist between them. The effect of rolling on corrosion protection arises from closing of the microchannels.

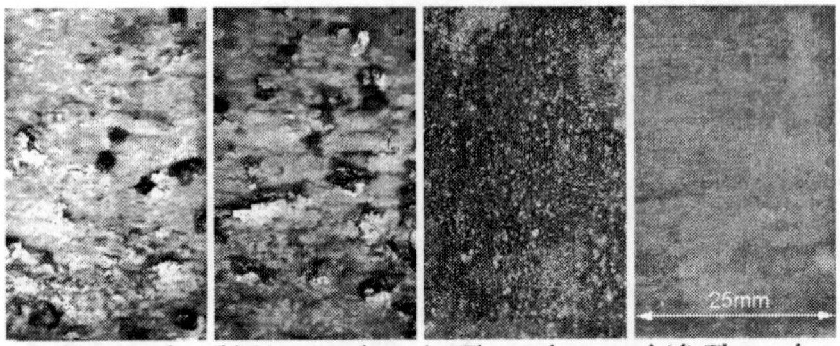

(a) Untreated (Emery-ground) (b) Untreated (Chemically polished) (c) Thermal sprayed (d) Thermal sprayed (+cold rolled)

Figure 4. Surfaces after salt-spray testing (test duration: 24 h).

The Mg-Li alloys with lithium contents of greater than 11 mass% have a BCC crystal structure. It allows these magnesium alloys to be cold-worked without cracking. Therefore, this slight cold-working process of plasma thermally sprayed pure aluminum layer is ideal for Mg-Li alloys. Furthermore, such clad plates will be a raw material for cold-press processing.

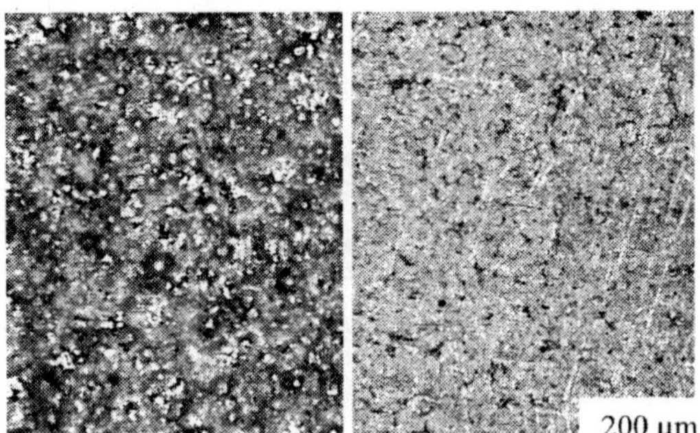

(a) As thermally sprayed (b) Rolled sur

Figure 5. Surface features: (a) as-sprayed, (b) cold-rolled 3%.

CONCLUSION

The Mg-Li alloy structural material is lightweight but has poor corrosion resistance. A protective pure aluminum surface layer was built up using plasma thermal spraying. Nevertheless, it was insufficient to prevent material degradation. Considerable staining occurred on the sprayed surface. Such staining was eliminated slightly by cold rolling of the sprayed plate. The effect of slight rolling follows the closing of microchannels that exist in the sprayed layer between the surface and substrate. This process is suitable for a cold-workable magnesium alloy with a BCC crystal structure.

REFERENCES

[1]M. Tsujikawa, Y. Abe, S.W. Chung, S. Oki, K. Higashi, M. Kamita, and I. Hiraki, "Cold Rolled Mg-14mass%Li-1mass%Al Alloy and Its Friction Stir Welding," *Mater. Trans.*, 47, 1077-1081 (2006).
[2]N. Yamauchi, N. Ueda, A. Okamoto, T. Sone, M. Tsujikawa, and S. Oki, "DLC coating on Mg–Li alloy," *Surface and Coatings Technology*, Article in Press, doi: 10.1016/ j.surfcoat. 2006.07.080, Available online 14 August 2006.
[3]N. Yamauchi, K. Demizu, N. Ueda, T. Sone, M. Tsujikawa, and Y. Hirose, "Effect of peening as pretreatment for DLC coatings on magnesium alloy," *Thin Solid Films*, 506-507, 378-383 (2005).
[4]Y. Zhang, C. Yan, F. Wang, and W. Li, "Electrochemical behavior of anodized Mg alloy AZ91D in chloride containing aqueous solution," *Corrosion Science*, 47, 2816-2831 (2005).
[5]Y. Zhang and C. Yan, "Development of anodic film on Mg alloy AZ91D," *Surface and Coatings Technology*, Article in Press, doi:10.1016/j.surfcoat.2006.04.015, Available online 8 June 2006.

BEHAVIOR OF SUPERFICIAL OXIDE AT DIFFUSION-BONDED INTERFACE OF TIN AND ITS INFLUENCE ON BOND STRENGTH

Shinji Koyama[1*], Makoto Takahashi[1], and Kenji Ikeuchi[1]
[1]Joining and Welding Research Institute, Osaka University
11-1 Mihogaoka
Ibaraki, Osaka, Japan, 567-0047

[1*] Now at Research Institute for Applied Science
49 Tanaka-Ohicho, Sakyoku
Kyoto, Japan, 606-8202

ABSTRACT

The microstructural evolution of the superficial oxide film during the diffusion bonding and its influence on the joint performance have been investigated mainly by TEM observations for a pure tin specimen having a faying surface finished by grinding on an emery paper of 800 grade. The tensile strength of joints increased with bonding temperature from 443 K to 483 K at a bonding pressure of 7 MPa, and a rise in bonding pressure lowered the temperature range where the joint tensile strength increased. Along the diffusion-bonded interface, a region of a few μm width was observed where a number of fine oxide particles and films were distributed. As the bonding temperature was increased, the oxide particle and film coalesced and their number density decreased. The rise in bonding pressure enhanced these changes in the morphology and distribution of the oxide. It can be considered that these changes in the interfacial oxide were accompanied with the expansion of the area of the metal-to-metal contact, which contributed to the increase in the joint strength. The interfacial oxide was identified as tetragonal SnO_2 on the basis of SAD pattern analyses.

INTRODUCTION

It is generally accepted that the oxide film covers almost all kinds of metals except for a few noble metals, and has significant influences on phenomena controlling the joint performance of solid-state bonding, brazing, and soldering[1, 2)]. For example, the oxide film, lying between the substrates of metals to be joined, interferes with the attainment of true metal-to-metal contact in the solid-state bonding, and impairs the wettability of the filler in brazing and soldering. To our knowledge, however, only little information has been reported about the microstructural change of the oxide film during the joining process except for those observed in diffusion-bonded joints of Al alloys[3)]. As well known, tin is the most common alloying element of solder, and the surface of most solders has been reported to be covered with tin oxide[4)]. However, only little is known about the behavior of tin oxide film during soldering as well as the solid state bonding. Therefore, we aimed this investigation at revealing the morphology and distribution of the oxide film at the diffusion-bonded interface of tin by direct observations with TEM (Transmission Electron Microscope), and grasping their influences on the bond strength of the joint interface.

EXPERIMENTAL DETAILS

The specimen to be bonded was a block 15×15×5 mm cut from a 99.999 % tin ingot. This tin block was bonded to a copper block for the convenience of the tensile test of joint[5)]. The faying surface (15×15 mm face) was finished by wet grinding on emery paper of 800 grade. The diffusion bonding has been carried out by holding a couple of specimens at bonding temperatures T_j from 373 to 503 K for a bonding time of 1.8 ks with a bonding pressure P_j applied to the interface between them. The bonding atmosphere was a vacuum of 3×10^{-2} Pa.

The surface (transverse section of the interface) for the metallographic observation was finished by grinding on emery paper of 1500 grade, followed by electrolytic polishing in a ethanol solution of perchloric acid, and was etched in aqueous solution of HCl. The thin film specimen for the observation with TEM was prepared by electrolytic polishing (in solution containing 5% perchloric acid in 10% ethylene glycol monobutyl and 85% ethyl alcohol) and subsequent Ar atom-beam milling. A FIB (Focused Ion Beam) system was also employed to prepare TEM specimens from the faying surface to observe its cross section and joints that could not withstand the preparation by the electrolytic polishing because of its poor strength. The bond strength was estimated from tensile strength of the joint. The tensile test was carried out at room temperatures at a displacement speed of 0.0017 mm/s.

EXPERIMENTAL RESULTS

As shown in **Fig. 1**, the tensile strength of the joint increased with bonding temperature, and became almost constant as it reached the level of the base metal strength. A rise in bonding pressure lowered the temperature range where the joint strength increased.

With a rise in bonding temperature, the microstructure of joint was changed as shown in **Fig. 2**. At a bonding temperature of 463 K, a bond line remained obviously, and many voids and/or inclusions were observed along it. As the bonding temperature was increased, the amounts of voids and/or inclusions reduced, and the bond line became almost disappeared at a

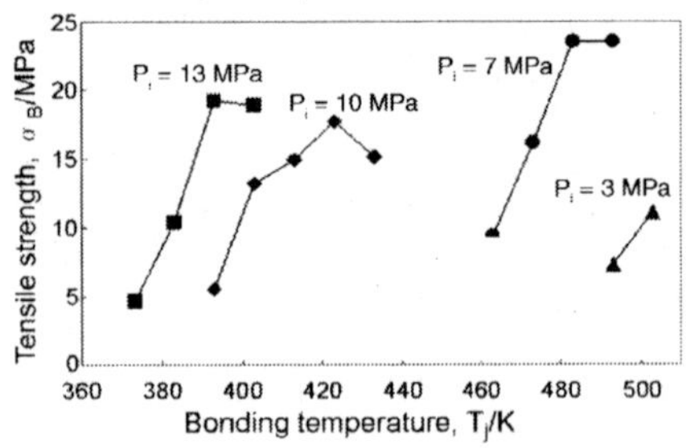

Figure. 1 Tensile strength vs. bonding temperature at bonding pressure $P_j = 3 - 13$ MPa.

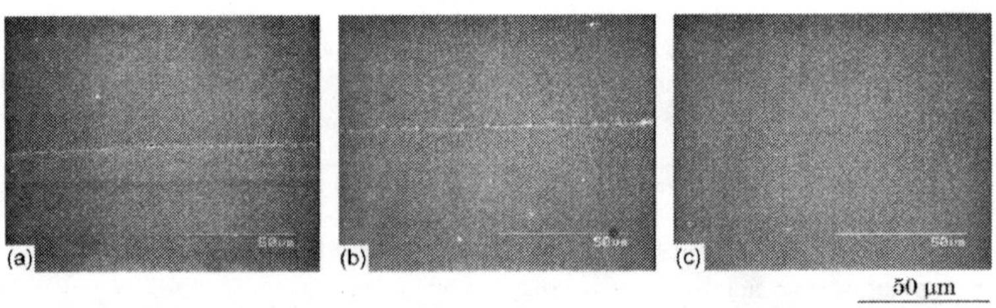

Figure. 2 SEM micrographs of joints bonded at a bonding temperature $T_j = 463$ K (a), 473 K (b), and 483 K (c) at $P_j = 7$ MPa.

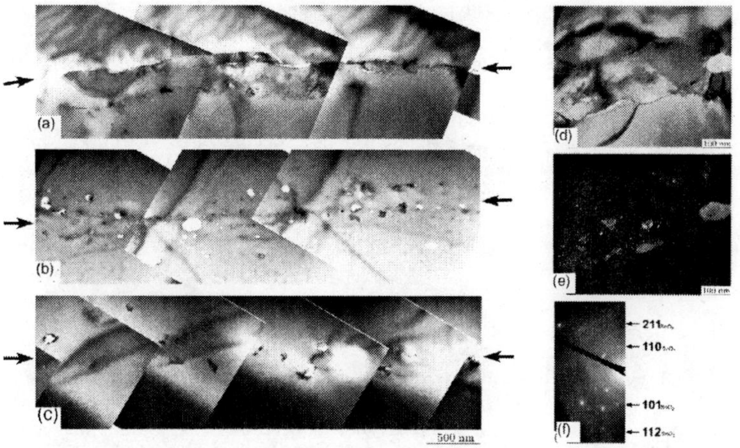

Figure. 3 TEM micrographs of joint interfaces (P_j = 7 MPa): (a) B.F.I. (Bright Field Image) at T_j = 463 K, (b) B.F.I. at T_j = 473 K, (c) B.F.I. at T_j = 483 K, (d) high magnification B.F. I. at T_j = 463 K, (e) D.F. I of (d) taken with reflection from SnO_2, and (f) SAD pattern from SnO_2.

bonding temperature of 483 K, when the bonding pressure was 7 MPa. The bond line tended to disappear in considerable areas of the interface, when the joint showed tensile strength comparable with the base metal strength.

TEM observations revealed that corresponding to these metallographic changes, the morphology and distribution of the oxide at the interface varied as shown in **Figs. 3(a), 3(b) and 3(c)**. At a bonding temperature of 463 K, a great number of granular and film inclusions a few 100 nm or less in size were distributed within a zone a few μm in width along the bond interface. As the bonding temperature was increased, these inclusions were coalesced to form coarser and more granular inclusions, reducing their number density. These inclusions were identified as tetragonal SnO_2 based on analyses of SAD (Selected Area Diffraction) patterns and dark field

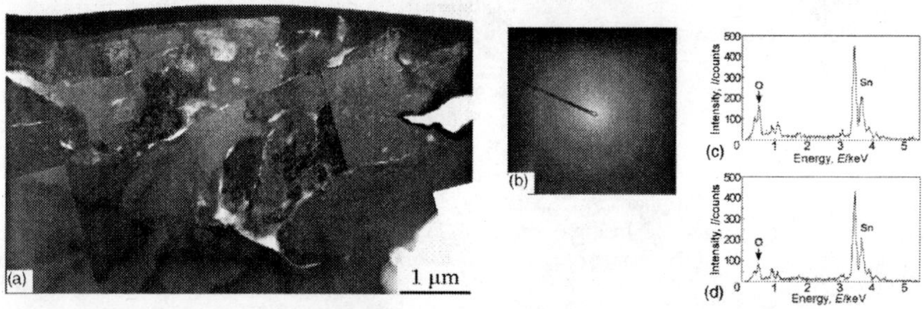

Figure. 4 (a) Section of a surface finished by grinding on the emery paper. (b) SAD pattern from an inclusion observed in (a). (c) and (d) EDX spectra from inclusions.

images (D.F.I.) as shown in **Figs. 3(d), 3(e), and 3(f)**. At a bonding temperature of 463 K, the oxide distributing in the interfacial region trapped a grain boundary as shown in **Fig. 3(a)**, while almost no grain boundary was observed in the interfacial region at a bonding temperature of 483 K as shown in Fig. 3(c). The disappearance of the bond line in the SEM micrograph at 483 K (see Fig. 2(c)) probably reflected the release of the grain boundary from the trap by the oxide inclusion in the interfacial region. When the bonding pressure was increased to 10 MPa, similar changes in the morphology and distribution of the oxide inclusion occurred at lower bonding temperatures. On the other hand, when the bonding pressure was decreased down to 3 MPa, many porosities were observed in the interfacial region even at a bonding temperature of 493 K (lower than the melting point of tin by only 15 K).

DISCUSSION

Behavior of oxide at bond interface

The interfacial region that involved a number of oxide inclusions similar to those observed in Fig. 3 was also observed in the diffusion-bonded joint of aluminum alloy when the faying surface was finished by mechanical processes such as wire brushing, turning in a lathe, and grinding on emery paper[6)]. It was suggested that the interfacial region came from a surface layer into which heavy plastic deformation and oxide particles were introduced during the mechanical process. The interfacial region observed in the joint of tin may also be considered to originate from the surface layer introduced during the grinding on the emery paper. In order to confirm the presence of the surface layer, the vertical section of the surface finished by grinding on the emery paper was observed with a TEM. As shown in **Fig. 4(a)**, a layer that consisted of grains much finer than those of the base metal and involved numerous inclusions was observed. The width of the layer, scattering quite widely, was a few μm or less in average. EDX analyses of these inclusions detected significant amounts of oxygen (see **Figs. 4(c) and (d)**), suggesting that they were tin oxides. From SAD patterns as shown in **Fig. 4(b)**, these inclusions are regarded as amorphous. It can be considered that the interfacial region involving a number of oxide inclusions observed in Fig. 3 came from the surface layer formed during the grinding on the emery paper.

Although the oxide inclusion observed in the interfacial region was SnO_2, the binary phase diagram of the Sn-O system indicates that the oxide in equilibrium with tin is Sn_3O_4. Because of its technological importance, the oxidation of tin has been investigated in many studies[7-9]. Since it is not easy to distinguish between the two oxides SnO and SnO_2, no comprehensive understanding has been obtained about the oxidation behavior of tin or the structure of the oxide film[4, 7-9]. Most of them, however, reported on the basis of results from AES, XPS, and EELS that the oxide film was composed of SnO and SnO_2, but not of Sn_3O_4. Since SnO can react with H_2O to form SnO_2 and the faying surface used in the present investigation was ground with running water on emery papers, the oxide observed in the surface layer may be considered to be mostly amorphous SnO_2. Probably, the amorphous SnO_2 crystallized during the diffusion bonding to form the inclusion observed in Fig. 3.

With a rise in bonding temperature, the tin oxide coalesced to reduce the interfacial energy between the oxide and the tin substrate, accompanied with an increase in areas where no oxide inclusion was involved as shown in Figs. 3(c), viz., areas where the metal-to-metal contact was achieved at the interface.

Influence of oxide inclusion on bond strength

It is generally accepted that the bond strength of a diffusion-bonded joint is controlled by the attainment of intimate contact between the surfaces to be bonded and the elimination of the superficial oxide film (contaminant on the surface) in an intimately contacted area[1]. The intimate contact can be achieved through the plastic deformation of the macro- and micro-asperities on the surface, and so it can be considered that the attainment of intimate contact can be roughly estimated from the degree of the plastic deformation that the specimens underwent during the diffusion bonding. In **Fig. 5**, the tensile strength of joint is plotted against the degree of the plastic deformation of the specimens (reduction in the total thickness of a couple of specimens). As shown in this figure, the joint strength increased with the degree of the deformation of specimens, suggesting the significant contribution of the attainment of intimate contact to the bond strength. However, joints bonded at lower bonding pressures tended to have higher bond strength when the degrees of the deformation were the same. As the bonding pressure is reduced, a higher bonding temperature is necessary in order to deform the specimen by a given degree. Therefore, the results shown in Fig. 5 suggest that the bond strength of joint increases with bonding temperature when the intimately contacted areas are the same. It follows from this that a factor other than the intimately contacted area also made a significant contribution to the bond strength of joint.

As can be seen in Fig. 3, oxide particles and films were distributed within an interfacial region ~1 μm in width, and they coalesced to form coarser and more granular ones, leaving areas almost free from the oxide. The interfacial energy, driving force of these changes in the oxide morphology, generally decreases with a rise in bonding temperature, whereas the diffusivity of atoms controlling the mass transport rate for the changes in the oxide morphology and distribution increases with temperature. Thus, it is conceivable that the increased diffusivity contributes to the expansion of the intimately contacted area with a rise in bonding temperature. The contribution of the area free from the oxide inclusion, viz., elimination of the superficial oxide, to the bond strength can probably explain the effect of bonding pressure on the relation between the bond strength and the degree of the deformation of specimens shown in Fig. 5.

With respect to the effect of the oxide in the interfacial region, it should be noted that submicron size particles of a nonmetallic phase impede dislocation movement at elevated temperature and strengthen the alloy. It is conceivable that the oxide in the interfacial region also has this effect, so-called dispersion strengthening. If this is the case, the deformation rate of the interfacial region at high bonding temperature is reduced in comparison with the bulk region and

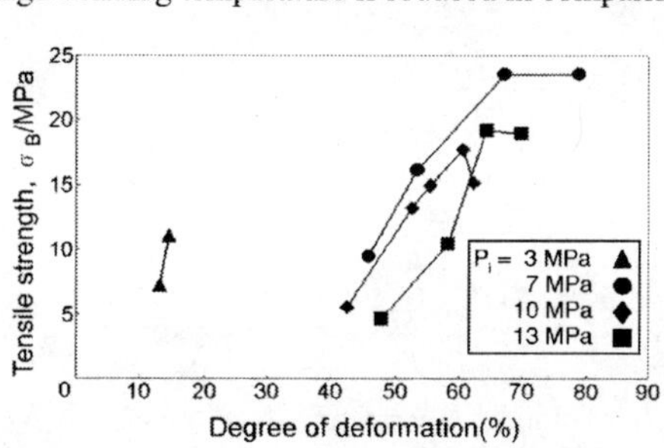

Figure. 5 Tensile strength vs. degree of deformation of specimens during diffusion bonding.

the elimination of porosities or uncontacted areas is reduced. The dispersion strengthening may be considered to be responsible for the low tensile strength of joints bonded at Pj = 3 MPa and Tj $\geq$ 493 K, though the bulk region of specimens is deformed by more than 10% during diffusion bonding, since many porosities were detected in these joints.

CONCLUSIONS

The morphology and distribution of the oxide inclusion in the interfacial zone of diffusion-bonded joint have been investigated mainly by TEM observations for a pure tin specimen whose faying surface was finished by grinding on an emery paper of 800 grade. The results obtained can be summarized as follows:

1) The tensile strength of joints increased with bonding temperature, and a rise in bonding pressure lowered the temperature range where the joint tensile strength increased.
2) Along the diffusion-bonded interface, a region of a few μm width was observed where a number of fine oxide particles and films were distributed. As the bonding temperature was increased, the oxide particle and film coalesced and their number density decreased. The rise in bonding pressure enhanced these changes in the morphology and distribution of the oxide. The interfacial oxide was identified as tetragonal SnO_2 on the basis of SAD pattern analyses.
3) It can be considered that these changes in the interfacial oxide described in 2) were accompanied with the expansion of the area of the metal-to-metal contact, which contributed to the increase in the joint strength.
4) Under the surface finished by grinding on the emery paper of 800 grade, a layer of a few μm width was observed which consisted of much finer grains than the bulk region and involved abundant film and granular inclusions of amorphous tin oxide. This surface layer is considered to be responsible for the formation of the interfacial zone described in 2).

REFERENCES

[1] R.F. Tylecote, The Solid Phase Welding of Metals, Edward Arnold, London, 35- 46 (1968).

[2] Welding Handbook 9th Ed. Vol. 2, Welding Processes, Part 1, AWS, Miami, 560-96 (2004).

[3] K. Kotani, K. Ikeuchi, F. Matsuda, "Interfacial Phases in Diffusion-Bonded Joints of Al-Mg Alloys," *J. Japan Weld. Soc.*, **14**, 551-562 (1996). (in Japanese).

[4] E.E. de Kluizenaar, "Surface oxidation of molten soft solder: An Auger study," *J. Vac. Sci. Technol. A*, **1**, 1480- 85 (1983).

[5] S. Koyama, M. Takahashi, and K. Ikeuchi, "Influence of surface treatment on interfacial microstructure and bond strength of diffusion-bonded tin," *Solid State Ionics*, **172**, 397-401 (2004).

[6] T. Enjo, K. Ikeuchi, and K. Furukawa, "Behavior of Oxide Film in Diffusion Bonding of Aluminum," *Trans. JWRI*, **14**, 115- 22 (1985).

[7] G.B. Hoflund and G.R. Crallo, "Electron-energy-loss study of the oxidation of polycrystalline tin," *Phys. Rev. B*, **46**, 7110- 20 (1992).

[8] C.L. Lau and G.K. Wertheim, "Oxidation of tin: An ESCA study," *J. Vac. Sci. Tecnl.*, **15**, 8622- 24 (1978).

[9] D.A. Asbury and G.B. Hoflund, "A surface study of the oxidation of polycrystalline tin," *J. Vac. Sci. Technol. A*, **5**, 1132- 35 (1987).

THE EFFECTS OF FUNCTIONAL GROUPS OF SURFACE MODIFIER ON INTERNAL FRICTION AND YOUNG'S MODULUS OF CERAMIC COMPOSITES

T. Shimazu[1,2], N. Isu[1], M. Miura[1], and E. H. Ishida[2]

[1] General Research Institute of Technology, INAX Corp., 3-77 Minatomachi, Tokoname, Aichi, 479-8588, Japan

[2] Graduate School of Environmental Studies Tohoku University, 6-6-20 Aoba, Aramaki, Sendai, Miyagi, 980-8579, Japan

ABSTRACT

The effect of surface modification on mechanical properties of Al_2TiO_5-$MgTi_2O_5$ high damping composites is discussed. Depending on the functional group of the silane coupling agent, using a hard type polymer surface modification treatment, Young's modulus increased but internal friction decreased. It was verified that the roughness of the ground surface is reduced after the surface treatment. The results suggest that the bonding strength between ceramic particles was increased.

INTRODUCTION

Many systems in the field of mechanical engineering are facing the noise filtering and the vibration control. The vibration can not be controlled completely by the introduction of the mechanical methods, such as the conventional air damper, an insulator of polymer, and computer controlled noise reduction system[1,2]. Structural materials having high damping capacities are required for precise vibration control for a fair price. Although damping polymers have been widely used as vibration reducing materials, their rigidity is not so high for some mechanical use. We have found new materials having high rigidity and high damping capacities by the impregnation of acrylic resin into porous ceramics such as Al_2TiO_5-$MgTi_2O_5$ solid solution ceramics. Al_2TiO_5 is known as low-thermal expansion ceramics has many microcracks along grain boundaries formed by anisotropic thermal expansion coefficient of grains under the cooling process[3]. The microstructure of Al_2TiO_5-$MgTi_2O_5$ was formed in similar process and composed of strongly anisotropic grains and pores. A lot of thick elongated columnar grains shaped complex tangled crystal structure and narrow pores were formed along almost all the grain boundaries. This unique microstructure gave high damping capacity to the ceramics, but Young's modulus was too low to use for structural material. To improve this problem, combining polymer was effective method and Young's modulus increased dramatically (about 10-80 times as large as before processing). The ceramics-polymer composite has high damping capacity (Q^{-1}=0.01-0.04) with high rigidity (E=50-100GPa) [4, 5]. Appropriate microstructure control of the base ceramics and the effective selection of polymer are the keys for the progress of properties. In this study, we focused on the interaction at boundary between ceramics and polymer. The effects of pore surface modification to increase the mutual force of ceramic and polymer boundary on mechanical property and damping properties of composites were investigated.

EXPERIMENT

Al_2TiO_5-$MgTi_2O_5$ solid solution (Al_2TiO_5:$MgTi_2O_5$=0.65:0.35(molar ratio)) was prepared by following methods. The raw powders of Al_2O_3 (AL-160SG-4: Showa Denko Co.), TiO_2 (30-2740-5: Sigma Aldrich Japan K.K.) and MgO (19-0375-5: Sigma Aldrich Japan K.K.) in appropriate amount were mixed with ethanol in a plastic ball mill for 20 hours. After drying the slurry, the obtained powders were die-pressed under 50MPa with $80\times12\times5(mm^3)$ shape. The formed body was fired at 1500 °C for 2 hours. Pore surface were modified by vacuum impregnation method using selected three silane coupling agents with different types of functional group (epoxy group(EP), amino group(AM), acryloxy group(AC)). two mass percent diluted agents by ethanol were used. After surface modification, two types of acrylic resin with difference in hardness (A: hard type acrylic resin; storage modulus E'=2110MPa, B:soft type acrylic resin; storage modulus E'=51.5MPa) depending on side chain were combined with the solid solution.

Damping capacities (internal friction) and Young's modulus were measured by free resonance method (Nippon Techno-plus Co., Ltd.: JE-RT). The specimens ground by waterproof abrasive paper(from #100 to #1200). Microstructure of polished surface was observed by SEM. Surface roughness of polished surface was measured (Tokyo Seimitsu Co., Ltd. : Surfcom590A)

RESULTS AND DISCUSSION

YOUNG'S MODULUS AND INTERNAL FRICTION

Fig.1 shows the change of Young's modulus and internal friction of Al_2TiO_5-$MgTi_2O_5$ solid solution - polymer A composite with pore surface modification treatment. Young's modulus of base ceramics (without polymer) was a fair bit low 2GPa as ceramics. However, by combining polymer increased Young's modulus about thirty five times as large as base ceramics and internal friction increased slightly. Furthermore Young's modulus increased by the surface modification while internal friction decreased depending on the functional group of the surface modifier. It is supposed that impregnating polymer A into the pore bonded mutual displacement of the grains and

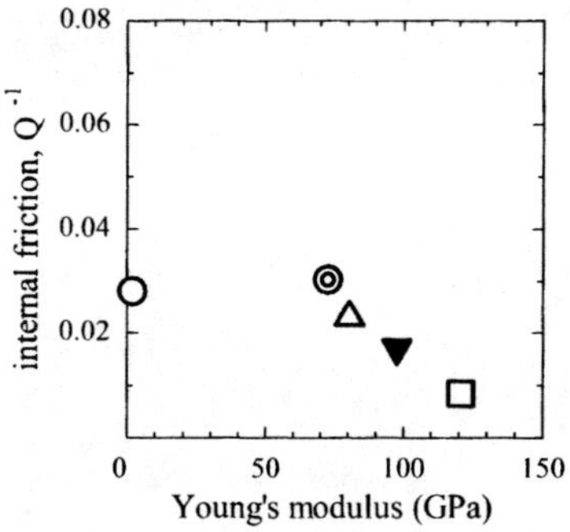

Fig.1 The effect of surface modification on Young's modulus and internal friction of ceramics - polymer A composites. Al_2TiO_5:$MgTi_2O_5$=0.65:0.35 (molar ratio). ○:base ceramics. Composite A; ◎:without surface modifier, △:EP, ▼:AM, □:AC.

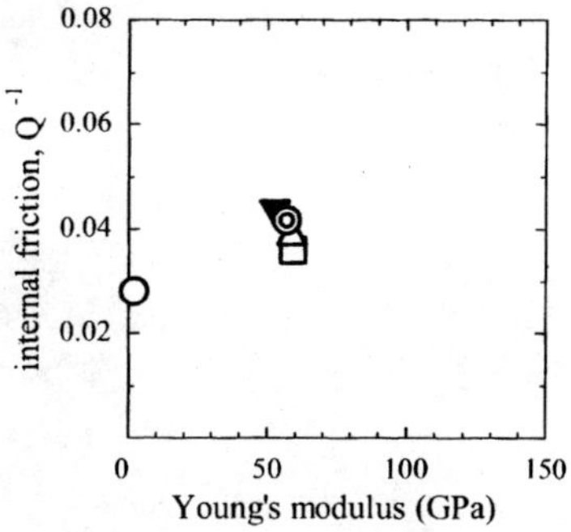

Fig.2 The effect of surface modification on Young's modulus and internal friction of ceramics - polymer B composites. Al_2TiO_5:$MgTi_2O_5$=0.65:0.35 (molar ratio). ○:base ceramics. Composite B; ◎:without surface modifier, △:EP, ▼:AM, □:AC.

bonding strength was increased with surface modification. On the other hand, using the polymer B, combining the polymer increased both Young's modulus and internal friction, but remarkable change of Young's modulus and internal friction was not observed with the surface modification (Fig.2). The effect of surface modification depended on combined polymer and functional groups of surface modifier.

MICROSTRUCTURE AND SURFACE ROUGHNESS

POLYMER A COMPOSITE

The effect of surface modification on microstructure and polished surface were discussed. Reflection electron images of Al_2TiO_5-$MgTi_2O_5$ solid solution-polymer A composites were shown in Fig.3. In this system, the particles of Al_2TiO_5-$MgTi_2O_5$ solid solution showed discontinuous grain growth because of MgO ingredient as flux[6]. In the figure, there are several areas in the pores with dark color which means polymer as light element. The polymer was filled in the pores over 10μm in size of all specimens. Though surface modification was considered to affect impregnating behavior, remarkable difference of impregnating polymers with the modifier was not found. Polished surface was rough without surface modification; On the other hand, it was smoother with the modification. Fig.4 shows the profile curves of surface roughness measurement. The surface was smoother with using coupling agent AC. The surface modification decreased the surface roughness depending on functional groups of coupling agents.

These results indicate that pore surface modification increased bonding strength between ceramics and polymer enough to prevent the particles from chipping away while grinding process. Without surface modification, ceramics particles would be coming off at the boundary because of weakness of bonding between ceramics and polymer. Surface modification would improve bonding strength tightly at the boundary.

Fig.3 Reflection electron images of Al_2TiO_5-$MgTi_2O_5$ – polymer A composites polished surface. Al_2TiO_5:$MgTi_2O_5$=0.65:0.35 (molar ratio).

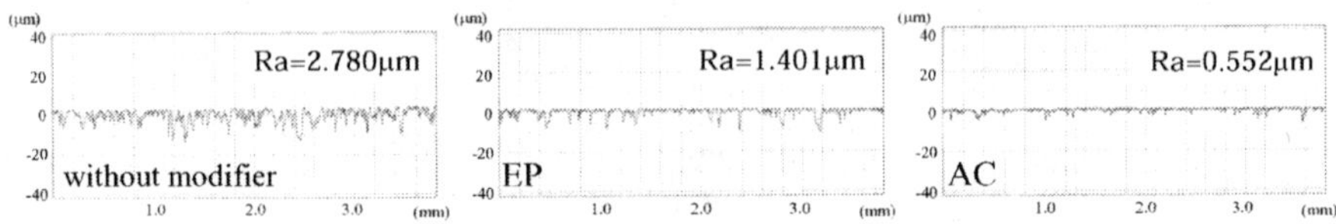

Fig.4 Profile curves of surface roughness measurement and Ra values of Al_2TiO_5-$MgTi_2O_5$ - polymer A composites. Al_2TiO_5:$MgTi_2O_5$=0.65:0.35 (molar ratio).

Fig.5 Reflection electron images of Al_2TiO_5-$MgTi_2O_5$ – polymer B composites polished surface. Al_2TiO_5:$MgTi_2O_5$=0.65:0.35 (molar ratio).

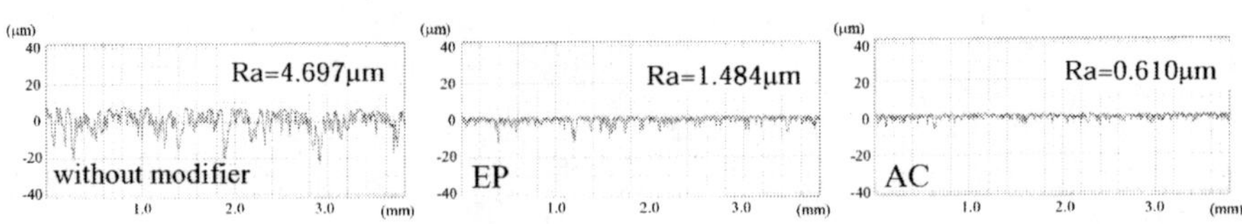

Fig.6 Profile curves of surface roughness measurement and Ra values of Al_2TiO_5-$MgTi_2O_5$ - polymer B composites. Al_2TiO_5:$MgTi_2O_5$=0.65:0.35 (molar ratio).

POLYMER B COMPOSITE

Fig.5 shows the reflection electron images of polished surface of Al_2TiO_5-$MgTi_2O_5$ solid solution - polymer B composites. Fig. 6 shows the profile curves of surface roughness measurement. Same as in the case of using polymer A, the surface modification treatment was effective to improve surface smoothness. Without the surface modification, polished surface was rough, on the other hand, polished surface was smoother with the surface modification depending on functional groups of the modifier, and most effective coupling agent was AC. Judging from Ra value of surface roughness, the surface of polymer A composites tended to slightly smoother than that of polymer B composites.

YOUNG'S MODULUS AND SURFACE ROUGHNESS

The relation between surface roughness and Young's modulus was shown in Fig. 7. In the case of using polymer A, Young's modulus decreased with surface roughness. On the other hand, in the case of using polymer B, Young's modulus did not change so much. The surface modification would improve chipping away of ceramic particles while grinding process, but about Young's modulus, the effect of surface modification depended on hardness of combined polymer. It is thought to be difficult for a soft type polymer B to restrain the displacement of the particles against applied stress.

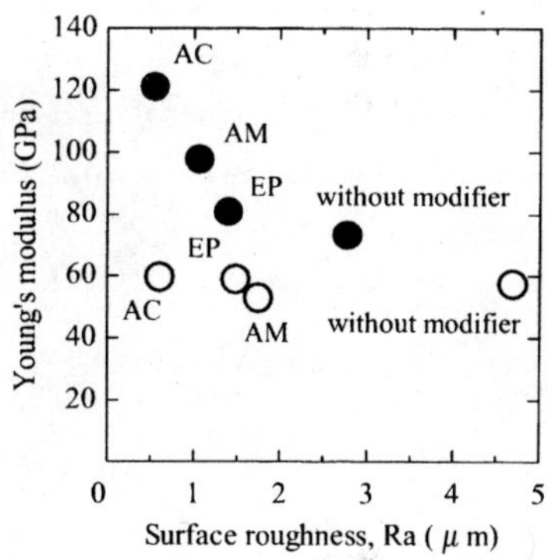

Fig. 7 The relation between surface roughness and Young's modulus. ●:polymer A, ○:polymer B.

CONCLUSION

The effect of surface modification on the mechanical properties of Al_2TiO_5-$MgTi_2O_5$ -polymer composites;

Surface roughness decreased, which is an advantage for materials in the mechanical field. Using a hard type polymer, Young's modulus increased but internal friction decreased, depending on the functional group of the modifier. Using a soft type polymer, Young's modulus and internal friction do not change significantly. Being able to control the mechanical properties of the polymer and the bonding strength between ceramics can provide flexibility to design and control the composites properties.

REFERENCES

[1] T. Shimazu, M. Miura, N.Isu, H.Kuno, E. H. Ishida, "Damping properties of resin - porous ceramics composites", Jpn. Ceram. Soc. 2002, Annual meeting abstract, 40 (2002).

[2] T. Shimazu, M. Miura, N.Isu, H.Kuno, E. H. Ishida, "Damping properties of porous alumina ceramics", Jpn. Ceram. Soc. 2002, Fall meeting abstract, 132 (2002).

[3] W.R. Buessem, N. R. Thielke, R. V. Sarakauskas, "Thermal Expansion Hysteresis of Alminum Titanate", Ceramic Age, **60**, 38-40 (1952).

[4] T. Shimazu, M. Miura, N. Isu, T.Ogawa, and E. H. Ishida, "Synthesis of Novel High Damping Ceramic-Polymer Composites and its Application as Ceramic Musical Instruments"Key Engineering Materials, **319** 173-180, (2006)

[5] T. Shimazu, M. Miura, H. Kuno, N. Isu, and E. H. Ishida, "High Damping Ceramic-Polymer Composites and its Application as Musical Instruments", Proceedings of the 22nd International Japan-Korea Seminar on Ceramics, 463-466(2005)

[6] K. Daimon, "Preparation and Thermal Expansion Characteristics of Sintered Bodies of Al_2TiO_5-$MgTi_2O_5$ Solid Solution with Preferred Orientation", J. Ceram. Soc. Jpn, **98** [4] , 365-69(1990)

Author Index